ABELIAN CATEGORIES WITH APPLICATIONS TO RINGS AND MODULES

L.M.S. MONOGRAPHS

Editors: P. M. COHN *and* G. E. H. REUTER

1. Surgery on Compact Manifolds *by* C. T. C. Wall, F.R.S.
2. Free Rings and Their Relations *by* P. M. Cohn.
3. Abelian Categories with Applications to Rings and Modules *by* N. Popescu.

Published for the London Mathematical Society
by Academic Press Inc. (London) Ltd.

Abelian Categories with Applications to Rings and Modules

N. POPESCU

Institut de Mathématique,
Académie Republique Socialiste de Roumanie,
Bucarest, Romania

1973

ACADEMIC PRESS · LONDON & NEW YORK

ACADEMIC PRESS INC. (LONDON) LTD.
24/28 Oval Road
London, N.W.1

U.S. Edition published by
ACADEMIC PRESS INC.
111 Fifth Avenue
New York, New York 10003

Second printing 1975

Library of Congress Catalog Card Number: 72–12276

ISBN: 0–12–561550–7

REPRODUCED PHOTOLITHO IN GREAT BRITAIN BY
J. W. ARROWSMITH LTD., BRISTOL

Preface

The theory of abelian categories begins almost with the first studies on abelian groups. A first systematization of the notions and results is made by Cartan and Eilenberg [33] for categories of modules and subsequently generalized by Buchsbaum [28] through the introduction of the notion of exact category.

An important advance was made by Grothendieck [79] who exhibited new aspects and results in the theory of abelian categories. A few years later, Gabriel [59] developed the theory of localization and laid the foundation of the theory of duality for abelian categories.

Another important contribution was made by Mitchell [133], to whom we owe the full imbedding theorem. Freyd [57], through his book and results, contributed much to the spreading of the theory of abelian categories.

Important results have also been obtained by Roos [199], Oberst [155], and others.

The present book attempts on the one hand, to systemize the principle results of the theory of abelian categories and, on the other, to show how these results participate in the theory of modules and rings.

Chapter 1 is devoted to the categorical language, which will be of frequent use in subsequent chapters.

Chapter 2 contains the basic notions and results from the theory of abelian categories: the fundamental lemmas, the isomorphism theorems, direct sums in abelian categories and the *Ab* conditions.

Chapter 3 is devoted to the study of additive functors. Here the exactness of functors, the tensor product, injective and projective objects are defined and discussed.

In Chapter 4 localization theory is developed. With the help of localization some important rings are constructed in the complete ring of quotients, the regular ring of quotients of a given ring, etc. Here also the Mitchell imbedding theorem is proved, flat epimorphisms and rings of fractions are studied and also, the theorem of Goldie, the general spectrum of a ring and bilocalizing subcategories are considered.

In Chapter 5 some types of Krull–Remak–Schmidt decompositions are studied, and also classes of abelian categories in which these decompositions hold

The theory of duality, in the form in which it has recently been developed by Gabriel, Roos and Oberst, is studied in Chapter 6.

Besides the authors cited above there are many others who contributed to the development of abelian categories, and their contribution cannot be diminished by the fact that their names do not appear in this book except in the bibliography (and, maybe, not even there). We hope that these omissions will be explained by the author's ignorance.

During the preparation of this book the author received support from various sources. Therefore thanks are owed to:

Professor Paul M. Cohn, who kindly recommended the book for publication to the publisher, and made some very useful remarks about its contents; Academic Press (London) Inc. which accepted the book for publication; the authors cited in the bibliography, who by their works made it possible for this book to appear; Mrs. Elena Stroescu, and Messrs. Ion Băianu and Silviu Teleman, who contributed to the English translation; Mr. C. Vraciu, who made an essential contribution to the writing of Chapter 6; Mrs. Ionica Stănculescu and Mr. Paul Cernovodeanu, who typed the manuscript; and all who by word, deed or thought contributed in some way to its publication.

February, 1973 N. Popescu

Contents

Note to the Reader

Chapter 1 consists of background material from category theory. The main subject matter of the book is introduced in Chapter 2 and the reader may wish to start here, referring back to Chapter 1 only when necessary.

The other chapters are closely related, thus it is very desirable that the reader study the book chapter by chapter.

The references at the end of each section should help the reader interested in other aspects of the problems which the section deals with.

All theorems, propositions, lemmas and corollaries are numbered consecutively in a single series in each section, the end (or absence) of a proof is indicated by ∎.

Generally even terms in current use have been defined, but not always on their first occurrence. In all such cases the definition can be traced through the subject index. For some of the more usual symbols and notations the following section (Some Terminology, Notations and Conventions used Throughout this Book) should be consulted.

Some Terminology, Notations and Conventions Used Throughout this Book

All rings occurring are associative, but not necessarily commutative. Every ring has a unit element denoted by 1, which is preserved by homomorphisms and inherited by subrings.

The letters **N**, **Z**, $\mathbf{Z}_n$, **Q** stand as usual for the set (respectively ring) of non-negative integers, all integers, all integers modulo n and rational numbers, respectively.

The following is a list of some important symbols and notations occurring in the book.

$\mathcal{A}b$: the category of abelian groups (Chapter 1, Section 1.1).

$\mathcal{S}et$: the category of sets (Chapter 1, Section 1.1).

h^X (respectively h_X): $\mathcal{C} \to \mathcal{S}et$, the functor $h^X(Y) = \mathrm{Hom}_{\mathcal{C}}(X, Y)$ (respectively $h_X(Y) = \mathrm{Hom}_{\mathcal{C}}(Y, X)$). (Chapter 1, Section 1.3).

$\varinjlim F$: the inductive limit of a functor (Chapter 1, Section 1.4).

$\varprojlim F$: the projective limit of a functor (Chapter 1, Section 1.4).

$\mathrm{Id}\,\mathcal{C}$: the identity functor of the category $\mathcal{C}$ (Chapter 1, Section 1.5).

$E(X)$: the injective envelope of an object (Chapter 3, Section 3.10).

$X_{\mathcal{F}}$: the greatest subobject of X in a localizing subcategory $\mathcal{F}$ (Chapter 4, Section 4.6).

$\mathcal{A}(X)$: the localizing subcategory cogenerated by X (Chapter 4, Section 4.6).

$\mathcal{L}(X)$: the smallest localizing subcategory which contains X (i.e. generated by X). (Chapter 4, Section 4.6).

$L(\mathcal{C})$: the family of all localizing subcategories of a category $\mathcal{C}$ (Chapter 4, Section 4.6).

$P(A)$: the set of all (proper) prelocalizing systems of a ring A (Chapter 4, Section 4.9).

$L(A)$: the set of all (proper) localizing systems of a ring A.

If $F \in P(A)$, then $\mathcal{F}$ is the associated prelocalizing sub-category (Chapter 4, Section 4.9).

$F_{\mathfrak{a}}$: the localizing system cogenerated by $A/_{\mathfrak{a}}$.

D: system of left-dense ideals (Chapter 4, Section 4.9).

E: the Goldie prelocalizing system.

$\bar{\mathscr{F}}$: closure of $\mathscr{F}$, i.e. the smallest localizing subcategory which contains a prelocalizing subcategory $\mathscr{F}$ (Chapter 4, Section 4.8).

$\bar{F}$: closure of F, i.e. the smallest localizing system containing a prelocalizing system F.

$\mathfrak{r}_F$: the smallest left ideal of a ring A in a localizing subcategory $\mathscr{F}$ of Mod A.

X_F: the localization module associated with X, relatively to a localizing system F (Chapter 4, Section 4.10).

$\mathfrak{R}(A)$: the Jacobson radical of a ring A (Chapter 3, Section 3.4).

$J(X)$: the singular submodule of X (Chapter 4, Section 4.9).

$\mathscr{C}_E$: the spectral category of $\mathscr{C}$ (Chapter 4, Section 4.2).

Sim($\mathscr{C}$): the set of types of simple objects of $\mathscr{C}$ (Chapter 5, Section 5.6).

So($\mathscr{C}$): the socle prelocalizing subcategory (Chapter 5, Section 5.6).

So(X): the socle of X (Chapter 5, Section 5.6).

Spel(A): the set of all left primes of A (Chapter 4, Section 4.20).

Spel($\mathscr{C}$): the set of all primes localizing subcategories of $\mathscr{C}$ (Chapter 4, Section 4.20).

Spep(A): the set of all two-sided prime ideals of A (Chapter 5, Section 5.11).

Sp($\mathscr{C}$): the spectrum of $\mathscr{C}$ (Chapter 5, Section 5.3).

$\mathscr{N}(\mathscr{C})$: the full subcategory of $\mathscr{C}$ generated by all noetherian objects.

$\mathscr{A}(\mathscr{C})$: the full subcategory of $\mathscr{C}$ generated by all artinian objects

$\mathscr{F}(\mathscr{C})$: the full subcategory of $\mathscr{C}$ generated by all finite objects.

Lex($\mathscr{C}$): (Chapter 4, Section 4.11).

$Q_\mathfrak{p}$: the injective envelope of $A/\mathfrak{p}$, where $\mathfrak{p}$ is a prime two-sided ideal (Chapter 5, Section 5.11).

$Q_\mathfrak{P}$: (Chapter 5, Section 5.10).

$\subseteq$: Inclusion.

$\subset$: Strict inclusion.

1. Categories and Functors

In this chapter we give the basic definitions of the general theory of categories and also some already classical results, which will be freely used in the rest of the book.

The results presented here are not proved; for details we refer the reader to references [31], [133], [162] or to any other reference in this domain.

1.1 The notion of a category. Examples

We denote by $\mathscr{C}$ and we shall call a *category* the following concept:

(a) a class Ob $\mathscr{C}$ of elements called *objects* of $\mathscr{C}$;

(b) for each ordered pair (X, Y) of objects of $\mathscr{C}$ a set $\mathrm{Hom}_{\mathscr{C}}(X, Y)$ (called the set of *morphisms* of X into Y) is given such that $\mathrm{Hom}_{\mathscr{C}}(X, Y)$ and $\mathrm{Hom}_{\mathscr{C}}(X', Y')$ are disjoint for $(X, Y) \neq (X', Y')$. [We shall frequently write $f : X \to Y$ or $X \xrightarrow{f} Y$ instead of $f \in \mathrm{Hom}_{\mathscr{C}}(X, Y)$]. We shall say that X is the *domain* and Y the *codomain* of f:

(c) for each triple (X, Y, Z) of objects of $\mathscr{C}$ a map:

$$\theta(X, Y, Z) : \mathrm{Hom}_{\mathscr{C}}(X, Y) \times \mathrm{Hom}_{\mathscr{C}}(Y, Z) \to \mathrm{Hom}_{\mathscr{C}}(X, Z)$$

is given (if $f \in \mathrm{Hom}_{\mathscr{C}}(X, Y)$, $g \in \mathrm{Hom}_{\mathscr{C}}(Y, Z)$, we shall denote by gf or $g \circ f$ the image $\theta(X, Y, Z)\ (f, g)$ of the pair (f, g) under $\theta(X, Y, Z)$; we shall say that θ is the *composition map* of morphisms);

(d) The composition map is associative: if X, Y, Z, T are objects of $\mathscr{C}$, then the following diagram of canonically constructed sets and maps is commutative:

$$\begin{array}{ccc}
\begin{array}{c}\mathrm{Hom}_{\mathscr{C}}(X, Y) \\ \times \mathrm{Hom}_{\mathscr{C}}(Y, Z) \times \mathrm{Hom}_{\mathscr{C}}(Z, T)\end{array} & \xrightarrow{\theta(X, Y, Z) \times 1} & \mathrm{Hom}_{\mathscr{C}}(X, Z) \times \mathrm{Hom}_{\mathscr{C}}(Z, T) \\
\Big\downarrow {\scriptstyle 1 \times \theta(Y, Z, T)} & & \Big\downarrow {\scriptstyle \theta(X, Z, T)} \\
\mathrm{Hom}_{\mathscr{C}}(X, Y) \times \mathrm{Hom}_{\mathscr{C}}(Y, T) & \xrightarrow{\theta(X, Y, T)} & \mathrm{Hom}_{\mathscr{C}}(X, T)
\end{array}$$

(i.e. if $X \xrightarrow{f} Y \xrightarrow{g} Z \xrightarrow{h} T$, then $h(gf) = (hg)f$);

(e) For each object X of $\mathscr{C}$ there is a morphism $1_X : X \to X$ (called the *identity morphism* of X) such that if $f : Y \to X$ and $g : X \to Z$ then $1_X f = f$, $g 1_X = g$.

EXAMPLES OF CATEGORIES

1. The category $\mathscr{S}et$. The objects of the category $\mathscr{S}et$ are the sets. If X and Y are two sets we shall define $\mathrm{Hom}_{\mathscr{S}et}(X, Y)$ to be the set of all maps defined on the set X and taking values in the set Y. The composition of morphisms is the usual composition of maps.

2. The category $\mathscr{T}op$. The objects of $\mathscr{T}op$ are topological spaces. If X, Y are two topological spaces we shall define $\mathrm{Hom}_{\mathscr{T}op}(X, Y)$ as the set of all continuous maps defined on the space X and taking values in Y. The composition of morphisms in $\mathscr{T}op$ is the usual composition of continuous maps.

3. The category $\mathscr{G}$. The objects of $\mathscr{G}$ are groups. If X and Y are two arbitrary groups, we shall define $\mathrm{Hom}_{\mathscr{G}}(X, Y)$ as the set of all homomorphisms of the group X into the group Y. The composition of morphisms is the usual composition of homomorphisms.

4. The category $\mathscr{A}b$. The objects of the category $\mathscr{A}b$ are abelian groups, and the morphisms are again homomorphisms of abelian groups.

5. *Duality*. Given a category $\mathscr{C}$, a new category $\mathscr{C}^0$—called the *dual category* of $\mathscr{C}$—is obtained in the following manner:

(i) The objects of the category $\mathscr{C}^0$ coincide with the objects of the category $\mathscr{C}$, that is $\mathrm{Ob}\,\mathscr{C} = \mathrm{Ob}\,\mathscr{C}^0$. If X is an object of $\mathscr{C}$, then the same object regarded as being in the category $\mathscr{C}^0$ is denoted by X^0 (X^0 is called the *dual of* X).

(ii) The set of morphisms $\mathrm{Hom}_{\mathscr{C}^0}(X^0, Y^0)$ is by definition identical with $\mathrm{Hom}_{\mathscr{C}}(Y, X)$. If $f \in \mathrm{Hom}_{\mathscr{C}}(Y, X)$ then the same morphism regarded as being in $\mathrm{Hom}_{\mathscr{C}^0}(X^0, Y^0)$ will be denoted by f^0 (f^0 is called the *dual of* f).

(iii) The composition map:

$$\theta^0(X^0, Y^0, Z^0) : \mathrm{Hom}_{\mathscr{C}^0}(X^0, Y^0) \times \mathrm{Hom}_{\mathscr{C}^0}(Y^0, Z^0) \to \mathrm{Hom}_{\mathscr{C}^0}(X^0, Z^0)$$

is defined as follows: if $f^0 \in \mathrm{Hom}_{\mathscr{C}^0}(X^0, Y^0)$, $g^0 \in \mathrm{Hom}_{\mathscr{C}^0}(Y^0, Z^0)$ then

$$\theta^0(X^0, Y^0, Z^0)\,(f^0, g^0) = g^0 f^0 = (\theta(Z, Y, X)\,(g, f))^0 = (fg)^0.$$

Clearly, $(\mathscr{C}^0)^0 \equiv \mathscr{C}$.

The possibility of associating with each category $\mathscr{C}$ its dual category $\mathscr{C}^0$ enables one to dualize each notion or statement with respect to a category $\mathscr{C}$, to a corresponding notion or statement with respect to the dual category $\mathscr{C}^0$ (from a practical point of view, this is the procedure of "reversing the arrows").

6. Let $\mathscr{C}$ be a category. By a *subcategory* of $\mathscr{C}$ we mean a category $\mathscr{C}'$ which is subject to the following conditions:

(i) $\mathrm{Ob}\,\mathscr{C}' \subseteq \mathrm{Ob}\,\mathscr{C}$;

(ii) $\mathrm{Hom}_{\mathscr{C}'}(X, Y) \subseteq \mathrm{Hom}_{\mathscr{C}}(X, Y)$;

(iii) The composition of morphisms in $\mathscr{C}'$ is induced by the composition of morphisms in $\mathscr{C}$.

(iv) The identity morphisms in $\mathscr{C}'$ are also identity morphisms in $\mathscr{C}$.

A subcategory $\mathscr{C}'$ of the category $\mathscr{C}$ is said to be *full* if, for each pair (X, Y) of objects of $\mathscr{C}'$, we have:

$$\mathrm{Hom}_{\mathscr{C}'}(X, Y) = \mathrm{Hom}_{\mathscr{C}}(X, Y)$$

7. If $(\mathscr{C}_i)_{i\in I}$ is a set of categories we define the *product category* $\prod_{i\in I}\mathscr{C}_i$ as follows:

The objects of $\prod_{i\in I}\mathscr{C}_i$ are all the families of the form $(X_i)_{i\in I}$ where for each $i \in I$, X_i is an object of $\mathscr{C}_i$;

$$\mathrm{Hom}_{\prod_i\mathscr{C}_i}((X_i)_i, (Y_i)_i) = \textstyle\prod_{i\in I} \mathrm{Hom}_{\mathscr{C}_i}(X_i, Y_i).$$

The composition of morphisms is defined componentwise.

Let $\mathscr{C}$ be any category. If $\mathrm{Ob}\,\mathscr{C}$ is a set, then $\mathscr{C}$ will be called a *small category*.

1.2 Special objects and morphisms

Let $f : X \to Y$ be a morphism of the category $\mathscr{C}$.

(a) f is called a *monomorphism* if $fu = fv$ implies that $u = v$ for all pairs of morphisms u, v with codomain X.

(b) f is called an *epimorphism* if f^0, the dual of f, is a monomorphism.

(c) f is called a *bimorphism* if f is at the same time a monomorphism and an epimorphism.

(d) f is called a *retraction* if there is a morphism $f' : Y \to X$ such that $ff' = 1_Y$.

(e) f is called a *section* if f^0, the dual of f, is a retraction, i.e., there is a morphism $f'' : Y \to X$ such that $f''f = 1_X$.

(f) f is called an *isomorphism* if f is both a retraction and a section. In this case we have $f' = 1_X f' = (f''f)\,f' = f''(ff') = f''\,1_Y = f''$. We call $f' = f''$ the *inverse* of f and we denote it by f^{-1}. Then from the definition we have $(f^{-1})^{-1} = f$.

PROPOSITION 2.1. (a) *Any retraction is an epimorphism.*

(b) *Any section is a monomorphism.*

(c) *If $f: X \to Y$ is a retraction (respectively, section) and is also a monomorphism (respectively, epimorphism) then it is an isomorphism.*

Let X be an object of $\mathscr{C}$. We shall say that X is an *initial object* in $\mathscr{C}$ if for any object Y, the set $\mathrm{Hom}_{\mathscr{C}}(X, Y)$ has a single element. Also, we shall say that X is a *final object* of $\mathscr{C}$ if X^0, the dual of X, is an initial object in $\mathscr{C}^0$. Finally, an object X will be called a *null object* if it is at the same time an initial and a final object.

Two objects X and Y of $\mathscr{C}$ are called *isomorphic* if there exists an isomorphism $f: X \to Y$. Any two initial (respectively, final or null) objects of $\mathscr{C}$ are isomorphic. Two isomorphic objects have essentially the same properties. Obviously, the "isomorphism" relation is in fact an equivalence relation on $\mathrm{Ob}\,\mathscr{C}$. The class of objects which are isomorphic to X will be denoted by $[X]$ and will be called the *type of* X. A new category can be constructed from the types of objects of a category $\mathscr{C}$, and this category will be called the *type category* of $\mathscr{C}$ or the *skeleton of* $\mathscr{C}$. Plainly, the type category of $\mathscr{C}$ has the same properties as $\mathscr{C}$.

Let X be an object in $\mathscr{C}$. We shall denote by $M(X)$ the category whose objects are couples (X', f') where $f': X' \to X$ is a monomorphism in $\mathscr{C}$. A morphism $g: (X', f') \to (X'', f'')$ of $M(X)$ is defined as a morphism $g: X' \to X''$ such that $f''g = f'$. A type of object of $M(X)$ will be called a *subobject of* X. Obviously, in order to get a subobject of X one has to use a representative of the class of equivalent objects from $M(X)$ which defines the subobject. In this way, a subobject of X will be represented by a couple (X', f) with monomorphism $f: X' \to X$. Sometimes, in order to emphasize that X' is a subobject of X' we shall write $X' \subseteq X$. The dual concept of a subobject is that of *quotient object*. This dualization is readily obtained: a quotient object of X will be a subobject of X^0 in the dual category $\mathscr{C}^0$. In other words a quotient object of X will be represented by a couple (p, X''), where $p: X \to X''$ is an epimorphism in $\mathscr{C}$.

A set $\{U_i\}_i$ of objects from $\mathscr{C}$ will be called a *set of generators of* $\mathscr{C}$ if for any couple (X, Y) of objects from $\mathscr{C}$ and for any two distinct morphisms $f, g \in \mathrm{Hom}_{\mathscr{C}}(X, Y)$, there is an index i_0 and a morphism $h \in \mathrm{Hom}_{\mathscr{C}}(U_{i_0}, X)$ such that $fh \neq gh$. Also, we shall say that the family given above is a *set of cogenerators of* $\mathscr{C}$ if the family $\{U_i^{\,0}\}_i$ is a set of generators of $\mathscr{C}^0$. We shall say that the given set of objects is a *set of S-generators of* $\mathscr{C}$ if for any monomorphism $f: X' \to X$ such that f is not an isomorphism, there is an index i_0 and a morphism $h \in \mathrm{Hom}_{\mathscr{C}}(U_{i_0}, X)$ such that for any morphism

$$g \in \mathrm{Hom}_{\mathscr{C}}(U_{i_0}, X'), \qquad fg \neq h.$$

Dually, one can define the concept of a *set of S-cogenerators*. More specifically, a category has a *generator* (respectively, *cogenerator*) if it has a set of generators (respectively, *cogenerators*) consisting of a single object. Analogously we get the concept of an *S-generator* (respectively *S-cogenerator*).

Let X be an object of $\mathscr{C}$; we say that X is a *locally small object* if the category $M(X)$ has a small skeleton. If each object of $\mathscr{C}$ is locally small, then $\mathscr{C}$ is called a *locally small* category. We shall say that X is *colocally small* if the dual X^0 of X is locally small; $\mathscr{C}$ is a *colocally small category* if $\mathscr{C}^0$ is locally small.

PROPOSITION 2.2. *If the category $\mathscr{C}$ has a set of S-generators and fibred products, then $\mathscr{C}$ is locally small.*

1.3 Functors

Let $\mathscr{C}$ and $\mathscr{C}'$ be two categories. A *covariant* (respectively, *contravariant*) *functor F* from $\mathscr{C}$ to $\mathscr{C}'$ consists of the following entities:

(a) a map $X \rightsquigarrow F(X)$ which associates to each object X of C an object $F(X)$ of $\mathscr{C}'$:

(b) for each pair (X, Y) of objects of $\mathscr{C}$, a map

$$F(X, Y): \mathrm{Hom}_{\mathscr{C}}(X, Y) \to \mathrm{Hom}_{\mathscr{C}'}(F(X), F(Y))$$

(respectively, $F(X, Y): \mathrm{Hom}_{\mathscr{C}}(X, Y) \to \mathrm{Hom}_{\mathscr{C}'}(F(Y), F(X))$); such that, if we write simply $F(f)$ instead of $F(X, Y)(f)$, then $F(1_X) = 1_{F(X)}$ and $F(gf) = F(g)F(f)$ (respectively, $F(gf) = F(f)F(g)$).

Clearly, a covariant functor from $\mathscr{C}$ to $\mathscr{C}'^0$ and a covariant functor $\mathscr{C}^0$ to $\mathscr{C}'$ are naturally associated with each contravariant functor from $\mathscr{C}$ to $\mathscr{C}'$ and conversely. Thus, the general study of contravariant functors can be reduced to the study of covariant functors.

EXAMPLES

8. We denote by $S: \mathscr{T}\!op \to \mathscr{S}\!et$ the functor obtained as follows: for any topological space X, $S(X)$ is the underlying set; for any continuous map f, $S(f)$ is the corresponding map. Obviously, S is a covariant functor, often called the *forgetful functor*.

9. Let G be a group; we denote by $c(G)$ the commutator of G (i.e. the normal subgroup generated by the elements $xyx^{-1}y^{-1}$, $x, y \in G$). If we put $A(G) = G/c(G)$, and $f: G \to G'$ is a homomorphism of groups, then

$$f(c(G)) \subseteq c(G')$$

and f induces a unique homomorphism $A(f): A(G) \to A(G')$. Thus, we obtain a covariant functor $A: \mathscr{G} \to \mathscr{Ab}$.

10. Let $\mathscr{C}$ be any category and let X be an object of $\mathscr{C}$. We denote by $h^X: \mathscr{C} \to \mathscr{Set}$ the functor obtained as follows: for any $Y \in \mathrm{Ob}\,\mathscr{C}$, $h^X(Y) = \mathrm{Hom}_{\mathscr{C}}(X, Y)$; if $f: Y \to Y'$ is a morphism of $\mathscr{C}$ then $h^X(f): \mathrm{Hom}_{\mathscr{C}}(X, Y) \to \mathrm{Hom}_{\mathscr{C}}(X, Y')$ is the map $h^X(f)\,(g) = fg$. Sometimes we shall denote h^X by $\mathrm{Hom}_{\mathscr{C}}(X, ?)$.

11. We denote by $h_X: \mathscr{C} \to \mathscr{Set}$ the functor:

$$h_X(Y) = \mathrm{Hom}_{\mathscr{C}}(Y, X)$$

for the sake of simplicity we shall denote h_X by $\mathrm{Hom}_{\mathscr{C}}(?, X)$.

Let $\mathscr{C}$ and $\mathscr{C}'$ be two categories and let F, G be two covariant functors from $\mathscr{C}$ to $\mathscr{C}'$. We say that a *functorial morphism* u from the functor F to the functor G is given, if for each object X in $\mathscr{C}$ we are given a morphism $u_X: F(X) \to G(X)$ in $\mathscr{C}'$, such that the diagram

$$\begin{array}{ccc} F(X) & \xrightarrow{u_X} & G(X) \\ {\scriptstyle F(f)}\downarrow & & \downarrow{\scriptstyle G(f)} \\ F(Y) & \xrightarrow{u_Y} & G(Y) \end{array}$$

is commutative, for any morphism $f: X \to Y$ in $\mathscr{C}$.

If u_X is a monomorphism (respectively epimorphism, isomorphism) for each X in $\mathscr{C}$, then f is called a *functorial monomorphism* (respectively, *epimorphism, isomorphism*).

If the functors F, G are contravariant, then in the definition of the functorial morphism the above diagram has of course to be replaced by the following diagram

$$\begin{array}{ccc} F(X) & \xrightarrow{u_X} & G(X) \\ {\scriptstyle F(f)}\uparrow & & \uparrow{\scriptstyle G(f)} \\ F(Y) & \xrightarrow{u_Y} & G(Y) \end{array}$$

EXAMPLES

12. Let $u: X \to X'$ be a morphism in the category $\mathscr{C}$. We associate to this morphism a functorial morphism $h(u)$ from the functor h_X to the functor $h_{X'}$, in the following way. Let Y be an arbitrary object in $\mathscr{C}$. We denote by $h(u)_Y$ the map from the set $\mathrm{Hom}_{\mathscr{C}}(Y, X)$ into the set $\mathrm{Hom}_{\mathscr{C}}(Y, X')$ which sends $f \in \mathrm{Hom}_{\mathscr{C}}(Y, X) = h_X(Y)$ to $uf \in \mathrm{Hom}_{\mathscr{C}}(Y, X') = h_X\,(Y)$. To conclude that

we have thus defined a functorial morphism from h_X to $h_{X'}$, we have to check the commutativity of the diagram

$$\begin{array}{ccc} h_X(Y) & \xrightarrow{h(u)_Y} & h_{X'}(Y) \\ {\scriptstyle h_X(f)}\uparrow & & \uparrow{\scriptstyle h_{X'}(f)} \\ h_X(Y') & \xrightarrow{h(u)_{Y'}} & h_{X'}(Y') \end{array}$$

for any morphism $f : Y \to Y'$ in $\mathscr{C}$. But this follows immediately, since if $g \in h_X(Y') = \mathrm{Hom}_{\mathscr{C}}(Y', Y)$, we have

$$\big(h(u)_Y h_X(f)\big)(g) = h(u)_Y(gf) = u(gf) = (ug)f = \big(h_{X'}(f)h(u)_{Y'}\big)(g)$$

13. Let F, G, H be three covariant functors from the category $\mathscr{C}$ to $\mathscr{C}'$ and let $u : F \to G$ and $v : G \to H$ be two functorial morphisms. We shall denote by $vu : F \to H$ and call the *composition* of the two functorial morphisms, a functorial morphism which is defined as follows: for any $X \in \mathrm{Ob}\,\mathscr{C}$, one has $(vu)_X = v_X u_X$. It may easily be noticed that the composition of functorial morphisms, as defined above, is associative.

14. Let $X \in \mathrm{Ob}\,\mathscr{C}$ and let $F : \mathscr{C} \to \mathscr{S}et$ be a covariant functor. Also, let $x \in F(X)$. We shall denote by $u_x : h^X \to F$ the functorial morphism defined as follows: if $Y \in \mathrm{Ob}\,\mathscr{C}$, then $(u_x)_Y : h^X(Y) \to F(Y)$ is the mapping defined by the equality $(u_x)_Y(f) = F(f)(x)$.

Moreover, we have the following:

LEMMA 3.1. (The Yoneda–Grothendieck Lemma). *Let $X \in \mathrm{Ob}\,\mathscr{C}$ and let $F : \mathscr{C} \to \mathscr{S}et$ be a covariant functor The assignment $x \rightsquigarrow u_x$ defines a bijection between the set $F(X)$ and the set of functorial morphisms from h^X to F.*

We shall say that u_x is the *functorial morphism associated to x.*

15. The above results are also valid for a contravariant functor F and h_X. In this case, for any $x \in F(X)$ we shall denote also by $u_x : h_X \to F$ the functorial morphism associated to x.

Let $F : \mathscr{C} \to \mathscr{C}'$, $G : \mathscr{C}' \to \mathscr{C}''$ be two functors. We shall denote by GF the functor from $\mathscr{C}'$ to $\mathscr{C}''$ which is defined as follows: for any $X \in \mathrm{Ob}\,\mathscr{C}$ we have $(GF)(X) = G(F(X))$, and for any f from $\mathscr{C}$ we also have that $(GF)(f) = G(F(f))$. The functor GF is called the *composition of G with F*. Obviously, GF is a covariant (respectively, contravariant) functor if both G and F are covariant or contravariant, (respectively, if one is covariant and the other contravariant). As may easily be checked the composition of functors is associative.

Let $F, F' : \mathscr{C} \to \mathscr{C}'$ be two functors, and let $u : F \to F'$ be a functorial morphism. If $G : \mathscr{C}' \to \mathscr{C}''$ is a functor, then we shall denote by Gu: $GF \to GF'$ the functorial morphism defined as follows: for any $X \in \operatorname{Ob} \mathscr{C}$, one has $(Gu)_X = G(u_X)$. Also, if $H : \mathscr{C}''' \to \mathscr{C}$ is a functor, then we shall denote by $uH : FH \to F'H$ the functorial morphism defined by the equality $(uH)_{X'''} = u_{H(X''')}$, which holds for any $X''' \in \operatorname{Ob} \mathscr{C}'''$. It will readily be seen that $(Gu)H = G(uH)$. This morphism is sometimes denoted by GuH.

Let $F : \mathscr{C} \to \mathscr{C}'$ be a functor. We shall say that F is a *faithful functor* if for any two morphisms f, g of $\mathscr{C}$ for which $F(f) = F(g)$ one also has $f = g$. A functor F will be called *full* if for any X, $Y \in \operatorname{Ob} \mathscr{C}$ and for any morphism $g \in \operatorname{Hom}_{\mathscr{C}'}(F(X), F(Y))$ there is an $f \in \operatorname{Hom}_{\mathscr{C}}(X, Y)$ such that $g = F(f)$.

F will be called *faithfully full* if it is at the same time a full functor and a faithful one.

If $\mathscr{C}'$ is a subcategory of $\mathscr{C}$ then there exists a canonically defined functor $I : \mathscr{C}' \to \mathscr{C}$. It is plain that $\mathscr{C}'$ is a full subcategory of $\mathscr{C}$ if and only if I is full.

A covariant (respectively, contravariant) functor $F : \mathscr{C} \to \mathscr{S}et$ is called *representable*, if there exists an object X of $\mathscr{C}$ such that F is isomorphic with h^X (respectively, with h_X).

1.4 Inductive and projective limits

Let $\mathscr{I}$, $\mathscr{C}$ be two categories and also let $X \in \operatorname{Ob} \mathscr{C}$.

We shall denote by $X_{\mathscr{I}} : \mathscr{I} \to \mathscr{C}$ the functor which is defined as follows: for any $i \in \operatorname{Ob} \mathscr{I}$ we have $X_{\mathscr{I}}(i) = X$ and if $f : i \to j$ is a morphism in $\mathscr{I}$, then we have $X_{\mathscr{I}}(f) = 1_X$.

We shall say that $X_{\mathscr{I}}$ is the *constant functor associated to* X. Sometimes, we shall write simply X instead of $X_{\mathscr{I}}$ if there is no risk of confusion.

Let $F : \mathscr{I} \to \mathscr{C}$ be a functor. We shall tacitly assume that F is covariant although this is inessential. A functorial morphism $u : F \to X_{\mathscr{I}}$ (respectively, $v : X_{\mathscr{I}} \to F$) will be called an *inductive cone* (respectively, *projective cone*) *over* F. Sometimes we shall also say that "a morphism is given from the functor F to the object X", (respectively "from X to F"). Obviously, to define a morphism from F to X, (respectively, from X to F) means to give the morphism, $u_i : F(i) \to X$ for any $i \in \operatorname{Ob} \mathscr{I}$ (respectively, $v_i : X \to F(i)$), such that if $f : i \to j$ is a morphism in $\mathscr{I}$, then $u_j F(f) = u_i$ (respectively, $F(f)\, v_i = v_j$. We shall denote by $F/\mathscr{C}$ (respectively, by $\mathscr{C}/F$), the category whose objects are the inductive (respectively, projective) cones over F (that is, these objects are couples (u, X), (respectively (X, v)), with $X \in \operatorname{Ob} \mathscr{C}$ and with $u : F \to X$. (respectively, $v : X \to F$), being functorial morphisms.

A morphism $f : (u, X) \to (u', X')$, (respectively, $g : (X, v) \to (X', v')$) is by definition a morphism $f : X \to X'$ of $\mathscr{C}$ such that $fu = u'$, (respectively $(v'f = v)$) .We shall say that $F/\mathscr{C}$ (respectively, $\mathscr{C}/F$) is the *category of inductive* (respectively, *projective*) *cones over* F. Also, we shall say that the functor F has an *inductive* (respectively, *projective*) *limit* if the category $\mathscr{C}/F$ (respectively, $F/\mathscr{C}$) has an initial (respectively, final) object. Obviously, if an inductive (respectively, projective) limit exists, it is unique up to an isomorphism. Let us denote an inductive (respectively, projective) limit of F by $(u, \varinjlim F)$, (respectively, by $(\varprojlim F, v)$) or simply, by $\varinjlim F$ (respectively, $\varprojlim F$), tacitly assuming the presence of the morphism u (respectively, v). The morphism u (respectively, v) is called the *structural morphism of the inductive* (respectively, *projective*) *limit*. Obviously, u stands for a family $\{u_i\}_{i \in \mathrm{Ob}\,\mathscr{I}}$ of morphisms of $\mathscr{C}$, and $u_i : F(i) \to \varinjlim F$ is such that if $h : i \to j$ is a morphism of $\mathscr{I}$ then we have: $u_j F(h) = u_i$.

We shall say that u_i are the *structural components* (or the *structural morphism*) of the inductive limit. The same terminology will also be used in the case of the projective limit since the concepts of inductive and projective limit are dual to each other.

Examples

16. A category is called *discrete* if all its morphisms are identity morphisms. A discrete category is in fact a class, and any class can be considered as a discrete category.

To define a covariant (or contravariant) functor from a discrete small category I to a category $\mathscr{C}$ is to give a set $\{X_i\}_{i \in I}$ of objects of C which are indexed by the set I.

An inductive (respectively, projective) limit of the above defined functor $\{X_i\}_{i \in I}$ will be denoted $\coprod_{i \in I} X_i$ (respectively $\prod_{i \in I} X_i$) and will be called the *direct sum* (respectively, *direct product*) *of the set* $\{X_i\}_{i \in I}$.

The disjoint union in the category of sets and the cartesian product in the same category are, respectively, typical examples of direct sum and direct product in the category $\mathscr{S}et$.

17. Let $\mathscr{I}$ be the category whose only objects are 1 and 2, and whose morphisms are $u, v : 1 \to 2$ and the identity morphisms. A functor from $\mathscr{I}$ to $\mathscr{C}$ will be called *a couple of morphisms.* To give such a couple is to give two objects X, Y in $\mathscr{C}$ and two morphisms $f, g \in \mathrm{Hom}_{\mathscr{C}}(X, Y)$. Then, we shall write $(f, g) : X \rightrightarrows Y$.

A projective limit of the couple (f, g) is called a *kernel of this couple.* Dually, an inductive limit of the couple will be called a *cokernel* of it. We shall denote by $\ker(f, g)$ the kernel of the couple (f, g), (respectively, the cokernel will be denoted by $\mathrm{coker}(f, g)$). Obviously, a kernel of (f, g) consists of a couple (Z, i) with $Z \in \mathrm{Ob}\,\mathscr{C}$ and $i : Z \to X$ being a morphism such that $fi = gi$; also, for any object T and any $h : T \to X$ for which $fh = gh$, there exists a unique morphism $h' : T \to Z$ such that $ih' = h$.

Dually, one can easily define the cokernel of the couple (f, g).

18. A diagram of the form

$$\begin{matrix} & X \\ & \downarrow \\ Z \to & Y \end{matrix} \quad \left(\text{respectively} \quad \begin{matrix} Y \to X \\ \downarrow \quad \\ Z \quad \end{matrix} \right)$$

will be called an *angle* (respectively, *co-angle*). Obviously, an angle (respectively, co-angle) of $\mathscr{C}$ can be viewed as a functor from a conveniently chosen category to $\mathscr{C}$. A projective limit of an angle will be called a *fibred product of the angle*. An inductive limit of a co-angle will be called a *fibred sum of the co-angle*.

A fibred product (respectively, fibred sum) of the above angle (respectively, co-angle) will be denoted by $X \prod_Y Z$ (respectively, $X \coprod_Y Z$) and it will be read as follows: the fibred product (respectively, the fibred sum) of X with Z, over Y.

Let $\mathscr{C}$ be a category. A functor $F : \mathscr{I} \to \mathscr{C}$, where $\mathscr{I}$ is a small category, will be called *small functor*. We shall say that $\mathscr{C}$ is a *category with projective* (respectively, *inductive*) *limits* if any small functor F with codomain $\mathscr{C}$ has a projective (respectively, inductive) limit.

It will be said that $\mathscr{C}$ is a *category with direct sums* (respectively, *direct products*) if any set of objects from $\mathscr{C}$ has a direct sum (respectively, direct product).

Also, it will be said that $\mathscr{C}$ is a category with *finite direct sums* (respectively, with *finite direct products*) if any finite set of objects $\mathscr{C}$ has a direct sum (respectively, direct product).

We shall say that $\mathscr{C}$ is a *category with kernels*, (respectively, *cokernels*) if each couple of morphisms of $\mathscr{C}$ has a kernel (respectively, cokernel).

THEOREM 4.1. *Let $\mathscr{C}$ be a category. The following assertions are equivalent*:

(a) *$\mathscr{C}$ is a category with inductive limits*;

(b) *$\mathscr{C}$ is a category with direct sums and cokernels*;

(c) *$\mathscr{C}$ is a category with direct sums in which any co-angle has a fibred sum.*

Obviously, this theorem may be dualized by replacing each concept by its dual.

Let I be a set which is partially ordered under the relation "$\leqslant$". We shall say that I is a *directed set* (respectively, *inverse set*), if for any $i, j \in I$ there is a $k \in I$ such that $i \leqslant k$, $j \leqslant k$, (respectively, $k \leqslant i$, $k \leqslant j$).

Any partially ordered set can be organized as a category by means of the following equalities:

(a) For any $i, j \in I$, $\mathrm{Hom}_I(i, j) = \emptyset$ if we do not have $i \leqslant j$;

(b) $\mathrm{Hom}_I(i, j) =$ the set with one morphism $i \to j$, (which stands for $i \leqslant j$), otherwise.

Let I be a directed (respectively, inverse) set. A covariant functor $F:I \to \mathscr{C}$ will be called a *direct* (respectively, *inverse*) *system* of objects of $\mathscr{C}$ or an *inductive* (respectively, *projective*) *system*. An inductive (respectively, projective) system will be denoted by $\{X_i, t_{ij}\}_{i \in I}$, with $X_i = F(i)$, and where $t_{ij} = X_i \to X_j$ is the unique morphism associated to the relation $i \leqslant j$.

It will be said that $\mathscr{C}$ is a category with *direct inductive limits* if any inductive system from $\mathscr{C}$ has an inductive limit. Dually, one can define a *category with inverse projective limits.*

1.5 Adjoint functors. Equivalences of categories

Let $F:\mathscr{C} \to \mathscr{C}'$ and $G:\mathscr{C}' \to \mathscr{C}$ be two covariant functors. We shall say that F is a *left adjoint of* G, and that G is a *right adjoint of* F if for any object X of $\mathscr{C}$ and any object X' of $\mathscr{C}'$, there exists a bijection $t(X, X')$: $\mathrm{Hom}_{\mathscr{C}}(X, G(X')) \to \mathrm{Hom}_{\mathscr{C}'}(F(X), X')$ such that for any morphism $f: X \to Y$ of $\mathscr{C}$ and any morphisms $f': X' \to Y'$ of C', the following diagrams of sets and canonically constructed mappings are commutative:

$$\begin{array}{ccc} \mathrm{Hom}_{\mathscr{C}}(Y, G(X')) & \xrightarrow{t(Y, X')} & \mathrm{Hom}_{\mathscr{C}'}(F(Y), X') \\ {\scriptstyle h_{G(X')}(f)}\downarrow & & \downarrow{\scriptstyle h_{X'}(F(f))} \\ \mathrm{Hom}_{\mathscr{C}}(X, G(X')) & \xrightarrow{t(X, X')} & \mathrm{Hom}_{\mathscr{C}'}(F(X), X') \end{array}$$

$$\begin{array}{ccc} \mathrm{Hom}_{\mathscr{C}}(X, G(X') & \xrightarrow{t(X, X')} & \mathrm{Hom}_{\mathscr{C}'}(F(X), X') \\ {\scriptstyle h^{X}(G(f'))}\downarrow & & \downarrow{\scriptstyle h^{F(X)}(f')} \\ \mathrm{Hom}_{\mathscr{C}}(X, G(Y')) & \xrightarrow{t(X, Y')} & \mathrm{Hom}_{\mathscr{C}'}(F(X), Y') \end{array}$$

In particular, we shall denote by $u_X: X \to GF(X)$ the morphism

$$t(X, F(X))^{-1}\,(1_{F(X)}).$$

Also, we shall denote by $v_{X'}: FG(X') \to X'$ the morphism $t(G(X'), X')(1_{G(X')})$.

It can be easily checked that the following diagram

$$\begin{array}{ccc} X & \xrightarrow{u_X} & GF(X) \\ {\scriptstyle f}\downarrow & & \downarrow{\scriptstyle GF(f)} \\ Y & \xrightarrow{u_Y} & GF(Y) \end{array} \qquad \text{resp.} \qquad \begin{array}{ccc} FG(X') & \xrightarrow{v_{X'}} & X' \\ {\scriptstyle FG(f)}\downarrow & & \downarrow{\scriptstyle f'} \\ FG(Y') & \xrightarrow{v_{Y'}} & Y' \end{array}$$

(which is canonically constructed), is commutative in $\mathscr{C}$ (respectively in $\mathscr{C}'$) for any morphism $F: X \to Y$ of $\mathscr{C}$, (respectively, for any morphism

$$f': X' \to Y' \text{ of } \mathscr{C}').$$

This implies that morphisms $\{u_X\}_{X \in \mathrm{Ob}\,\mathscr{C}}$ define a functorial morphism $u : \mathrm{Id}\mathscr{C} \to GF$ ($\mathrm{Id}\mathscr{C}$ is the identity functor of $\mathscr{C}$ and has an obvious definition). Also, the morphisms $\{v_{X'}\}_{X' \in \mathrm{Ob}\,C'}$ define a functorial morphism $v : FG \to \mathrm{Id}\mathscr{C}'$.

We shall say that u is an *adjunction arrow* of F with G, and that v is a *quasi-inverse of this arrow*. v is also called an adjunction arrow of G with F and u is its quasi-inverse. It may be readily checked that the two adjunction arrows, which are associated with the two functors, are subject to the following conditions

$$Gv \circ uG = 1_G \quad \text{and} \quad vF \circ Fu = 1_F,$$

(1_G being the identity functorial morphism of G).

The adjunction morphisms determine specifically the particular adjunction in a sense which is made clear in the next sentence.

Let us assume that the functorial morphisms $u : \mathrm{Id}\mathscr{C} \to GF$, $v : FG \to \mathrm{Id}\mathscr{C}'$ were given and that they fulfil the two relations:

$$Gv \circ uG = 1_G \quad \text{and} \quad vF \circ Fu = 1_F.$$

Then, F is the left adjoint of G and G is the right adjoint of F.

Let $F : \mathscr{C} \to \mathscr{C}'$ be a covariant functor and let X' be an object of $\mathscr{C}'$. We shall denote by F/X' the category whose objects are couples (X, f), X being an object of $\mathscr{C}$ and $f : F(X) \to X'$ being a morphism of $\mathscr{C}'$.

A morphism $g : (X, f) \to (X_1, f_1)$ of F/X' is defined as a morphism $g : X \to X_1$ of $\mathscr{C}$ such that $f_1\, F(g) = f$. Dually, one can define the category X'/F.

THEOREM 5.1. *Let $F : \mathscr{C} \to \mathscr{C}'$ be a covariant functor. The following assertions are equivalent.*

(1) *The functor F has a right adjoint;*

(2) *For any object X' of $\mathscr{C}'$ the category F/X' has a final object;*

(3) *For any object X' of $\mathscr{C}'$, the functor $h_{X'}F$ is representable.*

This theorem implies that the right adjoint of a functor is uniquely defined up to an isomorphism. Obviously, theorem 5.1 has a dual statement.

PROPOSITION 5.2. *Let $F : \mathscr{C} \to \mathscr{C}'$ be a covariant functor, G a right adjoint of F and $u : \mathrm{Id}\mathscr{C} \to GF$ an adjunction arrow. Also, let v be a quasi-inverse of u. Then, the following assertions are equivalent.*

(1) *The functor G is full and faithful;*

(2) *v is a functorial isomorphism.*

THEOREM 5.3. *Let $F : \mathscr{C} \to \mathscr{C}'$ be a covariant functor. The following assertions are equivalent.*

(1) *F is full and faithful, and any object X' of $\mathscr{C}'$ is isomorphic to an object $F(X)$, X being an object of $\mathscr{C}$;*

(2) *F is full and faithful, and has a full and faithful left adjoint;*

(3) *F is full and faithful, and has a full and faithful right adjoint.*

We shall say that two categories $\mathscr{C}$ and $\mathscr{C}'$ are *equivalent* if there is a covariant functor $F : \mathscr{C} \to \mathscr{C}'$ which satisfies any of the assertions of Theorem 5.3. The functor F will be called an *equivalence* from $\mathscr{C}$ to $\mathscr{C}'$.

Two categories are called *anti-equivalent* if one of them is equivalent to the dual category of the other. Obviously, any category is equivalent to each of its skeletons.

Let $S : \mathscr{I} \to \mathscr{C}$ be a functor such that $\varinjlim S$ (respectively, $\varprojlim S$) exists in $\mathscr{C}$. We shall say that the covariant functor $F : \mathscr{C} \to \mathscr{C}'$ *commutes with the inductive limit* of S (respectively, *with the projective limit* of S) if $F(\varinjlim S)$, (respectively, $F(\varprojlim S)$), is an inductive (respectively, projective) limit of FS. Here, we have tacitly assumed that if $f : S \to \varinjlim S$ (respectively, $f : \varprojlim S \to S$) is the structural morphism, then Ff is the structural morphism of the inductive (respectively, projective) limit of FS.

We shall say that the covariant functor $F : \mathscr{C} \to \mathscr{C}'$ *commutes with the inductive* (respectively *projective*) *limits* if F commutes with the inductive (respectively projective) limit of any functor $S : \mathscr{I} \to \mathscr{C}$. In particular, we shall say that F *commutes with direct sums* (respectively, *with direct products*) if F commutes with the inductive (respectively, projective) limit of any functor $S : I \to \mathscr{C}$, where I is a set.

THEOREM 5.4. *Let $F : \mathscr{C} \to \mathscr{C}'$ be a covariant functor and let G be its right adjoint. Then*

(1) *F commutes with the inductive limit in $\mathscr{C}$ of any functor,*

(2) *G commutes with the projective limit in $\mathscr{C}'$ of any functor.*

COROLLARY 5.5. *Let $F : \mathscr{C} \to \mathscr{C}'$ be a functor and let G be a full and faithful right adjoint of F. Then*

(1) *If $\mathscr{C}$ is a category with inductive (respectively, projective) limits then $\mathscr{C}'$ is also a category with inductive (respectively, projective) limits,*

(2) *If $\{U_i\}_\iota$ is a set of generators of $\mathscr{C}$, then $\{F(U_i)\}_\iota$ is a set of generators of $\mathscr{C}'$.*

EXAMPLES OF ADJOINT FUNCTORS

17. Let $I : \mathscr{Ab} \to \mathscr{G}$ be the inclusion functor, (obviously, $\mathscr{Ab}$ may be conceived as a full subcategory of $\mathscr{G}$). The functor $A : \mathscr{G} \to \mathscr{Ab}$, which was defined in Example 9, is a left adjoint of I. Obviously, $AI \equiv \mathrm{Id}\mathscr{Ab}$ and consequently, the identity morphism of $\mathrm{Id}\mathscr{Ab}$ can be taken for an adjunction morphism from A to I.

18. Let $\mathscr{Top}c$ be the category of compact topological spaces.

We shall denote by $U : \mathscr{Top}c \to \mathscr{Top}$ the inclusion functor. Also, we shall denote by $C : \mathscr{Top} \to \mathscr{Top}c$ the Stone–Čech compactification functor (see [24]). It will be noticed that C is a left adjoint of U and that U is full and faithful.

Exercises

1. A category with only one object is a *monoid.* A functor from one monoid to another is a *homomorphism.*

2. A monoid in which every element has an inverse is a *group.* Let F and G be two functors, each from a group A to a group B, and let $u : F \to G$ be a functorial morphism. There then is an $x \in B$ such that for all $y \in A$, $F(y) = xG(y)x^{-1}$ i.e., F and G are "conjugate" homomorphisms. An inner automorphism is a functor isomorphic to the identity functor.

3. Consider the categories $\mathscr{Set}$, $\mathscr{Ab}$, $\mathscr{Top}$, $\mathscr{G}$. In these categories a morphism is a monomorphism if and only if it is injective and an epimorphism if and only if it is surjective. All four categories are locally small, colocally small, have inductive and projective limits. Every non-empty set is a generator and every set containing a subset of two elements is a cogenerator of $\mathscr{Set}$, What are the generators and cogenerators in $\mathscr{Top}$? Let $\mathbf{Z}$ be the additive group of integers. Then $\mathbf{Z}$ is a generator of $\mathscr{G}$ and $\mathscr{Ab}$. Let $\mathbf{Q}$ the group of rational numbers. Then the quotient group $\mathbf{Q}/\mathbf{Z}$ is a cogenerator of $\mathscr{Ab}$.

4. In the category of all Hausdorff spaces, a morphism can be an epimorphism without being surjective as a function. In fact, in order that a continuous function $f : X \to Y$ be an epimorphism it suffices that $f(X)$ be a dense subspace of Y.

5. Let $\mathscr{C}$ be a category with kernels (cokernels) and $\{U_i\}_i$ a set of S-generators (resp. S-cogenerators). Then $\{U_i\}_i$ is a set of generators (cogenerators) of $\mathscr{C}$.

6. Let $\mathscr{C}$ be a category and 0 a null object of $\mathscr{C}$. For any object X of $\mathscr{C}$ there is a unique morphism $0_{0X} : 0 \to X$, and a unique morphism $0_{X0} : X \to 0$. For $X, Y \in \mathrm{Ob}\,\mathscr{C}$ we write $0_{XY} = 0_{0Y}0_{X0}$. We assume that for every morphism

$f: X \to Y$ the couple $(f, 0_{XY})$ has a cokernel (respectively, kernel). Then a set $\{U_i\}_i$ of generators (respectively, cogenerators) of $\mathscr{C}$ is a set of S-generators (respectively, S-cogenerators).

7. We denote by $\mathscr{C}at$ the category of small categories: the objects are small categories and the morphisms are covariant functors. $\mathscr{C}at$ is a category with inductive and projective limits. The category consisting of two objects 1, 2 and three morphisms: $u: 1 \to 2$ and identity morphisms, is a generator of $\mathscr{C}at$.

8. Let $\mathscr{C}$ be a category and $\mathscr{I}$ a small category. We denoted by $[\mathscr{I}, \mathscr{C}]$ the category of covariant functors from $\mathscr{I}$ to $\mathscr{C}$, and morphisms, the functorial morphisms. If $\mathscr{C}$ is a category having kernels, cokernels, inductive or projective limits, then the category $[\mathscr{I}, \mathscr{C}]$ has the same property. Particularly, we write $[\mathscr{I}, \mathscr{S}et] = \mathscr{I}^{\wedge}$. If $F: \mathscr{I} \to \mathscr{I}'$ is a covariant functor we shall denote by $F_*: \mathscr{I}'^{\wedge} \to \mathscr{I}^{\wedge}$ the restriction functor: for $X \in \operatorname{Ob} \mathscr{I}'^{\wedge}$, we have $F_*(X) = XF$. Then F_* has a left adjoint F^*, and a right adjoint F^+.

9. Let $\mathscr{I}$ be a small category; we denote by $h: \mathscr{I} \to \mathscr{I}^{\wedge}$ the functor: $h(i) = h_i$, $i \in \operatorname{Ob} \mathscr{I}$. Then h is a fully faithful functor and it commutes with projective limits. If $\mathscr{C}$ is a category with inductive limits and $t: \mathscr{I} \to \mathscr{C}$ a covariant functor, then a unique functor $t^*: \mathscr{I}^{\wedge} \to \mathscr{C}$ exists (up to isomorphism) such that:

(a) $t^* h \simeq t$;

(b) t^* has a right adjoint.

Every covariant functor $F: \mathscr{I}^{\wedge} \to \mathscr{C}$ which commutes with inductive limits has right adjoint. (Hint. A right adjoint of t^* is the functor $t_*: \mathscr{C} \to \mathscr{I}^{\wedge}$ such that for any object $X \in \operatorname{Ob} \mathscr{C}$, $t_*(X): \mathscr{I} \to \mathscr{C}$ is the functor : $t_*(X)(i) = \operatorname{Hom}_{\mathscr{C}}(t(i), X)$. For the construction of t^* we observe that any object of $\mathscr{I}^{\wedge}$ is a canonical limit of representable functors).

10. Let $\mathscr{M}$ be the category of monoids and $V: \mathscr{G} \to \mathscr{M}$ the inclusion functor. Then V has a left adjoint, denoted by R. If $\mathbf{N}^+$ is the additive monoid of natural numbers, then $R(\mathbf{N}^+)$ is canonically isomorphic with the group $\mathbf{Z}$ of integers. Also, if $\mathbf{N}^{\times}$ is the multiplicative monoid of non-zero natural numbers, then $R(\mathbf{N}^{\times})$ is canonically isomorphic with the group of all strictly positive rational numbers. Finally, if $\mathbf{Q}_+$ is the additive monoid of all non-negative rational numbers then $R(\mathbf{Q}_+)$ is the additive group of rational numbers.

2. Abelian categories

The theory of abelian categories is closely related to the theory of abelian groups, which represents the most important model for abelian categories.

This chapter has a technical character, containing the basic notions and results of the theory of abelian categories; we shall use them freely in the next chapters. In 2.8 we introduce Grothendieck's conditions Ab (see [79]) and we prove some results concerning them, the main one being Theorem 8.6 which gives equivalent forms for condition Ab5.

2.1 Preadditive and additive categories

We shall say that a category $\mathscr{C}$ is *preadditive* if the following condition is satisfied:

($Ad1$) For any couple of objects (X, Y) in $\mathscr{C}$, $\mathrm{Hom}_{\mathscr{C}}(X, Y)$ has a structure of an abelian group such that the composition map:

$$\theta(X, Y, Z): \mathrm{Hom}_{\mathscr{C}}(X, Y) \times \mathrm{Hom}_{\mathscr{C}}(Y, Z) \to \mathrm{Hom}_{\mathscr{C}}(X, Z)$$

is bilinear, that is, for any, $f, f' \in \mathrm{Hom}_{\mathscr{C}}(X, Y)$ and, $g, g' \in \mathrm{Hom}_{\mathscr{C}}(Y, Z)$ we have: $\theta(X, Y, Z)\ (f + f', g) = \theta(X, Y, Z)\ (f, g) + \theta(X, Y, Z)\ (f', g)$ and $\theta(X, Y, Z)\ (f, g + g') = \theta(X, Y, Z)\ (f, g) + \theta(X, Y, Z)\ (f, g')$, X, Y, Z being arbitrary objects in $\mathscr{C}$.

NOTES

(1) If we put $\theta(X, Y, Z)(f, g) = gf$, then the last equalities take the following form:

$$g(f + f') = gf + gf',$$
$$(g + g')f = gf + g'f.$$

(2) For any couple (X, Y) of objects we shall denote by 0_{XY} (or simply by 0, if there is no confusion) the null element of the group $\mathrm{Hom}_{\mathscr{C}}(X, Y)$. Then for any three objects (X, Y, Z) of $\mathscr{C}$ we have: $0_{YZ}0_{XY} = 0_{XZ}$. Indeed, for any $f \in \mathrm{Hom}_{\mathscr{C}}(X, Y)$, we get:

$$0_{YZ}(0_{XY} + f) = 0_{YZ}f = 0_{XZ} = 0_{YZ}0_{XY} + 0_{YZ}f.$$

(3) Sometimes it is convenient to assume that a null object 0 exists in $\mathscr{C}$. Then, for any object X, the group $\mathrm{Hom}_{\mathscr{C}}(0, X)$ (respectively the group $\mathrm{Hom}_{\mathscr{C}}(X, 0)$) reduces to the element 0_{0X} (respectively 0_{X0}). Obviously, for any couple (X, Y) of objects of $\mathscr{C}$ one has: $0_{XY} = 0_{0Y}0_{X0}$.

EXAMPLES

1. The category $\mathscr{Ab}$ of abelian groups is a preadditive category. Indeed, if X, Y are two abelian groups, then the set $\mathrm{Hom}_{\mathscr{Ab}}(X, Y)$ has canonically an abelian group structure: for $f, g \in \mathrm{Hom}_{\mathscr{Ab}}(X, Y)$ we put $f + g : X \to Y$ for the homomorphism of groups defined by $(f + g)(x) = f(x) + g(x)$. We see immediately that the composition of morphisms is bilinear.

2. Let A be a ring (with unit element). Then A is a preadditive category and will be denoted by the same symbol A. The category A has a single object *, and $\mathrm{Hom}_{\mathscr{A}}(*, *) = A$. The composition of morphisms is the multiplication of elements of A. The group structure of $\mathrm{Hom}_{\mathscr{A}}(*, *)$ is that of the underlying additive group A. Conversely, a preadditive category $\mathscr{C}$, with a single object, is in fact a ring. Indeed, if X is an object in a preadditive category $\mathscr{C}$, then the set $\mathrm{Hom}_{\mathscr{C}}(X, X)$ is a ring. This ring will be frequently denoted by $\mathrm{End}_{\mathscr{C}}(X)$.

3. Canonically, the category $\mathscr{C}^0$, the dual of a preadditive category, is also preadditive. If A is a ring, we denote by A^0 the dual or the opposite ring of A, that is, the dual category of A.

4. Let $\mathscr{C}$ be a preadditive category and let X be an object of $\mathscr{C}$. We denote by $X/\mathscr{C}$ the category whose objects are couples (f, Y), $Y \in \mathrm{Ob}\,\mathscr{C}$ and $f \in \mathrm{Hom}_{\mathscr{C}}(X, Y)$ and whose morphisms $g : (f, Y) \to (f', Y')$ are in fact morphisms $g : Y \to Y'$ such that $gf = f'$. $X/\mathscr{C}$ is a preadditive category. Analogously, we have the category $\mathscr{C}/X$.

5. Let $\mathscr{C}$ be a preadditive category. A subcategory $\mathscr{C}'$ of $\mathscr{C}$ is preadditive if for any two objects X, Y of $\mathscr{C}'$, the set $\mathrm{Hom}_{\mathscr{C}'}(X, Y)$ is a subgroup of $\mathrm{Hom}_{\mathscr{C}}(X, Y)$. In fact $\mathscr{C}'$ is a preadditive category if and only if, for any two objects X, Y of $\mathscr{C}'$ and for any $f, g \in \mathrm{Hom}_{\mathscr{C}'}(X, Y)$ the difference $f - g$ in $\mathrm{Hom}_{\mathscr{C}}(X, Y)$ is also in $\mathrm{Hom}_{\mathscr{C}'}(X, Y)$. If A is a ring, a preadditive subcategory of A is in fact a subring.

Let $\mathscr{C}$ be a preadditive category. We call *left* (respectively, *right*) *ideal of* $\mathscr{C}$, a preadditive subcategory $\mathscr{I}$ of $\mathscr{C}$ which is subject to the following condition: if $f : X \to Y$ is a morphism of $\mathscr{I}$ and $g : Y \to Z$ (respectively, $g : Z \to X$) is a morphism of $\mathscr{C}$ then gf (respectively, fg) is a morphism of $\mathscr{I}$. A *two-sided ideal of* $\mathscr{C}$ is a subcategory of $\mathscr{C}$ which is at the same time a left and right ideal.

Let X be an object of $\mathscr{C}$. A left (respectively, right) ideal in the category $X/\mathscr{C}$ (respectively, $\mathscr{C}/X$) is called *a left* (respectively, *right*) *ideal of* X.

Let A be a ring; we denote by $\mathfrak{a}, \mathfrak{b}, \mathfrak{c}, \ldots$ the left, right or two-sided ideals of A.

Let $\mathscr{I}$ be a two-sided ideal in $\mathscr{C}$. Then a new category $\mathscr{C}/\mathscr{I}$ may be constructed in the following manner. The objects of $\mathscr{C}/\mathscr{I}$ are the object of $\mathscr{C}$. If X is an object of $\mathscr{C}$, then the same object when considered in $\mathscr{C}/\mathscr{I}$ will be denoted by $\bar{X}$. If $\bar{X}$, $\bar{Y}$ are two objects in $\mathscr{C}/\mathscr{I}$, then $\mathrm{Hom}_{\mathscr{C}/\mathscr{I}}(\bar{X}, \bar{Y})$ is the quotient group $\mathrm{Hom}_{\mathscr{C}}(X, Y)/\mathrm{Hom}_{\mathscr{I}}(X, Y)$. We leave to the reader the task of verifying that $\mathscr{C}/\mathscr{I}$ is a preadditive category. $\mathscr{C}/\mathscr{I}$ is called the *factor category* of $\mathscr{C}$ by $\mathscr{I}$. Particularly, if A is a ring and $\mathfrak{a}$ an ideal of A we shall have the factor ring of A by $\mathfrak{a}$.

We shall always denote by $\mathscr{C}$ a preadditive category. Also, let $f : X \to Y$ be morphism of $\mathscr{C}$; we shall say that f has a *kernel* (respectively, *cokernel*) if the couple $(f, 0)$, has a kernel (respectively, cokernel). In other words, there is a morphism $t : X' \to X$ such that $ft = 0$ and for any morphism $g : Z \to X$ such that $fg = 0$ there exists a unique morphism $g' : Z \to X'$ such that $tg' = g$. The dual condition is valid for a cokernel. From these considerations we derive the following result:

PROPOSITION 1.1. *Let $(f, g) : X \to Y$ be a couple of morphisms of $\mathscr{C}$. The following assertions are equivalent*:

(1) *The couple (f, g) has a kernel (respectively, cokernel)*;

(2) *The morphism $f - g$ has a kernel (respectively, cokernel)*;

(3) *The morphism $g - f$ has a kernel (respectively, cokernel).* ∎

This proposition implies that if the morphism $f : X \to Y$ has a kernel (respectively, cokernel), then $-f$, the opposite morphism of f in the group $\mathrm{Hom}_{\mathscr{C}}(X, Y)$ has the same kernel (respectively, cokernel).

We have the following result concerning finite direct sums and finite direct products in a preadditive category.

THEOREM 1.2. *Let $\{X_i\}_{1 \leqslant i \leqslant n}$ be a finite set of objects of a preadditive category $\mathscr{C}$. The following assertions are equivalent*:

(1) *The given set of objects has a direct sum*;

(2) *The given set of objects has a direct product*;

(3) *There exist an object X of $\mathscr{C}$ and morphisms $u_i : X_i \to X$, $p_i : X \to X_i$, $1 \leqslant i \leqslant n$, such that*

$$\left.\begin{aligned} &\text{(a)}\ \sum_{i=1}^{n} u_i p_i = 1_X, \\ &\text{(b)}\ p_i u_j = \begin{cases} 0 & \text{if } i \neq j, \\ 1_{X_i} & \text{if } i = j. \end{cases} \end{aligned}\right\} \tag{1}$$

Moreover, the direct sum and the direct product of the above set of objects are canonically isomorphic.

Proof. (1) ⇒ (3). Let

$$X = \coprod_{i=1}^{n} X_i$$

and let $u_i : X_i \to X$, $1 \leqslant i \leqslant n$ be the structural morphisms. Then, for any $1 \leqslant i \leqslant n$ there is a unique morphism $p_i : X \to X_i$ such that: $p_i u_j = 0$ if $i \neq j$ and $p_i u_i = 1_{X_i}$. These morphisms are subject to the conditions (1). Let

$$f = \sum_i u_i p_i.$$

Then for any $1 \leqslant j \leqslant n$ we have:

$$fu_j = \left(\sum_i u_i p_i\right) u_j = \sum_i u_i p_i u_j = u_j.$$

From the definition of the direct sum we derive that $f = 1_X$.

(3) ⇒ (2). We shall prove that X and the morphisms $p_i : X \to X_i$, $1 \leqslant i \leqslant n$ define a direct product of the above set of objects. In order to do this, let Y be an object of $\mathscr{C}$ and let us consider for any $1 \leqslant i \leqslant n$, a morphism $g_i : Y \to X_i$. Also, let

$$g = \sum_i u_i g_i.$$

Then

$$p_j g = p_j \sum_i u_i g_i = \sum_i p_j u_i g_i = g_j,$$

according to the relations (a). Now we shall prove the uniqueness of g. Let $g' : Y \to X$ be another morphism such that $p_j g' = g_j$, $1 \leqslant j \leqslant n$. Then we have: $u_j p_j g' = u_j g_j$. Summarizing the last relations and conforming to relations (1) we get:

$$g' = \sum_j u_j p_j g' = \sum_j u_j g_j = g.$$

(2) ⇒ (1). The proof is now completed by reasoning in $\mathscr{C}^0$ the dual category of $\mathscr{C}$ and taking note of the fact that the relations (1) are self-dual. ■

Let

$$X = \coprod_i X_i, \qquad Y = \prod_i X_i \quad \text{and} \quad \text{let } u_i : X_i \to X$$

$p_i' : Y \to X_i$, $1 \leqslant i \leqslant n$ be the structural morphisms. There is a unique morphism $t : X \to Y$ such that $p_i' t = p_i$, where p_i are defined in (a). From Theorem 1.2 we derive that t is an isomorphism, and we may assume that $X = Y$ and $p_i = p_i'$, $i \leqslant 1 \leqslant n$. In the sequel, we shall usually identify in the same manner the direct sum with the direct product and this will be done for any finite set of objects from a preadditive category (whenever these notions have a meaning).

NOTES

1. In the relations (1) p_i is called the *projection associated with* u_i and dually u_i is called the *injection associated with* p_i.

A preadditive category $\mathscr{C}$ is said to be *additive* if the following condition is fulfilled:

($Ad\,2$). For any pair of objects (X, Y) in $\mathscr{C}$, there exists the direct sum $X \amalg Y$.

From the Theorem 1.2 we deduce that an additive category is a preadditive category with finite direct sums and finite direct products.

2. We shall tacitly assume that an additive category has a null object.

EXAMPLES

6. The category $\mathscr{Ab}$ of abelian groups is an additive category. Indeed if X, Y are two abelian groups we shall consider the cartesian product $X \Pi Y$ where the addition of elements is componentwise. Let $p : X \Pi Y \to X$, $q : X \Pi Y \to Y$ be the natural projections. Then $X \Pi Y$, p and q define a direct product of X and Y.

Let $\mathscr{C}$ be an arbitrary category with a null object 0 and let X, Y be two objects of $\mathscr{C}$. We shall denote by $0 : X \to Y$ the null morphism from X to Y. We shall assume that $X \amalg Y$ and $X \Pi Y$ exist in $\mathscr{C}$. Let $u : X \to X \amalg Y$, $v : Y \to X \amalg Y$ and also let $p : X \Pi Y \to X$, $q : X \Pi Y \to Y$ be the structural morphisms. Then, a unique morphism $t(X, Y) : X \amalg Y \to X \Pi Y$ exists, such that $pt(X, Y)\,u = 1_X$, $qt(X, Y)\,u = 0$, $qt(X, Y)\,v = 1_Y$, $pt(X, Y)\,v = 0$.

THEOREM 1.3. *Let $\mathscr{C}$ be a category. The following assertions are equivalent*:

(1) *$\mathscr{C}$ is an additive category*;

(2) *The following conditions are satisfied*;

(A_1) *$\mathscr{C}$ has a null object*;

(A_2) *$\mathscr{C}$ has finite direct sums and products*;

(A_3) *For any couple (X, Y) of objects of $\mathscr{C}$ the morphism $t(X, Y)$: $X \amalg Y \to X \Pi Y$ defined above is an isomorphism*;

(A_4) *For any object X of $\mathscr{C}$ there is a morphism $c(X) : X \to X$ such that the following diagram is commutative*:

$$\begin{array}{ccc} X & \xrightarrow{c(X)} & X \\ {\scriptstyle \Delta_X}\downarrow & & \uparrow{\scriptstyle \nabla_X} \\ X \amalg X & \xrightarrow{c(X) \amalg 1_X} & X \amalg X \end{array}$$

where the morphisms Δ_X and ∇_X are canonically defined below.

Proof. (1) $\Rightarrow$ (2). The proof results from the above, by taking $c(X) = -1_X$.

(2) $\Rightarrow$ (1). The hypothesis tells us that $X \amalg Y = X \Pi Y$ and $t(X, Y)$ is the identity morphism for any couple (X, Y) of morphisms of $\mathscr{C}$.

Let $\Delta_X : X \to X \Pi X$ the unique morphism such that $p_i \Delta_X = 1_X$, $i = 1, 2$, p_i: $X \Pi X \to X$ being the structural morphism; Δ_X is called the *diagonal morphism* of X. Dually we define the *codiagonal morphism* $\nabla_X : X \amalg X \to X$.

Let $f, g \in \mathrm{Hom}_{\mathscr{C}}(X, Y)$. We shall define $f + g$ to be the composition of the following morphisms

$$X \xrightarrow{\Delta_X} X \amalg X \xrightarrow{f \amalg g} Y \amalg Y \xrightarrow{\nabla_Y} Y$$

We may derive that the above addition of the morphisms in $\mathrm{Hom}_{\mathscr{C}}(X, Y)$ is commutative, associative, 0 is a neutral element and composition of morphism is distributive with respect to the addition. For an element

$$f \in \mathrm{Hom}_{\mathscr{C}}(X, Y)$$

we get $-f = fc(X) = c(Y)f$; $-f$ is the opposite element of f in the above addition. The details are left to the reader. ∎

Exercises

1. Let $\mathscr{C}$ be a preadditive category and $\mathscr{I}$ a two-sided ideal of $\mathscr{C}$. We consider the factor category $\mathscr{C}/\mathscr{I}$ and the functor $T : \mathscr{C} \to \mathscr{C}/\mathscr{I}$ defined as follows: $T(X) = \bar{X}$ for any object X of $\mathscr{C}$, and for any morphism

$$f \in \mathrm{Hom}_{\mathscr{C}}(X, Y)$$

we get $T(f) = \bar{f}$ the class of f modulo $\mathrm{Hom}_{\mathscr{I}}(X, Y)$. Then

(a) For any couple (X, Y) of objects of $\mathscr{C}$ the map $T(X, Y) : \mathrm{Hom}_{\mathscr{C}}(X, Y) \to \mathrm{Hom}_{\mathscr{C}/\mathscr{I}}(\bar{X}, \bar{Y})$ is a homomorphism of groups.

(b) Let $\mathscr{I}'$ be a left (respectively, right) ideal of $\mathscr{C}/\mathscr{I}$. We denote by $T^{-1}(\mathscr{I}')$ the subcategory of $\mathscr{C}$ defined as follows. The objects of $T^{-1}(\mathscr{I}')$ are the objects X of $\mathscr{C}$ such that $T(X)$ is an object in $\mathscr{I}'$. If (X, Y) are two objects of $T^{-1}(\mathscr{I}')$, then $\mathrm{Hom}_{T^{-1}(\mathscr{I}')}(X, Y)$ is the subset of $\mathrm{Hom}_{\mathscr{C}}(X, Y)$ consisting of those elements f such that $T(f) \in \mathrm{Hom}_{\mathscr{I}'}(\bar{X}, \bar{Y})$. We derive that $T^{-1}(\mathscr{I}')$ is a left (respectively, right) ideal of $\mathscr{C}$. The map $\mathscr{I}' \rightsquigarrow T^{-1}(\mathscr{I}')$ is a bijection from the left (respectively, right) ideals of $\mathscr{C}/\mathscr{I}$ to the left (respectively, right) ideals of $\mathscr{C}$ which contain $\mathscr{I}$. In this correspondence the two-sided ideals are preserved.

2. For any category $\mathscr{C}$ define a category $\mathrm{Add}(\mathscr{C})$ as follows. The objects of $\mathrm{Add}(\mathscr{C})$ are the same as the objects of $\mathscr{C}$. The set of morphisms from X to Y in $\mathrm{Add}(\mathscr{C})$ is the free abelian group generated by the elements of

$\mathrm{Hom}_{\mathscr{C}}(X, Y)$; that is, the set of all finite formal linear combinations of the form

$$\sum_i n_i f_i$$

where n_i is an integer and $f_i \in \mathrm{Hom}_{\mathscr{C}}(X, Y)$. Composition in $\mathrm{Add}(\mathscr{C})$ is defined by the rule:

$$(\sum n_i f_i)(\sum m_j g_j) = \sum_{i,j} (n_i m_j) f_i g_j.$$

If $\mathrm{Hom}_{\mathscr{C}}(X, Y)$ is empty, then $\mathrm{Hom}_{\mathrm{Add}(\mathscr{C})}(X, Y)$ contains a single element: the null morphism.

Then $\mathrm{Add}(\mathscr{C})$ is an preadditive category which contains $\mathscr{C}$ as a subcategory. If $\mathscr{C}$ is a category with a null object, then we can factor $\mathrm{Hom}_{\mathrm{Add}(\mathscr{C})}(X, Y)$ by the subgroup generated by 0_{XY}, and in this way we obtain a factor category $\overline{\mathrm{Add}(\mathscr{C})}$ which is preadditive, contains $\mathscr{C}$ as a subcategory, and has the same set of zero morphisms as $\mathscr{C}$. $\mathrm{Add}(\mathscr{C})$ is the factor category of $\overline{\mathrm{Add}(\mathscr{C})}$ by a suitable two-sided ideal.

3. Let $f : X \to Y$ be a morphism in the preadditive category $\mathscr{C}$. We assume that the following conditions are satisfied: (a) the morphism f possesses a kernel; (b) there is a morphism $g : Y \to X$ such that $fg = 1_Y$. Then X is the direct sum of Y and a suitable object. Dualize this result.

4. Let $f : X \to Y$ be a morphism in a preadditive category $\mathscr{C}$. The following assertions are equivalent:

(a) f is a monomorphism (epimorphism).

(b) 0 is a kernel (cokernel) for f.

5. An *idempotent* in a preadditive category $\mathscr{C}$ is a morphism $f : X \to X$ such that $f^2 = f$. We say that the idempotent f *splits* in $\mathscr{C}$ if there is an object Y and the morphisms $g : X \to Y$, $h : Y \to X$ such that $hg = f$ and $gh = 1_Y$. We say that idempotents split in $\mathscr{C}$ if every idempotent of $\mathscr{C}$ splits. Then:

(a) If every idempotent may be factored into an epimorphism followed by a monomorphism, then idempotents split.

(b) Let $\overline{\mathscr{C}}$ be the category whose objects are pairs (X, f) where $X \in \mathrm{Ob}\,\mathscr{C}$ and $f : X \to X$ is an idempotent. The morphisms from (X_1, f_1) to (X_2, f_2) are defined to be those morphisms $g : X_1 \to X_2$ such that $f_2 g f_1 = g$. Prove that $\overline{\mathscr{C}}$ is a preadditive category in which idempotents split. If $\mathscr{C}$ is an additive category then $\overline{\mathscr{C}}$ is additive.

6. Let $\mathscr{C}$ be an additive category, $\{X_i\}_{1 \leqslant i \leqslant n}$ and $\{Y_j\}_{1 \leqslant j \leqslant m}$ two families of objects of $\mathscr{C}$. Then a morphism

$$f : \coprod_i X_i \to \coprod_j Y_j$$

is canonically defined by an $n \times m$ morphism $f_{ij}: X_i \to Y_j$, that is by a $n \times m$ matrix $(f_{ij})_{\substack{1 \leq i \leq n \\ 1 \leq j \leq m}}$.

7. Let $f = (f_{ij}): X_1 \amalg X_2 \to X_1 \amalg X_2$ be an idempotent morphism in an additive category $\mathscr{C}$ with kernels, such that $f_{21} = 0$. Then f_{11} and f_{22} are idempotents, and so we can find a factorization of f_{11} and f_{22} of the form:

$$X_1 \xrightarrow{g_1} Y_1 \xrightarrow{g_2} X_1$$

$$X_2 \xrightarrow{h_1} Y_2 \xrightarrow{h_2} X_2$$

such that $g_1 g_2 = 1_{Y_1}$, and $h_1 h_2 = 1_{Y_2}$. Find morphisms $m: X_1 \amalg X_2 \to Y_1 \amalg Y_2$ and $n: Y_1 \amalg Y_2 \to X_1 \amalg X_2$ such that $nm = f$ and $mn = 1_{Y_1 \amalg Y_2}$. Prove that if $\mathrm{Hom}_{\mathscr{C}}(X_1, X_2) = 0$, then any retract of $X_1 \amalg X_2$ is isomorphic to an object of the form $Y_1 \amalg Y_2$, where Y_1 is a retract of X_1 and Y_2 is a retract of X_2.

References

Buchsbaum [28]; Bucur [30]; Bucur–Deleanu [31]; Burmistrovich [32]; Cartan–Eilenberg [33]; Freyd [57]; Gabriel [59]; Grothendieck [79]; Heller [88]; MacLane [120]; Mitchell [133]; Pareigis [163].

2.2 The canonical factorization of a morphism. Preabelian categories

Let $\mathscr{C}$ be a preadditive category. Let $f: X \to Y$ be a morphism of $\mathscr{C}$ and let $i: X' \to X$ (respectively, $p: Y \to Y'$) be a kernel (cokernel) of f. Then a cokernel (respectively, kernel) of i (respectively, of p), (if it exists), is called a *coimage* (respectively, *image*) of f. We shall write this as coim f (respectively im f).

PROPOSITION 2.1. *Let $\mathscr{C}$ be a preadditive category and let $f: X \to Y$ be a morphism in $\mathscr{C}$. Assume that there exist* $\ker f$, $\operatorname{coker} f$, $\operatorname{im} f$, *and* $\operatorname{coim} f$. *Then there exists a unique morphism $\bar{f}$*: $\operatorname{coim} f \to \operatorname{im} f$ *such that f is the composition of the following morphisms*:

$$X \xrightarrow{p} \operatorname{coim} f \xrightarrow{\bar{f}} \operatorname{im} f \xrightarrow{j} Y,$$

where p and j are canonically defined.

Proof. The uniqueness is checked as follows: If $j \bar{f} p = j \bar{\bar{f}} p$, then from the fact that j is a monomorphism it follows that $\bar{f} p = \bar{\bar{f}} p$ and from the fact that p is an epimorphism it follows $\bar{f} = \bar{\bar{f}}$.

In order to prove the existence of $\bar{f}$, consider the following diagram:

$$\begin{array}{ccccc} \ker f & \xrightarrow{i} & X & \xrightarrow{p} & \operatorname{coker} i = \operatorname{coim} f \\ & & \downarrow{\scriptstyle f} & & \\ \operatorname{coker} f & \xleftarrow{q} & Y & \xleftarrow{j} & \ker q = \operatorname{im} f. \end{array}$$

Since $f\,i = 0$, there exists a morphism $f' : \operatorname{coim} f \to Y$ such that $f'\,p = f$. But $q f' = 0$. Since p is an epimorphism, it is enough to prove that $(q\,f')p = 0$, which is immediate, since $q(f'\,p) = q\,f = 0$.

Then there exists $\bar{f} : \operatorname{coim} f \to \operatorname{im} f$ such that $j\bar{f} = f'$. ■

We shall say that $\bar{f}$ is the *parallel of* f, and that the equality $f = j\,\bar{f}p$ is the *canonical factorization* of f.

An additive category $\mathscr{C}$ is said to be *preabelian* if for any morphism f in $\mathscr{C}$ there exist $\ker f$ and $\operatorname{coker} f$.

It follows immediately that in a preabelian category, any morphism possesses an image, a coimage and a canonical factorization.

EXAMPLES

1. The category $\mathscr{Ab}$ is preabelian. Indeed if $f : X \to Y$ is a morphism of abelian groups, we denote by $\ker f$ the subset of X consisting of elements x for which $f(x) = 0$. Also we denote the natural inclusion by $i : \ker f \to X$. $\operatorname{Ker} f$ is a subgroup of X and is obviously a kernel of f. The cokernel of f is defined as follows: we consider the subgroup $f(X)$ of Y and we put $\operatorname{coker} f = Y/f(X)$. If $p : Y \to \operatorname{coker} f$ is the canonical morphism, then $(p, \operatorname{coker} f)$ is a cokernel of f.

2. Consider the category $\mathscr{C}$ defined as follows:

(i) the objects of $\mathscr{C}$ are Hausdorff abelian topological groups,

(ii) the morphisms of $\mathscr{C}$ are the continuous homomorphisms.

Obviously, $\mathscr{C}$ is an additive category (with the direct product of two objects being the direct product of abelian groups with the product topology). However, if $f : X \to Y$ is a morphism in the category $\mathscr{C}$, then $\ker f$ is the kernel of f, considered as a morphism of abelian groups, with induced topology, and $\operatorname{coker} f = Y/\overline{f(X)}$, (where $\overline{f(X)}$ is the closure of $f(X)$ in Y), and has the quotient topology.

In this category there exist some bijections which are not isomorphisms. For instance, the natural continuous homomorphism $i : \mathbf{Q} \to \mathbf{R}$ ($\mathbf{Q}$ is the topological additive group of rational numbers and $\mathbf{R}$ the topological additive group of real numbers) is such that $i(\mathbf{Q}) \neq \mathbf{R}$ and $\overline{i(\mathbf{Q})} = \mathbf{R}$.

3. The dual of a preabelian category is preabelian.

In order to study the functorial properties of kernels, cokernels, images and coimages in a preabelian category $\mathscr{C}$, we define a new category $\mathscr{C}_1$, as follows:

(i) the objects of $\mathscr{C}_1$ are triples (X, f, Y) where $f : X \to Y$ is a morphism of $\mathscr{C}$, in other words the diagrams in $\mathscr{C}$ of the form $X \xrightarrow{f} Y$.

(ii) a morphism from the diagram $X \xrightarrow{f} Y$ to the diagram $X' \xrightarrow{f'} Y'$ is by definition a pair of morphisms $u : X \to X'$, $v : Y \to Y'$ such that $f' u = v f$.

In other words, if we consider a category $\mathscr{I}$ with two objects 1 and 2 and three morphisms: the identical morphisms and a morphism $1 \to 2$, then $\mathscr{C}_1$ is the category $[\mathscr{I}, \mathscr{C}]$ of covariant functors of $\mathscr{I}$ into $\mathscr{C}$ (See Chapter 3, Section 3.4). Obviously, $\mathscr{C}_1$ is an additive category.

We define a functor $\ker : \mathscr{C}_1 \to \mathscr{C}_1$ as follows. To the diagram $X \xrightarrow{f} Y$ we associate the diagram $\ker f \xrightarrow{i} X$, $(\ker f, i)$ being a kernel of f. If the pair (u, v) defines a morphism from the diagram $f : X \to Y$ to the diagram $f' : X' \to Y'$, then there exists a unique morphism $\ker f \to \ker f'$ such that the diagram:

$$\begin{array}{ccc} \ker f & \xrightarrow{i} & X \\ \downarrow & & \downarrow u \\ \ker f' & \xrightarrow{i'} & X' \end{array}$$

is commutative. In other words, we obtain a uniquely determined morphism from the diagram $i : \ker f \to X$ to the diagram $i : \ker f' \to X'$.

One defines similarly coker, im, coim: $\mathscr{C}_1 \to \mathscr{C}_1$.

The canonical factorization of a morphism in a preabelian category also leads to a functor defined on the category $\mathscr{C}_1$, but taking its values in a more complicated category $\mathscr{C}_2$ which is defined as follows:

(i) the objects of the category $\mathscr{C}_2$ are diagrams in $\mathscr{C}$ of the form:

$$X_1 \xrightarrow{f_1} X_2 \xrightarrow{f_2} X_3 \xrightarrow{f_3} X_4$$

(ii) a morphism from the diagram

$$X_1 \xrightarrow{f_1} X_2 \xrightarrow{f_2} X_3 \xrightarrow{f_3} X_4$$

to the diagram

$$Y_1 \xrightarrow{g_1} Y_2 \xrightarrow{g_2} Y_3 \xrightarrow{g_3} Y_4$$

is by definition an ordered system (u_1, u_2, u_3, u_4) of morphisms $u_i : X_i \to Y_i$ $(i = 1, 2, 3, 4)$ such that:

$$u_2 f_1 = g_1 u_1, \qquad u_3 f_2 = g_2 u_2, \qquad u_4 f_3 = g_3 u_3.$$

We define a functor $F : \mathscr{C}_1 \to \mathscr{C}_2$ as follows:

To the diagram $f : X \to Y$ we associate the diagram

$$X \xrightarrow{p} \operatorname{coim} f \xrightarrow{\bar{f}} \operatorname{im} f \xrightarrow{j} Y.$$

If $f' : X' \to Y'$ is another object of $\mathscr{C}_1$ and $u : X \to X'$, $v : Y \to Y'$ are such that $v f = f' u$, then there exist $u' : \operatorname{coim} f \to \operatorname{coim} f'$, $v' : \operatorname{im} f \to \operatorname{im} f'$ such that $u' p = p' u$, $j' v' = v j$, $p' : X' \to \operatorname{coim} f'$ and $j' : \operatorname{im} f' \to Y'$ being canonically defined.

It remains to show that $v' \bar{f} = \bar{f}' u'$. We have: $j' \bar{f}' u' p = j' \bar{f}' p' u = f' u$ and $j' v' \bar{f} p = v j \bar{f} p = v f$. In other words $j' \bar{f}' u' p = j' v' \bar{f} p$, whence $\bar{f}' u' = v' \bar{f}$ because j' is a monomorphism and p is an epimorphism.

Proposition 2.2. *In a preabelian category $\mathscr{C}$ there exist fibred products and sums.*

Proof. Since the dual of a preabelian category is preabelian, we shall confine ourselves to proving the existence of fibred products. Consider the diagram:

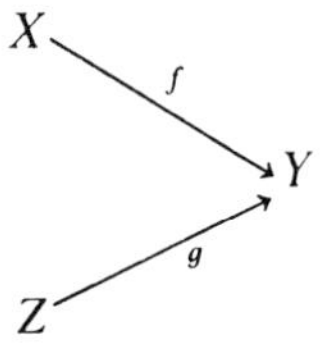

Consider the direct sum $X \amalg Z$ and the canonical projection $p : X \amalg Z \to X$, $q : X \amalg Z \to Z$; let $r : X \amalg Z \to Y$ be the morphism $r = f p - g q$ and let (T, i) be the kernel of r. We assert that the triple $(T, p i, q i)$ is a fibred product of the above diagram.

The relation $r i = 0$ implies $f(p i) = g(q i)$. However, if $t : U \to X$, $s : U \to Z$ are such that $f t = g s$, then there exists a unique $h : U \to T$ such that $p i h = t$, $q i h = s$. For, consider the morphism $ut + v s : U \to X \amalg Z$ (u, v are injections associated with projection p, q). We have $r(u t + v s) = (f p - g q)\,(u t + v s) = f t - g s = 0$. It follows that there exists a unique $m : U \to T$ such that $im = u t + v s$. Hence it follows immediately that $p i m = t, q i m = s$. The uniqueness follows from the fact that $p i m = t$ and $q i m = s$ imply $i m = u t + v s$. ■

Exercises

1. Let $f : X \to Y$ be a morphism in a preabelian category $\mathscr{C}$, $(\ker f, i)$ its kernel, $X \Pi_Y X$ the fibred product associated to the diagram

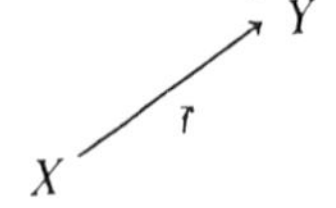

and let p_1, p_2 be the canonical projections of this fibred product. Then $\operatorname{coker}(p_1 - p_2) = \operatorname{coim} f$.

2. Let f be as above. We shall say that f is a *strict epimorphism* if it is an epimorphism and (f, Y) is a cokernel for $p_1 - p_2$. An epimorphism f of $\mathscr{C}$ is a strict epimorphism if and only if $\bar{f}$, the parallel of f, is an isomorphism.

3. Let $\mathscr{C}$ be a category with a null object. We shall say that $\mathscr{C}$ is a *normal category* if any monomorphism $i: X' \to X$ of $\mathscr{C}$ is the kernel of some morphism. $\mathscr{C}$ is said be a *conormal category* if $\mathscr{C}^0$ is normal.

Let $\mathscr{C}$ be a normal preabelian category. Then, there is an injective function from the class of subobjects of an object X to the class of quotient objects of X. In particular, if $\mathscr{C}$ is colocally small, then it is locally small. If $\mathscr{C}$ is normal and conormal, then the above function is a bijection between the class of subobjects of X and the class of quotient objects of X.

4. Let $\mathscr{C}$ be a preabelian category. Then the categories $\mathscr{C}_1$ and $\mathscr{C}_2$ considered above, are preabelian.

5. Let $\mathscr{C}$ be a preabelian category and $\mathscr{C}'$ a full subcategory. Then $\mathscr{C}'$ is a preabelian subcategory if and only if for every morphism $f: X \to Y$ in $\mathscr{C}'$, there is a kernel of f, a cokernel of f, and a direct sum of X and Y in $\mathscr{C}'$, all being in $\mathscr{C}$.

References

Buchsbaum [28]; Bănică–Popescu [15]; Bucur–Deleanu [31]; Freyd [57]; Gabriel [59]; Grothendieck [79]; Heller [88]; Mitchell [133]; Pareigis [163]; Raikov [179].

2.3 Abelian categories

A preabelian category $\mathscr{C}$ is called *abelian* if for any morphism $f: X \to Y$ in $\mathscr{C}$, the morphism $\bar{f}: \operatorname{coim} f \to \operatorname{im} f$, the parallel of f, is an isomorphism.

EXAMPLES

1. The category $\mathscr{Ab}$ is abelian.
2. The dual of an abelian category is abelian.

NOTE. The category considered in Example 2 of the previous paragraph is not abelian. This shows that there exist preabelian categories which are not abelian.

Proposition 3.1. *Let $\mathscr{C}$ be a preabelian category. The following assertions are equivalent:*

(1) *$\mathscr{C}$ is an abelian category;*

(2) *The parallel of any morphism is a bimorphism and every bimorphism of $\mathscr{C}$ is an isomorphism.*

Proof. If $\mathscr{C}$ is abelian and if $f : X \to Y$ is a bimorphism, then $\bar{f} = f$ (see Proposition 2.1). The converse assertion is obvious. ∎

Let $\mathscr{C}$ be an abelian category and X an object of $\mathscr{C}$. We denote by S the class of subobjects of X and by Q the class of quotient objects of X. These two classes are provided with natural partial ordering relations (we shall say that $(X', i') \leqslant (X'', i'')$ if and only if there is a morphism $u : X' \to X''$ such that $i'' u = i'$). Consider the function which associates to each subobject (X', i) of X the quotient object defined by the couple (coker i, p), where $p : X \to \operatorname{coker} i$ is the canonical epimorphism. Dually, consider the function which associates to each quotient object (X'', p) of X the subobject X defined by the pair $(\ker p, i)$, where $i : \ker p \to X$ is the canonical monomorphism. The aim of the following proposition is to show that these two functions are inverse to each other. Thus, it will follow in particular that there exists a natural bijection between the classes S and Q; this correspondence reverses the ordering relations on the two classes.

Proposition 3.2. *Let X be an object of an abelian category $\mathscr{C}$ and (X', i) a subobject of X. Let* $(\operatorname{coker} i, q)$ *be the cokernel of i. Then, the kernel of q coincides with (X', i). Dually, let (q, X'') be a quotient object of X, and* $(\ker p, i)$ *its kernel. Then (q, X'') is the cokernel of i.*

Proof. By definition $\ker(\operatorname{coker} i) = \operatorname{im} \mathrm{i}$. Let $j : \operatorname{im} i \to X$ be the canonical morphism. It will be sufficient to exhibit an isomorphism $f : X' \to \operatorname{im} i$ such that $j f = i$. To do this, consider the canonical factorization of

$$i : X' \xrightarrow{p} \operatorname{coim} i \xrightarrow{\bar{\imath}} \operatorname{im} i \xrightarrow{j} X.$$

Since i is a monomorphism, $\operatorname{coim} i = X'$ and $p = 1_{X'}$, $\mathscr{C}$ being abelian, $\bar{\imath}$ is an isomorphism. The proof is complete. ∎

Corollary 3.3. *Any monomorphism (respectively, epimorphism) in an abelian category is the kernel (respectively, cokernel) of a morphism.* ∎

If X' is a subobject of the object X in the abelian category $\mathscr{C}$, we shall denote by X/X' the quotient object of X which corresponds to X' under the bijection described above. Dually, if X'' is a quotient object of X, we shall sometimes denote by $X \setminus X''$ the subobject of X which corresponds to X'' under the above-mentioned bijection.

PROPOSITION 3.4. *Let $\mathscr{C}$ be an abelian category and let $\{U_i\}_{i \in I}$ be a set of objects of $\mathscr{C}$. The following assertions are equivalent:*

(1) *$\{U_i\}_i$ is a set of generators;*

(2) *$\{U_i\}_i$ is a set of S-generators.*

Proof. (1) ⇒ (2). Let $f: X \to Y$ be a monomorphism which is not an isomorphism, that is, the canonical morphism $p: Y \to \operatorname{coker} f$ is not null. There is an $i \in I$ and a morphism $g: U_i \to Y$ such that $p\,g \neq 0\,g = 0$.

(2) ⇒ (1). Let $f,\ g \in \operatorname{Hom}_{\mathscr{C}}(X, Y)$ be two distinct morphisms, that is, $f - g \neq 0$. Then, the canonical morphism $j: \ker(f - g) \to X$ is not an isomorphism. Then there is an $i \in I$ and a morphism $h: U_i \to X$ such that for any morphism $t: U_i \to \ker(f - g)$ we have that $j\,t \neq h$, that is $(f - g)$ $h \neq 0$, or $f\,h \neq g\,h$. ∎

Let $\mathscr{C}$ be an abelian category. We shall say that the sequence:

$$X \xrightarrow{\quad f \quad} Y \xrightarrow{\quad g \quad} Z$$

of objects and morphisms of $\mathscr{C}$, is *exact at* Y if $\ker g = \operatorname{im} f$. In other words, in the canonical factorization of the morphism f:

$$\begin{array}{ccccc} X & \xrightarrow{\quad f \quad} & Y & \xrightarrow{\quad g \quad} & Z \\ {\scriptstyle p}\downarrow & & \uparrow{\scriptstyle j} & & \\ \operatorname{coim} f & \xrightarrow{\quad \bar{f} \quad} & \operatorname{im} f & & \end{array}$$

we derive that $(\operatorname{im} f, j)$ is a kernel of g. Obviously, we have that $g\,f = 0$. We shall say that the sequence of objects and morphisms of $\mathscr{C}$

$$\cdots \longrightarrow X_{n-2} \longrightarrow X_{n-1} \longrightarrow X_n \longrightarrow X_{n+1} \longrightarrow \cdots$$

is *exact* if it is exact at X_n for every n. Particularly, the following sequence is exact at X':

$$0 \longrightarrow X' \xrightarrow{\quad f \quad} X$$

if and only if f is a monomorphism. Dually, the sequence:

$$X \xrightarrow{\quad p \quad} X'' \longrightarrow 0$$

is exact at X'' if and only if p is an epimorphism. Finally, we shall say that the sequence:

$$0 \longrightarrow X' \xrightarrow{\ i\ } X \xrightarrow{\ p\ } X'' \longrightarrow 0$$

is a *short exact sequence*, if it is exact at each object of the sequence. Equivalently, if X' is a subobject of X and $X'' = X/X'$, or X'' is a quotient object and $X' = X \diagdown X''$.

Sometimes, we shall utilize the so called *left short exact* sequence, that is an exact sequence of the form:

$$0 \longrightarrow X' \longrightarrow X \longrightarrow X''$$

Dually we have the *right short exact* sequence, i.e. an exact sequence of the form:

$$X' \longrightarrow X \longrightarrow X'' \longrightarrow 0$$

Sometimes, in order to emphasize that a morphism $f : X \to Y$ is a monomorphism (epimorphism), we shall write simply

$$0 \longrightarrow X \longrightarrow Y \text{ (respectively, } X \longrightarrow Y \longrightarrow 0)$$

This expression will usually be used in connection with diagrams.

For a morphism $f : X \to Y$ of C, we have the following exact sequence:

$$0 \longrightarrow \ker f \longrightarrow X \xrightarrow{f} Y \longrightarrow \operatorname{coker} f \longrightarrow 0$$

which is called the *exact sequence associated* with f.

PROPOSITION 3.5. *Let X_1, X_2 be two objects in an abelian category. Let $u_i : X_i \to X_1 \amalg X_2$, and $p_i : X_1 \amalg X_2 \to X_i$, $i = 1, 2$ the structural injections and projections. Then we have the exact sequences*:

$$0 \longrightarrow X_1 \xrightarrow{u_1} X_1 \amalg X_2 \xrightarrow{p_2} X_2 \longrightarrow 0$$

$$0 \longrightarrow X_2 \xrightarrow{u_2} X_1 \amalg X_2 \xrightarrow{p_1} X_1 \longrightarrow 0$$

Proof. Let $f : Y \to X_1 \amalg X_2$ be a morphism such that $p_2 f = 0$. Let us put $f' = p_1 f$. Then we have:

$$f = 1_{X_1 \amalg X_2} f = (u_1 p_1 + u_2 p_2) f = u_1 p_1 f + u_2 p_2 f = u_1 f',$$

according to the relations (1). We deduce that (X_1, u_1) is the kernel of p_2. ∎

We shall make some observations on morphisms in the category $\mathscr{Ab}$.

A morphism $f : X \to Y$ in the category $\mathscr{Ab}$ is a monomorphism if and only if it is injective, an epimorphism if and only if it is surjective and an isomorphism if and only if it is a bijection. If f is a monomorphism, then it is easy to see that f is a kernel for $p : Y \to \operatorname{coker} f$. Similarly, if f is an epimorphism, then it is a cokernel for $i : \ker f \to X$.

Exercises

1. If $\mathscr{C}$ is an arbitrary category, the following assertions are equivalent:

(a) $\mathscr{C}$ is an abelian category;

(b) $\mathscr{C}$ has a null object, kernels, cokernels, finite direct sums, finite direct products, and $\mathscr{C}$ is normal and conormal;

(c) $\mathscr{C}$ has a null object, fibred products and sums and $\mathscr{C}$ is normal and conormal.

(Hint: See Mitchell's book, [133]).

2. Let $\mathscr{C}$ be an abelian category. Then the categories $\mathscr{C}_1$ and $\mathscr{C}_2$ which were considered in 2.2 are abelian.

3. Suppose that the following diagram is commutative and has exact rows in $\mathscr{Ab}$.

$$\begin{array}{ccccccccc}
X_1 & \xrightarrow{u_1} & X_2 & \xrightarrow{u_2} & X_3 & \xrightarrow{u_3} & X_4 & \xrightarrow{u_4} & X_5 \\
\downarrow{\scriptstyle f_1} & & \downarrow{\scriptstyle f_2} & & \downarrow{\scriptstyle f_3} & & \downarrow{\scriptstyle f_4} & & \downarrow{\scriptstyle f_5} \\
Y_1 & \xrightarrow{v_1} & Y_2 & \xrightarrow{v_2} & Y_3 & \xrightarrow{v_3} & Y_4 & \xrightarrow{v_4} & Y_5
\end{array}$$

(i) If f_1 is an epimorphism and if f_2 and f_4 are monomorphisms, then f_3 is a monomorphism.

(ii) If f_5 is a monomorphism and f_2 and f_4 are epimorphisms, then f_3 is an epimorphism.

(iii) If f_1 is an epimorphism, f_5 is a monomorphism, and f_2 and f_4 are isomorphisms, then f_3 is an isomorphism.

4. Consider the following commutative diagram with exact rows in the category $\mathscr{Ab}$.

$$\begin{array}{ccccccccc}
 & & X' & \xrightarrow{u_1} & X & \xrightarrow{u_2} & X'' & \longrightarrow & 0 \\
 & & \downarrow{\scriptstyle f'} & & \downarrow{\scriptstyle f} & & \downarrow{\scriptstyle f''} & & \\
0 & \longrightarrow & Y' & \xrightarrow{v_1} & Y & \xrightarrow{v_2} & Y'' & &
\end{array}$$

Then, the induced sequences: $\ker f' \to \ker f \to \ker f''$ and $\operatorname{coker} f' \to \operatorname{coker} f \to \operatorname{coker} f''$ are exact. Define a function $d\colon \ker f'' \to \operatorname{coker} f'$ as follows: for any $x \in \ker f''$, let $u_2(y) = x$. Then $v_2 f(y) = 0$ so $f(y) = v_1(z)$ for some $z \in Y'$. Define $d(x) = \bar{z} \in \operatorname{coker} f'$. Then, d is a well defined morphism of groups and the sequence $\ker f \to \ker f'' \xrightarrow{d} \operatorname{coker} f' \to \operatorname{coker} f$ is exact.

5. If an abelian category has a set of generators, then it is locally small and colocally small.

6. Let $\mathscr{C}$ be an abelian category. Let $\mathscr{E}$ be the class of short exact sequences in $\mathscr{C}$. A morphism $u: E \to F$ of short exact sequences is a commutative diagram:

$$\begin{array}{ccccccccc} E: 0 & \longrightarrow & X' & \longrightarrow & X & \xrightarrow{p} & X'' & \longrightarrow & 0 \\ & & \downarrow{f'} & & \downarrow{f} & & \downarrow{f''} & & \\ F: 0 & \longrightarrow & Y' & \longrightarrow & Y & \longrightarrow & Y'' & \longrightarrow & 0 \end{array}$$

Then $\mathscr{E}$ is a preabelian category. The kernel of u is the sequence:

$$K: 0 \longrightarrow K' \longrightarrow K \longrightarrow p(K) \longrightarrow 0$$

where $K' = \ker f'$, $K = \ker f$. The cokernel K' is given by the the dual construction. Hence u is a monomorphism (epimorphism) if and only if f is a monomorphism (epimorphism) and so u is a bimorphism if and only if f is an isomorphism. Let I' be the cokernel of the morphism $K \to E$, and I the kernel of the morphism $F \to K'$. Then the induced morphism $I' \to I$ is a bimorphism.

References

Buchsbaum [28]; Bucur–Deleanu [31]; Freyd [57]; Gabriel [59]; Grothendieck [79]; MacLane [120]; Mitchell [133]; Oort [157]; Pareigis [163].

2.4 Fibred products and fibred sums

Let $\mathscr{C}$ be an abelian category. We shall say that the square of $\mathscr{C}$:

$$\begin{array}{ccc} X & \xrightarrow{f} & Y \\ \downarrow{k} & & \downarrow{h} \\ T & \xrightarrow{g} & Z \end{array} \qquad (2)$$

is *cartesian* (respectively, *cocartesian*) if the triple (X, f, k) (respectively the triple (h, g, Z)) is a fibred product (respectively, fibred sum) for the angle $Y \xrightarrow{h} Z \xleftarrow{g} T$ (respectively for the coangle $T \xleftarrow{k} X \xrightarrow{f} Y$). We shall say that diagram (2) is an *exact square* if it is simultaneously cartesian and cocartesian.

The result which follows is valid in any category.

LEMMA 4.1. *If in the cartesian (respectively, cocartesian) square (2) g is a monomorphism (respectively, f is an epimorphism), then f is a monomorphism (respectively, g is an epimorphism).*

Proof. Let $u, v: U \to X$ be morphisms such that $f u = f v$; then, $h f u = h f v$. Also, $g k u = g k v$, i.e. $k u = k v$. Then the hypothesis implies that $u = v$. We proceed analogously for h and k. ∎

Let us consider the square (2) in $\mathscr{C}$. There is a unique morphism

$$t : X \to Y \amalg T$$

such that $p\,t = f, q\,t = k$, where $p : Y \amalg T \to Y, q : Y \amalg T \to T$ are structural projections. Also, there is a unique morphism $s : Y \amalg T \to Z$ such that $s\,u = -h$ and $s\,v = g$, $u : Y \to Y \amalg T$ $v : T \to Y \amalg T$ being structural injections.

PROPOSITION 4.2. *With the above notations, the following assertions concerning the square* (2) *are true*:

(a) (2) *is commutative if and only if* $s\,t = 0$, (*that is, the sequence*

$$X \xrightarrow{t} Y \amalg T \xrightarrow{s} Z$$

is null).

(b) (2) *is cartesian if and only if the sequence*:

$$0 \longrightarrow X \xrightarrow{t} Y \amalg T \xrightarrow{s} Z$$

is exact.

(c) (2) *is cocartesian if and only if the sequence*

$$X \xrightarrow{t} Y \amalg T \xrightarrow{s} Z \longrightarrow 0$$

is exact;

(d) (2) *is exact if and only if the sequence*:

$$0 \longrightarrow X \xrightarrow{t} Y \amalg T \xrightarrow{s} Z \longrightarrow 0$$

is exact.

Proof. If the square (2) is commutative, then $h\,f = g\,k$, that is $g\,k + (-h)\,f = 0$. Also, the above considerations imply that $s\,v\,q\,t + s\,u\,p\,t = 0$, or $s(v\,q + u\,p)\,t = 0$. On the other hand, we have $v\,q + u\,p = 1_{Y \amalg T}$. Consequently $s\,t = 0$. Conversely, if $s\,t = 0$, then $s\,1_{Y \amalg T}t = s(v\,q + u\,p)\,t = s\,v\,q\,t + s\,u\,p\,t = g\,k - h\,f = 0$, that is $g\,k = h\,f$.

Let (2) be a cartesian square and let $r : U \to Y \amalg T$ be a morphism such that $sr = 0$. Then, $r = (up + vq)\,r$, and $0 = s\,r = s\,u\,p\,r + s\,v\,q\,r = -h\,p\,r + g\,q\,r$, i.e. $h\,p\,r = g\,q\,r$. Then, there is a unique morphism $m : U \to X$ such that $f\,m = p\,r$, $k\,m = q\,r$. Now, from the definition of t we have $p\,t\,m = p\,r$ and $q\,t\,m = q\,r$. Then, according to the definition of the direct product, we must have $t\,m = r$.

Conversely, if (X, t) is a kernel of s, then we have $s\,t = 0$, that is, according to (a) the square is commutative. Let $m : U \to Y$ and $n : U \to T$ be two morphisms such that $h\,m = g\,n$. Also, let $r : U \to Y \amalg T$ be the unique morphism for which $p\,r = m$ and $q\,r = n$. Then $s\,r = 0$. In fact, according to the definition of s, we shall have $s u m = - s v n$, that is, $0 = s u p r + s v q r = s(u p + v q) r = s r$.

Let $j : U \to X$ be the unique morphism such that $t j = r$. Then $p t j = p r$ and $q t j = q r$. Finally, we get $k j = n$ and $f j = m$. The other assertions are analogously derived or they follow by dualization. ■

COROLLARY 4.3. *If in the cartesian (respectively, cocartesian) square* (2), *g is an epimorphism (respectively, f is a monomorphism), then f is an epimorphism (respectively, g is a monomorphism).*

Proof. Let (2) be cocartesian and f monomorphism. According to assertion (c) of Proposition 4.2, we shall have that (s, Z) is a cokernel of t. Also, we have $p t = f$, hence t is a monomorphism. According to assertion (2) of the previous proposition, we shall have that (2) is cartesian. Now if $m : U \to T$ is such that $gm = 0$, then there is a unique morphism $n : U \to X$ such that $f n = 0$ and $kn = m$. But then $n = 0$, so that $m = 0$. ■

NOTE. If (2) is a cartesian (respectively, cocartesian) square and if g is an epimorphism (respectively, f a monomorphism), then (2) is exact.

Exercises

1. Let $X \xrightarrow{u} Y \xleftarrow{v} Z$ be an angle in the category $\mathscr{Ab}$. Then $X \Pi_Y Z$ is the subset of $X \Pi Z$ which consists of the pairs (x, z) for which $u(x) = v(z)$. The structural morphisms are the canonical projections.

2. Let $\mathscr{C}$ be a category and $f : X \to Y$ a morphism of $\mathscr{C}$. Then f is a monomorphism (respectively, epimorphism) if and only if the following square

$$\begin{array}{ccc} X & \xrightarrow{1_X} & X \\ {\scriptstyle 1_X}\downarrow & & \downarrow{\scriptstyle f} \\ X & \xrightarrow{f} & Y \end{array} \quad \text{resp.} \quad \begin{array}{ccc} X & \xrightarrow{f} & Y \\ {\scriptstyle f}\downarrow & & \downarrow{\scriptstyle 1_Y} \\ Y & \xrightarrow{1_Y} & Y \end{array}$$

is cartesian (respectively, cocartesian).

References

Buchsbaum [28]; Bucur–Deleanu [31]; Freyd [57]; Gabriel [59]; Grothendieck [79]; Heller [88]; MacLane [120]; Mitchell [133].

2.5 Basic lemmas on abelian categories

Let $\mathscr{C}$ be an abelian category.

LEMMA 5.1. *Let $X \xrightarrow{f} Y \xrightarrow{g} Z$ be two morphisms of $\mathscr{C}$. If g is a monomorphism (respectively, f is an epimorphism), then* $\ker f = \ker(g f)$ *(respectively.* $\operatorname{coker} g = \operatorname{coker}(g f)$*).* ■

LEMMA 5.2. *Let us consider the commutative diagram*

$$\begin{array}{ccccccc} & & X & \xrightarrow{f} & Y & & \\ & & {\scriptstyle u}\downarrow & & \downarrow{\scriptstyle v} & & \\ 0 & \longrightarrow & X' & \xrightarrow{g} & Y' & \xrightarrow{h} & Z \end{array}$$

the last sequence being exact. Then, the square is cartesian if and only if the sequence

$$0 \longrightarrow X \xrightarrow{f} Y \xrightarrow{hv} Z \tag{3}$$

is exact.

Proof. Assume that the square is cartesian. Then, $hvf = hgu = 0$. Let $s: U \to Y$ be a morphism such that $hvs = 0$. There is a unique morphism $t: U \to X'$ such that $gt = vs$. Then, there is a unique morphism $r: U \to X$ such that $fr = s$, that is (X,f) is a kernel for hv, or equivalently, (3) is exact.

Conversely, assume that (3) is exact and let $s: U \to Y$, $t: U \to X'$ be two morphisms such that $vs = gt$. Then, $hvs = hgt = 0$ and there is a unique morphism $r: U \to X$ for which $fr = s$. Also, $gur = vfr = vs = gt$. Finally one has $ur = t$, since g is a monomorphism. Consequently, the square is cartesian. ∎

LEMMA 5.3. *Let us consider the following commutative diagram*

$$\begin{array}{ccccccccc} & & X & \xrightarrow{f} & Y & \xrightarrow{p} & \operatorname{coker} f & \longrightarrow & 0 \\ & & {\scriptstyle g}\downarrow & & \downarrow{\scriptstyle k} & & \downarrow{\scriptstyle s} & & \\ 0 & \longrightarrow & T & \xrightarrow{h} & Z & \xrightarrow{q} & \operatorname{coker} h & \longrightarrow & 0 \end{array}$$

in which the first square is cartesian. Then s is a monomorphism. If k is an epimorphism, then s is an isomorphism.

Proof. Let $r: U \to \operatorname{coker} f$ be a morphism for which $sr = 0$. Let (K, u, v) be a fibred product for the angle $Y \xrightarrow{p} \operatorname{coker} f \xleftarrow{r} U$, $u: K \to U$, $v: K \to Y$. Then, $sp = qk$, and from Lemma 5.2 we derive that (X,f) is a kernel of qk, that is, (X,f) is a kernel of sp. Now we have $sru = spv = 0$. Then there is a unique morphism $t: K \to X$ such that $ft = v$. Hence $pv = pft = ru = 0$ and since u is an epimorphism, we may derive that $r = 0$. It follows that s is a monomorphism.

If k is an epimorphism, then $s\,p$ is an epimorphism and finally s is an epimorphism. ■

LEMMA 5.4. *Consider the following commutative diagram*

$$\begin{array}{ccccccc}
0 & \longrightarrow & \ker f & \xrightarrow{i} & X & \xrightarrow{f} & Y \\
 & & \downarrow{\scriptstyle u} & & \downarrow{\scriptstyle k} & & \downarrow{\scriptstyle h} \\
0 & \longrightarrow & \ker g & \xrightarrow{j} & Z & \xrightarrow{g} & T
\end{array}$$

the last square being cartesian. If h is a monomorphism, then u is a monomorphism. If h is a monomorphism and g an epimorphism, then u is an isomorphism.

Proof. If h is a monomorphism, then k is a monomorphism according to Lemma 4.1. Then, $j\,u$ is a monomorphism, that is u is a monomorphism. Assume that g is an epimorphism. Then the square is cocartesian (see the Note at the end of Section 4). Furthermore, we can use the dual of the previous Lemma. ■

LEMMA 5.5. *Consider the following commutative diagram*

$$\begin{array}{ccccccccc}
0 & \longrightarrow & X' & \xrightarrow{f'} & X & \xrightarrow{g} & X'' & \longrightarrow & 0 \\
 & & \downarrow{\scriptstyle 1_{X'}} & & \downarrow{\scriptstyle 1_X} & & \downarrow{\scriptstyle u} & & \\
0 & \longrightarrow & X' & \xrightarrow{f'} & X & \xrightarrow{g'} & Y & &
\end{array}$$

in which the upper sequence is exact. Then the last sequence is also exact if and only if u is a monomorphism.

Proof. Assume that the last sequence is exact. Let (k, i) a kernel of u and (U, t, s) a fibred product of the angle $X \to X'' \leftarrow K$, $t : U \to K$, $s : U \to X$. According to Corolloary 4.3, t is an epimorphism and s is a monomorphism. Now we have: $u\,i\,t = u\,g\,s = 0$, i.e. $g'\,s = 0$. Then there is a unique morphism $r : U \to X'$ such that $f'\,r = s$.

From this we derive that $i\,t = g\,s = g\,f'\,r = 0$; and $i = 0$, t being an epimorphism. Finally, we derive that u is a monomorphism.

Conversely, if u is a monomorphism, and if $r : U \to X$ is a morphism such that $g'\,r = 0$, then $u\,g\,r = 0$ and $g\,r = 0$. Furthermore, a unique morphism $h : U \to X'$ exists such that $f'\,h = r$, that is, f' is a kernel of g'. ■

Lemma 5.6. *We consider the following commutative diagram*

$$\begin{array}{ccccccc}
 & & 0 & & 0 & & 0 \\
 & & \downarrow & & \downarrow & & \downarrow \\
0 & \longrightarrow & X' & \xrightarrow{f'} & X & \xrightarrow{f} & X'' \\
 & & \downarrow{\scriptstyle u'} & & \downarrow{\scriptstyle u} & & \downarrow{\scriptstyle u''} \\
0 & \longrightarrow & Y' & \xrightarrow{g'} & Y & \xrightarrow{g} & Y'' \\
 & & \downarrow{\scriptstyle v'} & & \downarrow{\scriptstyle v} & & \\
0 & \longrightarrow & Z' & \xrightarrow{h} & Z & & \\
 & & \downarrow & & & & \\
 & & 0 & & & &
\end{array}$$

in which the all columns and the middle sequence are exact. Then, the upper sequence is also exact, if and only if the last sequence is exact (that is, if h is a monomorphism).

Proof. Assume that the the upper sequence is exact. Then $\ker(u'' f) = \ker f$ and it follows that $\ker(g\,u) = \ker f$, that is, the sequence

$$0 \longrightarrow X' \xrightarrow{f'} X \xrightarrow{gu} Y''$$

is exact. According to Lemma 5.2, we derive that the square

$$\begin{array}{ccc}
X' & \xrightarrow{f'} & X \\
\downarrow{\scriptstyle u'} & & \downarrow{\scriptstyle u} \\
Y' & \xrightarrow{g'} & Y
\end{array} \tag{4}$$

is cartesian. We use the Lemma 5.3 to check that h is a monomorphism.

Conversely, if h is a monomorphism, then we have the exact sequence

$$0 \longrightarrow X' \xrightarrow{f'} Y' \xrightarrow{h\,v'} Z$$

and it follows that the square (4) is cartesian.

Let $r : U \to X$ be a morphism such that $f\,r = 0$. Then, $g\,u\,r = 0$ and a unique morphism $t : U \to Y'$ exists such that $g'\,t = u\,r$. Since the square (4) is cartesian, a unique morphism $s : U \to X'$ exists such that $f'\,s = r$. Finally, (X', f') is a kernel of f. ∎

Lemma 5.7. (3 × 3 Lemma). *Let us consider the commutative diagram*

$$\begin{array}{ccccccccc}
 & & 0 & & 0 & & 0 & & \\
 & & \downarrow & & \downarrow & & \downarrow & & \\
0 & \longrightarrow & X' & \longrightarrow & X & \longrightarrow & X'' & \longrightarrow & 0 \\
 & & \downarrow & & \downarrow & & \downarrow & & \\
0 & \longrightarrow & Y' & \longrightarrow & Y & \longrightarrow & Y'' & \longrightarrow & 0 \\
 & & \downarrow & & \downarrow & & \downarrow & & \\
0 & \longrightarrow & Z' & \longrightarrow & Z & \longrightarrow & Z'' & \longrightarrow & 0 \\
 & & \downarrow & & \downarrow & & \downarrow & & \\
 & & 0 & & 0 & & 0 & &
\end{array}$$

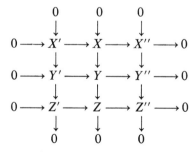

in which the sequence in the middle and the columns are exact. Then, the upper sequence is exact if and only if the last sequence is exact.

Proof. If one makes use of the preceding Lemma and of its dual, then the above Lemma becomes obvious. ■

References

Buchsbaum [28]; Bucur–Deleanu [31]; Freyd [57]; Grothendieck [79]; Mitchell [133].

2.6 The isomorphism theorems

Let (X_1, i_1), (X_2, i_2) be two subobjects of an object X in an abelian category $\mathscr{C}$. We denote by $X_1 + X_2$ and we call the *sum* of X_1 and X_2 the image of the unique morphism $s: X_1 \amalg X_2 \to X$ defined by the conditions $s u_1 = i_1$, $s u_2 = i_2$ where u_j is the structural morphism. We shall say that s is the *sum morphism.*

Dually, we denote by $X_1 \cap X_2$ and we call the *intersection* of X_1 and X_2 the kernel of the unique morphism $t: X \to X/X_1 \Pi X/X_2$ defined by the conditions $q_j t = p_j$, $j = 1, 2$ where $p_j: X \to X/X_j$ are the canonical epimorphisms and q_j the structural projections. We shall say that t is the *product morphism.*

PROPOSITION 6.1. *The sum $X_1 + X_2$ of any two subobjects of an object X is the upper bound of these subobjects in the ordered class of subobjects of X.*

Dually, $X_1 \cap X_2$ is the lower bound of these subobjects in the above ordered class.

Proof. Obviously, we have $X_j \subseteq X_1 + X_2$, $j = 1, 2$.

Let (X_3, i_3) be an subobject of X such that $X_j \subseteq X_3$, $j = 1, 2$. Then there exist certain morphisms $f_j: X_j \to X_3$, such that $i_3 f_j = i_j$, $j = 1, 2$. Let $u: X_1 \amalg X_2 \to X_3$ be the unique morphism such that $u u_j = f_j$, $j = 1, 2$. Then $i_3 u = s$ and according to Lemma 5.1, $\operatorname{im} s \subseteq \operatorname{im} i_3 = X_3$. ■

More generally, we may define the sum and intersection of any finite set of subobjects of an object in an obvious manner.

LEMMA 6.2. (First isomorphism theorem). *Let X' be a subobject of X and Y a subobject of X'. Then we have the following commutative diagram with exact rows:*

$$\begin{array}{ccccccccc}
0 & \longrightarrow & X' & \xrightarrow{i} & X & \xrightarrow{p} & X/X' & \longrightarrow & 0 \\
 & & \downarrow{\scriptstyle u'} & & \downarrow{\scriptstyle u} & & \downarrow{\scriptstyle u''} & & \\
0 & \longrightarrow & X'/Y & \xrightarrow{j} & X/Y & \xrightarrow{q} & X/X' & \longrightarrow & 0
\end{array}$$

such that u'' is an isomorphism and all morphisms are canonically constructed.

Proof. Indeed, we have the following commutative diagram:

$$\begin{array}{ccccccccc}
 & & 0 & & 0 & & & & \\
 & & \downarrow & & \downarrow & & & & \\
0 & \longrightarrow & Y & \xrightarrow{1_Y} & Y & \longrightarrow & 0 & \longrightarrow & 0 \\
 & & \downarrow & & \downarrow & & \downarrow & & \\
0 & \longrightarrow & X' & \xrightarrow{i} & X & \xrightarrow{p} & X/X' & \longrightarrow & 0 \\
 & & \downarrow{\scriptstyle u'} & & \downarrow{\scriptstyle u} & & \downarrow{\scriptstyle u''} & & \\
0 & \longrightarrow & X'/Y & \xrightarrow{j} & X/Y & \xrightarrow{q} & X/X' & \longrightarrow & 0 \\
 & & \downarrow & & \downarrow & & \downarrow & & \\
 & & 0 & & 0 & & 0 & &
\end{array}$$

The proof results from Lemma 5.7. ∎

PROPOSITION 6.3. *Let X_1, X_2 be two subobjects of X. Then we have the commutative diagram canonically constructed with exact columns and rows:*

$$\begin{array}{ccccccccc}
 & & 0 & & 0 & & 0 & & \\
 & & \downarrow & & \downarrow & & \downarrow & & \\
0 & \longrightarrow & X_1 \cap X_2 & \longrightarrow & X_2 & \longrightarrow & X_2/X_1 \cap X_2 & \longrightarrow & 0 \\
 & & \downarrow & & \downarrow & & \downarrow & & \\
0 & \longrightarrow & X_1 & \longrightarrow & X & \longrightarrow & X/X_1 & \longrightarrow & 0 \\
 & & \downarrow & & \downarrow & & \downarrow & & \\
0 & \longrightarrow & X_1/X_1 \cap X_2 & \longrightarrow & X/X_2 & \longrightarrow & X/(X_1 + X_2) & \longrightarrow & 0 \\
 & & \downarrow & & \downarrow & & \downarrow & & \\
 & & 0 & & 0 & & 0 & &
\end{array} \qquad (5)$$

The proof results from Lemma 5.7. ∎

PROPOSITION 6.4. (The second isomorphism theorem). *Let X_1, X_2 be two subobjects of X. Then we have the following commutative diagram, canonically constructed, with exact rows and u'' an isomorphism:*

$$\begin{array}{ccccccccc}
0 & \longrightarrow & X_1 \cap X_2 & \longrightarrow & X_2 & \longrightarrow & X_2/X_1 \cap X_2 & \longrightarrow & 0 \\
 & & \downarrow{\scriptstyle u'} & & \downarrow{\scriptstyle u} & & \downarrow{\scriptstyle u''} & & \\
0 & \longrightarrow & X_1 & \longrightarrow & X_1 + X_2 & \longrightarrow & (X_1 + X_2)/X_1 & \longrightarrow & 0
\end{array}$$

Proof. The assertion is a consequence of the last Proposition, if we put $X = X_1 + X_2$. ∎

Let X be an object of $\mathscr{C}$. We shall denote as above by $S(X)$ the class of subobjects of X and by $Q(X)$ the class of quotient objects of X.

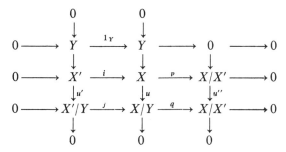

Let now $f : X \to Y$ be a morphism. We intend to show that one associates to this morphism in a natural way the maps:

$$S(X) \longrightarrow S(Y)$$
$$Q(X) \longrightarrow Q(Y)$$
$$S(Y) \longrightarrow S(X)$$
$$Q(Y) \longrightarrow Q(X)$$

which we denote somewhat improperly (to comply with tradition) as follows: the first two by f, the last two by f^{-1}. The definitions of these maps are the following:

Let (X', i) be a subobject of X. Then, by definition $f(X') = \operatorname{im}(f\,i)$. Moreover, there exists a morphism $X' \to f(X')$ defined to be the composition $X' \to \operatorname{coim}(f\,i) \to \operatorname{im}(f\,i)$.

If (X'', p) is a quotient object of X, then we denote by (X', i) the subobject of X which corresponds to it under the bijection defined in Section 2.3, that is $(X', i) = \ker p$. We define $f(X'') = \operatorname{coker}(f\,i)$. Clearly there exists a morphism $X'' \to f(X'')$ such that the diagram

$$\begin{array}{ccc} X & \xrightarrow{f} & Y \\ {\scriptstyle p}\downarrow & & \downarrow \\ X'' & \longrightarrow & f(X'') \end{array}$$

is commutative.

We obtain the maps f^{-1} by dualization. The explicit definition is as follows:

Let (Y'', q) be a quotient object of Y. By definition $f^{-1}(Y'') = \operatorname{coim}(qf)$. If (Y', j) is a subobject of Y, we denote by (Y'', q) the quotient object such that $(Y'', q) = \operatorname{coker} j$. We set $f^{-1}(Y') = \ker(q\,f)$. There exist morphisms $f^{-1}(Y') \xrightarrow{t} Y'$, $f^{-1}(Y'') \xrightarrow{s} Y''$ such that the diagram:

$$\begin{array}{ccccc} f^{-1}(Y') & \xrightarrow{i} & X & \xrightarrow{p} & f^{-1}(Y'') \\ \downarrow{\scriptstyle t} & & \downarrow{\scriptstyle f} & & \downarrow{\scriptstyle s} \\ Y' & \xrightarrow{j} & Y & \xrightarrow{q} & Y'' \end{array}$$

i, j being defined canonically, is commutative.

LEMMA 6.5. *The first square of the last diagram is cartesian.*

Proof. We shall consider the commutative diagram in which the square is cartesian:

$$\begin{array}{ccccc} X\Pi_Y Y' & \xrightarrow{c} & X & & \\ \downarrow & & \downarrow{\scriptstyle f} & & \\ Y' & \xrightarrow{j} & Y & \xrightarrow{q} & Y/Y' \end{array}$$

and we shall prove that c is a kernel of $q\,f$. Indeed, let $d: U \to X$ be a morphism such that $q\,f\,d = 0$. There is a unique morphism $g: U \to Y'$ such that $j\,g = f\,d$. Also there is a unique morphism $h: U \to X\Pi_Y Y'$ such that $c\,h = d$. ■

For the subobject Y' of Y we have the following commutative diagram:

$$\begin{array}{ccccccc} \ker f & \xrightarrow{r} & X & \xrightarrow{\qquad f \qquad} & & Y & \xrightarrow{q} Y/Y' \\ \| & & \uparrow{\scriptstyle i} & & & \uparrow{\scriptstyle j} & \\ \ker f & \xrightarrow{r'} & f^{-1}(Y') & \xrightarrow{k} & \operatorname{coker} r' & \xrightarrow{u} Y' & \end{array} \qquad (6)$$

such that $u\,k = t$ and r, r', i, q, j being canonically constructed.

Proposition 6.6. (Third isomorphism theorem). *Let $f: X \to Y$ be a morphism and (Y', j) a subobject of Y. Then:*

(a) *the morphism u in the diagram* (6) *is a monomorphism*;

(b) *if f is an epimorphism, then u is an isomorphism*;

(c) *Let us denote by $f'': X/f^{-1}(Y') \to Y/Y'$ the unique morphism which makes commutative the diagram*:

$$\begin{array}{ccc} X & \xrightarrow{f} & Y \\ {\scriptstyle p}\downarrow & & \downarrow{\scriptstyle q} \\ X/f^{-1}(Y') & \xrightarrow{f''} & Y/Y' \end{array}$$

Then f'' is a monomorphism.

(d) *If f is an epimorphism, then f'' is an isomorphism.*

Proof. (a) Let $h: Z \to \operatorname{coker} r'$ be a morphism such that $u\,h = 0$. Let (K, m, n) be a fibred product for the angle $f^{-1}(Y') \xrightarrow{k} \operatorname{coker} r' \xleftarrow{h} Z$, $m: K \to Z$ $n: K \to f^{-1}(Y')$. According to Corollary 4.3, m is an epimorphism, k being an epimorphism. We have $j\,u\,k\,n = j\,u\,h\,m = 0$. Now, $f\,i = j\,u\,k$, and thus $f\,i\,n = 0$. There is a unique morphism $w: K \to \ker f$ such that $r\,w = i\,n$, also, $i\,r'\,w = i\,n$, that is $r'\,w = n$, i being a monomorphism. Then, $h\,m = k\,n = k\,r'\,w = 0$, in the other words $h = 0$, m being an epimorphism. Hence u is a monomorphism.

(b) Assume that f is an epimorphism. Let $h: Y' \to U$ be a morphism such that $hu = 0$. Let (m, n, K) be a fibred sum of the coangle $Y \xleftarrow{j} Y' \xrightarrow{h} U$, $m: U \to K$, $n: Y \to K$. According to Corollary 4.3 we deduce that m is a monomorphism, j being a monomorphism. Then $n f i = n j u k = m h u k = 0$. There exists a unique morphism w: $Y/Y' \to K$ such that $wqf = nf$. $(q f, Y/Y')$ being a cokernel of i, according to Lemma 5.1. Then $wq = n$, f being an epimorphism. Also, $mh = nj = wqj = 0$ and thus $h = 0$, m being a monomorphism. Hence u is an epimorphism, that is an isomorphism, according to (a).

(c) We shall consider the following commutative diagram with exact rows:

$$\begin{array}{ccccccccc} 0 & \longrightarrow & f^{-1}(Y') & \xrightarrow{i} & X & \xrightarrow{p} & X/f^{-1}(Y') & \longrightarrow & 0 \\ & & \downarrow{\scriptstyle uk=t} & & \downarrow{\scriptstyle f} & & \downarrow{\scriptstyle f''} & & \\ 0 & \longrightarrow & Y' & \xrightarrow{j} & Y & \xrightarrow{q} & Y/Y' & \longrightarrow & 0 \end{array}$$

According to Lemma 6.5, and Lemma 5.3 we deduce that f'' is a monomorphism.

(d) Again, using the Lemma 6.5 and Lemma 5.3 it is easy to check the result. ■

NOTE 1. For an object X we shall denote by $\phi_X: S(X) \to Q(X)$ the bijection defined in Section 2.3. Let $f: X \to Y$ be a morphism. We shall show that we can go over from one of the maps to the other and analogously from one of the maps f^{-1} to the other by using the above bijections ϕ_X. Precisely, the following diagrams of classes and maps are commutative:

$$\begin{array}{ccc} Q(X) & \xrightarrow{f} & Q(Y) \\ \uparrow{\scriptstyle \phi_X} & & \uparrow{\scriptstyle \phi_Y} \\ S(X) & \xrightarrow{f} & S(Y) \end{array} \qquad \begin{array}{ccc} Q(Y) & \xrightarrow{f^{-1}} & Q(X) \\ \uparrow{\scriptstyle \phi_Y} & & \uparrow{\scriptstyle \phi_X} \\ S(Y) & \xrightarrow{f^{-1}} & S(X) \end{array}$$

To verify the commutativity of the first diagram, let (X', i) be a subobject of X, then f associates to this subobject of X the subobject $\operatorname{im}(f\, i)$ of Y, and ϕ_Y associates to this subobject of Y its quotient object $\operatorname{coker}(\operatorname{im}(f\, i))$. Under ϕ_X the subobject (X', i) is carried onto coker i, and the latter is carried under f onto $\operatorname{coker}(f\, i)$. Thus we must show that $\operatorname{coker}(\operatorname{im}(f\, i)) = \operatorname{coker}(f\, i)$. But we have:

$$\operatorname{coker}(\operatorname{im}(f\, i)) = \operatorname{coker}(\ker(\operatorname{coker}(f\, i))) = \operatorname{coker}(f\, i).$$

The commutativity of the second diagram follows in a similar way from the relation:

$$\ker(\operatorname{coim}(qf)) = \ker(qf).$$

These remarks enable us to reduce the study of the four maps described above to the study of one of them.

Direct and inverse images verify certain conditions of "universality".

PROPOSITION 6.7. *Let $f : X \to Y$ be a morphism, (X', i) a subobject of X and (Y', j) a subobject of Y.*

(a) *$f(X')$ is a subobject of Y such that for any subobject (Y'', t) of Y for which there is a morphism $u : X' \to Y''$ with $f\,i = t\,u$, then $f(X') \subseteq Y''$.*

(b) *$f^{-1}(Y')$ is a subobject of X such that for any subobject (X'', r) of X for which there is a morphism $v : X'' \to Y'$ with $f\,r = j\,v$, then $X'' \subseteq f^{-1}(Y')$.*

The proof is left to the reader. ∎

PROPOSITION 6.8. *Let $f : X \to Y$ be a morphism, $X_1 \subseteq X_2 \subseteq X$ and*

$$Y_1 \subseteq Y_2 \subseteq Y$$

subobjects of X and Y respectively. Then:

(a) $f(X_1) \subseteq f(X_2)$;

(b) $f^{-1}(Y_1) \subseteq f^{-1}(Y_2)$;

(c) $X_1 \subseteq f^{-1}(f(X_1))$;

(d) $Y_1 \supseteq f(f^{-1}(Y_1))$;

(e) $f(X_1) = f(f^{-1}(f(X_1)))$;

(f) $f^{-1}(Y_1) = f^{-1}(f(f^{-1}(Y_1)))$.

Proof. (a) and (b) are immediate consequences of the definition of direct and inverse images, while the rest are consequences of Proposition 6.7 and (a) and (b). ∎

NOTE 2. We may consider the class $S(X)$ of the subobjects of X a category in an evident manner. From Proposition 6.8 we deduce that direct and inverse images are in fact two functors $f_* : S(X) \to S(Y)$ and $f^* : S(Y) \to S(X)$, with $f_*(X') = f(X')$, $f^*(Y') = f^{-1}(Y')$ and f_* is a left adjoint of f^*. In fact, the above inclusions (c) and (d) are morphisms of adjunctions (see Chapter 1, Section 1.5).

From Proposition 6.1 we deduce that the sum of any finite set of subobjects of X is a direct sum in $S(X)$. Also, the intersection of any finite set of subobjects of X is the direct product in $S(X)$. Then $S(X)$ is a category with finite direct sums and products. According to above considerations and by Theorem 5.4, Chapter 1, we deduce the following result:

PROPOSITION 6.9. *Let $f: X \to Y$ be a morphism.*

(a) *For any two subobjects X_1, X_2 of X we have $f(X_1 + X_2) = f(X_1) + f(X_2)$.*

(b) *For any two subobjects Y_1, Y_2 of Y we have:*

$$f^{-1}(Y_1 \cap Y_2) = f^{-1}(Y_1) \cap f^{-1}(Y_2). \blacksquare$$

PROPOSITION 6.10. *Let $X \xrightarrow{f} Y \xrightarrow{g} Z$ be morphisms.*

(a) *For any subobject X' of X we have $(gf)(X') = g(f(X'))$;*

(b) *For any subobject Z' of Z we have $(g\,f)^{-1}(Z') = f^{-1}(g^{-1}(Z'))$.*

The proof is left to the reader. ■

Exercises

1. For the morphisms $X \xrightarrow{f} Y \xrightarrow{g} Z$ in an abelian category, we have the following relations:

(a) $\operatorname{im} f \cap \ker g = f(\ker(g\,f))$;

(b) $\operatorname{im} f + \ker g = g^{-1}(\operatorname{im}(g\,f))$;

(c) $\operatorname{coim} g \cap \operatorname{coker} f = g^{-1}(\operatorname{coker}(g\,f))$;

(d) $\operatorname{coim} g + \operatorname{coker} f = f(\operatorname{coim}(g\,f))$.

2. Let $0 \to X' \to X \xrightarrow{g} X'' \to 0$ be a short exact sequence in an abelian category. If X_1, X_2 are subobjects of X, satisfying the relation $X' \subseteq X_1 \subseteq X_2$, then in order that $X_1 = X_2$ it is necessary and sufficient that $g(X_1) = g(X_2)$.

3. Formulate and prove Propositions 6.7, 6.8 and 6.9 for quotient objects.

4. Let $\{X_i\}_{1 \leqslant i \leqslant n}$ be a finite set of subobjects of an object X of $\mathscr{Ab}$. Show that $\sum X_i$ is the set of elements in X of the form $\sum_i a_i$ where $a_i \in X_i$. Moreover, if $\{X_i\}_{i \in I}$ is a certain set of subobjects of X, then $\sum_{i \in I} X_i$ is the set of elements in X of the form $\sum_i a_i$ where $a_i \in X_i$ for all $i \in I$ and $a_i = 0$ for all but a finite number of i. The intersection $\bigcap_i X_i$ is evident.

5. (Modular law). Let X_1, X_2, X_3 be the subobjects of an object X in an abelian category, such that $X_1 \subseteq X_2$. Then

$$X_2 \cap (X_1 + X_3) = X_1 + (X_2 \cap X_3).$$

6. Let $f : X \to Y$ be a morphism, X' and X'' the subobjects of X. Then we have the following isomorphisms:

$$f(X') \simeq X'/(X' \cap \ker f) \simeq (X' + \ker f)/\ker f$$

$$X'/(X' \cap X'') \simeq (X' + X'')/X''.$$

7. Let $f : X \to Y$ be an epimorphism, and Y' be a subobject of Y. Then, $f(f^{-1}(Y')) = Y'$. Dually, if f is a monomorphism and X' a subobject of X, then $f^{-1}(f(X')) = X'$.

8. Let $f : X \to Y$ be a morphism and Y' a subobject of Y. Then, we have a canonical isomorphism:

$$X/f^{-1}(Y') \simeq \operatorname{im} f/(\operatorname{im} f \cap Y').$$

9. Let $f : X \to Y$ be a morphism, X' a subobject of X and Y' a subobject of Y. Then:

$$f(f^{-1}(Y') \cap X') = f(X') \cap Y'.$$

10. Prove Exercise 3 of Section 2.3 as follows in an abelian category. In proving part (i) replace X_1 by $\operatorname{im} u_1$ and Y_1 by $\operatorname{im} v_1$ and show that the induced morphism $\operatorname{im} u_1 \to \operatorname{im} v_1$ is an isomorphism. Then, by Lemma 6.2 the induced morphism $\operatorname{im} u_2 \to \operatorname{im} v_2$ is a monomorphism. Hence we may assume that u_2, v_2 and f_2 are all monomorphisms. From the fact that f_4 is a monomorphism, it follows that f_3 is a monomorphism. Part (ii) now follows by duality.

References

Buchsbaum [28]; Bucur–Deleanu [31]; Freyd [57]; Grothendieck [79]; Mitchell [133].

2.7 Direct sums of subobjects

Let $\mathscr{C}$ be an abelian category.

LEMMA 7.1. *Let X_1, X_2 be two subobjects of an object X of $\mathscr{C}$.*

(a) *The following assertions are equivalent:*

1. $X_1 \cap X_2 = 0$;
2. *The sum morphism $s : X_1 \amalg X_2 \to X$ is a monomorphism.*
3. *The product morphism $t : X \to X/X_1 \prod X/X_2$ is a monomorphism.*

(b) *Dually, the following assertions are equivalent:*

1. $X_1 + X_2 = X$;
2. *The product morphism $t : X \to X/X_1 \prod X/X_2$ is an epimorphism.*
3. *The sum morphism $s : X_1 \amalg X_2 \to X$ is an epimorphism.*

Proof. It suffices to prove the assertion (a).

(1) ⇒ (2). Let $r : Y \to X_1 \amalg X_2$ be a morphism such that $sr = 0$. Let $u_i : X_i \to X_1 \amalg X_2$ be structural morphisms and p_i, $i = 1, 2$ associated projections (such that the relations (1) are satisfied). Then: $0 = sr = s(u_1 p_1 + u_2 p_2) r = s u_1 p_1 r + s u_2 p_2 r$.

But $s u_i = v_i$, $i = 1, 2$, where $v_i : X_i \to X$ are canonical inclusions. Then we have:

$$(v_1 p_1) r = (-v_2 p_2) r,$$

hence

$$v_1(p_1(r(Y))) = v_2(p_2(r(Y))) = 0,$$

and obviously, $p_1 r = p_2 r = 0$ that is $r(Y) \subseteq \ker p_1$, $r(Y) \subseteq \ker p_2$. But $\ker p_1 \cap \ker p_2 = 0$, according to Proposition 3.5 and thus $r(Y) = 0$ i.e. $r = 0$.

(2) ⇒ (1). Let $X_3 = X_1 \cap X_2$. There is a unique morphism

$$h : X_3 \to X_1 \amalg X_2,$$

such that $r_1 = p_1 h$, $r_2 = p_2 h$, where $r_i : X_3 \to X_i$, $i = 1, 2$ are canonical inclusions. Then we have:

$$s u_1 r_1 = s u_1 p_1 h$$

and

$$s u_2 r_2 = s u_2 p_2 h.$$

But $s u_i = v_i$, $i = 1, 2$, so that $v_1 r_1 = v_2 r_2$, that is $s u_1 p_1 h = s u_2 p_2 h$ or $u_1 p_1 h = u_2 p_2 h$, s being a monomorphism. Then $p_1 u_1 p_1 h = p_1 u_2 p_2 h = p_1 h = 0$ and similarly, $p_2 h = 0$. Obviously then $h = 0$ and $X_3 = 0$.

All remaining equivalences follow by dualization. ■

We call *comaximal* two subobjects X_1, X_2 of X which satisfy the condition (b) of the above lemma. Two subobjects are called *supplementary* if they satisfy the conditions (a) and (b) simultaneously. We shall say that X_1 is a *supplement* of X_2 if X_1 and X_2 are supplementary.

THEOREM 7.2. *Let X_1 be a subobject of X. The following assertions are equivalent*:

(1) *X_1 has a supplement*;

(2) *There is a subobject X_2 of X such that the sum morphism*

$$s : X_1 \amalg X_2 \to X$$

is an isomorphism.

(2′) *There is a subobject* X_2 *of* X *such that the product morphism*

$$X \to X/X_1 \amalg X/X_2$$

is an isomorphism.

(3) *The canonical inclusion* $i_1 : X_1 \to X$ *is a section.*

(4) *There is a morphism* $q_1 : X \to X_1$ *such that* $u = i_1 q_1$ *is an idempotent.*

Proof. According to Lemma 7.1 we have that (1) $\Leftrightarrow$ (2) $\Leftrightarrow$ (2′).

(2) $\Rightarrow$ (3). Let s^{-1} be an inverse of s, and $q_1 = p_1 s^{-1}$, p_1 being the canonical projection associated with the structural morphism $u_1 : X_1 \to X_1 \amalg X_2$. Then:

$$q_1 i_1 = p_1 s^{-1} i_1 = p_1 s^{-1} s u_1 = p_1 u_1 = 1_{X_1}.$$

(3) $\Rightarrow$ (4). Obvious.

(4) $\Rightarrow$ (2). Let (X_2, i_2) be a kernel of u. We consider the following diagram:

$$\begin{array}{ccccc} X_2 & \xrightarrow{i_2} & X & \xrightarrow{u} & X \\ & & \uparrow{\scriptstyle 1_X - u} & & \\ & & X & & \end{array}$$

We see that $u(1_X - u) = u - u = 0$. Then, there is a unique morphism $q_2 : X \to X_2$, such that $i_2 q_2 = 1_X - u$, hence $i_1 q_1 + i_2 q_2 = 1_X$. But $i_2 q_2 i_2 = (1_X - u) i_2 = i_2$, hence $q_2 i_2 = 1_{X_2}$, i_2 being a monomorphism. On the other hand, $i_1 q_1 i_2 = u i_2 = 0$ and $q_1 i_2 = 0$, i_1 being a monomorphism. We have $q_2 i_1 q_1 + q_2 i_2 q_2 = q_2$ and then $q_2 i_1 q_1 + q_2 = q_2$, or $q_2 i_1 q_1 = 0$ that is $q_2 i_1 = 0$, q_1 being an epimorphism. According to relations (1) we deduce that X with canonical morphisms i_1, i_2 is a direct sum of X_1 and X_2. Obviously, the sum morphism is an isomorphism. ∎

A subobject X_1 of X which verifies the equivalent conditions of the above theorem is called a *direct summand of X*.

COROLLARY 7.3. *Let* $\{X_i\}_{i \in I}$ *be a finite set of objects and* I' *a subset of* I. *Then* $X' = \coprod_{i' \in I'} X_{i'}$ *is a direct summand of* $X = \coprod_{i \in I} X_i$.

Proof. Let $u'_{i'} : X_{i'} \to X'$, $u_i : X_i \to X$ be structural morphisms and $u : X' \to X$ the unique morphism such that $u u_{i'}' = u_{i'}$, $i' \in I'$. Also, let $p : X \to X'$ be the unique morphism such that $p_{i'}' p = p_{i'}$, $p_{i'}$ and $p_{i'}'$ being the projections associated with $u_{i'}$ and $u_{i'}'$ respectively. Then we have $p u_{i'} = u_{i'}'$. In fact, for any $i'' \in I'$ we deduce that $p_{i''}' p u_{i'} = p_{i''} u_{i'} = 0$ if $i' \neq i''$ or $= 1_{X_{i'}}$ if $i' = i''$. From the uniqueness of the injections associated with the given projections we may deduce that $p\,u_{i'} = u'_{i'}$. Then obviously, we have $p\,u\,u'_{i'} = p\,u_{i'} = u_{i'}'$ and from the definition of direct sum, we may deduce that $p\,u = 1_{X'}$, that is, X' is a direct summand of X according to Theorem 7.2. ■

COROLLARY 7.4. *Let* $0 \to X' \xrightarrow{i} X \xrightarrow{q} X'' \to 0$ *be a short exact sequence. The following assertions are equivalent*:

(1) *i is a section*;

(2) *q is a retraction*;

(3) *X' is a direct summand of X.*

Then X is canonically isomorphic with the direct sum $X' \amalg X''$.

Proof. (1) ⇒ (2). Let p be a retraction of i, and $u = i\,p$. Then $u^2 = u$ and $(1_X - u)\,i = i - u\,i = 0$. There is a unique morphism $j: X'' \to X$ such that $jq = 1_X - u$, that is, $jq + i\,p = 1_X$. Also $q\,j\,q = q$, and thus $q\,j = 1_{X''}$, q being an epimorphism.

The other implications are obvious. ■

Using the above notations, we say that X with the monomorphisms i, j is a direct sum of X' and X'', according to relations (1).

An exact sequence $0 \to X' \to X \to X''$ is called *left-split* if i is a section. Dually, we have the notion of *right-split*. From the above corollary, a short exact sequence which is left-split is also right-split and conversely. Then we shall speak of a *split short exact sequence.*

COROLLARY 7.5. *Let u be an endomorphism of X. The following assertions are equivalent*:

(1) *u is an idempotent,*

(2) ker *u is a direct summand in X,*

(3) im *u is a direct summand in X.*

Proof. We consider the exact sequence

$$0 \to \ker u \xrightarrow{i} X \xrightarrow{q} \operatorname{im} u \to 0,$$

canonically constructed. Let $j: \operatorname{im} u \to X$ be the canonical inclusion, then $jq = u$.

(1) ⇒ (3). We have $0 = u(1_X - u) = jq(1_X - u)$, and thus $q(1_X - u) = 0$, j being a monomorphism. There is a unique morphism $p: X \to \ker u$, such that $ip = 1_X - u$, or $1_X = ip + jq$. Then $q = q\,i\,p + q\,j\,q$, hence $q = q\,j\,q$ and finally, $q\,j = 1$, that is im u is a direct summand according to Corollary 7.4.

The other implications are obvious. ■

COROLLARY 7.6. *Let $X_1, \ldots X_n$ be subobjects of an object X. The following assertions are equivalent:*

(1) *The sum morphism:*

$$s : \sum_{i=1}^{n} X_i \to X$$

is an isomorphism.

(2) *The product morphism*

$$t : X \to \prod_{i=1}^{n} X/X_i$$

is an isomorphism.

(3) $\sum_i X_i = X$, *and for any* $1 \leqslant i_0 \leqslant n$ *we have:*

$$X_{i_0} \cap (\textstyle\sum_{i \neq i_0} X_i) = 0.$$

This corollary is a direct consequence of Theorem 7.2, by induction over n. ■

Let X_1 be a direct summand of X. According to Theorem 7.2, we deduce that any two supplements of X_1 are isomorphic. Generally, any two supplements of X, will be distinct, as is shown by the following example.

Let X be a non-null object, $\Delta : X \to X \amalg X$, and $u_i : X \to X \amalg X$ $i = 1, 2$ be respectively diagonal and structural morphisms. Also let $X_3 = \Delta(X)$, $X_i = u_i(X)$, $i = 1, 2$. Then X_1 and X_2 are non-null supplements of X_3 and $X_1 \cap X_2 = 0$.

Exercises

1. Consider a commutative diagram:

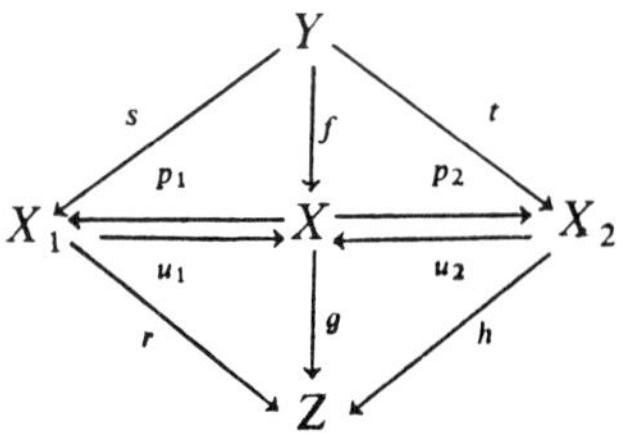

where $g\,f = 0$ and the sequences $X_1 \xrightarrow{u_1} X \xrightarrow{p_2} X_2$, $X_2 \xrightarrow{u_2} X \xrightarrow{p_1} X_1$ are exact. Then $u_1 p_1 + u_2 p_2 = 1_X$, that is, X is the direct sum of X_1 and X_2. Furthermore, $r\,s + h\,t = 0$.

References

Bourbaki [21]; Bucur–Deleanu [31]; Cartan–Eilenberg [33]; Freyd [57]; Gabriel [59]; Mitchell [133].

2.8 Inductive limits. The conditions Ab

In order to obtain more precise results, we are often obliged to impose additional conditions on an abelian category. The most manageable such conditions considered until now are those introduced by Grothendieck [79] under the name of conditions Ab 3, Ab 4, Ab 5, Ab 6 and their duals which are usually denoted by Ab 3*, Ab 4*, Ab 5*, Ab 6*. We now formulate and comment on these conditions.

We shall say that an abelian category $\mathscr{C}$ verifies the *condition* Ab 3 or that $\mathscr{C}$ is an Ab 3-*category* if $\mathscr{C}$ has arbitrary direct sums. Dually, $\mathscr{C}$ verifies *conditions* Ab 3* or is an Ab 3*-*category* if $\mathscr{C}^0$, the dual of $\mathscr{C}$, verifies Ab 3, that is $\mathscr{C}$ has arbitrary direct products.

Sometimes the condition Ab 3 is formulated as a function of a cardinal For example, $\mathscr{C}$ verifies Ab 3 countable, if $\mathscr{C}$ has countable direct sum.

According to Theorem 4.1 of Chapter 1, if $\mathscr{C}$ is an Ab 3-category (resp. Ab 3*-category), then $\mathscr{C}$ has arbitrary inductive (resp. projective) limits.

Let $\mathscr{C}$ be an Ab 3 (resp. Ab 3*)-category and $\{X_i\}_{i \in I}$ a set of objects of $\mathscr{C}$. Let $u_i : X_i \to \coprod_i X_i$ (resp. $q_i : \prod_i X_i \to X_i$) be structural morphisms. Then for any $i_0 \in I$ there exists a unique morphism $p_{i_0} : \coprod_i X_i \to X_{i_0}$ (resp. $v_{i_0} : X_{i_0} \to \prod_i X_i$) such that $p_{i_0} u_{i_0} = 1_{X_{i_0}}$, and $p_{i_0} u_i = 0$ (respectively $q_{i_0} v_{i_0} = 1_{X_{i_0}}$ and $q_{i_0} v_i = 0$) if $i \neq i_0$. The morphisms p_i (resp. v_i) are also called canonical *projections* (resp. *injections*) *associated with* u_i (resp. q_i). (See Section 1).

Let us assume that there exist $\coprod_i X_i$ and $\prod_i X_i$ in $\mathscr{C}$ (for example if $\mathscr{C}$ is an Ab 3 and Ab 3*-category). Then there is a unique morphism

$$t : \coprod_i X_i \to \prod_i X_i$$

such that $q_i t = p_i$ for any $i \in I$. If the set I is finite, we find (see Section 2.1) that t is an isomorphism. Generally t is not a monomorphism or epimorphism. Later we shall prove that under certain additional conditions t is a monomorphism.

Let $\mathscr{C}$ be an Ab 3-category (respectively Ab 3*-category), X an object in $\mathscr{C}$ and I a set. Then by $X^{(I)}$ (respectively by X^I) we shall denote the direct sum (product) of any set of objects equal to X (or equivalently isomorphic with X).

Let $\{X_i\}_{i \in I}$ be a set of subobjects of X in an Ab 3-category. Then there exists a unique morphism $s : \coprod_i X_i \to X$ such that $s u_i = r_i$, $r_i : X_i \to X$ being canonical monomorphisms and u_i the structural morphisms of the direct sum. We denote by $\sum_i X_i$ the subobject $\operatorname{im} s$ and we shall call this subobject the *sum* of the given set of subobjects.

We shall assume that $\mathscr{C}$ is an Ab 3*-category, and we shall denote by

$t : X \to \prod_i X/X_i$ the unique morphism such that $q_i t = h_i$, $h_i : X \to X/X_i$

being the canonical epimorphisms and q_i the structural morphisms of the direct product. Also we shall denote by $\bigcap_i X_i$ the subobject ker t, and we shall call this subobject the *intersection* of the given set of subobjects.

Using the observations made at the end of Section 2.6, we deduce that if $f: X \to Y$ is a morphism, $\{X_i\}_i$ a set of subobjects of X and $\{Y_j\}_j$ a set of subobjects of Y, then

$$f(\sum_i X_i) = \sum_i f(X_i)$$

and

$$f^{-1}(\bigcap_j Y_j) = \bigcap_j f^{-1}(Y_j).$$

Proposition 6.1 is actually valid for the sum (intersection) of any set of subobjects.

The following result will be used later.

LEMMA 8.1. *Let* $F: \mathscr{I} \to \mathscr{C}$ *be a small covariant functor in an* Ab 3-*category. Let* (f, X) *be an inductive limit of* F, $g: F \to Y$ *a morphism and* $u: X \to Y$ *the unique morphism such that* $u f = g$. *Then* $u(X) = \sum_i g_i(F(i))$, $i \in \operatorname{Ob} \mathscr{I}$.

Proof. For any $i \in \operatorname{Ob} \mathscr{I}$ we have: $u f_i = g_i$ thus $\operatorname{im}(u f_i) = \operatorname{im} g_i \subseteq \operatorname{im} u$. Then $Y' = \sum_i g(F(i)) \subseteq \operatorname{im} u$. Let $k: Y' \to Y$ be the canonical inclusion, and $g'_i: F(i) \to Y'$ be the unique morphism such that $k g'_i = g_i$. For any morphism $t: i \to j$ of $\mathscr{I}$, we have $g_j F(t) = g_i$, that is $k g_j'(F(t)) = k g_i'$ and $g_j' F(t) = g_i'$, k being a monomorphism.

There is a unique morphism $u': X \to Y'$ such that $u' f_i = g_i'$, or $k u' f_i = k g_i' = g_i$, in other words, $u f_i = k u' f_i$ for any $i \in \operatorname{Ob} \mathscr{I}$. Finally, $k u' = u$, and thus $\operatorname{im} u \subseteq Y'$. ∎

PROPOSITION 8.2. *Let* $\mathscr{C}$ *be an* Ab 3-*category and* $\{U_i\}_i$ *a set of objects of* $\mathscr{C}$. *The following assertions are equivalent*:

(1) *The given set is a set of generators of* $\mathscr{C}$;

(2) *The object* $U = \coprod_i U_i$ *is a generator of* $\mathscr{C}$.

(3) *For any object* X, *there is a set* A *and an epimorphism*: $U^{(A)} \to X$.

Proof. (1) ⇒ (2). Let $t_i: U_i \to \coprod_i U_i = U$ be the structural morphism and $f: X \to Y$ a morphism. Let $i_0 \in I$ and $g: U_{i_0} \to X$ a morphism such that $f g \neq 0$. Then $f g p_{i_0} \neq 0$, p_{i_0} being the projection associated with t_{i_0}. It is clear that U is a generator of $\mathscr{C}$.

(2) ⇒ (3). Let $A = \operatorname{Hom}_{\mathscr{C}}(U, X)$. We denote by $p: U^{(A)} \to X$ the unique morphism such that $p i_f = f$, $i_f: U \to U^{(A)}$ being the structural morphism associated with $f \in A$. Then p is an epimorphism. Indeed, let $q: X \to \operatorname{coker} p$

be the cokernel of p. If $q \neq 0$, then there is a morphism $h\colon U \to X$ such that $q\,h \neq 0$. But $q\,h = q\,p\,i_h = 0$, a contradiction. Thus $q = 0$, and p is an epimorphism.

(3) ⇒ (1). Let $f : X \to Y$ be a non null morphism of $\mathscr{C}$ and $p : U^{(A)} \to X$ an epimorphism. Then $f\,p \neq 0$. There exists $a \in A$ such that $f\,p\,i_a \neq 0$. From the definition of U there is an index $i_0 \in I$, such that $f\,p\,i_a\,t_{i_0} \neq 0$. In other words, the considered set is a set of generators, according to Proposition 3.4. ∎

Let $\{f_i : X_i \to Y_i\}_{i \in I}$ be a set of morphisms of $\mathscr{C}$. We denote by

$$\coprod_i f_i : \coprod_i X_i \to \coprod_i Y_i$$

the unique morphism such that $(\coprod f_i)\,u_i = v_i f_i$, u_i resp. v_i being the structural morphisms of the direct sum $\coprod_i X_i$, respectively $\coprod_i Y_i$. We shall say that $\coprod_i f_i$ is the *sum of the given set of morphisms.* Dually, we have the *product of any set of morphisms.*

Let $\mathscr{C}$ be an Ab 3-category and I a set. Assume that for any $i \in I$ we have a right short exact sequence of $\mathscr{C}$:

$$X_i' \xrightarrow{f_i} X_i \xrightarrow{g_i} X_i'' \longrightarrow 0 \tag{7}$$

Then we have the commutative diagram:

$$\begin{array}{ccccccc} X_i' & \xrightarrow{f_i} & X_i & \xrightarrow{g_i} & X_i'' & \longrightarrow & 0 \\ \downarrow{\scriptstyle u'_i} & & \downarrow{\scriptstyle u_i} & & \downarrow{\scriptstyle u''_i} & & \\ \coprod_i X'_i & \xrightarrow{\coprod f_i = f} & \coprod_i X_i & \xrightarrow{\coprod g_i = g} & \coprod_i X_i'' & \longrightarrow & 0 \end{array} \tag{8}$$

in which the vertical morphisms are structural morphisms. Then $(g, \coprod_i X_i'')$ is a cokernel of f. Indeed, $g\,f\,u_i' = g\,u_i f_i = u_i''\,g_i f_i = 0$ for any i, and thus $g\,f = 0$. Let $h : \coprod_i X_i \to Z$ be a morphism such that $h\,f = 0$. Then $h\,f\,u_i' = h\,u_i f_i = 0$, and a unique morphism $h_i : X_i'' \to Z$ may be calculated such that $h_i g_i = h\,u_i$. Also a unique morphism $\bar{h} : \coprod_i X_i'' \to Z$ is defined by the condition that $\bar{h}\,u_i'' = h_i$ for all $i \in I$. Hence $\bar{h}\,u_i''\,g_i = \bar{h}\,g\,u_i = h\,u_i$, and finally, $\bar{h}\,g = h$, that is, the last row of (8) is exact.

We shall say in such a situation that the direct sum is right exact, understanding by this that for any set of right short exact sequences (7), the lower row of the diagram (8) is a right short exact sequence.

Dually, if $\mathscr{C}$ is an Ab 3-category, then the direct product is left exact. In other words, for any set of left short exact sequences:

$$0 \longrightarrow X_i' \xrightarrow{f_i} X_i \xrightarrow{g_i} X_i''$$

We have the left short exact sequence:

$$0 \longrightarrow \prod_i X_i' \xrightarrow{\Pi f_i} \prod_i X_i \xrightarrow{\Pi g_i} \prod_i X_i''$$

We derive that if $\{0 \to X_i' \to X_i \to X_i'' \to 0\}_i$ is a finite set of short exact sequences, then we have the short exact sequence:

$$0 \to \prod_i X_i' \to \prod_i X_i \to \prod_i X_i'' \to 0$$

We shall say that an Ab 3-category is an Ab 4-*category* if for any set

$$\{f_i : X_i' \to X_i\}_i$$

of monomorphisms, the sum morphisms $\coprod f_i$ is also a monomorphism. That is, an Ab 3-category is Ab 4 if and only if the direct sums are left exact.

Dually, an Ab 3*-category is an Ab 4*-*category* if the direct product of any set of epimorphisms is also an epimorphism, that is, if and only if the direct products are right exact.

PROPOSITION 8.3. *Let $\mathscr{C}$ be an* Ab 3-*category* (*respectively an* Ab 3*-*category*). *The following assertions are equivalent*:

(1) *$\mathscr{C}$ is an* Ab 4-*category* (*resp. or* Ab 4*-*category*).

(2) *For any set $X_i' \to X_i \to X_i''$ of exact sequences, we have the exact sequence*:

$$\coprod_i X_i' \to \coprod_i X_i \to \coprod_i X_i''$$

$$(\textit{respectively}\colon \prod_i X_i' \to \prod_i X_i \to \prod_i X_i'').$$

The proof is left to the reader (compare with Chapter 3, Section 3.2). ∎

PROPOSITION 8.4. *Let $\mathscr{C}$ be an* Ab 3 *and* Ab 3*-*category. We suppose that for any set $\{X_i\}_i$ of objects of $\mathscr{C}$, the canonical morphism $t : \coprod_i X_i \to \prod_i X_i$ is a monomorphism. Then $\mathscr{C}$ is an* Ab 4-*category*.

Proof. Let $\{f_i : X_i' \to X_i\}_i$ be a set of monomorphisms. Then we have the commutative diagram:

$$\begin{array}{ccc} \coprod_i X_i' & \xrightarrow{\Pi f_i} & \coprod_i X_i \\ {\scriptstyle t'}\downarrow & & \downarrow{\scriptstyle t} \\ \prod_i X'_i & \xrightarrow{\Pi f_i} & \prod_i X_i \end{array}$$

Then $\coprod f_i$ is a monomorphism, $\prod f_i$ being a monomorphism. ∎

NOTE 1. A category $\mathscr{C}$ verifying the conditions of the above proposition is called by Mitchell [133] a C_2-*category*.

PROPOSITION 8.5. *Let $\mathscr{C}$ be an* Ab 3 *and* Ab 3*-*category. The following assertions are equivalent*:

(1) *$\mathscr{C}$ is a C_2-category*;

(2) *If $\{X_i\}_i$ is any set of objects of $\mathscr{C}$, and $f : Y \to \coprod_i X_i$ is a morphism such that $p_i f = 0$ for any i, p_i being the canonical projection, then $f = 0$.*

Proof. (1) $\Rightarrow$ (2). Let $q_i : \prod_i X_i \to X_i$ be the structural projections

$$u_i : X_i \to \coprod_i X_i$$

structural injections and p_i the projections associated with u_i. Then $q_i t = p_i$. In fact $q_i t\, u_i = q_i v_i = 1_{X_i}$, and $q_i t\, u_{i'} = q_i v_{i'} = 0$ if $i \neq i'$, v_i being the injections associated with the projections q_i.

If $p_i f = 0$, then $q_i t\, f = 0$, and thus $t\, f = 0$, that is $f = 0$, t being a monomorphism.

(2) $\Rightarrow$ (1). Let (Y, j) the kernel of t. First we notice that $q_i t j = 0$ and $p_i j = 0$ for any i. Then $j = 0$, hence t is a monomorphism. ∎

Let I be a directed set. Assume that for any $i \in I$ a short exact sequence

$$0 \longrightarrow X_i' \xrightarrow{f_i} X_i \xrightarrow{g_i} X_i'' \longrightarrow 0$$

is given, such that if $i \leqslant j$, then we have the morphisms:

$$f_{ij}' : X_i' \to X_j', \qquad f_{ij} : X_i \to X_j$$

and $f_{ij}'' : X_i'' \to X_j''$ making commutative the diagram:

$$\begin{array}{ccccccccc}
0 & \longrightarrow & X_i' & \xrightarrow{f_i} & X_i & \xrightarrow{g_i} & X_i'' & \longrightarrow & 0 \\
 & & \downarrow{\scriptstyle f'_{ij}} & & \downarrow{\scriptstyle f_{ij}} & & \downarrow{\scriptstyle f'_{ij}} & & \\
0 & \longrightarrow & X_j' & \xrightarrow{f_j} & X_j & \xrightarrow{g_j} & X_j'' & \longrightarrow & 0
\end{array}$$

Next, we suppose that $f_{ii}' = 1_{X'_i}$, $f_{ii} = 1_{X_i}$, $f_{ii}'' = 1_{X''_i}$ and if $i \leqslant j \leqslant k$ are elements of I, then the following compatibility relations are verified:

$$f_{jk}' f_{ij}' = f_{ik}', \qquad f_{jk} f_{ij} = f_{ik}, \qquad f_{jk}'' f_{ij}'' = f_{ik}''.$$

In this case, we shall say that a *direct system of short exact sequences in* $\mathscr{C}$ is given.

NOTE 2. A direct system of short exact sequences in $\mathscr{C}$ is in fact a covariant functor of I into the category $\mathscr{E}$ considered in Exercise 6 of Section 2.3.

Now, we have in $\mathscr{C}$ the commutative diagram:

$$\begin{array}{ccccccc} 0 \longrightarrow & X_i' & \xrightarrow{f_i} & X_i & \xrightarrow{g_i} & X_i'' & \longrightarrow 0 \\ & \downarrow{\scriptstyle u'_i} & & \downarrow{\scriptstyle u_i} & & \downarrow{\scriptstyle u''_i} & \\ & \varinjlim_i X_i' & \xrightarrow{f} & \varinjlim_i X_i & \xrightarrow{g} & \varinjlim_i X_i'' & \longrightarrow 0 \end{array} \tag{9}$$

in which f and g are canonically constructed.

It is easy to prove that last row of the above diagram is exact, precisely

$$(g, \varinjlim_i X_i'')$$

is a cokernel of f.

We shall say that the inductive direct limits are right exact. Dually, we deduce that projective inverse limits are left exact.

A set $\{X_i\}_{i\in I}$ of subobjects of an object X is called a *direct set of subobjects* if for any $i, j \in I$ there is $k \in I$ such that $X_i \subseteq X_k$ and $X_j \subseteq X_k$. Obviously, a direct set of subobjects may be considered as a direct system of objects.

THEOREM 8.6. *Let $\mathscr{C}$ be an* Ab 3-*category. The following assertions are equivalent*:

(1) *For any direct system of short exact sequences in $\mathscr{C}$, the canonical morphism f in the diagram* (9) *is a monomorphism* (*that is, the direct inductive limits are exact*).

(2) *For any direct set $\{X_i\}_i$ of subobjects of any object X, the unique morphism $j : \varinjlim X_i \to X$ defined such that $j u_i = j_i$,* (*where $u_i : X_i \to \varinjlim X_i$ are structural morphisms and $j_i : X_i \to X$ are canonical inclusions*) *is an isomorphism of $\varinjlim_i X_i$ onto $\sum_i X_i$.*

(3) *Let X be any object, $\{X_i\}_i$ a direct set of subobjects, and X' a subobject of X. Then*:

$$\textstyle\sum_i (X_i \cap X') = (\sum_i X_i) \cap X' \tag{10}$$

(4) *Let $f : Y \to X$ be a morphism and $\{X_i\}_i$ a direct set of subobjects of X. Then*:

$$f^{-1}(\textstyle\sum_i X_i) = \sum_i f^{-1}(X_i)$$

(5) *Let $\{X_i\}_{i\in I}$ be a set of objects. For any finite subset F of I, we denote by X_F image of canonical morphism $u_F : \coprod_{i'\in F} X_{i'} \to \coprod_i X_i$, defined such that $u_F u_{i'}' = u_{i'}$, where $u_{i'}'$ and $u_{i'}$ are respectively structural morphisms. Then for any subobject X of $\coprod_i X_i$ we have*:

$$X = \textstyle\sum_{F\in T} (X \cap X_F)$$

T being the set of all finite subsets of I.

Proof. (1) $\Rightarrow$ (2). Let $\{X_i\}_i$ be a direct set of subobjects of X, and $j_i : X_i \to X$ canonical inclusions. Then, we have the following direct set of short exact sequences canonically defined:

$$0 \longrightarrow X_i \xrightarrow{j_i} X \xrightarrow{p_i} X/X_i \longrightarrow 0$$

If $X_i \subseteq X_{i'}$, then we have the commutative diagram:

$$\begin{array}{ccccccccc} 0 & \longrightarrow & X_i & \longrightarrow & X & \longrightarrow & X/X_i & \longrightarrow & 0 \\ & & \downarrow \cap & & \| & & \downarrow f_{ii'}'' & & \\ 0 & \longrightarrow & X_{i'} & \longrightarrow & X & \longrightarrow & X/X_{i'} & \longrightarrow & 0 \end{array}$$

$f_{ii'}''$ being canonically constructed.

According to (1) we have the exact sequence:

$$0 \to \varinjlim_i X_i \xrightarrow{j} X \xrightarrow{p} \varinjlim_i (X/X_i) \to 0$$

For any i, we have $j u_i = j_i$. According to Lemma 8.1, we have

$$\operatorname{im} j = \textstyle\sum_i \operatorname{im}(j u_i) = \sum_i X_i.$$

Obviously, j is an isomorphism of $\varinjlim X_i$ to $\sum_i X_i$.

(2) $\Rightarrow$ (3). Let $\{X_i\}_i$ be a direct set of subobjects of X. Then, for any subobject X' of X, we have $X_i \cap X' \subseteq X_i$ and thus $\sum_i(X_i \cap X') \subseteq \Sigma X_i$. On the other hand, $X_i \cap X' \subseteq X'$, and consequently $\sum_i(X_i \cap X') \subseteq X'$. Finally, $\sum_i(X_i \cap X') \subseteq (\sum_i X_i) \cap X'$.

By the definition of sum of any set of subobjects, we have:

$$\textstyle\sum_i(X_i + X') = (\sum_i X_i) + X'.$$

We shall consider the following direct set of short exact sequences, canonically defined, t_i being the canonical inclusion:

$$0 \longrightarrow X' \xrightarrow{t_i} (X_i + X') \longrightarrow (X_i + X')/X' \longrightarrow 0$$

According to (2) we have

$$\varinjlim_i (X_i + X') = \textstyle\sum_i (X_i + X') = (\sum_i X_i) + X'.$$

The inductive direct limits being right exact, we deduce the following exact sequence:

$$X' \xrightarrow{t} \varinjlim_i (X_i + X') \to \varinjlim_i ((X_i + X')/X') \to 0$$

t being a monomorphism. (In fact, the structural morphisms

$$v_i : (X_i + X) \to \varinjlim_i (X_i + X')$$

are monomorphisms).

Then we have the canonical isomorphisms:

$$\varinjlim_i ((X_i + X')/X') \simeq (\varinjlim_i (X_i + X'))/X' \simeq ((\textstyle\sum_i X_i) + X')/X'. \quad (11)$$

On the other hand, we have the direct set of short exact sequences:

$$0 \longrightarrow X_i \cap X' \xrightarrow{k_i} X_i \longrightarrow (X_i + X')/X' \longrightarrow 0$$

canonically constructed. This set gives an exact direct limit sequence:

$$\textstyle\sum_i (X_i \cap X') \xrightarrow{k} \sum_i X_i \to \varinjlim_i ((X_i + X')/X') \to 0.$$

According to (2), k is a monomorphism and thus we have the isomorphism:

$$\varinjlim_i ((X_i + X')/X') \simeq (\textstyle\sum_i X_i)/(\sum_i (X_i \cap X')). \quad (12)$$

According to Proposition 6.4 we have:

$$((\textstyle\sum_i X_i) + X')/X' \simeq (\sum_i X_i)/((\sum_i X_i) \cap X').$$

From the isomorphisms (11) and (12) we deduce that

$$(\textstyle\sum_i X_i) \cap X' \subseteq \sum_i (X_i \cap X').$$

(3) ⇒ (2). Let $\{X_i\}_i$ be a direct set of subobjects of X,

$$X' = \varinjlim_i X_i, \qquad u_i : X_i \to X'$$

structural morphisms and $j : X' \to X$ the unique morphism such that $j u_i = j_i$, $j_i : X_i \to X$ being canonical inclusions. According to Lemma 8.1, we have $j(X') = \sum_i X_i$. We shall prove that j is a monomorphism, hence an isomorphism of X' to $\sum_i X_i$.

Let $K = \ker j$ and $X_i' = u_i(X_i)$. The set $\{X_i'\}$ of subobjects of X' is direct and $\sum_i X_i' = X'$. Then $K = (\sum_i X_i') \cap K = \sum_i (X_i' \cap K)$. If $K \neq 0$, $u_i^{-1}(K) = u_i^{-1}(X_i' \cap K) \neq 0$ for an i, and thus $u_i^{-1}(K) = \ker(j u_i) = \ker j_i = 0$, a contradiction. Finally $K = 0$ and j is a monomorphism.

(3) ⇒ (4). Let $\{X_i\}_i$ be a direct set of subobjects of X and $f : Y \to X$ a morphism. Then the set $\{f^{-1}(X_i)\}_i$ of subobjects of Y is direct. For any i we have the exact sequence:

$$0 \to f^{-1}(X_i) \to Y \to (\operatorname{im} f/(X_i \cap \operatorname{im} f)) \to 0$$

according to Exercise 8 of Section 2.6.

Passing to the inductive limit, we shall obtain the exact sequence:

$$\varinjlim_i f^{-1}(X_i) \xrightarrow{u} Y \to \varinjlim_i (\operatorname{im} f/(X_i \cap \operatorname{im} f)) \to 0 \qquad (13)$$

According to equivalence (2) ⇔ (3) we deduce that u is a monomorphism. Now we shall utilize the following lemma:

LEMMA 8.7. *Let $\mathscr{C}$ be an* Ab 3-*category and $\{X_i\}_i$ a direct set of subobjects of X. Then $\varinjlim_i (X/X_i)$ is canonically isomorphic with $X/(\sum_i X_i)$.*

Proof. We consider the exact sequences:

$$0 \to X_i \to X \to X/X_i \to 0.$$

Passing to the inductive limit, we have the exact sequence:

$$\varinjlim_i X_i \xrightarrow{u} X \to \varinjlim_i (X/X_i) \to 0.$$

According to Lemma 8.1 we see that $\operatorname{im} u = \sum_i X_i$, that is

$$X/(\textstyle\sum_i X_i) \simeq \varinjlim_i (X/X_i). \blacksquare$$

Applying the Lemma 8.7 to the exact sequence (13) we have:

$$\varinjlim_i (\operatorname{im} f/(X_i \cap \operatorname{im} f)) \simeq \operatorname{im} f/(\textstyle\sum_i (X_i \cap \operatorname{im} f)) \simeq \operatorname{im} f/(\operatorname{im} f \cap (\sum_i X_i)).$$

By Exercise 8 of Section 2.6 and (13), we deduce:

$$\operatorname{im} f/(\operatorname{im} f \cap (\textstyle\sum_i X_i)) \simeq Y/(f^{-1}(\sum_i X_i)) \simeq Y/(\sum_i f^{-1}(X_i)).$$

The assertion (4) of Theorem 8.6 now follows immediately.

The implication (4) ⇒ (3) is now obvious. Indeed, if $i: Y \to X$ is a subobject of X, then for any subobject X' of X we have: $i^{-1}(X') = X' \cap Y$.

It is clear that (5) is a special case of (3), that is, (3) ⇒ (5)

(5) ⇒ (3). Let $\{X_i\}_{i \in I}$ be a direct set of subobjects of X, and X' a subobject of X. Let $s: \coprod_i X_i \to X$ be the sum morphism. According to Exercise 7 of Section 2.6 we have:

$$X' \cap (\textstyle\sum_i X_i) = X' \cap \operatorname{im} s = s(s^{-1}(X')) = s(s^{-1}(X') \cap (\coprod_i X_i)). \qquad (14)$$

Let T be the set of finite subsets of I. Obviously, $\{X_F\}_{F \in T}$ is a direct set of $\coprod_i X_i$ and $\sum_F X_F = \coprod_i X_i$.

Then, according to Exercise 9 of Section 2.6, we have:

$$\begin{aligned} s\big(s^{-1}(X') \cap \textstyle\coprod_i X_i\big) &= s\big(s^{-1}(X') \cap \textstyle\coprod_F X_F\big) \\ &= s\big(\textstyle\sum_F (s^{-1}(X') \cap X_F)\big) = \textstyle\sum_F s\big(s^{-1}(X') \cap X_F\big) \\ &= \textstyle\sum_F \big(X' \cap s(X_F)\big). \end{aligned} \tag{15}$$

Now, for any $F \in T$, there is $i \in I$ such that $s(X_F) \subseteq X_i$, and thus:

$$\textstyle\sum_F \big(X' \cap s(X_F)\big) \subseteq \textstyle\sum_i (X' \cap X_i).$$

The converse relation being obvious, we deduce that (3) is valid, using the relations (14) and (15).

We must prove that (2)+(3)+(4)+(5)⇒(1). For that, we shall utilise the following result:

LEMMA 8.8. *Let $\mathscr{C}$ be an* Ab 3-*category in which one of the assertions* (2) *to* (5) *holds. Let $(X_i, f_{ij})_i$ be a direct system in $\mathscr{C}$. Denote by K_{ij} the kernel of $f_{ij}: K_i \to K_j$, $i \leqslant j$, and let K_i be the kernel of u_i, $u_i : X_i \to \varinjlim X_i$ being the structural morphisms. Then $K_i = \sum_{i \leqslant j} K_{ij}$.*

Proof. Firstly it is clear that $\sum_{i \leqslant j} K_{ij} \subseteq K_i$, so that we need only prove the reverse inclusion. Let R be subset of $I \times I$ consisting of all ordered pairs (i, j) such that $i \leqslant j$. Let $X = \coprod_i X_i$ and v_i be the structural morphisms. If S is any subset of R, let:

$$\textstyle\sum_{(i,\,j) \in S} \operatorname{im}(v_i - v_j f_{ij}) = X_S \subseteq X.$$

Then, by Exercise 2 we have:

$$\varinjlim_i X_i = X/X_R$$

and $u_i = p\, v_i$, $p : X \to X/X_R$ canonically.

Now $X_R = \sum_F X_F$ where F runs through all finite subsets of R. Obviously, $\{X_F\}_F$ is a direct set of subobjects of X. Hence by assertion (4) of Theorem 8.6 we have:

$$K_i = v_i^{-1}(X_R) = \textstyle\sum_F v_i^{-1}(X_F),$$

and so it suffices to show that for each finite subset F of R we have

$$v_i^{-1}(X_F) \subseteq K_{ij}$$

for some $j \geqslant i$. Given F, let p be any index which follows i and all indices which appear either in the first position or in the second position in a member of F. Now, $v_i^{-1}(X_F)$ is the kernel of the composition:

$$X_i \xrightarrow{v_i} X \longrightarrow X/X_F.$$

Define a morphism $f : X \to X_p$ by taking $f v_j = f_{jp}$ for $j \leqslant p$ and $f v_j = 0$ otherwise. Then for $(k, q) \in F$ we have

$$f(v_k - v_q f_{kq}) = f_{kp} - f_{qp} f_{kq} = f_{kp} - f_{kp} = 0.$$

Consequently, f factors through $X \to X/X_F$ and so we have:

$$0 = f(v_i(v_i^{-1}(X_F)) = f_{ip}(v_i^{-1}(X_F)).$$

This shows that $v_i^{-1}(X_F) \subseteq K_{ip}$ as required. ■

Now suppose that $\mathscr{C}$ satisfies conditions (2) to (5) and let

$$\{f_i : X_i' \to X_i\}_{i \in I}$$

be a set of monomorphisms defining a morphism from the direct system $\{X_i', f_{ij}'\}_{i \in I}$ to the direct system $\{X_i, f_{ij}\}_{i \in I}$. Consider the commutative diagram:

$$\begin{array}{ccccccc}
 & & & X_i' & \xrightarrow{f_i} & X_i \\
 & & & \downarrow{\scriptstyle u'_i} & & \downarrow{\scriptstyle u_i} \\
0 & \longrightarrow & K \longrightarrow & \varinjlim_i X_i' & \xrightarrow{f} & \varinjlim_i X_i
\end{array}$$

where K is kernel of f, u_i', u_i being the structural morphisms. Let $Y_i = \operatorname{im} u_i'$, so that by Lemma 8.1 we have $\varinjlim_i X_i' = \sum_i Y_i$ and $\{Y_i\}_i$ is a direct set of subobjects. Then by (10) we have $K = (\sum_i Y_i) \cap K = \sum_i (K \cap Y_i)$ and so if $K \neq 0$, then $K \cap Y_{i_0} \neq 0$ for some $i_0 \in I$. Writing $M = u_{i_0}'^{-1}(K \cap Y_{i_0})$ we see from the definition of inverse image that $M \neq 0$, and $u_{i_0}'(M) \neq 0$. On the other hand we have:

$$\begin{aligned}
u_{i_0}(f_{i_0}(M)) &= f(u_{i_0}'(M)) \\
&= f(u_{i_0}'(u_{i_0}'^{-1}(K \cap Y_{i_0}))) \subseteq f(K \cap Y_{i_0}) \subseteq f(K) = 0.
\end{aligned}$$

Therefore, by Lemma 8.8, $f_{i_0}(M)$ is a subobject of $\sum_{i_0 \leqslant j} \ker f_{i_0 j}$. Again using (10) we then have:

$$f_{i_0}(M) = \textstyle\sum_{i_0 \leqslant j} (\ker f_{i_0 j} \cap f_{i_0}(M)).$$

Hence

$$\begin{aligned}
M = f_{i_0}^{-1}(f_{i_0}(M)) &= f_{i_0}^{-1}(\textstyle\sum_{i_0 \leqslant j} (\ker f_{i_0 j} \cap f_{i_0}(M)) \\
&= \textstyle\sum_{i_0 \leqslant j} f_{i_0}^{-1}(\ker f_{i_0 j} \cap f_{i_0}(M))
\end{aligned} \tag{16}$$

the first equality being true because f_{i_0} is a monomorphism and the third equality being true by condition (4) of the theorem. It follows from the fact that f_j is a monomorphism that $f_{i_0 j}'(f_{i_0}^{-1}(\ker f_{i_0 j} \cap f_{i_0}(M))) = 0$ for

$j \geqslant i_0$, and so $u_{i_0}'(f_{i_0}^{-1}(\ker f_{i_0 j} \cap f_{i_0}(M))) = 0$ for all $j \geqslant i_0$. But then, using (16) and Proposition 6.9 we see that $u_{i_0}'(M) = 0$. This contradiction proves that $K = 0$, and so f is a monomorphism. ■

An Ab 3-category which satisfies the conditions of the above theorem is called an Ab 5-*category*. Dually we have an Ab 5*-*category*. this is an Ab 3*-category which satisfies the duals of above conditions.

Sometimes we shall say that $\mathscr{C}$ is *locally an* Ab 5-*category* if it is not necessarily an Ab 3-category, but the condition (3) of above theorem is fulfilled. In this category, the sum of any direct set of subobjects need not be defined. For example, the category of finitely generated abelian groups is a locally Ab 5-category.

COROLLARY 8.9. *An* Ab 5-*category* $\mathscr{C}$ *is* Ab 4.

Proof. Let $\{f_i : X_i' \to X_i\}_{i \in I}$ be a set of monomorphisms of $\mathscr{C}$ and

$$f : \coprod_i X_i' \to \coprod_i X_i$$

the sum morphism. Let T be the set of finite subsets of I; for any $F \in T$ we denote by X_F' the canonical image of $\coprod_{j \in F} X_j'$ in $\coprod_i X_i'$. Then $\{X_F'\}_{F \in T}$ is a direct set of subobjects and $\sum_F X_F' = \coprod_i X_i'$. Let $K = \ker f$. Then $K = (\sum_F X_F') \cap K = \sum_F' \cap K)$. If $K \neq 0$, then $X_F' \cap K \neq 0$ for some F. Now, let $u_F : \coprod_{j \in F} X_j' \to \coprod_i X_i'$ be the canonical monomorphism. Obviously, $u_F^{-1}(X_F' \cap K) \neq 0$. Also, let $v_F : \coprod_{j \in F} X_j \to \coprod_i X_i$ and

$$f_F : \coprod_{j \in F} X_j' \to \coprod_{j \in F} X_j$$

be canonically constructed. Then $f\,u_F = v_F f_F$ and $u_F^{-1}(X_F' \cap K) = \ker(f\,u_F) = \ker(v_F f_F) = 0$, v_F and f_F being monomorphisms. This is a contradiction, hence f is a monomorphism. ■

COROLLARY 8.10. *Let* $\mathscr{C}$ *be an* Ab 5 *and an* Ab 3*-*category Then* $\mathscr{C}$ *is a* C_2-*category.*

Proof. Let $\{X_i\}_{i \in I}$ be a set of objects of $\mathscr{C}$. Let T be a set of finite subsets of I, and for any $F \in T$, $X_F = \operatorname{im} u_F$, $u_F : \coprod_{j \in F} X_j \to \coprod_i X_i$ the canonical map. If $K = \ker t \neq 0$, $t : \coprod_i X_i \to \prod_i X_i$, then for some F, we have

$$K \cap X_F \neq 0,$$

that is, $0 \neq u_F^{-1}(K \cap X_F) = \ker(t\,u_F) = 0$, $t\,u_F$ being a monomorphism, a contradiction. Hence $K = 0$. ■

COROLLARY 8.11. *An* Ab 5-*category which is also* C_2* *consists only of zero objects.*

Proof. By Proposition 8.5, $\mathscr{C}$ is C_2 and so since it is also $C_2{}^*$, we find that for any set $\{X_i\}_i$ of objects the canonical morphism $t : \coprod_i X_i \to \prod_i X_i$ is an isomorphism. In particular, given $X \in \operatorname{Ob} \mathscr{C}$, then $X^{(N)} = X^N$. Let $\Delta : X \to X^N$ be the diagonal morphism, that is the unique morphism such that $p_i \Delta = 1_X$ for all structural projections p_i. For any natural n, X^n is canonically a subobject of X^N and $\{X^n\}_n$ is a direct set of subobjects such that $\sum_n X^n = X^N$. Then from Theorem 8.6, we have $X = \sum_{n \in N} \Delta^{-1}(X^n)$. We show that $\Delta^{-1}(X^n) = 0$ for all n. Let q be the composition $X \xrightarrow{\Delta} X^N \xrightarrow{p} X^N/X^n$. Then the exact sequence

$$0 \to \Delta^{-1}(X^n) \to X \xrightarrow{q} X^N/X^n$$

defines $\Delta^{-1}(X^n)$. Also, if $u_n : X^n \to X^N$ is the canonical inclusion, then $p_{n+1} u_n = 0$, so that we have a morphism $\bar{p}_{n+1} : X^N/X^n \to X_{n+1} = X$ such that $\bar{p}_{n+1} p = p_{n+1}$. Then $\bar{p}_{n+1} q = 1_X$ and so q must be a monomorphism. Therefore $\Delta^{-1}(X^n) = \ker q = 0$ and so $X = 0$. ∎

PROPOSITION 8.12. *Let $\mathscr{C}$ be an* Ab 5*-category and $\{X_i\}_{i \in I}$ a set of subobjects of an object X. The following assertions are equivalent*:

(1) *The sum morphism $s : \coprod_i X_i \to X$ is an isomorphism*;

(2) *The following conditions are fulfilled*:

(a) $\sum_i X_i = X$;

(b) *For any $i_0 \in I$, we have* $(\sum_{i \neq i_0} X_i) \cap X_{i_0} = 0$.

Proof. (1) ⇒ (2). Let $v : \coprod_{i \neq i_0} X_i \to \coprod_i X_i$ be the unique morphism canonically defined. If $u_i : X_i \to \coprod_i X_i$ are structural morphisms, then $p_{i_0} v = 0$, where p_i is the projection associated with u_i. Then, $u_{i_0}(X_{i_0}) \cap \operatorname{im} v = 0$. But $s u_{i_0}$ is the canonical inclusion of X_{i_0} in X and $\operatorname{im}(s v) = \sum_{i \neq i_0} X_i$. Then $s\big(u_{i_0}(X_{i_0}) \cap \operatorname{im} v\big) = s\big(u_{i_0}(X_{i_0})\big) \cap s(\operatorname{im} v) = 0$, s being an isomorphism.

(2) ⇒ (1). According to (a), s is an epimorphism. We shall prove that $\ker s = 0$. Let $K = \ker s$ and F a finite subset of I. Again we denote by X_F the image of canonical morphism $u_F : \coprod_{j \in F} X_j \to \coprod_i X_i$. According to (b) and Corollary 7.6 we deduce that $s u_F$ is a monomorphism. Let T be set of finite subsets of I. Then $\sum_{F \in T} X_F = \coprod_i X_i$ and $\{X_F\}_F$ is a direct set of subobjects. Then: $K = \sum_F (X_F \cap K)$ and for any F, $u_F{}^{-1}(X_F \cap K) = \ker(s u_F) = 0$. Hence $K = 0$, and finally s is an isomorphism. ∎

We shall say that an Ab 3-category is an Ab 6-*category* if for any object X and any set J such that for each $j \in J$ a direct set $\{X_{j(i)}\}_{i \in I_j}$ of subobjects of X is given, we have:

$$\bigcap_{j \in J} \Big(\sum_{i \in I_j} X_{j(i)}\Big) = \sum_{(j(i)) \in \prod_{j \in J} I_j} \Big(\bigcap_{(j(i))} X_{j(i)}\Big).$$

COROLLARY 8.13. *An* Ab 6-*category is* Ab 5. ∎

Dually we have the condition Ab 6*.

According to Corollaries 8.10 and 8.11 we may deduce that the conditions Ab 5 and Ab 6* are independent. (In fact the category $\mathcal{Ab}$ is an Ab 6-category). In the following chapters we shall give some examples which prove the independence of all conditions Ab. (See Chapter 4, Section 4.21, Note 3 and Chapter 5, Section 5.3, Note 1).

NOTE 3. An Ab 5-category with a generator is sometimes called a *Grothendieck-category*.

We illustrate these notions within the category $\mathcal{Ab}$. If $\{X_i\}_{i\in I}$ is any set of abelian groups, then a direct product is defined by the cartesian product $\prod_i X_i$ with addition defined by the rule: $(x_i) + (y_i) = (x_i + y_i)$. The projection $p : \prod_i X_i \to X_i$ is given by $p_i\left((x_i)\right) = x_i$. A direct sum for the above set of objects is obtained by taking the subgroup of $\prod_i X_i$ consisting of all those elements (x_i) such that $x_i = 0$ for all but a finite number of i. The structural injection $u_i : X_i \to \coprod_i X_i$ is given by $u_i(x) = (x_i)$ where $x_j = 0$ for $j \neq i$ and $x_i = x$, for any $x \in X_i$. It is easy to prove that $\mathcal{Ab}$ is an Ab 6 and Ab 4* category.

Exercises

1. In an Ab 3-category the intersection of any set of subobjects of an object has a sense.

Dually, in an Ab 3*-category the sum of any set of subobjects has a sense.

2. Let $(X_i, f_{ij})_{i\in I}$ be a direct system in an Ab 3-category. Let R be the subset of $I \times I$ consisting of all ordered pairs (i, j) such that $i \leqslant j$. Let $X = \coprod_i X_i$ with structural morphisms v_i, and put: $X_R = \sum_{(i,j)\in R} \operatorname{im}(v_i - v_j f_{ij})$. Then X/X_R with canonical morphisms $u_i = p\, v_i (p : X \to X/X_R$ being canonical) is an inductive limit of the system considered.

Dualize this exercise.

3. Let $\mathbf{Z}$ be the additive group of integers. Find three subgroups X_1, X_2 and Y in $\mathbf{Z} \amalg \mathbf{Z}$ such that $(X_1 + X_2) \cap Y \neq (X_1 \cap Y) + (X_2 \cap Y)$.

4. An abelian group X is called a *torsion group* if for each $x \in X$ there is a non-zero integer n such that $n\,a = a + a + \ldots + a$ (n-fold) $= 0$. Let $\mathcal{Tors}$ be the full subcategory of $\mathcal{Ab}$ consisting of all torsioin groups. Then $\mathcal{Tors}$ is an abelian subcategory of $\mathcal{Ab}$ and the direct sums in $\mathcal{Tors}$ are the same as in $\mathcal{Ab}$.

However the product in $\mathscr{T}ors$ of a family $\{X_i\}_i$ of torsion groups is given by the subgroup of the product in $\mathscr{A}b$ consisting of all elements of the form (x_i) such that $x_i \in X_i$ have bounded orders; that is there is a non-zero integer n such that $n\,x_i = 0$ for all i. For each positive integer m, let $f_m : Z_{2m} \to Z_2$ be the morphism which takes the coset r modulo 2^m into the coset r modulo 2. Then, each f_m is an epimorphism, but $\prod_m f_m$ is not an epimorphism. Thus an Ab 3 and Ab 3* need not be an Ab 4-category, that is, the independence of Ab 3 and Ab 4 is proved.

5. $\mathscr{C}$ is an Ab 5-category if and only if, for any direct system of exact sequences:

$$X_i' \xrightarrow{f_i} X_i \xrightarrow{g_i} X_i''$$

We have the exact sequence, canonically constructed:

$$\varinjlim_i X_i' \xrightarrow{f} \varinjlim_i X_i \xrightarrow{g} \varinjlim_i X_i''$$

(compare with Chapter 3, Section 3.2.)

6. Consider a direct system $\{X_i, f_{ij}\}_{i \in I}$ in an Ab 5-category and let $f : Y \to X_i$ be a morphism for some $i \in I$. Then: $\ker(u_i\, f) = \sum_{j \geqslant i} \ker(f_{ij} f)$, $u_i : X_i \to \varinjlim_i X_i$ being structural morphisms.

7. Let $\mathscr{C}$ be an Ab 3-category (respectively an Ab 3*-category). Then the category $\mathscr{E}$ defined as in Exercise 6 of Section 2.3 has direct sums (respectively, direct products).

8. Let $\{X_i\}_i$ be a set of subobjects of any abelian group X. Show that $\sum_i X_i$ is the set of elements in X of the form $\sum_i x_i$, when $x_i \in X_i$, for all $i \in I$ and $x_i = 0$ for all but a finite number of i.

References

Bucur–Deleanu [31]; Freyd [57]; Gabriel [59]; Grothendieck [79]; Mitchell [133]; Pareigis [163].

3. Additive Functors

The notion of additive functor is a fundamental tool in the theory of abelian categories. The aim of this chapter is to expose the fundamental results about additive functors and to study some particular classes of additive functors which are of special interest.

In the theory of abelian categories an important rôle is played by the representable additive functors, which in this context have special properties in connection with certain types of objects. An important class of functors is that of the exact functors and these are in close connection with the injective and the projective objects, by which the exactness of the representable additive functors may be expressed. This connection is yet more apparent in homological algebra, where the principal problem is the study of the derived functors of a given additive functor (functors by which, in fact, the degree of non-exactness of the given functor is measured). This study is carried out by means of the injective and projective objects. (In connection with homological algebra, the reader is referred to [33], [79], or [133]).

An illustration of the close connection existing between the exactness of the functors and injective objects is expressed by Theorem 2.8, which has numerous consequences.

The notion of injective object in the category $\mathscr{Ab}$ was introduced by Baer [12]. Subsequently his results have been systematically generalized for a larger class of abelian categories.

As already mentioned, an important application of the injective (and projective) objects is made in homological algebra but their theory is important also for other reasons, as we shall see in the following chapters. Thus, of great importance are the abelian categories with sufficiently many injectives. Baer [12] has already shown that $\mathscr{Ab}$ is a category with sufficient injectives. Subsequently, Cartan and Eilenberg [33] have shown that the category Mod A, the category of left modules over a ring A, is a category with sufficiently many injectives. With the help of their method, Grothendieck [79] proved that an Ab 5-category which has a set of generators has sufficiently many injectives. These results are proved in the present chapter. First we show that $\mathscr{Ab}$ is a category with sufficiently many injectives (Theorem 3.3) and then, by using Theorem 2.8 we show that Mod A has sufficiently many injectives (Theorem 7.8).

In Section 3.4 categories of modules are defined. The study of some special classes of objects (small objects, objects of finite type, finitely presented objects, coherent objects) forms the contents of Section 3.5, where a characterization of the objects of finite type and finitely presented objects for categories of modules is given.

In Section 3.6 the tensor product is constructed; this is, in fact, an additive alternative to the Kan extension theorem (see [133]). Here we also give a characterization of the categories of modules, due to Freyd (Corollary 6.4).

In Section 3.7 we apply the results of the preceding section to the special case of the category of modules over a ring. An important result is Corollary 7.7, due to Morita [134], which gives necessary and sufficient conditions for the equivalence of categories of modules. An application of this result is given in Section 3.11 (Theorem 11.6) where semisimple rings are characterized.

An important result of Chapter 3 is Theorem 7.9, which gives a characterization of Grothendieck categories. The first proof of this theorem utilized localization theory. The present proof, which is partly due to Harada [86], does not appeal to localization theory, nor to the fact that a Grothendieck category has sufficiently many injectives. On the contrary, by using this theorem and Theorem 7.8 one can infer that any Grothendieck category has sufficiently many injectives (Theorem 10.10).

Section 3.8 is devoted to the exactness of the tensor product, by which an important class of modules is exhibited: the flat modules.

Section 3.9 lies somewhat outside the fundamental aim of this Chapter, but the results presented, most of them due to Kaplansky [102], are of great importance for the structure of the projective objects.

A study of injective envelopes is made in Section 3.10. The notion of injective envelope was introduced by Baer for the category $\mathscr{Ab}$. Eckmann and Schopf [48] generalized his result for categories of modules, whereas Gabriel [59] extended it to the case of Grothendieck categories. All these results are presented here in the general form given by Gabriel [59].

As a consequence of Morita's duality theorem (Corollary 7.7) and of Theorem 7.9, in Section 3.11 we establish the structure of semisimple rings (Theorem 11.6).

3.1 Additive functors

Let $\mathscr{C}, \mathscr{C}'$ be two preadditive categories and $F : \mathscr{C} \to \mathscr{C}'$ a functor. F is said to be *additive* if for any pair of morphisms $f, g \in \mathrm{Hom}_{\mathscr{C}}(X, Y)$ we have $F(f + g) = F(f) + F(g)$.

Unless otherwise stated, whenever we consider functors defined on preadditive categories and taking their values in preadditive categories, we will assume them to be additive.

NOTES (1) The functor F is additive if and only if for any couple (X, Y) of objects of $\mathscr{C}$, the map $F(X, Y)$ is a morphism of abelian groups.

(2) Let F be a covariant additive functor. Then, for any two morphisms f, g such that $f\,g = 0$, we have: $F(f\,g) = F(f)\,F(g) = 0$.

EXAMPLES. 1. Let $\mathscr{C}$ be a preadditive category and $X \in \operatorname{Ob}\mathscr{C}$. Then the functors h_X and h^X were defined in Chapter 1, Section 1.3, taking their values in the category $\mathscr{Ab}$. These functors are additive. They will be denoted by the same symbols: h_X and h^X.

2. We denote by $T : \mathscr{Ab} \to \mathscr{Tors}$ (see Chapter 2, Section 2.8, Exercise 4) the functor defined as follows. For any abelian group X, $T(X)$ is the subset of X consisting of the torsion elements x, that is such that $n\,x = 0$ for a non-null integer n. Then T is an additive functor.

3. Each of the functors ker, coker, im, coim defined in Chapter 2, Section 2.2 is additive.

4. Let A, B be two rings. Then an additive covariant functor from A to B is a morphism of rings.

PROPOSITION 1.1. *Let $F : \mathscr{C} \to \mathscr{C}'$ be an additive functor and $\{X_i\}_{1 \leqslant i \leqslant n}$ objects of $\mathscr{C}$. If the object $\coprod_i X_i$ exists in $\mathscr{C}$, then $\coprod_i F(X_i)$ exists in $\mathscr{C}'$ and $F(\coprod_i X_i)$ is canonically isomorphic to $\coprod_i F(X_i)$.*

Proof. Let u_i be structural injections and p_i associated projections, $i = 1, \ldots, n$. Then the relations (1) of Chapter 2 are satisfied. Obviously, the same relations are satisfied by the morphisms $F(u_i)$ and $F(p_j)$ of $\mathscr{C}'$. ■

The next result is in some sense a converse of the above proposition.

PROPOSITION 1.2. *Let $\mathscr{C}$ be an additive category, $\mathscr{C}'$ a preadditive category and $F : \mathscr{C} \to \mathscr{C}'$ a functor which commutes with finite direct sums. Then F is additive.*

Proof. Let $f, g \in \operatorname{Hom}_{\mathscr{C}}(X, Y)$. According to Chapter 2, Theorem 1.3, we have:

$$f + g = \nabla t \Delta,$$

$t : X \amalg X \to Y \amalg Y$ being canonically induced by f and g. Let $u_i : X \to X \amalg X$, and $v_i : Y \to Y \amalg Y$ be the structural morphisms. Then, by hypothesis, $F(u_i)$ and $F(v_i)$ are also structural morphisms. Since $t\,u_1 = v_1 f$, $t\,u_2 = v_2 g$, $F(\Delta) = \Delta$ and $F(\nabla) = \nabla$, we deduce that $F(t)$ is canonically induced by $F(f)$ and $F(g)$. On the basis of Chapter 2, Theorem 1.3, we deduce that $F(f + g) = \nabla F(t)\,\Delta = F(f) + F(g)$. ■

COROLLARY 1.3. *Let $\mathscr{C}$, $\mathscr{C}'$ be additive categories, $F : \mathscr{C} \to \mathscr{C}'$ a covariant functor and G a right adjoint to F. Then F and G are additive.*

Proof. We use Theorem 5.2 of Chapter 1 and the above proposition. ■

Exercises

1. The functor defined in Chapter 2, Section 2.1, Exercise 1 is additive.

2. Let **Z** be the "preadditive category" of integers and A any ring. There exists a unique additive functor $u_A : \mathbf{Z} \to A$.

3. Let $\mathscr{C}$ be any category and let Add ($\mathscr{C}$) be as in Chapter 1, Section 1.1, Exercise 1. If $\mathscr{B}$ is any preadditive category, then for any functor $T : \mathscr{C} \to \mathscr{B}$ there is a unique additive functor $\overline{T} : \text{Add}(\mathscr{C}) \to \mathscr{B}$ such that $\overline{T}I = T$, $I : \mathscr{C} \to \text{Add}(\mathscr{C})$ is canonically defined.

4. We say that a functor $T : \mathscr{C} \to \mathscr{C}'$ *reflects* a property of a diagram D in $\mathscr{C}$ if the condition that TD has the property implies that D itself has the property. Thus, for example, T reflects inductive limits if (f, X) is an inductive limit for any functor D in $\mathscr{C}$ whenever $(T(f), T(X))$ is an inductive limit for TD.

Let $\mathscr{C}$, $\mathscr{C}'$ be abelian categories and $T : \mathscr{C} \to \mathscr{C}'$ an additive covariant functor such that T reflects either inductive limits or exact sequences. Then T is faithful. Conversely, a faithful functor reflects exact sequences.

5. If $\mathscr{C}$ is an Ab 3 and Ab 3* category and $\mathscr{C}'$ is a C_2-category, then for any additive functor $T : \mathscr{C} \to \mathscr{C}'$ and any direct sum $\coprod_i X_i$ in $\mathscr{C}$ the canonical morphism

$$\coprod_i T(X_i) \to T(\coprod_i X_i)$$

is a monomorphism.

References

Buchsbaum [28]; Bucur–Deleanu [31]; Cartan–Eilenberg [33]; Freyd [57]; Gabriel [59]; Grothendieck [79]; MacLane [120]; Mitchell [133]; Pareigis [162].

3.2 Exactness of functors. Injective and projective objects

All categories considered here are abelian.

Let $F : \mathscr{C} \to \mathscr{C}'$ be a covariant (respectively contravariant) additive functor.

(a) We shall say that F is *left exact* if for any exact sequence of $\mathscr{C}$

$$0 \to X' \to X \to X'' \to 0 \tag{1}$$

we get an exact sequence in $\mathscr{C}'$:

$$0 \to F(X') \to F(X) \to F(X'') \quad (\text{resp. } 0 \to F(X'') \to F(X) \to F(X'))$$

(b) We shall say that F is *right exact* if for any exact sequence (1) of $\mathscr{C}$ we have the exact sequence in $\mathscr{C}'$:

$$F(X') \to F(X) \to F(X'') \to 0 \quad (\text{resp. } F(X'') \to F(X) \to F(X') \to 0)$$

(c) We shall say that F is *exact* if F is left and right exact.

Let us consider in the category $\mathscr{C}$ the following commutative square:

$$\begin{array}{ccc} X & \longrightarrow & Y \\ \downarrow & & \downarrow \\ Z & \longrightarrow & T \end{array} \tag{2}$$

PROPOSITION 2.1. *Let $F : \mathscr{C} \to \mathscr{C}'$ be a covariant additive functor:*

(a) *The following assertions are equivalent:*

(1) *F is left exact;*

(2) *For any exact sequence*

$$0 \to X' \to X \to X''$$

of $\mathscr{C}$, we have in $\mathscr{C}'$ the exact sequence:

$$0 \to F(X') \to F(X) \to F(X'').$$

(3) *For any cartesian square* (2) *of $\mathscr{C}$, the corresponding square in $\mathscr{C}'$ is also cartesian.*

(b) *The following assertions are equivalent:*

(1) *F is right exact;*

(2) *For any exact sequence of $\mathscr{C}$:*

$$X' \to X \to X'' \to 0$$

we have in $\mathscr{C}'$ the exact sequence:

$$F(X') \to F(X) \to F(X'') \to 0.$$

(3) *For any cocartesian square* (2) *in $\mathscr{C}$, the corresponding square in $\mathscr{C}'$ is also cocartesian.*

(c) *The following assertions are equivalent;*

(1) *F is exact;*

(2) *For any exact sequence*:

$$X' \to X \to X''$$

of $\mathscr{C}$ *we have in* $\mathscr{C}'$ *the exact sequence*:

$$F(X') \to F(X) \to F(X'').$$

(3) *For any exact square* (2) *in* $\mathscr{C}$, *the corresponding square in* $\mathscr{C}'$ *is also exact.*

Proof. (a), (1) $\Rightarrow$ (2). We shall consider in $\mathscr{C}$ the commutative diagram:

$$\begin{array}{ccccccc} 0 & \longrightarrow & X' & \xrightarrow{f} & X & \xrightarrow{g} & X'' \\ & & & & {\scriptstyle p}\searrow & & \nearrow{\scriptstyle j} \\ & & & & & \operatorname{im} g & \end{array}$$

p and j being defined canonically. Obviously, $\ker p = \ker(jp) = \ker g = \operatorname{im} f$. Then we have in $\mathscr{C}'$ the diagram:

$$\begin{array}{ccccccc} 0 & \longrightarrow & F(X') & \xrightarrow{F(f)} & F(X) & \xrightarrow{F(g)} & F(X'') \\ & & & & {\scriptstyle F(p)}\searrow & & \nearrow{\scriptstyle F(j)} \\ & & & & & F(\operatorname{im} g) & \end{array}$$

in which $\operatorname{im} F(f) = \ker F(p)$, and $F(j)$ is a monomorphism according to the hypothesis. Then $\ker F(g) = \ker\big(F(j)\,F(p)\big) = \operatorname{im} F(f)$.

(2) $\Rightarrow$ (3). Obviously, according to Chapter 2, Proposition 4.2.

The remaining implications are left to the reader to verify. ∎

COROLLARY 2.2. *For any object* X *of* $\mathscr{C}$ *the functors* h^X *and* h_X *are left exact.*

Proof. We consider in $\mathscr{C}$ the exact sequence:

$$0 \longrightarrow Y' \xrightarrow{i} Y \xrightarrow{p} Y''$$

Then we have in $\mathscr{Ab}$ the diagram

$$0 \longrightarrow \operatorname{Hom}_{\mathscr{C}}(X, Y') \xrightarrow{h^X(i)} \operatorname{Hom}_{\mathscr{C}}(X, Y) \xrightarrow{h^X(p)} \operatorname{Hom}_{\mathscr{C}}(X, Y'').$$

Obviously, $h^X(p)\,h^X(i) = h^X(p\,i) = 0$. Let $f \in \operatorname{Hom}_{\mathscr{C}}(X, Y)$ be a morphism such that $h^X(p)\,(f) = p\,f = 0$. There is a morphism $f' : X \to Y'$ such that $i\,f' = f$, that is $h^X(i)\,(f') = f$. Hence, the kernel of $h^X(p)$ is the image of $h^X(i)$. ∎

COROLLARY 2.3. *Let $F : \mathscr{C} \to \mathscr{C}'$ be a covariant functor and G a right adjoint of F. Then F is left exact and G is right exact.*

The proof follows according to Chapter 1, Theorem 5.2 and the above Proposition. ∎

We shall say that an object X of $\mathscr{C}$ is *injective* (resp. *projective*) if the functor h_X (resp. h^X) is exact.

COROLLARY 2.4. *Let X be an object of $\mathscr{C}$. The following assertions are equivalent:*

(1) *X is an injective (resp. projective) object;*

(2) *For any coangle (resp. angle) of $\mathscr{C}$:*

$$\begin{array}{ccc} 0 \to Y' \xrightarrow{i} Y & \qquad & X \\ {\scriptstyle f}\downarrow \quad\quad & & \downarrow{\scriptstyle g} \\ X \quad\quad & & Y \xrightarrow{p} Y'' \to 0 \end{array}$$

with an exact row, there is a morphism $\bar{f} : Y \to X$ (resp. $\bar{g} : X \to Y$) such that $\bar{f} i = f$ (resp. $p \bar{g} = g$). ∎

Obviously, the concepts of injective and projective objects are dual.

COROLLARY 2.5. *Let $f : X \to Y$ be a monomorphism and X injective. Then f is a section. If f is an epimorphism and Y projective, then f is a retraction.* ∎

PROPOSITION 2.6. *Let $\{X_i\}_i$ be a set of objects of $\mathscr{C}$ such that $\prod_i X_i$ (respectively $\coprod_i X_i$) exists in $\mathscr{C}$. Then $\prod_i X_i$ (resp. $\coprod_i X_i$) is an injective (projective) object if and only if for any i, X_i is an injective (projective) object.*

Proof. Assume that $\prod_i X_i$ is an injective object, and let p_i be the canonical projections. Let $j : X' \to X$ be a monomorphism and $f : X' \to X_{i_0}$ a morphism. Then there exists a unique morphism $g : X' \to \prod_i X_i$ such that $p_{i_0} g = f$ and $p_i g = 0$ if $i \neq i_0$. Also there is a morphism $\bar{g} : X \to \prod_i X_i$ such that $\bar{g} j = g$. It is clear that $p_{i_0} \bar{g} j = p_{i_0} g = f$. Hence X_{i_0} is injective.

Conversely, assume that X_i is an injective object for any i. Let $j : X' \to X$ be a monomorphism and $f : X' \to \prod_i X_i$ a morphism. Then for any i, there exists a morphism $g_i : X \to X_i$ such that $g_i j = p_i f$. Let $g : X \to \prod_i X_i$ be the unique morphism such that $p_i g = g_i$. Then $p_i g j = g_i j = p_i f$, hence $g j = f$ by the definition of direct product. ∎

We say that $\mathscr{C}$ has *sufficiently many injectives* (*respectively, projectives*) if any object of $\mathscr{C}$ is a subobject (respectively, quotient object) of an injective (respectively, projective) object of $\mathscr{C}$.

The injective and projective objects are intimately connected to the exactness of adjoint functors, as follows from the next results.

LEMMA 2.7. *Let $F:\mathscr{C}\to\mathscr{C}'$ be a covariant functor and G a right adjoint of F. If the functor F is exact, then G transforms any injective object of $\mathscr{C}'$ into an injective object of $\mathscr{C}$.*

Proof. Let Q be an injective object of $\mathscr{C}'$. We have to prove that $G(Q)$ is an injective object of the category $\mathscr{C}$. To do this, it suffices to show that for any exact sequence

$$0\to X'\to X\to X''\to 0$$

in the category $\mathscr{C}$, the sequence of abelian groups, canonically constructed:

$$0\to \mathrm{Hom}_{\mathscr{C}}(X'', G(Q))\to \mathrm{Hom}_{\mathscr{C}}(X, G(Q))\to \mathrm{Hom}_{\mathscr{C}}(X', G(Q))\to 0 \qquad (3)$$

is also exact. But the hypothesis implies that the sequence

$$0\to F(X')\to F(X)\to F(X'')\to 0$$

is exact. Hence, Q being injective, the sequence

$$0\to \mathrm{Hom}_{\mathscr{C}'}(F(X''), Q)\to \mathrm{Hom}_{\mathscr{C}'}(F(X), Q)\to \mathrm{Hom}_{\mathscr{C}'}(F(X'), Q)\to 0$$

is exact. Since we have the commutative diagram:

$$\begin{array}{ccccccccc} 0\to & \mathrm{Hom}_{\mathscr{C}'}(F(X''), Q) & \to & \mathrm{Hom}_{\mathscr{C}'}(F(X), Q) & \to & \mathrm{Hom}_{\mathscr{C}'}(F(X'), Q) & \to 0 \\ & \downarrow & & \downarrow & & \downarrow & \\ 0\to & \mathrm{Hom}_{\mathscr{C}}(X'', G(Q)) & \to & \mathrm{Hom}_{\mathscr{C}}(X, G(Q)) & \to & \mathrm{Hom}_{\mathscr{C}}(X', G(Q)) & \to 0 \end{array}$$

where the vertical arrows are isomorphisms, it follows that the sequence (3) is exact. ■

We prove the following theorem which is in a sense a converse of the above lemma.

THEOREM 2.8. *Let $F:\mathscr{C}\to\mathscr{C}'$ be a covariant functor and G a right adjoint of F. Assume that $\mathscr{C}'$ is a category with sufficiently many injectives. The following assertions are equivalent:*

(1) *F is an exact functor;*

(2) *G transforms any injective object of $\mathscr{C}'$ into an injective object of $\mathscr{C}$ (that is, G preserves injective objects).*

Proof. According to the above lemma it is sufficient to prove (2) ⇒ (1), that is, F preserves monomorphisms. Let $u : \mathrm{Id}\,\mathscr{C} \to GF$ and $v : FG \to \mathrm{Id}\,\mathscr{C}'$ be two arrows of adjunction (Chapter 1, Section 1.5) such that $v\,F \circ F\,u = 1_F$ and $G\,v \circ u\,G = 1_G$, and let $i : X' \to X$ be a monomorphism in $\mathscr{C}$. We have in $\mathscr{C}'$ the diagram:

$$\begin{array}{ccc} F(X') & \xrightarrow{F(i)} & F(X) \\ {\scriptstyle f}\downarrow & & \\ Q & & \end{array}$$

where f is a monomorphism and Q is an injective object. From this diagram we obtain the following commutative diagram in $\mathscr{C}$:

$$\begin{array}{ccccc} & & X' & \xrightarrow{i} & X \\ & & {\scriptstyle u_{X'}}\downarrow & & \downarrow{\scriptstyle u_X} \\ G(Q) & \xleftarrow{G(f)} & GF(X') & \xrightarrow{GF(i)} & GF(X) \end{array}$$

Let $g : X \to G(Q)$ be a morphism in $\mathscr{C}$ such that $g\,i = G(f)\,u_{X'}$. By applying the functor F to this diagram and taking into account the relations:

$$v_{F(X')}\,F(u_{X'}) = 1_{F(X')}, \qquad f\,v_{F(X')} = v_Q\,FG(f)$$

we get: $f = v_Q\,FG(f)\,F(u_{X'})$. But $F(g)\,F(i) = FG(f)\,F(u_{X'})$, hence

$$f = v_Q\,F(g)\,F(i),$$

whence we conclude that $F(i)$ is a monomorphism. ■

COROLLARY 2.9. *An* Ab 3-*category* $\mathscr{C}$ *with sufficiently many injectives is an* Ab 4-*category*.

Proof. Let I be a set and $\overline{\mathscr{C}}$ the product category of I copies of $\mathscr{C}$. Obviously, $\overline{\mathscr{C}}$ is an Ab 3-category. We denote by $F : \overline{\mathscr{C}} \to \mathscr{C}$ the functor which assigns to any object $(X_i)_{i \in I}$ of $\overline{\mathscr{C}}$ the object $\coprod_i X_i$ of $\mathscr{C}$. Let $G : \mathscr{C} \to \overline{\mathscr{C}}$ be the functor $G(X) = (X_i)_{i \in I}$ with $X_i = X$ for each i. We see that F is a left adjoint of G and G preserves injectives. Then F is exact, in other words, a direct sum of monomorphisms is also a monomorphism. ■

PROPOSITION 2.10. *Let* $T : \mathscr{C} \to \mathscr{C}'$ *be an additive covariant functor,* $\mathscr{C}$ *being an* Ab 3-*category. Then* T *commutes with inductive limits if and only if* T *commutes with direct sums and is right exact.*

Proof. A glance at Chapter 1, Theorem 4.1 gives us the result. ■

Obviously, the above proposition is valid for any covariant functor which commutes with projective limits and also for contravariant functors.

Later, we shall use the following result.

LEMMA 2.11. *Let $\mathscr{C}$ be a category with sufficiently many injectives. Then for any object X of $\mathscr{C}$ we may construct an exact sequence*:

$$0 \to X \to Q_0 \to Q_1 \to Q_2 \to \dots \to Q_n \to \dots$$

Q_i being injective for any i.

This sequence is called an *injective resolution* of X.

Proof. First, note that we have the exact sequence:

$$0 \longrightarrow X \longrightarrow Q_0 \xrightarrow{p_1} X_1 \longrightarrow 0$$

Q_0 being injective. Let $i_1 : X_1 \to Q_1$ be a monomorphism with Q_1 injective. Then the sequence:

$$0 \longrightarrow X \longrightarrow Q_0 \xrightarrow{i_1 p_1} Q_1$$

is obviously exact. In order to construct the objects $Q_2, \dots, Q^3, \dots$ we proceed in the same manner. ∎

Dually, we have the notion of a *projective resolution.*

Let $\mathscr{C}$ be an additive category and $\mathscr{C}_1$ the category defined in Chapter 2, Section 2.2. Recall that an object of $\mathscr{C}_1$ is a triple (X, f, Y) where $f : X \to Y$ is a morphism of $\mathscr{C}$. For brevity, an object of $\mathscr{C}_1$ will be denoted simply by the morphism f of $\mathscr{C}$. A morphism $(u, v) \in \mathrm{Hom}_{\mathscr{C}_1}(f, f')$ is said to be *homotopic* to zero if there exists a morphism $h : Y \to X'$ such that $hf = u$ (here $f : X \to Y$, $f' : X' \to Y'$, $u : X \to X'$ and $v : Y \to Y'$ are such that $vf = f'u$). Denote by $h(\mathscr{C}_1)$ the subcategory of $\mathscr{C}_1$ consisting of all morphisms homotopic to zero. More precisely, objects of $h(\mathscr{C}_1)$ are objects of $\mathscr{C}_1$ and for each $f, f' \in \mathrm{Ob}\,\mathscr{C}_1$, $\mathrm{Hom}_{h(\mathscr{C}_1)}(f, f')$ consists of the subgroup of $\mathrm{Hom}_{\mathscr{C}_1}(f, f')$ of all morphisms homotopic to zero. We shall write

$$\bar{\mathscr{C}} = \mathscr{C}_1 / h(\mathscr{C}_1)$$

and denote by $P : \mathscr{C}_1 \to \bar{\mathscr{C}}$ the canonical functor.

Now let $\mathscr{C}$ be an abelian category and $\mathscr{B}$ the full subcategory of $\mathscr{C}$ generated by all injective objects. Denote by $K : \mathscr{B}_1 \to \mathscr{C}$ the functor defined as follows. If $f : X \to Y$ is an object of $\mathscr{B}_1$, then $K(f) = \ker f$ (for each object f we choose a kernel). If $(u, v): f \to f'$ is a morphism of $\mathscr{B}_1$, then by the commutative diagram:

$$\begin{array}{ccccc} \ker f & \xrightarrow{i} & X & \xrightarrow{f} & Y \\ & & \downarrow{\scriptstyle u} & & \downarrow{\scriptstyle v} \\ \ker f' & \xrightarrow{i'} & X' & \xrightarrow{f'} & Y' \end{array} \qquad (*)$$

in which i and i' are canonical inclusions, we can calculate a morphism $K(u, v): \ker f \to \ker f'$ such that $ui = i'K(u, v)$.

Now it is clear that (u, v) is homotopic to zero, then $K(u, v) = 0$, so that K defines a unique functor $\overline{K}: \overline{\mathscr{B}} \to \mathscr{C}$ such that $\overline{K}P = K$, where $P: \mathscr{B}_1 \to \overline{\mathscr{B}}$ is canonical.

THEOREM 2.12. *Let $\mathscr{C}$ be an abelian category with sufficiently many injectives and $\mathscr{B}$ the full subcategory of $\mathscr{C}$, generated by injective objects. With the above notations, the functor $\overline{K}: \overline{\mathscr{B}} \to \mathscr{C}$ is an equivalence of categories.*

Proof. If X is an object $\mathscr{C}$, then by hypothesis one has the exact sequence:

$$0 \longrightarrow X \xrightarrow{i} Q_0 \xrightarrow{f} Q_1$$

where Q_0 and Q_1 are injective objects. Now, it is clear that $\overline{K}P(f) \simeq X$, in other words any object of $\mathscr{C}$ is isomorphic with an object of the form $\overline{K}P(f)$, $f \in \mathrm{Ob}\,\mathscr{B}_1$. Furthermore, if $(u, v) \in \mathrm{Hom}_{\mathscr{B}_1}(f, f')$ is such that $\overline{K}P(u, v) = 0 = K(u, v)$, then by diagram (*) we check that $ui = 0$, so that there exists a morphism $h': \mathrm{im} f \to X'$ with $h'p = u$, where $p: X \to \mathrm{im} f$ is canonical. Now, because X' is injective in $\mathscr{C}$, a morphism $h: Y \to X'$ can be found with $hj = h'$, where $j: \mathrm{im} f \to Y$ is also canonical. Finally, $hf = hjp = h'p = u$, hence (u, v) is homotopic to zero. Therefore $\overline{K}$ is a faithful functor.

To complete the proof, it will be enough to prove that $\overline{K}$ is full. In fact, let $g: X \to X'$ be a morphism of $\mathscr{C}$. Then one has the diagram

$$\begin{array}{ccccccc} 0 & \longrightarrow & X & \xrightarrow{i} & Q_0 & \xrightarrow{f} & Q_1 \\ & & \downarrow{\scriptstyle g} & & & & \\ 0 & \longrightarrow & X' & \xrightarrow{i'} & Q_0' & \xrightarrow{f'} & Q_1' \end{array}$$

with exact row, Q_0, Q_1, Q_0', Q_1' being injectives. Then by injectivity of Q_0', a morphism $u: Q_0 \to Q_0'$ can be calculated such that $ui = i'g$. Thus, $f'ui = f'i'g = 0$, so that as above, a morphism $v: Q_1 \to Q_1'$ can be found with $f'u = vf$. Now it is obvious that $\overline{K}P(u, v) = K(u, v) = g$. ∎

Exercises

1. Let $\mathscr{C}$ be an abelian category. Then an object Y is injective (respectively, projective) if and only if any monomorphism $f\colon X \to Y$ (resp. any epimorphism $g\colon Z \to X$) is a section (respectively, a retraction).

2. Dualize the results 2.7, 2.8 and 2.9.

3. Let $F:\mathscr{C} \to \mathscr{C}'$ be an additive functor. Then the following assertions are equivalent:

(a) F is faithful;

(b) F carries non-commutative diagrams in non-commutative diagrams;

(c) F carries non-exact sequences into non-exact sequences.

4. A projective object P of $\mathscr{C}$ is a generator if and only if for any non-null object X of $\mathscr{C}$, the abelian group $h^P(X)$ is non-null.

5. Let $T:\mathscr{C} \sqcap \mathscr{C}' \to \mathscr{C}''$ be an additive covariant functor, where $\mathscr{C}$, $\mathscr{C}'$ and $\mathscr{C}''$ are abelian categories. Then T is left exact if and only if for every pair of exact sequences:

$$0 \longrightarrow X' \xrightarrow{f_1} X \xrightarrow{f_2} X''$$

$$0 \longrightarrow Y' \xrightarrow{g_1} Y \xrightarrow{g_2} Y''$$

of $\mathscr{C}$ and $\mathscr{C}'$ respectively, the sequence

$$0 \to T(X', Y') \to T(X, Y) \to T(X'', Y) \amalg T(X, Y'')$$

is exact, where the first morphism is $T(f_1, g_1)$ and the second morphism is induced by $T(f_2, 1_Y)$ and $T(1_X, g_2)$.

6. Assume that $\mathscr{C}$ is an abelian category with sufficiently many injectives. If the category $\mathscr{B}$ generated by all injective objects is equivalent with a product $\prod_i \mathscr{B}_i$ of additive categories, then $\mathscr{C}$ is equivalent with $\prod_i \overline{\mathscr{B}}_i$.

7. Let $\mathscr{C}$ and $\mathscr{B}$ be as above, $\mathscr{C}'$ another abelian category and $F:\mathscr{B} \to \mathscr{C}'$ an additive functor. Then F can be canonically "extended" to an additive functor $\overline{F}:\overline{\mathscr{B}} \to \overline{\mathscr{C}}'$ so that one has the following diagram:

$$\mathscr{C} \xrightarrow{S} \overline{\mathscr{B}} \xrightarrow{\overline{F}} \overline{\mathscr{C}}' \xrightarrow{\overline{K}'} \mathscr{C}'$$

where S is a left (or right!) adjoint of the functor $\overline{K}:\overline{\mathscr{B}} \to \mathscr{C}$ and $\overline{K}'$ is defined in the same manner as $\overline{K}$. Let us put $F_1 = \overline{K}'\overline{F}S$. Conversely, if $G:\mathscr{C} \to \mathscr{C}'$ is an additive functor, we denote by $G':\mathscr{B} \to \mathscr{C}'$ the restriction of G to $\mathscr{B}$. Then the functor G is left exact if and only if it is isomorphic with $(G')_1$. Moreover, any additive functor $F:\mathscr{B} \to \mathscr{C}'$ is isomorphic with $(F_1)'$.

8. Let $\mathscr{C}$, $\mathscr{C}'$ be abelian categories with sufficiently many injectives. Then a functor $G:\mathscr{C} \to \mathscr{C}'$ defines an equivalence between $\mathscr{C}$ and $\mathscr{C}'$ if and only if G is left exact and the restriction of G to $\mathscr{B}$ defines an equivalence of $\mathscr{B}$ with $\mathscr{B}'$ (here, as above $\mathscr{B}$ [respectively, $\mathscr{B}'$] is the full subcategory of $\mathscr{C}$ [respectively, of $\mathscr{C}'$] generated by all injective objects).

9. Let $\mathscr{C}$, $\mathscr{C}'$ be abelian categories and $\mathscr{C}$ with sufficiently many injectives. Then, for any additive functor $G:\mathscr{C} \to \mathscr{C}'$, there exists a functor $R^0G:\mathscr{C} \to \mathscr{C}'$ and a functorial morphism $u: G \to R^0G$ such that:

(a) The functor R^0G is left exact;

(b) For each functorial morphism $v: G \to G'$ where G' is left exact, there exists a unique functorial morphism $\bar{v}: R^0G \to G'$ with $\bar{v}u = v$.

References

Baer [12]; Bănică–Popescu [13]; Buchsbaum [28]; Bucur–Deleanu [31]; Cartan–Eilenberg [33]; Eckmann–Schopf [48]; Gabriel [59]; Gabriel–Zisman [62]; Grothendieck [79]; Lambek [113]; MacLane [120]; Mitchell [133]; Pareigis [162].

3.3 The injective and projective objects in the category $\mathscr{A}b$

LEMMA 3.1. *Let $\mathscr{C}$ be an* Ab 5-*category, U a generator and Q an object of $\mathscr{C}$. The following assertions are equivalent*:

(1) *Q is an injective object*;

(2) *For any monomorphism $j: U' \to U$ and for any morphism $f: U' \to Q$ there exists a morphism $\bar{f}: U \to Q$* such that $\bar{f}j = f$.

Proof. The implication (1) $\Rightarrow$ (2) is clear by Corollary 2.4.

(2) $\Rightarrow$ (1). Let X' be a subobject of any object X and $f': X' \to Q$ a morphism. We denote by M the set of couples (X'', f''), where X'' is a subobject of X which contains X' and $f'': X'' \to Q$ a morphism such that $f''u = f'$, $u: X' \to X''$ being the canonical inclusion, that is f'' prolongs f'.

We put $(X'', f'') \leqslant (X_1'', f_1'')$ if $X'' \subseteq X_1''$ and $f_1''v = f''$, $v: X'' \to X_1''$ being the canonical inclusion. Then M is an inductive set. In fact, let

$$\{(X_i'', f_i'')\}_i$$

be a totally ordered set of elements of M. We put $X'' = \sum_i X_i''$. According to Chapter 2, Theorem 8.6, there is a morphism $f'': X'' \to Q$, such that $f''u_i = f_i''$, u_i being the structural morphisms. By Zorn's lemma, we deduce that M has a maximal element. We may assume that (X', f') is a maximal element of M. If $X' \neq X$, then there is a morphism $t: U \to X$ such that $t(U) \nsubseteq X'$ and $X'' = X' + t(U) \neq X'$. Let $U' = t^{-1}(X')$ and $g': U' \to X'$ the unique morphism such that $f'g' = tr$, where $r: U' \to \tilde{U}$ is the canonical inclusion. We shall consider the exact sequence.

$$U' \xrightarrow{h} U \coprod X' \xrightarrow{p} X'' \longrightarrow 0$$

where p is defined by canonical inclusion $X' \to X''$ and t; also h is defined by $-r$ and g'. Let $g: U \to Q$ be a morphism which prolongs g' (according to the hypothesis) and $v: U \amalg X' \to Q$ the morphism defined by g and f'. Then $vh = 0$ by the construction of h. Then there is a unique morphism $f'': X'' \to Q$, such that $f''p = v$, that is $f''pk = f'$, where $k: X' \to U \amalg X'$ is the structural morphism. Contradiction. ∎

We now characterize the injective and projective objects of the category $\mathscr{Ab}$.

Let X be an abelian group. We say that X is *divisible* if for any $x \in X$ and $n \in \mathbf{Z}$ there is an element $x_0 \in X$ such that $n x_0 = x$.

THEOREM 3.2. *An object X in $\mathscr{Ab}$ is injective if and only if it is divisible.*

Proof. Let X be an injective object of $\mathscr{Ab}$, $x \in X$ and $n \in \mathbf{Z}$. Let $f: \mathbf{Z} \to X$ be the unique morphism of groups such that $f(m) = mx$ for any $m \in \mathbf{Z}$. Let $u: n\mathbf{Z} \to \mathbf{Z}$ be the morphism $u(nm) = m$. Then there is a morphism $g: \mathbf{Z} \to X$ such that for any $m \in \mathbf{Z}$ we have $g(nm) = (fu)(nm) = f(m) = mx$. Particularly, $g(n) = ng(1) = x$, that is X is divisible.

Conversely, we suppose that X is a divisible abelian group. Let $n \in \mathbf{Z}$ and $f: n\mathbf{Z} \to X$ be a morphism; then there is an element $x_0 \in X$ such that $nx_0 = f(n)$. If $\bar{f}: \mathbf{Z} \to X$ is the morphism $\bar{f}(m) = mx_0$, then $\bar{f}(mn) = mnx_0 = m\,f(n) = f(mn)$. Thus X is injective, any subgroup of $\mathbf{Z}$ being principal. ∎

EXAMPLES. 1. The additive group of rational numbers $\mathbf{Q}$ is divisible, hence injective in $\mathscr{Ab}$.

2. Let p be a prime number. Consider the multiplicative group $\mathbf{Q}_p$ of all complex numbers which satisfy an equation of the form $z^{p^n} = 1, n > 0$. Then $\mathbf{Q}_p$ is an injective object of $\mathscr{Ab}$. Indeed, let $x \in \mathbf{Q}_p$. We have to check that for any integer $m > 0$ there exists $x_1 \in \mathbf{Q}_p$ such that $x_1^m = x$. We need obviously prove this only for prime numbers. If $m = p$, this is evident. Assume now that m is prime to p and $x^{p^r} = 1$. We have $p^r s + mq = 1$. Hence $x = x^{p^r s + mq} = x^{p^r s} x^{mq} = (x^q)^m$. Thus $\mathbf{Q}_p$ is divisible and therefore injective.

3. Any direct sum or product of divisible abelian groups is divisible. A quotient group of a divisible abelian group is divisible.

THEOREM 3.3. *The category $\mathscr{A}b$ has sufficiently many injectives.*

Proof. For any object X of $\mathscr{A}b$, we consider the following diagram:

$$\begin{array}{ccccccccc} 0 & \longrightarrow & \ker p & \xrightarrow{i} & \mathbf{Z}^{(I)} & \xrightarrow{p} & X & \longrightarrow & 0 \\ & & & & \downarrow u & & \downarrow v & & \\ & & & & \mathbf{Q}^{(I)} & \xrightarrow{q} & \mathbf{Q}^{(I)}/R & \longrightarrow & 0 \end{array}$$

with exact row, where p is defined according to Chapter 2, Proposition 8.2 and $R = u(\ker p)$. Furthermore there exists a morphism $v : X \to \mathbf{Q}^{(I)}/R$ such that $vp = qu$, q being the canonical epimorphism. We see easily that v is a monomorphism, and $\mathbf{Q}^{(I)}/R$ is divisible, hence injective. ■

We see that $\mathbf{Z}$ is a projective object in $\mathscr{A}b$, and according to Chapter 2, Proposition 8.2 we deduce:

PROPOSITION 3.4. *The category $\mathscr{A}b$ has sufficiently many projective objects.* ■

In 3.9 we shall prove that any projective abelian group is free, that is, of the form $\mathbf{Z}^{(I)}$ for a suitable set I.

Exercises

1. Let p be a prime number. For any two natural numbers n, m such that $n \leqslant m$, we denote by $f_{nm} : \mathbf{Z}_{p^n} \to \mathbf{Z}_{p^m}$ the morphism $f(x) = p^{m-n}\bar{x}$ where $x \in \mathbf{Z}$ and $\bar{x}$ is the class of x modulo p^n. Then $(\mathbf{Z}_{p^n}, f_{nm})_{n \in \mathbf{N}}$ is a direct system in $\mathscr{A}b$. Prove that $\mathbf{Q}_p$ is the inductive limit of this system.

2. Let P be the set of prime numbers. Then $\mathbf{Q}/\mathbf{Z} \simeq \coprod_{p \in P} \mathbf{Q}_p$.

References

Baer [12]; Bucur–Deleanu [31]; Freyd [57]; Godement [66]; Grothendieck [79]; Kuroš [110]; Mitchell [133]; Pareigis [162].

3.4 Categories of additive functors. Modules

For any two categories $\mathscr{C}$ and $\mathscr{C}'$ let $[\mathscr{C}, \mathscr{C}']$ denote the class of all covariant functors from $\mathscr{C}$ to $\mathscr{C}'$. For any $S, T \in [\mathscr{C}, \mathscr{C}']$ let $[S, T]$ denote the class of functorial morphisms from S to T. With the law of composition of functorial morphisms given in Chapter 1, Section 2, $[\mathscr{C}, \mathscr{C}']$ comes very close to being a category. The only requirement that is missing is the fact that $[S, T]$ may not be a set. However, if we assume that $\mathscr{C}$ is small, then the functorial morphisms from S to T may be regarded as a subclass of the cartesian product $\prod_{X \in \mathrm{Ob}\,\mathscr{C}} \mathrm{Hom}_{\mathscr{C}}(S(X), T(X))$ and since the latter is a set, so is $[S, T]$. In speaking of the functor category $[\mathscr{C}, \mathscr{C}']$, we shall always assume that $\mathscr{C}$ is small.

In general $[\mathscr{C}, \mathscr{C}']$ inherits the properties of $\mathscr{C}'$ and a morphism $u : S \to T$ in $[\mathscr{C}, \mathscr{C}']$ has the properties which are common to all morphisms

$$u_X : S(X) \to T(X)$$

in $\mathscr{C}'$. Thus u is a monomorphism (resp. epimorphism, isomorphism) whenever its components are monomorphisms (resp. epimorphisms, isomorphisms), that is, u_X is a monomorphism (epimorphism, isomorphism) in $\mathscr{C}'$ for any $X \in \mathrm{Ob}\,\mathscr{C}$.

THEOREM 4.1. *Let $\mathscr{C}'$ be an abelian category and let $\mathscr{C}$ be a small category. Then*

(1) *The category $[\mathscr{C}, \mathscr{C}']$ is abelian.*

(2) *If $\mathscr{C}'$ is an* Ab3 (*resp.* Ab3*, Ab4, Ab4*, Ab5, Ab5*, Ab6, Ab6*)—*category, then so is $[\mathscr{C}, \mathscr{C}']$.*

Proof. All constructions in $[\mathscr{C}, \mathscr{C}']$ are carried out componentwise.

Let T and S be two objects in $[\mathscr{C}, \mathscr{C}']$. Then the set $\mathrm{Hom}_{[\mathscr{C}, \mathscr{C}']}(T, S)$ is an abelian group. Namely, if u, v are two functorial morphisms from T to S, then $u + v$ is defined by the equality $(u + v)_X = u_X + v_X$, for any object X of $\mathscr{C}$.

For any functorial morphism $u : T \to S$ we shall define $\ker u$ as follows. For any $X \in \mathrm{Ob}\,\mathscr{C}$ let $(\ker u_X, j_X)$ be a kernel of $u_X : T(X) \to S(X)$. Then $(\ker u)(X) = \ker u_X$. If $r : X \to X'$ is a morphism of $\mathscr{C}$, then $u_{X'}\, T(r)\, j_X = S(r)\, u_X\, j_X = 0$, and a unique morphism $(\ker u)(r) : \ker u_X \to \ker u_{X'}$ exists such that $j_{X'}\, (\ker u)(r) = T(r)\, j_X$. Obviously, the morphisms $\{j_X\}_{X \in \mathrm{Ob}\,\mathscr{C}}$ define a functorial monomorphism from $\ker u$ to T which is denoted by j, and $(\ker u, j)$ is a kernel of u.

Similarly, we can define the cokernel of u and the finite direct sum. Now we can prove that $[\mathscr{C}, \mathscr{C}']$ is an abelian category.

Let us suppose that $\mathscr{C}'$ is an Ab 3 category and that $\{T_k\}_{k \in K}$ is a set of objects in $[\mathscr{C}, \mathscr{C}']$. Then $\coprod_k T_k$ is defined as follows: for any $X \in \mathrm{Ob}\,\mathscr{C}$, we have that $(\coprod_k T_k)(X) = \coprod_k T_k(X)$. Computing componentwise, we get the second assertion of the theorem. ■

Analogously, if we denote by $[\mathscr{C}^0, \mathscr{C}']$ the category of contravariant functors from $\mathscr{C}$ to $\mathscr{C}'$, we can make similar remarks concerning the category $[\mathscr{C}^0, \mathscr{C}']$.

EXAMPLES. 1. Let I be a set. Then I is a discrete category (see Chapter 1, Section 1.4). Consequently, $[I, \mathscr{C}]$ is the category $\prod_{i \in I} \mathscr{C}_i$ with $\mathscr{C}_i = \mathscr{C}$, for any $i \in I$. Obviously, we have $[I, \mathscr{C}] = [I^0, \mathscr{C}]$.

2. Let E be a topological space, and let $D(E)$ be the collection of open subsets of E ordered by inclusion and regarded as a category (see Chapter 1. Section 1.4). An object F in the category $[D(E)^0, \mathscr{C}]$ is called a *presheaf* in $\mathscr{C}$ over E. Sometimes we shall write $[D(E)^0, \mathscr{C}] = \mathscr{P}(E, \mathscr{C})$.

Let $\mathscr{C}'$ be an abelian category and let $\mathscr{C}$ be a small category. We define a functor $k_{\mathscr{C}} : \mathscr{C}' \to [\mathscr{C}, \mathscr{C}']$ as follows: for any $X \in \operatorname{Ob} \mathscr{C}'$, $k_{\mathscr{C}}(X) : \mathscr{C} \to \mathscr{C}'$ is the constant functor which is associated to X. (Chapter 1, Section 1.4). Then $\mathscr{C}$ is an Ab 3 (respectively, Ab 3*) category if and only if, for any set I, the functor k_I has a left (respectively, right) adjoint. Also, $\mathscr{C}$ is an Ab 4 (respectively, Ab 4*) category if and only if for any set I, the functor k_I has an exact left (respectively, right) adjoint. In the same manner, we find that $\mathscr{C}$ is an Ab 5 (resp. Ab 5*)-category, if and only if for any directed set I, the functor k_I has an exact left (respectively, right) adjoint. We leave to the reader the full development of these considerations.

Let $\mathscr{C}$ and $\mathscr{C}'$ be two preadditive categories and let $\mathscr{C}$ be small. We denote by $(\mathscr{C}, \mathscr{C}')$ the full subcategory of $[\mathscr{C}, \mathscr{C}']$, which consists of all additive functors from $\mathscr{C}$ to $\mathscr{C}'$. We have the same results for the category $(\mathscr{C}, \mathscr{C}')$ as those which were obtained for $[\mathscr{C}, \mathscr{C}']$ when computations are carried out component-wise.

THEOREM 4.2. *Let $\mathscr{C}$ be a small preadditive category and let $\mathscr{C}'$ be an abelian category. Then we have that:*

(1) *The category $(\mathscr{C}, \mathscr{C}')$ is abelian;*

(2) *If $\mathscr{C}'$ is an Abn-category* ($n = 3, 3^*, 4, 4^*, 5, 5^*, 6, 6^*$) *then $(\mathscr{C}, \mathscr{C}')$ is also an Abn-category.* ■

We shall denote by $(\mathscr{C}^0, \mathscr{C}')$ the category of additive contravariant functors from $\mathscr{C}$ to $\mathscr{C}'$.

Particularly, if $\mathscr{C}$ is a small preadditive category, then we shall define $\operatorname{Mod}\mathscr{C}$ as the category $(\mathscr{C}, \mathscr{Ab})$, and $\operatorname{Mod}\mathscr{C}^0$ as $(\mathscr{C}^0, \mathscr{Ab})$. An object of $\operatorname{Mod}\mathscr{C}$ is called a *left module over* $\mathscr{C}$. An object of $\operatorname{Mod}\mathscr{C}^0$ is called a *right module over* $\mathscr{C}$.

If T, S are two left (resp. right) $\mathscr{C}$-modules, then we shall write simply $\operatorname{Hom}_{\operatorname{Mod}\mathscr{C}}(T, S) = \operatorname{Hom}_{\mathscr{C}}(T, S)$ (resp. $\operatorname{Hom}_{\operatorname{Mod}\mathscr{C}^0}(T, S) = \operatorname{Hom}_{\mathscr{C}}(T, S)$. We shall say that we have defined a *$\mathscr{C}$-module* (resp. a *$\mathscr{C}^0$-module*) if it is an object of $\operatorname{Mod}\mathscr{C}$ (resp. an object of $\operatorname{Mod}\mathscr{C}^0$).

An important contravariant functor $h: \mathscr{C} \to \operatorname{Mod}\mathscr{C}$ is defined as follows: for any $X \in \operatorname{Ob}\mathscr{C}$, $h(X) = h^X$, such that $h^X(Y) = \operatorname{Hom}_{\mathscr{C}}(X, Y)$. Sometimes, we shall write $h(X) = X_l$ or simply X, if there is no danger of confusion, and we shall say that X_l is the *left module structure* of X. Similarly, a functor $h^0 : \mathscr{C} \to \mathrm{Mod}\,\mathscr{C}^0$ is defined by the equality $h^0(X) = h_X$, so that

$$h_X(Y) = \mathrm{Hom}_{\mathscr{C}}(Y, X).$$

We shall write $h^0(X) = X_r$ or simply X if there is no danger of confusion, and say that X_r is the *right module structure* of X.

LEMMA 4.3. *Let $\mathscr{C}$ be a small preadditive category. Then:*

(1) *For any left (resp. right) $\mathscr{C}$-module T and for any $X \in \mathrm{Ob}\,\mathscr{C}$, the groups* $\mathrm{Hom}_{\mathscr{C}}(h(X), T))$ *(resp.* $\mathrm{Hom}_{\mathscr{C}}(h^0(X), T))$ *and* $T(X)$ *are canonically isomorphic.*

(2) *The functors h and h^0 are full and faithful.*

Proof. Let $u : h(X) \to T$ be a functorial morphism. Then we define $\phi(u) = u_X(1_X)$. One may see that ϕ is a morphism of groups from $\mathrm{Hom}_{\mathscr{C}}(h(X), T)$ to $T(X)$. If $x \in T(X)$, let us define the functorial morphism $u_x : h(X) \to T$ as follows: for any object Y of $\mathscr{C}$, $(u_x)_Y : h^X(Y) \to T(Y)$ is the assignment $(u_x)_Y(r) = T(r)(x)$, for any $r \in h^X(Y) = \mathrm{Hom}_{\mathscr{C}}(X, Y)$. We leave to the reader the proof of the fact, that the assignment $x \rightsquigarrow u_x$ is the inverse mapping of ϕ. ■

We shall say that the u_x defined above is the *associated morphism* of x (see Chapter 1, Lemma 3.1).

COROLLARY 4.4. *For any two objects, X, Y of $\mathscr{C}$, we have* $\mathrm{Hom}_{\mathscr{C}}(h(X), h(Y)) = \mathrm{Hom}_{\mathscr{C}}(Y, X)$ *and* $\mathrm{Hom}_{\mathscr{C}}(h^0(X), h^0(Y)) = \mathrm{Hom}_{\mathscr{C}^0}(Y, X)$. *For any object X of $\mathscr{C}$, the ring* $\mathrm{Hom}_{\mathscr{C}}(h(X), h(X))$ *is canonically isomorphic to the ring* $(\mathrm{Hom}_{\mathscr{C}}(X, X))^0$, *the opposite (or dual) ring of* $\mathrm{Hom}_{\mathscr{C}}(X, X)$. ■

Let A be a ring. Then the categories $\mathrm{Mod}A$ and $\mathrm{Mod}A^0$ are here of particular interest.

An object X of $\mathrm{Mod}A$ (or a left A-module) is an abelian group X such that for any element a of A, a morphism $t_a : X \to X$ of abelian groups is given and for any a, $a' \in A$, we have $t_a t_{a'} = t_{aa'}$ and $t_{a+a'} = t_a + t_{a'}$. Also $t_1 = 1_X$, where 1 is the unit-element of A. (A right A-module is an abelian group X such that for any $a \in A$, a morphism $v_a : X \to X$ of abelian groups is given, and for any two $a, a' \in A$ one has: $v_a v_{a'} = v_{a'a}$ and $v_{a+a'} = v_a + v_{a'}$. Also $v_1 = 1_X$). If we put $t_a(x) = ax$ for an $a \in A$ and $a\,x \in X$, then one can see that a left A-module is subject to the following rules:

(1) $1x = x$ for all $x \in X$;

(2) $a(x_1 + x_2) = ax_1 + ax_2$;

(3) $(a + a')x = ax + a'x$;

(4) $(aa')(x) = a(a'x)$.

Conversely, given an abelian group X and an operation $A \times A \to X$ satisfying the rules (1) to (4), this defines a left A-module.

For a right A-module the first three rules are the same, but the fourth rule must be replaced by $a(a'x) = (a'a)\,x$. For this reason, we put xa instead of $v_a(x)$, when we are referring to right A-modules, in which case rule (4) becomes: $(xa')\,a = x(a'a)$. In the sequel, any left A-module will be simply called "A-module". Also, any right A-module is called simply an "A^0-module".

Note that a morphism of left (respectively, right) A-modules is also a morphism $f: X \to Y$ of abelian groups, such that $f(ax) = af(x)$ (respectively, $f(xa) = f(x)\,a$) for all $a \in A$ and $x \in X$.

Let $\{X_i\}_{i \in I}$ be a set of A-modules, (respectively A^0-modules). Then $\prod_i X_i$ is the direct product of abelian groups, the left (respectively right) A-module structure being defined component-wise. In the same manner one can define $\coprod_i X_i$.

Let $T: \mathscr{C} \to \mathscr{C}'$ be a covariant additive functor, $\mathscr{C}$ and $\mathscr{C}'$ being small. Then a covariant additive functor $T_*: (\mathscr{C}', \bar{\mathscr{C}}) \to (\mathscr{C}, \bar{\mathscr{C}})$ is defined for any preadditive category $\bar{\mathscr{C}}$ by the equality $T_*(S) = S\,T$ for $S \in \mathrm{Ob}(\mathscr{C}', \bar{\mathscr{C}})$.

T_* is called the *associated restriction functor* of T. If $T': \mathscr{C}' \to \mathscr{C}''$ is another covariant additive functor, $\mathscr{C}''$ being small, then $T_*\, T_*' \equiv (T'\,T)_*$.

LEMMA 4.5. *Let $T: \mathscr{C} \to \mathscr{C}'$ be an additive covariant functor where $\mathscr{C}$ and $\mathscr{C}'$ are small. Assume that for any object X' of $\mathscr{C}'$, there is an object X of $\mathscr{C}$ such that $T(X) \simeq X'$. Then, T_* is faithful for any preadditive category $\bar{\mathscr{C}}$.* ∎

Particularly, if $u: A \to B$ is a morphism of rings, then the restriction functor $u_*: \mathrm{Mod}\, B \to \mathrm{Mod}\, A$ is defined as follows: for any left B-module X, $u_*(X)$ is the abelian group X, and for any $a \in A$ and any $x \in X$ one has $ax = u(a)\,x$.

If $\bar{\mathscr{C}}$ is an abelian category, the restriction functor defined above is exact. Particularly, for any morphism of rings $u: A \to B$ the functor u_* is exact.

Let X be an abelian group. Then, X has a unique left $\mathbf{Z}$-module structure. In fact, for any positive integer n and any $x \in X$ one has $nx = \underbrace{x + \ldots + x}_{n\text{-times}}$.
Thus, the category $\mathscr{A}b$ is canonically equivalent with the category $\mathrm{Mod}\, \mathbf{Z}$.

Let A be a ring. Then there exists a unique morphism of rings $u_A: \mathbf{Z} \to A$ such that $u_A(n) = n.1 = \underbrace{1 + \ldots + 1}_{n\text{-times}}$, 1 being the unit-element of A. The functor $(u_A)_*: \mathrm{Mod}\, A \to \mathscr{A}b$ is sometimes denoted by S and is called the *underlying functor*. For any left A-module X, $S(X)$ is the underlying abelian group. Obviously S is an exact functor.

THEOREM 4.6. *Let $\mathscr{C}$ be a small preadditive category. Then* Mod $\mathscr{C}$ *is a category with generators and with sufficiently many projective objects.*

Proof. We shall prove first that the objects $\{X_l\}_{X \in \mathrm{Ob}\,\mathscr{C}}$ are projective generators of Mod $\mathscr{C}$. According to Lemma 4.3, $X_l(=h(X))$ is projective for any object X of $\mathscr{C}$. Now, let $u: S \to T$ be a non-null morphism of Mod $\mathscr{C}$. Then there exists an object X of $\mathscr{C}$ and $x \in S(X)$ such that $u_X(x) \neq 0$. Consequently, $uu_x \neq 0$ where $u_x: X_l \to S$ is the associated morphism of X. Hence,

$$\{X_l\}_{X \in \mathrm{Ob}\,\mathscr{C}}$$

is a set of generators of Mod $\mathscr{C}$. ■

Let A be a ring. We shall denote by A_l (respectively A_r) the object $h(A)$ respectively $h^0(A)$, of Mod A, respectively of Mod A^0. According to the above theorem, A_l is a projective generator of Mod A. Sometimes, we put $A_l = A$ (and $A_r = A$) if there is no danger of confusion. Let I be a set; then $A^{(I)}$ is called the *free* left (respectively, right) *module* associated to I. Obviously, any free left (right) module is projective. We denote by e_i the element $(x_j)_j$ of $A^{(I)}$ such that $x_j = 0$, for any $j \neq i$ and $x_i = 1$. Let $x = (x_i)_i$ be an arbitrary element of $A^{(I)}$ and let $i_1, \dots i_n$ be the elements of I such that $x_{i_1}, \dots x_{i_n}$ are non-null components of x. Then x may be uniquely written as $x = x_{i_1} e_{i_1} + \dots + x_{i_n} e_{i_n}$ with $x_{i_1}, \dots, x_{i_n}$ being, of course, elements of A. In the sequel, we shall say that $\{e_i\}_{i \in I}$ is the *canonical basis* of $A^{(I)}$.

LEMMA 4.7. Any morphism $f: A^{(I)} \to X$ is uniquely defined by the elements: $\{f(e_i)\}_{i \in I}$. ■

Particularly, if X is an A-module, then a unique morphism $p: A^{(X)} \to X$ exists such that $p(e_x) = x$ for any $x \in X$. Obviously p is an epimorphism.

Let X be an arbitrary A-module and let $\{x_i\}_i$ be a set of elements of X. We denote by $\langle x_i \rangle_i$ the smallest submodule of X which contains all the elements x_i. This is called the *submodule generated* by the elements $\{x_i\}_i$. We invite the reader to provide a full description of this submodule. Particularly if $x_1, \dots, x_n$ are elements of X, the submodule generated by this set of elements is denoted by $\langle x_1, \dots, x_n \rangle$. Also the submodule generated by an element x of X is sometimes denoted by Ax (if X is an A^0-module we put xA for the submodule generated by an element x). Ax is the *principal submodule* generated by x.

Any A-submodule of A_l is in fact a left ideal of A. Obviously, a left ideal $\mathfrak{a}$ of A consists of an additive subgroup $\mathfrak{a}$ such that for any $x \in A$ and $y \in \mathfrak{a}$ one has $xy \in \mathfrak{a}$. Dually, the right-ideals of A, are the submodules of A_r. For any element x of A, we shall denote by Ax (respectively by xA) the left (respectively, right) ideal generated by x. Ax (respectively xA) is the principal left (respectively, right) ideal generated by x.

For any element x of an A-module X we shall denote by $\mathrm{Ann}(x)$ the kernel of the canonical morphism $u_x : A \to X$. $\mathrm{Ann}(x)$ is called the *annihilator* of x. Of course, $\mathrm{Ann}(x)$ consists of all elements $a \in A$ for which $ax = 0$. An element x of A is called *left-regular* if the annihilator of x is null. Dually, we have the notion of a *right-regular* element. x is called *regular*, if it is at the same time left and right regular. A ring A is called a *domain*, if any non-null element of A is regular. A commutative domain is called an *integral domain*. Obviously, any non-null element x of A is left regular if and only if the canonical morphism u_x is a monomorphism.

On the other hand a non-null element of A is called *left* (*respectively, right*)*-zero-divisor* if it is not left (respectively, right) regular.

Let X' be a submodule of an A-module X and let $x \in X$. We shall denote by $(X' : x)$ the left ideal of A consisting of all elements $a \in A$ such that $ax \in X'$. If $\bar{x}$ is the image of x in the quotient module X/X', then

$$(X' : x) = \mathrm{Ann}(\bar{x}).$$

The term *proper left ideal* will be used for a left ideal of a ring which is different from the ring itself. A *maximal ideal* $\mathfrak{m}$ of A is a proper ideal which has the property that there is no proper ideal which strictly contains it. If $\mathfrak{a}$ is an arbitrary proper ideal, then a simple application of Zorn's lemma shows that $\mathfrak{a}$ is contained in at least one maximal ideal. Such considerations are also valid for right and two-sided ideals.

Let n, m be two natural numbers. A morphism $f : A^n \to A^m$ in $\mathrm{Mod}\,A$ is completely determined by the elements $f(e_i), i = 1, \ldots, n$, where $\{e_i\}_i$ is a canonical basis of A^n. Then we get

$$f(e_i) = \sum_{j=1}^{m} a_{ij}\, e_j'$$

$i = 1, \ldots n$, $\{e_j'\}_j$ being the canonical basis of A^m, and so the $n \times m$ elements (a_{ij}) in A determine the morphism f. The $n \times m$ rectangular table (a_{ij}) (with n rows and m columns) is called the *matrix* associated to f. Sometimes the abelian group $\mathrm{Hom}_A(A^n, A^n)$ is denoted by A_n. Obviously A_n has the canonical structure of a ring, which is called the *ring of* $n \times n$*-matrices* (or square matrices of order n) over A.

For any submodule X' of an A-module X and for any left ideal $\mathfrak{a}$ we shall denote by $\mathfrak{a}X'$ the set of all finite sums $\sum a_i x_i$ with $a_i \in \mathfrak{a}$ and $x_i \in X'$. $\mathfrak{a}X'$ is obviously a submodule of X, which is called the *product* of $\mathfrak{a}$ by X'. Particularly, if $\mathfrak{a}$ and $\mathfrak{a}'$ are left ideals of A, then $\mathfrak{a}\mathfrak{a}'$ is also a left ideal, called the product of $\mathfrak{a}$ and $\mathfrak{a}'$.

The same considerations are valid for right and two-sided ideals.

Exercises

1. If $T:\mathscr{C}\to\mathscr{C}'$ is an equivalence of categories, then for any small category $\mathscr{I}$ the functor $\bar{T}:[\mathscr{I},\mathscr{C}]\to[\mathscr{I},\mathscr{C}']$ defined by the equality $\bar{T}(S)=TS$ for any $S\in\mathrm{Ob}[\mathscr{I},\mathscr{C}]$ is also an equivalence. Likewise, if $T:\mathscr{I}\to\mathscr{I}'$ is an equivalence of small categories, then for any category $\mathscr{C}$ the functor

$$T_*:[\mathscr{I}',\mathscr{C}]\to[\mathscr{I},\mathscr{C}]$$

which has an obvious definition is an equivalence. We have the same situation for additive functors.

2. Let $\mathscr{C}$ be a preadditive category and let $T:\mathscr{C}\to\mathscr{Ab}$ be a covariant additive functor. For each $X\in\mathrm{Ob}\,\mathscr{C}$ let S_X be a subset (which may be empty!) of $T(X)$. The subfunctor of T which is generated by S is defined as the smallest subfunctor M of T such that $S_X\subseteq M(X)$ for all $X\in\mathrm{Ob}\,\mathscr{C}$ (that is, it is the intersection of all subfunctors). Prove that $M(X)$ is the subgroup of $T(X)$ consisting of all finite sums of the form $\sum_i T(f_i)\,(x_i)$ with

$$f_i:X_i\to X$$

and $x_i\in S_X$.

The subfunctor of T generated by a single element $x_0\in T(X_0)$ is given by $M(X)=\{T(f)\,(x_0), f\in\mathrm{Hom}_{\mathscr{C}}(X_0,X)\}$.

If S is a subset of a left A-module X, then $\langle S\rangle$ is defined as the smallest submodule of X containing each member of S. Show that $\langle S\rangle$ is the set of all finite sums of the form $\sum_i a_i x_i$, where $a_i\in A$ and $x_i\in S$. Interpret this as a special case of the above.

Show that for any $X\in\mathrm{Ob}\,\mathscr{C}$, $1_X\in h^X(X)$ generates all of h^X.

3. Let A be a ring and let X be an abelian group. X is a left (resp. right) A-module if and only if a ring homomorphism $f:A\to\mathrm{Hom}_{\mathbf{Z}}(X,X)$ (resp. $g:A^0\to\mathrm{Hom}_{\mathbf{Z}}(X,X)$) is defined. Moreover, it is a bijection from the set of structures of left (right) A-modules of X to the set of ring homomorphisms, from A into $\mathrm{Hom}_{\mathbf{Z}}(X,X)$ (respectively from A^0 into $\mathrm{Hom}_{\mathbf{Z}}(X,X)$).

4. If X is a left (right) A-module, then the group $\mathrm{Hom}_A(A,X)$ has a left (right) A-module structure. Show that $u_X:X\to\mathrm{Hom}_A(A,X)$ which is defined by $u_X(x)=u_x$ leads to an isomorphism of functors: from the functor $\mathrm{Id}\,\mathrm{Mod}\,A$ to h^A.

5. Let $\mathscr{C}$ be an arbitrary category and let $\mathrm{Add}(\mathscr{C})$ be as in Chapter 2, Section 2.1, Exercise 2. If $\mathscr{C}'$ is an arbitrary preadditive category, establish an equivalence of categories $[\mathscr{C},\mathscr{C}']\simeq(\mathrm{Add}\mathscr{C},\mathscr{C}')$.

6. We say that a ring A is *commutative* if for any a, $a'\in A$, we have $aa'=a'a$. Then, the categories $\mathrm{Mod}\,A$ and $\mathrm{Mod}\,A^0$ are identical.

7. Let X be a left A-module and let M be a subset of X. Denote by $p : A^{(M)} \to X$ the unique morphism of modules such that $p(e_x) = x$ for any $x \in M$, $\{e_x\}_x$ being the canonical basis of $A^{(M)}$. Then p is a monomorphism if and only if M is *free*, that is, for any elements $x_1, \dots x_n \in M$ and $a_1, \dots, a_n \in A$ such that $\sum_i a_i x_i = 0$ it follows that $a_i = 0$ for any $1 \leqslant i \leqslant n$. The null element is not in a free set. A subset of a free set is also a free set. A set M is free if and only if any finite subset of M is free. Denote by $F(X)$ the set of free subsets of X, ordered by inclusion. Then $F(X)$ has maximal elements.

Dually, we say that M is a *set of generators* of X if p is an epimorphism. Any super-set of a set of generators is also a set of generators. M is a set of generators if and only if for any $x \in X$ there exist $x_1, \dots, x_n \in M$ and $a_1, \dots, a_n \in A$ such that $x = \sum_i a_i x_i$.

8. A subset M of a left A-module X is a *basis* of X if M is a free set of generators. The module X has a basis if and only if X is free.

9. The following assertions concerning the ring A are equivalent:

(a) Any non-null element of A is invertible;

(b) Any non-null A-module is free;

(c) Any non-null A^0-module is free.

(Hint. A maximal element in the set $F(X)$, defined as in Exercise 7, is a set of generators of X).

A ring A which has the above properties is called a *skew field* or *division ring*. A commutative skew field is called a *field*.

10. If $\mathscr{C}$ is a small abelian category and if T is a projective object in Mod $\mathscr{C}$, then T is left exact.

11. Let $\mathscr{C}$ be a small additive category with kernels, and let T be a small projective object in Mod $\mathscr{C}$. Then T is isomorphic with X_l for an object X of $\mathscr{C}$.

(Hint: Find a composition $T \to X_l \to T$ which is the identity, for some $X \in \mathrm{Ob}\,\mathscr{C}$). This exercise is due to Freyd.

12. Let A be a ring. The following assertions are equivalent:

(a) Any non-null element of A is left-regular;

(b) Any non-null element of A is right-regular;

(c) Any element of A is regular.

13. A left ideal $\mathfrak{m}$ of A is maximal if and only if for any element $x \in A$, $x \notin \mathfrak{m}$ there exists an element $y \in A$ such that $1 - yx \in \mathfrak{m}$. If $\mathfrak{m}$ is a left maximal ideal, then $(\mathfrak{m} : x)$ is also maximal for any $x \notin A$, $x \notin \mathfrak{m}$.

14. For any element x of a ring A the following assertions are equivalent:

(a) x lies in the intersection of all maximal left ideals of A;

(b) x is in the intersection of all maximal right ideals of A;

(c) For any element $y \in A$, $1 - yx$ is an invertible element.

In conclusion, the intersection of all maximal left ideals is the same as the intersection of all maximal right ideals. This intersection is a two-sided ideal denoted by $\mathfrak{R}(A)$ and called the *Jacobson radical* of A.

15. A left ideal $\mathfrak{a}$ in A is a direct summand if and only if there exists an idempotent element e in A such that $\mathfrak{a} = Ae$.

16. Let $u \in \operatorname{Hom}_{\mathscr{C}}(X, X) = A$, $\mathscr{C}$ being an abelian category. Then the following assertions are equivalent:

(a) u is idempotent ($u^2 = u$);

(b) $\ker u$ is a direct summand in X;

(c) $\operatorname{im} u$ is a direct summand in X;

(d) The left ideal Au is a direct summand in A;

(e) The right ideal uA is a direct summand in A.

17. For any ring A, the abelian group $\operatorname{Hom}(A^n, A^m)$ is canonically a free A-module.

18. Let $\{X_i\}_i$ be a family of submodules of an A-module X. Then the sum $\sum_i X_i$ is direct if and only if the relation $\sum_i x_i = 0$, where $x_i \in X_i$ for any i (and of course, $x_i = 0$ for all but a finite number of i's) implies $x_i = 0$ for all i.

19. Let $\mathscr{C}$ be a small preadditive category, $X \in \operatorname{Ob} \mathscr{C}$. Then the subobjects of X_l (resp. X_r) in the category $\operatorname{Mod} \mathscr{C}$ (respectively, $\operatorname{Mod} \mathscr{C}^0$) are in one-to-one correspondence with left (respectively, right) ideals of X in $\mathscr{C}$.

References

Bourbaki [21]; Bucur–Deleanu [31]; Cartan–Eilenberg [33]; Jacobson [96]; Lambek [113]; MacLane [120]; Mitchell [133]; Northcott [152]; Pareigis [162].

3.5 Special objects in abelian categories

PROPOSITION 5.1. *Let X be an object in an additive category $\mathscr{C}$. The following assertions are equivalent:*

(1) *For any morphism $f : X \to \coprod_{i \in I} X_i$, from X into a direct sum there exists a factorization*

$$X \xrightarrow{f'} \coprod_{j \in F} X_j \xrightarrow{u_F} \coprod_{i \in I} X_i$$

where F is a finite subset of I and u_F the unique morphism such that $u_F u_j' = u_j$, for any $j \in F$, where u_i resp. u_j' are structural morphisms of the direct sum $\coprod_{i \in I} X_i$ and of $\coprod_{j \in F} X_j$ resp.

(2) *For any morphism $f : X \to \coprod_{i \in I} X_i$ from X into a direct sum, there exists a finite subset F of I such that $f = \sum_{j \in F} u_j p_j f$, u_j being the structural morphisms and p_j the associated projections.*

(3) *The functor $h^X : \mathscr{C} \to \mathscr{Ab}$ commutes with direct sums.*

Proof. (1) $\Rightarrow$ (2). Let p_i respectively, p_j' be the projections associated with structural morphisms u_i respectively, u_j'. Then $p_j' = p_j u_F$. Hence

$$f = u_F f' = u_F(\textstyle\sum_{j \in F} u_j' p_j') f' = \sum_{j \in F} u_j p_j' f' = \sum_{j \in F} u_j p_j u_F f' = \sum_{j \in F} u_j p_j f.$$

(2) $\Rightarrow$ (1). Let $f' = \sum_{j \in F} u_j' p_j f$. Then we have

$$u_F f' = \textstyle\sum_{j \in F} u_F u_j' p_j f = \sum_{j \in F} u_j p_j f = f.$$

(2) $\Rightarrow$ (3). Let $\coprod_{i \in I} X_i$ be a direct sum in $\mathscr{C}$. The set of morphisms $h^X(u_i)$ gives rise to a morphism of groups:

$$t : \textstyle\coprod_{i \in I} h^X(X_i) \to h^X(\coprod_{i \in I} X_i) \tag{3}$$

so that $tv_i = h^X(u_i)$. To say that h^X commutes with direct sums is to say that t is an isomorphism for every direct sum in $\mathscr{C}$. Now a member of the left-hand side of (3) can be considered as a family $f_i : X \to X_i$ of morphisms, such that $f_i = 0$ for all but a finite number of i's. Then, $t((f_i)_i) = \sum_i u_i f_i$. If $(f_i)_i \neq 0$, then $f_j \neq 0$ for some j and thus we have $p_j \sum_i u_i f_i = f_j \neq 0$. Therefore, $\sum_i u_i f_i \neq 0$ and this shows that in all cases (3) is a monomorphism. We must prove that t is an epimorphism. Let us consider a morphism f on the right of (3). Then there is a finite subset F of I such that

$$f = \textstyle\sum_{j \in F} u_j p_j f.$$

If we take $f_i = p_i f$ for $i \in F$ and $f_i = 0$ otherwise, we have $f = \sum_{i \in I} u_i f_i$.
This shows that t is an isomorphism.

(3) $\Rightarrow$ (2). Let $f : X \to \coprod_{i \in I} X_i$ be a morphism. Then $f = t(g)$, where $g = (f_i)_{i \in I} \in \coprod_{i \in I} \mathrm{Hom}_{\mathscr{C}}(X, X_i)$. Let F be the subset of I consisting of all indices j such that $f_j \neq 0$. It is clear that F is a finite set and

$$f = t((f_i)_i) = \textstyle\sum_{i \in I} u_i f_i = \sum_{j \in F} u_j f_j.$$

Now we compose both sides of the above equation with p_j; then we get: $p_j f = f_j$, for all $j \in F$. Hence $f = \sum_{j \in F} u_j p_j f$. ∎

An object X with the properties of the above proposition is called a *small* object.

COROLLARY 5.2. *Let $\mathscr{C}$ be a preadditive small category. Then for any $X \in \mathrm{Ob}\,\mathscr{C}$ the object $h(X)$ of* Mod $\mathscr{C}$ *is small. Hence* Mod $\mathscr{C}$ *has a set of preadditive small generators.*

Proof. Indeed, according to Lemma 4.3, any morphism $f : h(X) \to \coprod_j T_j$ is defined by $f_X(1_X)$. ∎

COROLLARY 5.3. *Let $\mathscr{C}$ be an abelian category.*

(1) *Any finite direct sum of small objects is a small object*;

(2) *Any quotient object of a small object is also a small object.*

Proof. (1) Let X, X' be two small objects in $\mathscr{C}$, and let $f : X \coprod X' \to \coprod_{i \in I} X_i$ be a morphism. Then, a finite subset F of I exists such that $fu = \sum_{j \in F} u_j p_j fu$ and $fu' = \sum_{j \in F} u_j p_j fu'$, u, u' and u_i being respectively, structural morphisms. If p, p' are the projections associated to u and u', then, according to relation (1) of Chapter 2, $f = fup + fu'p' = \sum_{j \in F} u_j p_j f(up + u'p')$. By induction, we get the same result for any finite set of small objects. ∎

Let X be a left A-module and let M be a subset of X. We denote by $\langle M \rangle$ the submodule generated by M. Any element of M is of the form $\sum_i a_i x_i$, where $x_i \in M$, $a_i \in A$ and $a_i = 0$ for all but a finite number of i's. M is a system of generators of X if and only if $\langle M \rangle = X$. X is called *finitely generated* if it has a finite system of generators.

COROLLARY 5.4. *An A-module X is finitely generated if and only if it is isomorphic with a quotient of A^n for some natural n. Particularly, any finitely generated module is small.* ∎

A directed set $\{X_i\}_i$ of subobjects of an object X is called *stationary* if there exists an index i_0 so that $X_i \subseteq X_{i_0}$ for any i, and it is *complete* if $\sum_i X_i = X$.

Let $\{Y_i\}_i$ be a directed set of subobjects of an object Y. Then, for any object X, there exists a unique morphism:

$$t_X : \varinjlim_i \mathrm{Hom}_{\mathscr{C}}(X, Y_i) \to \mathrm{Hom}_{\mathscr{C}}(X, Y) \tag{4}$$

such that $t_X u_i = h^X(j_i)$, $j_i : Y_i \to Y$ being the canonical inclusions and u_i being the structural morphisms of the inductive limit.

LEMMA 5.5. *For any object X the above morphism t_X is a monomorphism.* ∎

PROPOSITION 5.6. *Let X be an object in an* Ab 5-*category. The following assertions are equivalent*:

(1) *Any complete direct set $\{X_i\}_i$ of subobjects of X is stationary (that is, $X = X_{i_0}$ for an index i_0).*

(2) *For any complete direct set $\{Y_i\}_i$ of subobjects of an object Y, the morphism* (4) *is an isomorphism.*

Proof. (1) $\Rightarrow$ (2). Let $f : X \to Y$ be a morphism, then $X = f^{-1}(\sum_i Y_i) = \sum_i f^{-1}(Y_i)$. Hence, $X = f^{-1}(Y_{i_0})$, for an index i_0; that is, the morphism f has the factorization

$$X \xrightarrow{f'} Y_{i_0} \longrightarrow Y,$$

the last composition being the canonical inclusion. Obviously,

$$t_X(u_{i_0}(f')) = f.$$

(2) $\Rightarrow$ (1). Let $\{X_i\}_i$ be a complete direct set of subobjects of X. Thus, the morphism

$$t_X : \varinjlim_i \operatorname{Hom}_{\mathscr{C}}(X, X_i) \to \operatorname{Hom}_{\mathscr{C}}(X, X)$$

is an isomorphism. Let i_0 and $f : X \to X_{i_0}$ be such that $t_X(u_{i_0}(f)) = 1_X$, that is, $j_{i_0} f = 1_X$. Obviously, j_{i_0} is an isomorphism, hence $X_{i_0} \equiv X$. ■

An object X in an Ab 5-category, as in the above proposition, is called of *finite type*. In other words, an object X is of finite type, if for any *monodirect system* $(X_i, f_{ij})_i$ (that is, the f_{ij} are monomorphisms) one has the canonical isomorphism:

$$\varinjlim_i \operatorname{Hom}_{\mathscr{C}}(X, X_i) \xrightarrow{\sim} \operatorname{Hom}_{\mathscr{C}}(X, \varinjlim_i X_i).$$

(that is, the functor h^X commutes with monoinductive direct limits.

COROLLARY 5.7. *An A-module X is finitely generated if and only if it is of finite type.*

Proof. Let X be finitely generated, say $X = \langle x_1, \dots x_n \rangle$ and let $\{X_i\}$ be a complete directed set of subobjects of X.

Obviously, we have $X = \sum_i X_i = \bigcup_i X_i$ (by "$\bigcup$" we mean the union of sets). Then $x_1, \dots, x_n \in X_{i_0}$ for a suitable i_0, that is we have $X = X_{i_0}$.

Conversely, we see that the set of finitely generated submodules of any module is a complete direct set. ■

The same result is true in the category Mod $\mathscr{C}$, where $\mathscr{C}$ is a small preadditive category.

Let $\{U_i\}_i$ be a set of generators of $\mathscr{C}$. An object X is called *finitely generated with respect to* the set $\{U_i\}_i$ (or equivalently $\{U_i\}_i$*-finitely generated*) if X is the quotient object of $\prod_{j\in J} U_j$, J being a finite set and with $U_j \in \{U_i\}_i$ for all $j \in J$. If U_i is small for any i, then any $\{U_i\}_i$-finitely generated object is also small. If $\mathscr{C}$ is an Ab 5-category and if U_i is of finite type for all i, then any $\{U_i\}_i$-finitely generated object is also of finite type.

Sometimes we shall make use of the following result:

PROPOSITION 5.8. *Let $\mathscr{C}$ be an* Ab 5*-category and let $\{U_i\}_i$ be a set of generators of X. Then, for any object X, the set of $\{U_i\}_i$-finitely generated subobjects is direct and complete.*

The proof is left to the reader. ∎

Sometimes, a $\{U_i\}_i$-finitely generated object will be called simply "*finitely generated*" if no confusion can arise.

An object X is called *finitely presented* if it is of finite type and if for any epimorphism $p: Y \to X$, Y being of finite type, it follows that $\ker p$ is also of finite type.

LEMMA 5.9. *Let $0 \to X' \to X \to X'' \to 0$ be an exact sequence in $\mathscr{C}$.*

(1) *If X is finitely presented and if X' is of finite type, then X'' is finitely presented.*

(2) *If X' and X'' are finitely presented, then X is also finitely presented.*

The proof is a simple computation and is left to the reader. ∎

An abelian category $\mathscr{C}$ is called *locally of finite type* if any object X of $\mathscr{C}$ is the sum of its subobjects of finite type. It is not difficult to prove that any object of a locally small and locally of finite type Ab 5-category, is an inductive limit of the direct set of its subobjects of finite type.

The following result gives a characterization of finitely presented objects of some categories.

THEOREM 5.10. *Let $\mathscr{C}$ be a locally small and locally of finite type* Ab 5*-category. An object X of $\mathscr{C}$ is finitely presented if and only if for any direct system $\{Y_i, u_{ij}\}_i$ in $\mathscr{C}$, the canonical morphism of abelian groups*

$$t_X : \varinjlim_i \operatorname{Hom}_{\mathscr{C}}(X, Y_i) \to \operatorname{Hom}_{\mathscr{C}}(X, \varinjlim_i Y_i)$$

is an isomorphism (that is, the functor $\operatorname{Hom}_{\mathscr{C}}(X, ?)$ *commutes with inductive direct limits).*

Proof. Let us assume first, that t_X is an isomorphism for any direct system. Then, according to Proposition 5.6, it follows that X is of finite type. Now, let $p : Y \to X$ be an epimorphism with Y being of finite type and with $Y' = \ker p$. We must prove that Y' is of finite type. In order to do this, let $\{Y_i'\}_i$ be the direct set of subobjects of finite type of Y'. From the hypothesis we have $\sum_i Y_i' = Y'$. Also, for any i, we have the commutative diagram:

$$\begin{array}{ccccccccc} 0 & \longrightarrow & Y' & \xrightarrow{p} & Y & \longrightarrow & X & \longrightarrow & 0 \\ & & \uparrow{\scriptstyle u_i} & & \| & & \uparrow{\scriptstyle v_i} & & \\ 0 & \longrightarrow & Y_i' & \xrightarrow{t_i} & Y & \xrightarrow{p_i} & Y/Y_i' & \longrightarrow & 0 \end{array}$$

where the rows are exact, u_i, t_i are the canonical inclusions and where v_i is canonically constructed. Now, proceeding to the inductive limit, we obtain another commutative diagram

$$\begin{array}{ccccccccc} 0 & \longrightarrow & Y' & \longrightarrow & Y & \longrightarrow & X & \longrightarrow & 0 \\ & & \| & & \| & & \uparrow{\scriptstyle v} & & \\ 0 & \longrightarrow & \varinjlim_i Y_i' & \longrightarrow & Y & \longrightarrow & \varinjlim_i (Y/Y_i') & \longrightarrow & 0 \end{array}$$

which has exact rows and where v is an isomorphism. According to the conditions of the theorem, for some i, a morphism $g : X \to Y/Y_i'$ exists, such that $r_i g = v^{-1}$, i.e. $vr_i g = 1_X$,

$$r_i : Y/Y_i' \to \varinjlim_i (Y/Y_i')$$

being structural morphisms. More precisely, we can derive that X is a direct summand of Y/Y_i', that is, $\ker v_i$ is an object of finite type. But $\ker v_i = Y'/Y_i'$ and thus, from Exercise 5 we know that Y' is of finite type.

Conversely, let X be a finitely presented object and f an element of the group

$$\varinjlim_i \mathrm{Hom}_{\mathscr{C}}(X, Y_i),$$

such that $t_X(f) = 0$. Then, an index i exists, such that $f = h_i(g)$ with

$$h_i : \mathrm{Hom}_{\mathscr{C}}(X, Y_i) \to \varinjlim_i \mathrm{Hom}_{\mathscr{C}}(X, Y_i)$$

being structural morphisms. Then, necessarily, one has that $u_i g = 0$ with

$$u_i : Y_i \to \varinjlim_i Y_i$$

being also structural morphisms.

According to Exercise 6 of Chapter 2, Section 2.8, we have that

$$\ker(u_i\, g) = \textstyle\sum_{j \geqslant i} \ker(u_{ij}\, g).$$

But $\operatorname{im} g \subseteq \ker u_i\, g$ and thus an index $j \geqslant i$ exists such that $u_{ij}\, g = 0$, Since X is of finite type. To conclude: $f = 0$, and thus t_X is a monomorphism.

Now it remains to show that t_X is an epimorphism. In order to do this, let

$$f : X \to \varinjlim_i Y_i = Y$$

be a morphism. We know that $Y = \sum_i \operatorname{im} u_i$ and consequently an index i exists such that $\operatorname{im} f \subseteq \operatorname{im} u_i$. Then, from the cartesian diagram

$$\begin{array}{ccc} X & \xrightarrow{\quad f \quad} & Y \\ {\scriptstyle g}\uparrow & & \uparrow{\scriptstyle u_i} \\ X' = X \prod_Y Y_i & \xrightarrow{\quad f_i \quad} & Y_i \end{array}$$

we derive that g is an epimorphism. From the hypothesis, we obtain that X' is the inductive limit of its subobjects of finite type. Then a subobject X'' of finite type of X' exists such that $g(X'') = X$. In conclusion, we obtain the commutative diagram

$$\begin{array}{ccc} X & \xrightarrow{\quad f \quad} & Y \\ {\scriptstyle g'}\uparrow & & \uparrow{\scriptstyle u_i} \\ X'' & \xrightarrow{\quad f' \quad} & Y_i \end{array}$$

where g' is an epimorphism and where X'' is of finite type, f' being induced by f_i.

By hypothesis $\ker g'$ is of finite type and from Exercise 6 of Chapter 2, Section 2.8, we derive that an index $j \geqslant i$ exists such that $(u_{ij}\, f')(\ker g') = 0$. Then, a morphism $\bar{f} : X \to Y_j$ exists such that $\bar{f} g' = u_{ij}\, f'$. Furthermore, we get $u_j \bar{f} g' = u_j\, u_{ij}\, f' = u_i\, f' = f g'$ and finally $u_j \bar{f} = f$, g' being an epimorphism. We conclude that $f = t_X(\bar{f})$. Hence t_X is also an epimorphism. The proof is now complete. ∎

COROLLARY 5.11. *Let $\mathscr{C}$ be as in Theorem 5.10. Then, a finite direct sum of finitely presented objects is also finitely presented.* ∎

From Corollary 5.4 we have that for any small preadditive category $\mathscr{C}$, the category $\operatorname{Mod} \mathscr{C}$ is locally of finite type.

LEMMA 5.12. *Let $\mathscr{C}$ be a small preadditive category. Then, any object X of $\mathscr{C}$ is finitely presented in* Mod $\mathscr{C}$.

Proof. Let $\{Y_i, u_{ij}\}$ be a direct system in Mod $\mathscr{C}$. We must prove that the morphism

$$t_X : \varinjlim_i \operatorname{Hom}_{\mathscr{C}}(X, Y_i) \to \operatorname{Hom}_{\mathscr{C}}(X, \varinjlim_i Y_i)$$

is an isomorphism. Indeed, if $f : X \to Y_i$ is a morphism such that

$$(t_X v_i)(f) = u_i f = 0,$$

where v_i and u_i are respectively structural morphisms, then in particular, we get $(u_i f)_X (1_X) = 0$. Furthermore, there exists an index j such that $((u_{ij})_X f_X)(1_X) = 0$. In other words, $h^X(u_{ij})(f) = 0$, that is $v_j(f) = 0$. Hence, t_X is a monomorphism.

Now, let

$$X \to \varinjlim_i Y_i$$

be a morphism and let $x \in Y_i(X)$ be an element such that $(u_i)_X(x) = f_X(1_X)$. Then $(t_X v_i)(u_x) = f$. To conclude: t_X is an epimorphism, and so an isomorphism. ∎

Now we shall give a useful characterization of finitely presented objects of Mod $\mathscr{C}$.

PROPOSITION 5.13. *Let $\mathscr{C}$ be a small preadditive category and X an object in* Mod $\mathscr{C}$. *The following assertions are equivalent*:

(1) *X is finitely presented*;

(2) *There is an exact sequence*:

$$\coprod_{1 \leqslant r \leqslant m} X_r' \to \coprod_{1 \leqslant k \leqslant n} X_k \to X \to 0$$

where X_k and X_r' are objects in $\mathscr{C}$ for any k and r.

Proof. (1) ⇒ (2). This results directly from Corollary 5.4, and the very definition of finitely presented objects.

(2) ⇒ (1). Let $\{Y_i, u_{ij}\}$ be a direct system in Mod $\mathscr{C}$. For any i we have the exact sequence:

$$0 \to \operatorname{Hom}_{\mathscr{C}}(X, Y_i) \to \operatorname{Hom}_{\mathscr{C}}(\textstyle\coprod_k X_k, Y_i) \to \operatorname{Hom}_{\mathscr{C}}(\coprod_r X_r, Y_i).$$

Now, the result is obtained by calculating the inductive limit over i and making use of Corollary 5.11 and Lemma 5.12. ∎

Let $\mathscr{C}$ be a locally small Ab 5-category. An object X of $\mathscr{C}$ is called *coherent* if it is of finite type and if for any morphism $f : Y \to X$ with Y of finite type, $\ker f$ is of finite type. Let $\mathrm{Coh}(\mathscr{C})$ be the full subcategory of $\mathscr{C}$ which is generated by the coherent objects.

LEMMA 5.14. *If in the following exact sequence:*

$$0 \to X' \to X \to X'' \to 0$$

X is coherent and X' of finite type, then X'' is also coherent.

Proof. Let $f : Y \to X''$ a morphism such that Y is of finite type. Then we have the commutative diagram with exact rows:

$$\begin{array}{ccccccccc} 0 & \longrightarrow & X' & \longrightarrow & X \prod_{X''} Y & \longrightarrow & Y & \longrightarrow & 0 \\ & & \| & & \downarrow{\scriptstyle f'} & & \downarrow{\scriptstyle f} & & \\ 0 & \longrightarrow & X' & \longrightarrow & X & \longrightarrow & X'' & \longrightarrow & 0 \end{array}$$

By Exercise 5 we deduce that $X \prod_{X''} Y$ is of finite type and so $\ker f'$ is of finite type. Finally, we see that $\ker f$ is isomorphic with $\ker f'$. ∎

LEMMA 5.15. *Let $f : X \to Y$ be a morphism with X and Y coherent. Then* $\ker f$, $\mathrm{im}\, f$ *and* $\mathrm{coker}\, f$ *are also coherent.*

Proof. By definition of a coherent object $\ker f$ is of finite type. Then by Lemma 5.14 we deduce that $\mathrm{im}\, f$ is coherent. In addition by same result we have that $\mathrm{coker}\, f$ is coherent. To complete the proof, it is sufficient to apply Exercise 13. ∎

THEOREM 5.16. *Let $\mathscr{C}$ be a locally small* Ab 5-*category. Then* $\mathrm{Coh}(\mathscr{C})$ *is an abelian category and the inclusion functor* $\mathrm{Coh}(\mathscr{C}) \to \mathscr{C}$ *is exact. Moreover, if in the exact sequence*

$$0 \to X' \to X \to X'' \to 0$$

X' and X'' are in $\mathrm{Coh}(\mathscr{C})$ *we deduce that X is also in* $\mathrm{Coh}(\mathscr{C})$.

The result is a consequence of the above two lemmas and Exercise 13. ∎

Exercises

1. A projective module is small if and only if it is finitely generated.
2. Prove that any countably generated small A-module is finitely generated.

3. If X is a small A-module and if $X_1 \subset X_2 \subset \ldots$ is an increasing sequence of proper submodules such that X/X_i are not finitely generated, then $X/\sum_i X_i$ is not finitely generated.

4. Let U be a small projective object of an Ab 3-category. Then, show that U has the following property: if $\{X_i\}_i$ is a direct set of proper subobjects of U, then $\sum_i X_i$ is a proper subobject of U.

Conversely, prove that if X is any object in an Ab 5-category and if X has the above property with respect to subobjects, then X is small.

5. Let $0 \to X' \to X \to X'' \to 0$ be an exact sequence in an Ab 5-category. Show that if X is of finite type, then X'' is of finite type too. If X' and X'' are of finite type, then X is also of finite type.

6. Let $0 \to X' \to X \to X'' \to 0$ be an exact sequence in an Ab 5-category and let $\{U_i\}_i$ be a set of generators of $\mathscr{C}$. Show that if X is $\{U_i\}_i$-finitely generated, then X'' is $\{U_i\}_i$-finitely generated too. If X' and X'' are $\{U_i\}_i$-finitely generated, then X is also $\{U_i\}_i$-finitely generated.

7. Let $\mathscr{C}$ be an Ab 5-category. We say that a subobject X' of X is of *finite type relative* to X if for any complete direct set $\{X_i\}_i$ of subobjects of X, the direct set $\{X' \cap X_i\}_i$ is stationary. Assuming that $\mathscr{C}$ is locally small, prove that the following assertions are equivalent:

(a) $\mathscr{C}$ is an Ab 6-category;

(b) Any object X of $\mathscr{C}$ is a sum of subobjects which are of finite type relative to X.

This result is due to Roos [192].

8. (Bourbaki, [26], Chapter 1, p. 62). (a) Let A be a ring, I a set and X a submodule of $A^{(I)}$. Also, let T be the set of couples (J, S) where J is a finite subset of I, and where S is a submodule of finite type of $A^J \cap X$ (A^J is canonically considered as a submodule in $A^{(I)}$). We put $(J, S) \leqslant (J', S')$ if and only if $J \subseteq J'$ and $S \subseteq S'$. Then prove that T is a right filtered set and that the set of modules $\{A^J/S\}_{(J,S)\in T}$ is canonically a direct system. Also, show that a canonical isomorphism exists from

$$A^{(I)}/X \quad \text{to} \quad \varinjlim_{(J,S)} (A^J/S).$$

(b) Making use of (a), prove that any module is an inductive limit of finitely presented modules.

9. Let X be a countably generated A-module. Assume that any direct summand X' of X has the following property: any element of X' can be embedded in a finitely generated direct summand of X'. Then X is a direct sum of finitely generated modules.

10. We define a *proper submodule* as a submodule of a module which is different from the module itself. A *maximal submodule* X' of X is to be understood as a proper submodule which has the property that there is no other proper submodule which strictly contains it. Show that if X is a finitely generated module, and if X' is an arbitrary proper submodule of X, then X' is contained in at least one maximal submodule.

11. (Lemma of Nakayama). Let X be a finitely generated A-module and let $\mathfrak{a}$ be a left ideal of A which is contained in $\mathfrak{R}(A)$, such that $\mathfrak{a}X = X$. In these conditions, show that $X = 0$. (Hint: use the characterization of elements of $\mathfrak{R}(A)$ as stated in Exercise 14 of Section 3.4).

12. Let $\mathscr{C}$ be a Grothendieck category which is locally of finite type. If every finite object is injective, then every finitely presented object is projective.

13. Let $\mathscr{C}$ be a locally small Ab 5-category. If in the exact sequence

$$0 \to X' \to X \to X'' \to 0$$

X' and X are coherent, then X'' is also coherent. Also if X and X'' are coherent, then X' is coherent.

14. $\mathscr{C}$ is as above. If X_1, X_2 are two subobjects of finite type, (respectively coherent), of any coherent object X, then $X_1 \cap X_2$ is of finite type (respectively, coherent).

15. Let $\mathscr{C}$ be a preadditive small category. The following assertions are equivalent:

(1) Any object X of $\mathscr{C}$ is coherent in Mod $\mathscr{C}$.

(2) Any finitely presented object of Mod $\mathscr{C}$ is coherent.

References

Auslander [10]; Bourbaki [21]; [23], [26]; Gabriel [59]; Gabriel–Rentschler [63]; Gabriel–Ulmer [64]; Gruson [81]; Mitchell [133]; Northcott [152]; Rentschler [183]; Roos [192]; [197], Stenström [211], [214].

3.6 Tensor products. A characterization of functor categories

LEMMA 6.1. *Let us consider the following diagram in an abelian category* $\mathscr{C}$:

$$\begin{array}{ccccccc} P_1 & \xrightarrow{d} & P_0 & \xrightarrow{q} & X & & \\ & & & & \downarrow{\scriptstyle f} & & \\ P_1' & \xrightarrow{d'} & P_0' & \xrightarrow{q'} & X' & \longrightarrow & 0 \end{array} \qquad (5)$$

where P_1 and P_0 are projective, the bottom row is exact and $qd = 0$. Then, there exist morphisms $f_0 : P_0 \to P_0'$ and $f_1 : P_1 \to P_1'$ such that (5) *is commutative. Furthermore, let $T : \mathscr{C}_1 \to \mathscr{C}'$ be a covariant additive functor into an additive category with cokernels, where $\mathscr{C}_1$ is a full subcategory of $\mathscr{C}$ containing P_0, P_1, P_0' and P_1'. Then the induced morphism*

$$\operatorname{coker} T(d) \to \operatorname{coker} T(d')$$

does not depend on the choice of f_0 and f_1.

Proof. Making use of the projectivity of P_0, one can find $f_0 : P_0 \to P_0'$ such that $fq = q'f_0$. Then one has $q'f_0d = fqd = 0$, and thus the image of f_0d is a subobject of $\ker q' = \operatorname{im} d'$. Hence, by projectivity of P_1 we can find $f_1 : P_1 \to P_1'$ such that $f_0d = d'f_1$. Let $p : T(P_0) \to Y$ and $p' : T(P_0') \to Y'$ be the cokernels of $T(d)$ and $T(d')$, respectively. Suppose that $g_0 : P_0 \to P_0'$ and $g_1 : P_1 \to P_1'$ are another pair of morphisms making the diagram (5) commutative. Let $u, v : Y \to Y'$ be the morphisms induced by f_0, f_1 and g_0, g_1 respectively. Now, $q'(f_0 - g_0) = 0$ and again making use of the projectivity of P_0 and of the exactness of the bottom row, we obtain a morphism $h : P_0 \to P_1'$ such that $d'h = f_0 - g_0$. Then $(u - v)p = up - vp = p'T(f_0) - p'T(g_0) = p'T(f_0 - g_0) = p'T(d'h) = p'T(d')T(h) = 0$.

Since p is an epimorphism, we must have $u = v$. ∎

THEOREM 6.2. *Let $\mathscr{P}$ be a full subcategory of an* Ab 3*-category $\mathscr{C}$ and suppose that the objects of $\mathscr{P}$ form a set of small projective generators of $\mathscr{C}$. Let $T : \mathscr{P} \to \mathscr{C}'$ be an additive covariant functor into an* Ab 3*-category. Then, T can be uniquely extended (up to a functorial isomorphism) to a functor $\overline{T} : \mathscr{C} \to \mathscr{C}'$ such that $\overline{T}$ commutes with inductive limits.*

Proof. We first extend T to the subcategory of $\mathscr{C}$ which consists of all free objects, that is, the objects of the form $\coprod_{i \in I} P_i$, where $P_i \in \operatorname{Ob} \mathscr{P}$ for all $i \in I$. We define $\overline{T}(\coprod_{i \in I} P_i) = \coprod_{i \in I} T(P_i)$. For a morphism $t : \coprod_{i \in I} P_i \to \coprod_{j \in J} P_j$, let $t_i = tu_i$, $u_i : P_i \to \coprod_{i \in I} P_i$ be the structural morphisms. By Proposition 5.1 since P_i is small we can write $t_i = \sum_{j \in J} u_j p_j t_i$ where u_j are also structural morphisms and p_j associated projections of the direct sum $\coprod_{j \in J} P_j$. Then we define $\overline{T}(t) : \coprod_{i \in I} T(P_i) \to \coprod_{j \in J} T(P_j)$ as the morphism which when composed with $\bar{u}_i$ (the structural morphisms of $\coprod_{i \in I} T(P_i)$) gives $\sum_{j \in J} \bar{u}_j T(p_j t_i)$ where $\bar{u}_j$ denotes the structural morphisms of $\coprod_{j \in J} T(P_j)$. The additive functorial properties of $\overline{T}$ so far constructed follow from the additive functorial properties of T.

Let X be an object of $\mathscr{C}$. According to the dual of Lemma 2.11 we get the following exact sequence:

$$P_1 \xrightarrow{d} P_0 \xrightarrow{q} X \longrightarrow 0 \tag{6}$$

where P_0 and P_1 are free in the above sense.

Define $\bar{T}(X)$ as coker $\bar{T}(d)$. For a morphism $f : X \to X'$, let

$$P_1' \xrightarrow{d'} P_0' \xrightarrow{q'} X' \longrightarrow 0$$

be the exact sequence used to define $\bar{T}(X')$. Then, by Lemma 6.1 we can find the morphisms $f_0 : P_0 \to P_0'$, $f_1 : P_1 \to P_1'$ such that $fq = q'f_0$, $f_0 d = d'f_1$. Define $\bar{T}(f)$ as the morphism making commutative the diagram

$$\begin{array}{ccccccc}
\bar{T}(P_1) & \xrightarrow{\bar{T}(d)} & \bar{T}(P_0) & \xrightarrow{p} & \bar{T}(X) & \longrightarrow & 0 \\
\downarrow{\scriptstyle \bar{T}(f_1)} & & \downarrow{\scriptstyle \bar{T}(f_0)} & & \downarrow{\scriptstyle \bar{T}(f)} & & \\
\bar{T}(P_1') & \xrightarrow{\bar{T}(d')} & \bar{T}(P_0') & \xrightarrow{p'} & \bar{T}(X') & &
\end{array}$$

p and p' being the canonical epimorphisms.

Then again by Lemma 6.1 we know that $\bar{T}(f)$ is independent of the choice of f_0 and f_1 and in this way, $\bar{T}$ is seen to be an additive functor from $\mathscr{C}$ to $\mathscr{C}'$. Setting $X = X'$ and $f = 1_X$ in the above discussion, we see at this point that up to an isomorphism, $\bar{T}$ is independent of the choice of the sequence (6) used to define it, and furthermore that this isomorphism is natural with respect to morphisms to and from X.

We now show that $\bar{T}$ is right exact. Let

$$0 \to X' \to X \to X'' \to 0 \tag{7}$$

be an exact sequence in $\mathscr{C}$, and take free resolutions

$$P_1' \to P_0' \to X' \to 0$$

and

$$P_1'' \to P_0'' \to X'' \to 0$$

for X' and X''. We construct the following commutative diagram:

$$\begin{array}{ccccccc}
P_1' & \longrightarrow & P_0' & \longrightarrow & X' & \to & 0 \\
\downarrow & & \downarrow & & \downarrow{\scriptstyle i} & & \\
P_1' \amalg P_1'' & \longrightarrow & P_0' \amalg P_0'' & \longrightarrow & X & \to & 0 \\
\downarrow & & \downarrow & & \downarrow{\scriptstyle p} & & \\
P_1'' & \longrightarrow & P_0'' & \xrightarrow{p} & X'' & \to & 0
\end{array} \tag{8}$$

The middle column is a split exact sequence, and the morphism $P_0'' \to X$ is defined by the projectivity of P_0'' and the epimorphism p. Taking the kernel of the morphism from the middle column to the right-hand column, we obtain from Lemma 5.5 of Chapter 2 a short exact sequence. If we repeat functional properties of T so far constructed follow from the additive the above construction with the sequence of kernels, replacing P_0' and P_0'' by P_1' and P_1'' respectively, we obtain the commutative diagram (8), the rows and columns being short exact sequences. Hence we can use the middle row to define $\bar{T}(X)$. Applying $\bar{T}$ to (8) the left two columns will still be split exact sequences according to the additivity of $\bar{T}$. It follows from Chapter 2, Lemma 5.7 that the sequence

$$\bar{T}(X') \xrightarrow{\bar{T}(i)} \bar{T}(X) \xrightarrow{\bar{T}(p)} \bar{T}(X'') \longrightarrow 0$$

is exact. In other words, $\bar{T}$ is right exact.

Now, it follows from Proposition 2.10 that in order to show that $\bar{T}$ is inductive limit preserving it will suffice to prove that $\bar{T}$ commutes with direct sums. Given a set $\{X_i\}_{i\in I}$ of objects in $\mathscr{C}$, choose a free resolution:

$$P_1{}^i \xrightarrow{d_i} P_0{}^i \xrightarrow{q_i} X_i \longrightarrow 0$$

for each $i \in I$. Then, if one uses the fact that direct sums are right exact (Chapter 2, Section 2.8) one obtains the following sequence:

$$\coprod_{i\in I} P_1{}^i \xrightarrow{d} \coprod_{i\in I} P_0{}^i \xrightarrow{q} \coprod_{i\in I} X_i \longrightarrow 0$$

where $d = \coprod_i d_i$, $q = \coprod_i q_i$ is a free resolution of $\coprod_i X_i$. We derive directly from this that $\bar{T}$ preserves direct sums.

Since the construction of $\bar{T}$ was constrained at each stage by the requirement that it should be inductive limit preserving, we have the uniqueness of this construction. ■

THEOREM 6.3. *Let $\mathscr{C}$ be a preadditive small category and $T : \mathscr{C} \to \mathscr{C}'$ an additive contravariant functor into an* Ab 3-*category. Then, a unique functor $T^* : \text{Mod}\,\mathscr{C} \to \mathscr{C}'$ exists such that:*

(1) *$T^*h \simeq T$, with $h : \mathscr{C} \to \text{Mod}\,\mathscr{C}$ being the canonical functor;*

(2) *T^* has a right adjoint.*

Moreover, any covariant functor $F : \text{Mod}\,\mathscr{C} \to \mathscr{C}'$ which commutes with inductive limits has a right adjoint.

Proof. Let $\mathscr{P}$ be the full subcategory of Mod $\mathscr{C}$ generated by the objects of $\mathscr{C}$. We denote by $T' : \mathscr{P} \to \mathscr{C}'$ the functor: $T'(X) = T(X)$ and for any morphism $f : X \to Y$ of $\mathscr{C}$, $T'(f) = T(f)$. Obviously T' is a covariant functor and according to the above theorem, it has a unique extension which commutes with the inductive limit functor T^* to Mod $\mathscr{C}$. Let us define a functor $T_* : \mathscr{C}' \to \text{Mod}\,\mathscr{C}$ as follows. For any object X' of $\mathscr{C}'$, $T_*(X') : \mathscr{C} \to \mathscr{Ab}$ is the functor which associates to each object X of $\mathscr{C}$ the abelian group $\text{Hom}_{\mathscr{C}'}(T(X), X')$. If $f' : X' \to Y'$ is a morphism of $\mathscr{C}'$, then $T_*(f')$ has the components $(T_*(f'))_X = h^{T(X)}(f'), X \in \mathrm{Ob}\,\mathscr{C}$. Thus we find that T_* is a covariant additive functor.

For any object F in $\mathrm{Mod}\,\mathscr{C}$ and any X' in $\mathscr{C}'$, we shall define the following functorial isomorphism:

$$t(F, X') : \mathrm{Hom}_{\mathscr{C}'}(T^*(F), X') \to \mathrm{Hom}_{\mathscr{C}}(F, T_*(X')).$$

In order to define $t(F, X')$ we suppose that F is an object X of $\mathscr{C}$. Then, $\mathrm{Hom}_{\mathscr{C}'}(T^*(X), X') = \mathrm{Hom}_{\mathscr{C}'}(T(X), X') = T_*(X')(X)$. If $g : T^*(X) \to X'$ is a morphism of $\mathscr{C}'$, we put $t(X, X')(g) = u_g$ where $u_g : X \to T_*(X')$ is the morphism associated with g, according to Lemma 4.3. The functoriality of t is now obvious.

Now assume that $F = \coprod_{i \in I} X_i$ is a free object in $\mathrm{Mod}\,\mathscr{C}$. Then we have the canonically defined functorial isomorphisms:

$$u : \mathrm{Hom}_{\mathscr{C}'}(T^*(\textstyle\coprod_i X_i), X') \xrightarrow{\sim} \prod_i \mathrm{Hom}_{\mathscr{C}'}(T(X_i), X')$$
$$v : \mathrm{Hom}_{\mathscr{C}}(\textstyle\coprod_i X_i, T_*(X')) \xrightarrow{\sim} \prod_i \mathrm{Hom}_{\mathscr{C}}(X_i, T_*(X')).$$

Let us take $t(\coprod_i X_i, X') = v^{-1}(\prod_i t(X_i, X'))\, u$.

For any object F of $\mathrm{Mod}\,\mathscr{C}$, we get the resolution

$$F_1 \xrightarrow{d} F_0 \xrightarrow{q} F \to 0$$

F_0 and F_1 being free objects. Define $t(F, X')$ as the morphism rendering commutative the following canonically constructed diagram:

$$\begin{array}{ccccccc}
0 \to & \mathrm{Hom}_{\mathscr{C}'}(T^*(F), X') & \to & \mathrm{Hom}_{\mathscr{C}'}(T^*(F_0), X') & \to & \mathrm{Hom}_{\mathscr{C}'}(T^*(F_1), X') \\
 & \downarrow t(F, X') & & \Big\Vert\!\downarrow t(F_0, X') & & \Big\Vert\!\downarrow t(F_1, X') \\
0 \to & \mathrm{Hom}_{\mathscr{C}}(F, T_*(X')) & \to & \mathrm{Hom}_{\mathscr{C}}(F_0, T_*(X')) & \to & \mathrm{Hom}_{\mathscr{C}}(F_1, T_*(X'))
\end{array}$$

Finally, we have that T_* is a right adjoint of T^*.

If $T' : \mathrm{Mod}\,\mathscr{C} \to \mathscr{C}'$ is a covariant functor, which commutes with inductive limits, then $T = T'h$ is a contravariant functor and according to the above theorem one has that $T^* \simeq T'$. Obviously T_* is a right adjoint of T'. ∎

The above defined functor T^* is called sometimes the "*tensor product*" associated to T.

NOTE 1. The above corollary has a dual. Let $\mathscr{C}$ be a preadditive small category and let $T : \mathscr{C} \to \mathscr{C}'$ be a covariant additive functor into an Ab 3-category. Then a unique functor $T^* : \mathrm{Mod}\,\mathscr{C}^0 \to \mathscr{C}'$ exists such that:

(1) $T^* h^0 \simeq T$;

(2) T^* has a right adjoint.

Moreover, any functor $T^*: \text{Mod}\,\mathscr{C}^0 \to \mathscr{C}'$ which commutes with inductive limits has a right adjoint.

T^* is also called the tensor product associated with T.

The following result due to Freyd shows that Theorem 6.3 is in fact characteristic for the category Mod $\mathscr{C}$ where $\mathscr{C}$ is a small preadditive category.

COROLLARY 6.4. *For any abelian category $\mathscr{C}$ the following assertions are equivalent*:

(1) *$\mathscr{C}$ is an* Ab 3-*category and $\{U_i\}_{i\in I}$ is a set of projective small generators of $\mathscr{C}$*;

(2) *There is a small preadditive category $\mathscr{P}$ such that $\mathscr{C}$ is equivalent with* $\text{Mod}\,\mathscr{P}^0$.

Proof. (1) ⇒ (2). Let $\mathscr{P}$ be the full subcategory of $\mathscr{C}$ generated by the objects U_i for all $i \in I$, and $T: \mathscr{P} \to \mathscr{C}$ the canonical inclusion functor. Then, a unique functor $T^*: \text{Mod}\,\mathscr{P}^0 \to \mathscr{C}$ exists such that

$$T^*(h^0(U_i)) = T(U_i) = U_i,$$

for any i and T^* has a right adjoint T_*. We shall prove that T_* is an equivalence of categories. Remark first that for any i, we have that $T_*(U_i) = h^0(U_i)$ and according to Theorem 6.2 we can derive that T_* is inductive limit preserving.

Furthermore, let $u: \text{IdMod}\,\mathscr{P}^0 \to T_*T^*$ be an adjunction arrow and v its quasi-inverse. For any U_i we may take $u_{h^0(U_i)} = 1_{h^0(U_i)}$. Also, for any free object F of $\text{Mod}\,\mathscr{P}^0$ we may take $u_F = 1_F$.

Again, according to the construction of functors T^* and T_* it is easy to check that for any object X of $\text{Mod}\,\mathscr{P}^0$ we may take $u_X = 1_X$. Then, according to Proposition 5.2 of Chapter 1, we derive that T^* is full and faithful.

Now, we have only to prove that T_* is full and faithful, that is, we have to show that v is an isomorphism, and this results directly from the construction of T^* and T_*. The result follows from Theorem 5.3 of Chapter 1. The implication (2) ⇒ (1) is obvious according to the Corollary 5.2.

Further, we shall make use of the following result.

THEOREM 6.5. (Mitchell, [133] Chapter 4). *Let $\mathscr{C}$ be an* Ab 5-*category with a full subcategory $\mathscr{P}$ the objects of which form a set of projective generators of $\mathscr{C}$. Let $T, S: \mathscr{C} \to \mathscr{C}'$ be additive covariant functors with $\mathscr{C}'$ being an abelian category and suppose that T commutes with inductive limits. Let us denote the restrictions of T and S to $\mathscr{P}$ by T' and S', respectively. Then, any functorial morphism $u: T' \to S'$ can be uniquely extended to a functorial morphism $\bar{u}: T \to S$.*

Proof. For a free object $P' = \coprod_{i\in I} P_i, P_i \in \operatorname{Ob}\mathscr{P}$, we denote by $u'_{P'}$ the composition

$$T(\coprod_{i\in I} P_i) = \coprod_{i\in I} T(P_i) \to \coprod_{i\in I} S(P_i) \xrightarrow{r} S(\coprod_{i\in I} P_i).$$

Here, the equality is a result of the fact that T commutes with direct sums, the middle morphism being $\coprod_{i\in I} u_{P_i}$ and the last morphism r is defined such that $rv_i = S(u_i)$ with $u_i : P_i \to \coprod_{i\in I} P_i$, and $v_i : S(P_i) \to \coprod_{i\in I} S(P_i)$ being structural morphisms. Consider a morphism $f : \coprod_{i\in I} P_i \to \coprod_{j\in J} P_j = P''$ where J is finite. Making use of the fact that a morphism from a direct sum to a direct product is determined by its coordinate morphisms, we obtain the commutative diagram

$$\begin{array}{ccc} \coprod_{i\in I} T(P_i) = T(\coprod_{i\in I} P_i) & \to & T(\coprod_{j\in J} P_j) \\ \downarrow u'_{P'} & & \downarrow u'_{P''} \\ S(\coprod_{i\in I} P_i) & \to & S(\coprod_{j\in J} P_j) = \prod_{j\in J} S(P_j) \end{array} \tag{9}$$

Now let X be a $\mathscr{P}$-finitely generated object. This means that we can find an exact sequence

$$P_1 \longrightarrow P_0 \xrightarrow{p} X \longrightarrow 0$$

with P_1 free and P_0 finitely free (that is, P_0 is a direct sum of the form $\coprod_{j\in J} P_j$ with J finite). Making use of (9) we get another commutative diagram

$$\begin{array}{ccccccc} T(P_1) & \longrightarrow & T(P_0) & \xrightarrow{T(p)} & T(X) & \longrightarrow & 0 \\ \downarrow u'_{P_1} & & \downarrow u'_{P_0} & & & & \\ S(P_1) & \longrightarrow & S(P_0) & \longrightarrow & S(X) & & \end{array} \tag{10}$$

with the top row being exact because T is right exact and the bottom row being a null-sequence according to the additive functorial property of S. Hence we can find a unique morphism $\bar{u}_X : T(X) \to S(X)$ which renders (10) commutative.

Let $f : X \to X'$ be a morphism with X and X' $\mathscr{P}$-finitely generated. Consider the diagram

$$\begin{array}{ccccc} P_0 & \longrightarrow & X & \longrightarrow & 0 \\ & & \downarrow f & & \\ P_0' & \longrightarrow & X' & \longrightarrow & 0 \end{array} \tag{11}$$

where the top row was used to define $\bar{u}_X$ and where the bottom row was used to define $\bar{u}_{X'}$. Since P_0 is projective, we can find a morphism $P_0 \to P_0'$

which renders the above diagram commutative. Now $\bar{u}$ as so far constructed, induces a morphism from the diagram

$$\begin{array}{ccc} T(P_0) & \xrightarrow{T(p)} & T(X) \\ \downarrow & & \downarrow{\scriptstyle T(f)} \\ T(P_0') & \longrightarrow & T(X') \end{array}$$

to the diagram

$$\begin{array}{ccc} S(P_0) & \longrightarrow & S(X) \\ \downarrow & & \downarrow \\ S(P_0') & \longrightarrow & S(X') \end{array}$$

Furthermore, in the resulting cube one sees that all the faces are commutative, save possibly the face

$$\begin{array}{ccc} T(X) & \xrightarrow{T(f)} & T(X') \\ {\scriptstyle \bar{u}_X}\downarrow & & \downarrow{\scriptstyle \bar{u}_{X'}} \\ S(X) & \xrightarrow{S(f)} & S(X') \end{array} \tag{12}$$

According to the above considerations, we have

$$\bar{u}_{X'}\, T(f)\, T(p) = S(f)\, \bar{u}_X\, T(p).$$

But since $T(p)$ is an epimorphism, the commutativity follows. Thus the above constructed $\bar{u}$ is natural.

Let the object X be not necessarily $\mathscr{P}$-finitely generated and let $\{X_i\}_i$ be the direct set of all $\mathscr{P}$-finitely generated subobjects of X. We know by Proposition 5.8 that X is the direct inductive limit of this system, and since T commutes with inductive limits, one has that $T(X)$ is the inductive limit of the direct system $\{T(X_i)\}$ which was canonically constructed. The family of morphisms

$$T(X_i) \xrightarrow{\bar{u}_{X_i}} S(X_i) \xrightarrow{S(j_i)} S(X)$$

with $j_i : X_i \to X$ being the canonical inclusions, defines a morphism

$$\bar{u}_X : T(X) \to S(X).$$

Given a morphism $f : X \to X'$ of $\mathscr{C}$, we must prove the commutativity of (12). For each $\mathscr{P}$-finitely generated subobject X_i of X, let $X_i' \to X'$ be the image of the composition

$$X_i \to X \xrightarrow{f} X'.$$

Then X_i' is a $\mathscr{P}$-finitely generated subobject of X'. Hence it results from what we have already shown, that the diagram

$$\begin{array}{ccc} T(X_i) & \longrightarrow & T(X_i') \\ {\scriptstyle \bar{u}_{X_i}}\downarrow & & \downarrow{\scriptstyle \bar{u}_{X'_i}} \\ S(X_i) & \longrightarrow & S(X_i') \end{array} \tag{13}$$

exists, and is commutative for each i. Consider the morphism from the diagram (13) to the diagram (12) which is induced by the morphisms $X_i \to X$ and $X_i' \to X'$. Again, we have a cube in which all faces are commutative save possibly (12). From this we see that for each i the composition

$$T(X_i) \to T(X) \to T(X') \to S(X')$$

is the same as the composition

$$T(X_i) \to T(X) \to S(X) \to S(X').$$

Since $T(X)$ is the inductive limit of the direct system $\{T(X_i)\}_i$, this proves the commutativity of (12) and establishes the naturality of $\bar{u}$.

Uniqueness of $\bar{u}$ is clear since each step in the above construction was forced by the requirement that $\bar{u}$ should be natural. ∎

COROLLARY 6.6. *In the conditions of Theorem* 6.5, *if S commutes with inductive limits and u is a functorial isomorphism, then $\bar{u}$ is a functorial isomorphism, too.* ∎

Exercises

1. Let $\mathscr{C}$ be a small abelian category with sufficiently many projectives (for example, the category of finitely generated abelian groups). Suppose that $T : \mathscr{C} \to \mathscr{Ab}$ is an additive functor which preserves epimorphisms and consider the functorial morphism:

$$u : \coprod (P, v) \to T$$

which is defined such that $uu_{(P,v)} = v$ with P running through all projectives in $\mathscr{C}$, and v running through all functorial morphisms from P into T and with $u_{(P,v)}$ being structural morphisms. Show that this functorial morphism is an epimorphism in the category Mod $\mathscr{C}$.

2. The centre of a ring A is defined as the subset $z(A)$ which consists of all elements x such that $xy = yx$ for any $y \in A$. Prove that if X is an object in a C_2-category $\mathscr{C}$, then the centre of $\mathrm{Hom}_{\mathscr{C}}(X, X)$ is a ring isomorphic to

the centre of $\mathrm{Hom}_{\mathscr{C}}(X^{(I)}, X^{(I)})$ for any set I. Hence, show that if U and V are two projective generators of $\mathscr{C}$, then the rings $\mathrm{Hom}_{\mathscr{C}}(U, U)$ and

$$\mathrm{Hom}_{\mathscr{C}}(V, V)$$

have isomorphic centres. (Hint: write $U \amalg M \simeq V^{(I)}$ and $V \amalg N \simeq U^{(I)}$ for some infinite set I and show from this that $U^{(I)} \simeq V^{(I)}$.) Therefore, if Mod A is equivalent to Mod B, then A and B have isomorphic centres. In particular, show that if A and B are commutative rings, then Mod A is equivalent to Mod B if and only if A is isomorphic to B.

3. Prove that if the functor S from Theorem 6.5 commutes with projective limits, then it suffices to assume that $\mathscr{C}$ is a C_2 category. If S commutes with direct sums and if $\mathscr{C}'$ is a C_2-category, then it suffices to assume that $\mathscr{C}$ is an Ab 3-category. (Hint: In either case, show that the diagram (9) is commutative even when J is not a finite set. Hence, there is no need to restrict ourselves in these cases to inductive direct limits.)

4. Let $T : \mathscr{C} \to \mathscr{C}'$ be an additive covariant functor with $\mathscr{C}$ and $\mathscr{C}'$ being small preadditive categories. Show that the restriction functor

$$T_* : \mathrm{Mod}\,\mathscr{C}' \to \mathrm{Mod}\,\mathscr{C}$$

has a right adjoint $*T$ and a left adjoint T^*. Moreover prove that

$$T^* h \simeq h' T,$$

where $h : \mathscr{C} \to \mathrm{Mod}\,\mathscr{C}$ and $h' : \mathscr{C}' \to \mathrm{Mod}\,\mathscr{C}'$ are canonical.

5. Let $\mathscr{C}$ be a small preadditive category and $\mathscr{C}'$ and Ab 3-category. We denote by $\mathrm{Ad}(\mathscr{C}, \mathscr{C}')$ the category defined as follows. Any object in $\mathrm{Ad}(\mathscr{C}, \mathscr{C}')$ is a pair (T, S) where $T : \mathrm{Mod}\,\mathscr{C} \to \mathscr{C}'$ is a covariant functor and S a right adjoint of T. A morphism from (T, S) to (T', S') is a pair (t, s) with

$$t : T \to T'$$

and $s : S' \to S$ being functorial morphisms such that $St \circ u = sT' \circ u'$ and $v \circ Ts = v' \circ tS'$, where $u : \mathrm{Id}\,\mathrm{Mod}\,\mathscr{C} \to ST$ is an arrow of adjunction from S to T and v is a quasi-inverse of u. Analogously, the same is true for u' and v' with respect to S' and T'. Finally, prove that the categories $(\mathscr{C}^0, \mathscr{C}')$ and $\mathrm{Ad}(\mathscr{C}, \mathscr{C}')$ are canonically equivalent.

References

Bourbaki [21]; Cartan–Eilenberg [33]; Freyd [57], [58]; Gabriel [59]; Gabriel–Ulmer [64]; Mitchell [133]; Pareigis [162].

NOTE 2. The presentation of 3.6 follows Mitchell [133], Chapter 4, Section 5.

3.7 Tensor products of modules

We shall now apply the results of the preceding paragraph to the category of modules. Particularly, we shall give a characterization of Grothendieck categories. Also, some results concerning projective objects are presented here.

Let A be a ring and $\mathscr{C}$ any preadditive category. A covariant additive functor from A into $\mathscr{C}$ consists of an object X of $\mathscr{C}$ and a morphism of rings from A into $\mathrm{Hom}_{\mathscr{C}}(X, X)$. In this case, X is called a *left A-object of $\mathscr{C}$*. Particularly, any left A-module is a left A-object of $\mathscr{Ab}$. The definition of the notion of right A-object is left to the reader.

Theorem 6.3 and Corollary 6.4 have the following reformulation for modules.

THEOREM 7.1. *Let A be a ring, $\mathscr{C}$ an* Ab 3-*category and X a right A-object of $\mathscr{C}$. Then, a unique functor denoted by $X \otimes_A ? : \mathrm{Mod}\, A \to \mathscr{C}$ exists, such that*

(1) $X \otimes_A A = X$;

(2) $X \otimes_A ?$ *has a right adjoint.*

Particularly, any covariant functor from Mod A *into $\mathscr{C}$ which commutes with inductive limits has a right adjoint.* ■

In the above theorem, the right adjoint of $X \otimes_A ?$ is the functor denoted here by $\mathrm{Hom}_{\mathscr{C}}(X, ?)$ and which is defined as follows. For any object Y of $\mathscr{C}$, $\mathrm{Hom}_{\mathscr{C}}(X, ?)(Y)$ is the abelian group $\mathrm{Hom}_{\mathscr{C}}(X, Y)$ when it is considered as an A-module. Thus, if $f \in \mathrm{Hom}_{\mathscr{C}}(X, Y)$ and if $a \in A$, then af is the morphism $ft(a)$ with $t : A \to (\mathrm{Hom}_{\mathscr{C}}(X, X))^0$ being the morphism of rings which defines X as a right A-object of $\mathscr{C}$.

The dual of this theorem is also valid and is given below.

THEOREM 7.2. *Let A be a ring, $\mathscr{C}$ an* Ab 3-*category and X a left A-object of $\mathscr{C}$. Then, a unique functor $? \otimes_A X : \mathrm{Mod}\, A^0 \to \mathscr{C}$ exists such that:*

(1) $A \otimes_A X = X$;

(2) $? \otimes_A X$ *has a right adjoint.*

Particularly, any covariant functor from Mod A^0 into $\mathscr{C}$ which commutes with inductive limits has a right adjoint. ■

Notice that the right adjoint of $? \otimes_A X$ is denoted by $\mathrm{Hom}_{\mathscr{C}}(X, ?)$ and is defined as above.

The above functors $X \otimes_A ?$ and $? \otimes_A X$ define the *tensor product* associated with the right (resp. left) A object X.

COROLLARY 7.3. *Let $\mathscr{C}$ be an* Ab 3-*category, X an object of $\mathscr{C}$ and*

$$A = \mathrm{Hom}_{\mathscr{C}}(X, X).$$

Then X is a left A-object and a unique functor $? \otimes_A X : \mathrm{Mod}\, A^0 \to \mathscr{C}$ exists such that:

(1) $A \otimes_A X = X$;

(2) *The right adjoint of $? \otimes_A X$ is the canonically defined functor*

$$\mathrm{Hom}_{\mathscr{C}}(X, ?). \blacksquare$$

NOTE 1. According to the proof of Theorem 6.2, it will suffice to assume in Corollary 7.3 that $\mathscr{C}$ is an abelian category and for any set I, there will exist the object $X^{(I)}$.

COROLLARY 7.4. *Any* Ab 3-*category with a small projective generator U is equivalent with the category* $\mathrm{Mod}\, A^0$ *for a suitable ring A.*

Proof. According to Corollaries 6.4 and 7.3, we have that $A = \mathrm{Hom}_{\mathscr{C}}(U, U)$. $\blacksquare$

Let A, B be two rings. Any left A-object X in $\mathrm{Mod}\, B$ is called an (A, B)-*bimodule*. An object of this kind is the B-module X and a ring morphism $t: A \to \mathrm{Hom}_B(X, X)$. Then, for any $a \in A$ and for any $b \in B$, $x \in X$ we have that $t(a)(bx) = bt(a)(x)$. If we take $t(a)x = ax$ with X being an A-module in the obvious way, then we get $a(bx) = b(ax)$.

Conversely, let X be an abelian group such that X is simultaneously an A-module and a B-module such that for any $a \in A$, any $b \in B$ and any $x \in X$ we have: $a(bx) = b(ax)$. Then X is an (A, B)-bimodule.

Any right A-object of $\mathrm{Mod}\, B$ is called an (A^0, B)-bimodule. We define an (A, B^0)-bimodule as a left A-object of $\mathrm{Mod}\, B^0$. Finally an (A^0, B^0)-bimodule is defined as a right A-object in $\mathrm{Mod}\, B^0$.

COROLLARY 7.5. (Watts, [238]). *Let A and B be two rings and let X be an arbitrary (A, B)-bimodule. Then, a unique functor $? \otimes_A X : \mathrm{Mod}\, A^0 \to \mathrm{Mod}\, B$ exists such that:*

(1) $A \otimes_A X = X$;

(2) $? \otimes_A X$ *has a right adjoint.*

Moreover, for any covariant functor $T : \mathrm{Mod}\, A^0 \to \mathrm{Mod}\, B$ which commutes with inductive limits the object $T(A)$ is canonically an (A, B)-bimodule and one has that $T \simeq\, ? \otimes_A T(A)$. $\blacksquare$

NOTE 2. Let $T : \mathrm{Mod}\, A \to \mathrm{Mod}\, B$ be a covariant additive functor. Then $T(A)$ is an (A, B)-bimodule. Thus, for any $a \in A$ we denote by $u_a : A \to A$

the morphism of the A-module associated with a. Obviously,

$$T(u_a): T(A) \to T(A)$$

is a morphism of B-modules, that is, for any $b \in B$ and any $x \in T(A)$ we have $T(u_a)(bx) = bT(u_a)(x)$. If we take $T(u_a)(x) = ax$, then $a(bx) = b(ax)$.

NOTE 3. If X is an arbitrary (A^0, B)-bimodule, (A, B^0)-bimodule or (A^0, B^0)-bimodule, then a result analogous to Corollary 7.5 is obtained. We leave to the reader the explication of these results.

NOTE 4. Let $t: A \to B$ be a morphism of rings. Then B may be considered in an obvious manner as an (A^0, B)-bimodule and as an (A, B^0)-bimodule. The functor $B \otimes_A ?: \text{Mod } A \to \text{Mod } B$ (respectively

$$? \otimes_A B: \text{Mod } A^0 \to \text{Mod } B^0)$$

has as right adjoint the restriction functor $t_*: \text{Mod } B \to \text{Mod } A$ (respectively $t_*^0: \text{Mod } B^0 \to \text{Mod } A^0$).

THEOREM 7.6. *Let A and B be two rings. Also. let $S: \text{Mod } B \to \text{Mod } A$ be a covariant functor. Then, the following assertions are equivalent*:

(1) *S is an equivalence of categories*;

(2) *There exists a finitely generated projective generator U of $\text{Mod } B$ and an isomorphism of rings $t: A^0 \to \text{Hom}_B(U, U)$ such that $S = \text{Hom}_B(U, ?)$.*

Proof. (1) ⇒ (2). Let T be a left adjoint of S and $U = T(A)$. U is an (A, B)-bimodule, and according to the hypothesis, one has that the assignment $u_a \rightsquigarrow T(u_a)$ is an isomorphism from A^0 to $\text{Hom}_B(U, U)$. Furthermore, U is a small projective generator in Mod B, since A is a small projective generator in Mod A.

The implication (2) ⇒ (1) is a special case of Corollary 7.4. ∎

COROLLARY 7.7. (Morita, [134]). *Let A and B be two rings and U an (A, B^0)-bimodule. The following assertions are equivalent*:

(1) *The functor $? \otimes_A U: \text{Mod } A^0 \to \text{Mod } B^0$ is an equivalence of categories*;

(2) *The functor $\text{Hom}_B(U, ?): \text{Mod } B^0 \to \text{Mod } A^0$ is an equivalence of categories*;

(3) *U is a projective finitely generated A-module and a projective finitely generated B^0-module and the canonically morphisms of rings*

$$t: A \to \text{Hom}_{B^0}(U, U) \quad \textit{and} \quad s: B^0 \to \text{Hom}_A(U, U)$$

are isomorphisms. ∎

Let A be a ring and $t: \mathbf{Z} \to A$ the unique morphism of rings. Then, the "underlying" functor $t_*: \operatorname{Mod} A \to \mathscr{Ab}$ is exact, and according to Theorem 2.8 we derive that the functor $\operatorname{Hom}_{\mathbf{Z}}(A, ?): \mathscr{Ab} \to \operatorname{Mod} A$—the right adjoint of t_*—preserves injectives.

Remark that for any A-module X one has

$$(\operatorname{Hom}_{\mathbf{Z}}(A, ?) \circ t_*)(X) = \operatorname{Hom}_{\mathbf{Z}}(A, t_*(X)).$$

Let us denote by $u_X: X \to \operatorname{Hom}_{\mathbf{Z}}(A, t_*(X))$ the map $u_X(x) = u_x$. It may be easily checked that the maps $\{u_X\}_X$ define an adjunction arrow from

$$\operatorname{Hom}_{\mathbf{Z}}(A, ?)$$

to t_*, and that u_X is a monomorphism for any X. Thus one gets the following theorem.

THEOREM 7.8. *For any ring A the category* Mod A *has sufficiently many injectives.*

Proof. Let X be an arbitrary A-module and let $i: t_*(X) \to Q$ be a monomorphism, Q being injective in Ab (see Theorem 3.3). Then $\operatorname{Hom}_{\mathbf{Z}}(A, Q)$ is an injective object of Mod A and $\operatorname{Hom}_{\mathbf{Z}}(A, i)\, u_X: X \to \operatorname{Hom}_{\mathbf{Z}}(A, Q)$ is a monomorphism. ■

Now, we shall prove a basic result in the theory of abelian categories.

THEOREM 7.9. (Gabriel–Popescu [60]). *Let $\mathscr{C}$ be a Grothendieck category, U an object of $\mathscr{C}$ and $A = \operatorname{Hom}_{\mathscr{C}}(U, U)$. Also, let us denote by T the functor $? \otimes_A U : \operatorname{Mod} A^0 \to \mathscr{C}$ and denote by S its right adjoint. The following assertions are equivalent:*

(1) *U is a generator of $\mathscr{C}$;*

(2) *S is full and faithful, and T is exact.*

Proof. (1) ⇒ (2). We shall prove first, that S is full and faithful, U being a generator. Thus, let us prove that S is full.

Indeed, let X, Y be two objects of $\mathscr{C}$ and let

$$S(X, Y): \operatorname{Hom}_{\mathscr{C}}(X, Y) \to \operatorname{Hom}_{A^0}(S(X), S(Y))$$

being the canonical map: $f \rightsquigarrow S(f)$, $f \in \operatorname{Hom}_{\mathscr{C}}(X, Y)$. We shall prove that $S(X, Y)$ is an epimorphism. In order to do this, take $g: S(X) \to S(Y)$ to be a morphism in Mod A^0. Also denote by I the set $\operatorname{Hom}_{\mathscr{C}}(U, X)$. Let $\{U_f\}_{f \in I}$ be a family of objects of $\mathscr{C}$, each U_f being isomorphic with U. Let us denote by $p: \coprod_{f \in I} U_f \to X$ the unique morphism defined such that $pv_f = f$ for all $f \in I$, v_f being structural morphisms of the direct sum. Also, let

$$q: \coprod_{f \in I} U_f \to Y$$

be the canonical morphism such that the diagram:

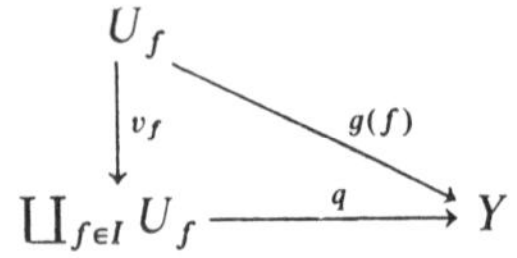

is commutative for any $f \in I$.

If we prove that $\ker q$ contains $\ker p$, then we have that $q = kp$ and therefore $g = S(k)$ (this is so because p is an epimorphism according to Proposition 8.2 in Chapter 2).

Let F be a finite subset of I, $v_F: \coprod_{f' \in F} U_{f'} \to \coprod_{f \in I} U_f$ the canonical inclusion and $K_F = \operatorname{im} v_F \cap \ker p$. Since $\mathscr{C}$ is an Ab 5-category, we derive that $\ker p = \sum_F K_F$ with F running through the family of finite subsets of I. In order to complete the proof it will be sufficient to show that K_F is contained in $\ker q$, for any F. Let $K_F' = v_F^{-1}(K_F)$ and let

$$h: K_F' \to \coprod_{f' \in F} U_{f'}$$

be the canonical inclusion. Also let $t: U \to K_F'$ be a morphism in $\mathscr{C}$, and $p_{f'}: \prod_{f' \in F} U_{f'} \to U_{f'}$, the canonical projections associated with structural morphisms $w_{f'}: U_{f'} \to \coprod_{f' \in F} U_{f'}$. Then one has that:

$$pv_F\, ht = pv_F(\textstyle\sum_{f' \in F} w_{f'}\, p_{f'}\, ht) = \sum_{f' \in F} f'\, r_{f'} = 0$$

with $r_{f'} = p_{f'}\, ht$ being an element of A, ($pv_F\, w_{f'} = pv_{f'} = f'$, according to the definition of p).

This implies that

$$\begin{aligned} qv_F\, ht = \textstyle\sum_{f' \in F}(qv_F\, w_{f'}\, p_{f'}\, ht) = \sum_{f' \in F} g(f')\, r_{f'} &= \textstyle\sum_{f' \in F} g(f'\, r_{f'}) \\ &= g(\textstyle\sum_{f' \in F} f'\, r_{f'}) = 0 \end{aligned}$$

for any $t \in \operatorname{Hom}_{\mathscr{C}}(U, K_F')$. Now we may conclude that $qv_F h = 0$, therefore K_F is contained in $\ker q$.

Thus, it remains to prove that T is exact.

In order to do this, we observe that $TS = \operatorname{Id}\mathscr{C}$ and an adjunction arrow from T to S is the identity of $\operatorname{Id}\mathscr{C}$. Then for any object Y of $\mathscr{C}$, we have that $TS(Y) = Y$. Also, for any morphism $f: S(Y) \to S(Y')$ in $\operatorname{Mod} A^0$ we have that $ST(f) = f$ and particularly, we get $STS(Y) = S(Y)$ for any object Y of $\mathscr{C}$.

We shall prove first that if X is an A^0-module which is contained in a free module F, and if $i: X \to F$ is the canonical inclusion, then $T(i)$ is a monomorphism. In order to do this, we assume that X is finitely generated, say $X = \langle x_1, \ldots, x_n \rangle$, and hence we may assume that F is also finitely generated, say $F = A^m$. Then we get the following commutative diagram

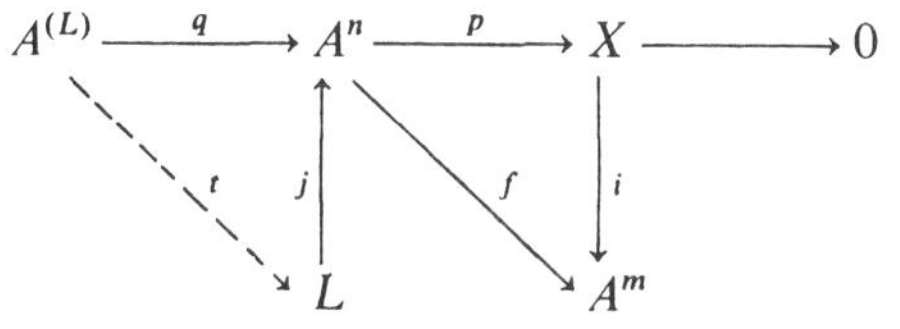

where i is the inclusion morphism, p is the natural morphism such that $p(e_i) = x_i$, $1 \leqslant i \leqslant n$, $\{e_i\}_{1 \leqslant i \leqslant n}$ is a basis of A^n, (L, j) is a kernel of p, t is the natural morphism such that $t(e_s) = s$ for any $s \in L$, $\{e_s\}_{s \in L}$ is a basis of $A^{(L)}$, $q = jt$ and $f = ip$.

If we operate with T on the above diagram, we obtain the following commutative diagram:

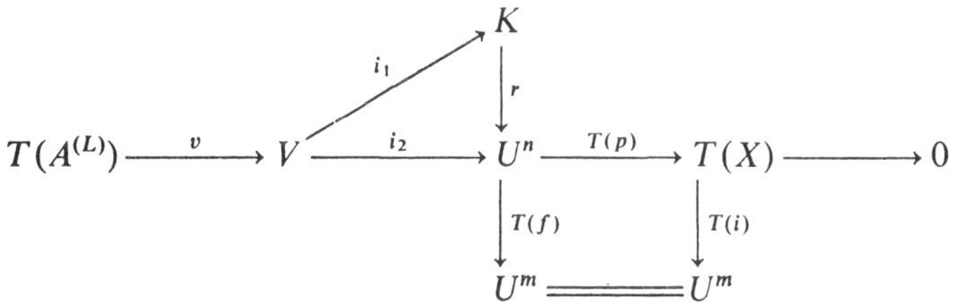

which is exact in $\mathscr{C}$, with $V = \operatorname{im} T(q)$, $i_2 v = T(q)$, and r being a kernel of $T(f)$. Obviously, there exists a morphism i_1 such that $r i_1 = i_2$. Acting now with S on the above diagram, we get

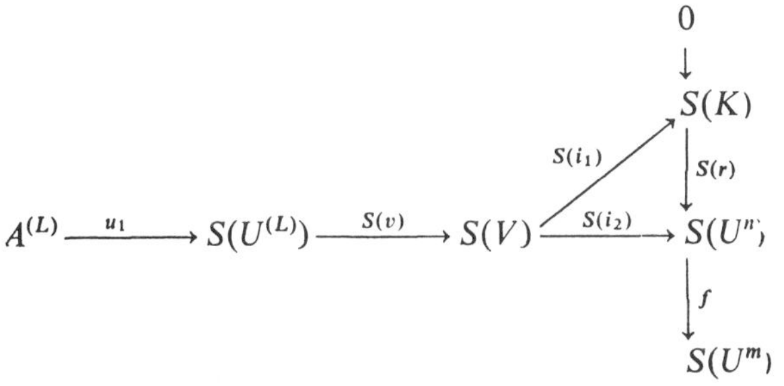

where the column is exact, $S(i_1)$ and $S(i_2)$ are monomorphisms,

$$S(i_2)\, S(v)\, u_1 = ST(q) u_1 = q$$

and where u_1 is the canonical morphism of adjunction. Since (L, j) is in fact $\ker f$, there exists a unique isomorphism $g : S(K) \to L$ such that $S(r) = jg$. Let $a \in S(K)$ and take $b = g(a)$. Then we find

$$\big(S(i_2)\,S(v)\,u_1\big)\,(e_b) = q(e_b) = j^{(b)} = S(r)\,(a) = \big(S(r)\,S(i_1)\,S(v)\,u_1\big)\,(e_b).$$

Now put $c = \big(S(v)\,u_1\big)\,(e_b)$, then we get $S(r)\,(a) = S(r)\,S(i_1)\,(c)$. Since $S(r)$ is monomorphism, we have that $a = S(i_1)\,(c)$. Hence $S(i_1)$ is an epimorphism (in fact an isomorphism). Since S is full and faithful, we have that i_1 is an isomorphism. Therefore, $\ker(T(i)\,T(p)) = \ker T(f) = \ker T(p)$, and since $T(p)$ is an epimorphism, we derive that $T(i)$ is a monomorphism.

Now, let X be a submodule of a free A^0-module F. Then X is the inductive limit of the family of finitely generated A-submodules $\{X_i\}_i$ and we get the commutative diagram

$$\begin{array}{ccc} X_i & \xrightarrow{t_i} & X \\ {\scriptstyle j_i}\downarrow & & \downarrow{\scriptstyle j} \\ F_i & \xrightarrow{r_i} & F \end{array}$$

with t_i and j inclusion morphisms, F_i being a free finitely generated module and r_i being a section.

$T(r_i j_i)$ is a monomorphism for any i and thus $T(j)$ is a monomorphism, because T commutes with inductive limits ($T(t_i)$ is a monomorphism for all i).

Finally, let $f : X \to Y$ be a monomorphism in $\operatorname{Mod} A^0$. Then we obtain the following commutative diagram

$$\begin{array}{ccccccccc} 0 & \longrightarrow & \ker p & \xrightarrow{i'} & F \prod_Y X & \xrightarrow{p'} & X & & \\ & & \| & & \downarrow{\scriptstyle f'} & & \downarrow{\scriptstyle f} & & \\ 0 & \longrightarrow & \ker p & \xrightarrow{i} & F & \xrightarrow{p} & Y & \longrightarrow & 0 \end{array}$$

which is canonically constructed, and where the bottom row is exact and F is free. Since f is a monomorphism and the last square is cartesian, f' is a monomorphism. Then $T(f')$ and $T(i)$ are monomorphisms. Hence $T(i')$ is a kernel of $T(p')$ and of $T(f)\,T(p') = T(p)\,T(f')$. Thus, one has that $T(f)$ is a monomorphism.

The implication (2) $\Rightarrow$ (1) is obvious. ∎

NOTE 6. A part of the above proof is due to Harada [86, Theorem 0]. Note that in the above proof we have not assumed that a Grothendieck category has sufficiently many injectives. Another proof of the above theorem will be presented in the next chapter. This new proof will be based upon a localization technique.

COROLLARY 7.10. *Any Grothendieck category is an* Ab 3*-*category.*

The proof results from Corollary 5.5 of Chapter 1. ∎

Later on, we shall prove that any Grothendieck category has sufficiently many injectives. We remark that this result was obtained directly in [79].

At the end of this paragraph, we shall make some observations concerning the construction of tensor products of modules.

Let A and B be two rings, X—an (A^0, B)-bimodule and Y—an A-module. Let us look more closely at the B-module $X \otimes_A Y$. According to Theorem 6.2 this module can be constructed as follows. First, notice that $X \otimes_A A = X$ and for any $a \in A$ one has that $X \otimes_A u_a : X \to X$ is the morphism

$$(X \otimes_A u_a)(x) = xa.$$

For any A-module Y consider the exact sequence:

$$A^{(Y)} \xrightarrow{p} Y \longrightarrow 0$$

with p being defined such that $p(e_y) = y$, for any $y \in Y$. From the construction of $X \otimes_A Y$ we have the following exact sequence in Mod B:

$$X^{(Y)} = (X \otimes_A A)^{(Y)} \xrightarrow{X \otimes p} X \otimes_A Y \longrightarrow 0.$$

Let $u_y : X \to X^{(Y)}$, $y \in Y$, be the structural morphisms. We write

$$X \otimes_A p = X \otimes p$$

and $((X \otimes p)\, u_y)(x) = x \otimes y$ for any $y \in Y$ and any $x \in X$. If $x, x' \in X$, then:

$$((X \otimes p)\, u_y)(x + x') = (x + x') \otimes y = x \otimes y + x' \otimes y.$$

Also, if $y, y' \in Y$, then $((X \otimes p)\, u_{y+y'})(x) = x \otimes (y + y')$.

But $p(e_{y+y'}) = y + y' = p(e_y) + p(e_{y'}) = p(e_y + e_{y'})$. A glance at the construction of $X \otimes_A Y$ shows that $u_{y+y'}(x) - u_y(x) - u_{y'}(x) \in \ker(X \otimes p)$, and thus

$$x \otimes (y + y') = x \otimes y + x \otimes y'.$$

Finally, we derive as above that for any $a \in A$, any $x \in X$ and any $y \in Y$, one has that $u_{ay}(x) - u_y(xa) \in \ker(X \otimes p)$, that is:

$$xa \otimes y = x \otimes ay.$$

Any A^0-module X may be considered as an $(A^0, \mathbf{Z})$-bimodule, and thus, we obtain the canonical functor:

$$X \otimes_A ? : \text{Mod}\, A \to \mathcal{Ab}.$$

PROPOSITION 7.11. *Let U be an (A^0, B)-bimodule, X—an A-module and Y—a B^0-module. Then, there exists a functorial isomorphism of abelian groups*:

$$\theta : (Y \otimes_B U) \otimes_A X \to Y \otimes_B (U \otimes_A X).$$

Proof. Consider the category $\mathscr{C} = \operatorname{Mod} B^0 \amalg \operatorname{Mod} A$. Obviously $\mathscr{C}$ is an Ab 5-category, and the object (B^0, A) is a projective generator of $\mathscr{C}$. Let $T, T' : \mathscr{C} \to \mathscr{Ab}$ be two functors:

$$T(Y, X) = (Y \otimes_B U) \otimes_A X$$

$$T'(Y, X) = Y \otimes_B (U \otimes_A X).$$

Then, we have:

$$T(B^0, A) = (B^0 \otimes_B U) \otimes_A A \simeq U \otimes_A A \simeq U$$

$$T'(B, A) = B^0 \otimes_B (U \otimes_A A) \simeq B^0 \otimes_B U \simeq U.$$

Also, let $\theta_U = 1_U$. Obviously, there is a functorial isomorphism from the restriction of T and T' to the full subcategory generated by (B^0, A) in $\mathscr{C}$. According to Corollary 6.6 we find that θ_U has a unique extension θ from T to T', and θ is an isomorphism. ∎

We remark that θ is defined as $\theta((y \otimes u) \otimes x) = y \otimes (u \otimes x)$, for any $x \in X$, $y \in Y$ and $u \in U$.

LEMMA 7.12. *Let $\mathscr{C}$ be an* Ab 3 *category with generators and sufficiently many injectives. Then, $\mathscr{C}$ has an injective cogenerator.*

Proof. Let U be a generator of $\mathscr{C}$, X— the direct sum of all quotient objects U/U' with U' running through the set of subobjects of U and V an injective object which contains X. We show that V is a cogenerator. According to the dual of Proposition 3.4 in Chapter 2, it will suffice to show that for each non-zero object Y of $\mathscr{C}$, there is a non-zero morphism $f : Y \to V$.

We already know that there is a non-zero morphism $U \to Y$ because U is a generator. The image of this morphism is isomorphic to U/U' for some subobjects $U' \neq U$. The structural morphism $U/U' \to X$ composed with the canonical inclusion $X \to V$ is not zero since each of these morphisms is a monomorphism. Then, from the injectivity of V, this composition can be extended to a morphism $Y \to V$ as required. ∎

COROLLARY 7.13. *The category* Mod A *has an injective cogenerator.* ∎

This corollary has the following important consequence which is due to Watts.

THEOREM 7.14. (Watts, [238]). *Let A be a ring and $T: \text{Mod}\, A \to \mathcal{Ab}$ a covariant functor which commutes with projective limits. Then, there exists an A-module M and a functorial isomorphism*:

$$u: \text{Hom}_A(M, ?) \xrightarrow{\sim} T.$$

Proof. Let C be an injective cogenerator of Mod A and let $V = C^{T(C)}$ (that is, V is a direct product of $T(C)$-copies, each of them being equal to C). Then, $T(V) = T(C^{T(C)}) = T(C)^{T(C)}$. Let $e \in T(C)^{T(C)}$ be the element defined as follows: for any $x \in T(C)$ the xth component of e is x. Let X be an A-module and

$$t_X: \text{Hom}_A(V, X) \to T(X)$$

the morphism of groups defined as follows: if $f: V \to X$ is a morphism, then $t_X(f) = T(f)(e)$. We shall prove that t_C is an epimorphism. Indeed, let $x \in T(C)$ and let $p_x: V \to C$ be the xth structural projection. Then $T(p_x)(e) = t_C(p_x) = x$ because T commutes with the direct product.

We have to compute the kernel of t_C. In order to do this let M be the intersection of all submodules Y of V such that $e \in T(Y)$. Then $T(M)$ is the intersection of the corresponding subgroups $T(Y)$ (because T commutes with projective limits). Obviously, one has that $e \in T(M)$. If

$$t_C(f) = T(f)(e) = 0,$$

we derive that $\ker f \supseteq M$ and conversely, if $\ker f \supseteq M$, then

$$e \in T(\ker f) = \ker T(f),$$

(that is $t_C(f) = 0$). Thus, we get $\ker t_C = \text{Hom}_A(V/M, C)$. At this stage we shall define a functorial morphism:

$$u: \text{Hom}_A(M, ?) \to T$$

as follows: if X is an A-module and if $f \in \text{Hom}_A(M, X)$, then $u(f) = T(f)(e)$.

Since C is injective, from the exact sequence:

$$0 \to M \to V \to V/M \to 0$$

we calculate another exact sequence:

$$0 \to \text{Hom}_A(V/M, C) \to \text{Hom}_A(V, C) \to \text{Hom}_A(M, C) \to 0$$

On the other hand, we also get the exact sequence:

$$0 \to \text{Hom}_A(V/M, C) \to \text{Hom}_A(V, C) \xrightarrow{t_C} T(C) \to 0$$

Consequently, t_C is an isomorphism of groups.

Now, let X be an A-module and consider the exact sequence:

$$0 \to X \xrightarrow{j} C^I \to C^J$$

From this sequence, we can derive the following commutative diagram:

$$\begin{array}{ccccccc} 0 \longrightarrow & \mathrm{Hom}_A(M, X) & \xrightarrow{\bar{j}} & (\mathrm{Hom}_A(M, C))^I & \longrightarrow & (\mathrm{Hom}_A(M, C))^J \\ & \downarrow u_X & & \downarrow u_{C^I} & & \downarrow u_{C^J} \\ 0 \longrightarrow & T(X) & \xrightarrow{T(j)} & T(C)^I & \longrightarrow & T(C)^J \end{array}$$

with $\bar{j}$ and $T(j)$ being induced by j. Obviously, u_{C^I} and u_{C^J} are isomorphisms and thus u_X is an isomorphism. ∎

Exercises

1. Let $t : A \to B$ be a ring morphism. Then show that the restriction functor $t_* : \mathrm{Mod}\, B \to \mathrm{Mod}\, A$ which is associated to t has a right adjoint. Particularly, prove that t_* is exact and faithful.

2. Let A be a ring, X an A^0-module, Y—an A-module and G—an abelian group. A map $f : X \Pi Y \to G$ is called *balanced* if for any x, $x' \in X$, any y, $y' \in Y$ and any $a \in A$ we have:

$$f(x + x', y) = f(x, y) + f(x', y)$$
$$f(x, y + y') = f(x, y) + f(x, y')$$
$$f(xa, y) = f(x, ay)$$

Prove that a balanced map $k : X \Pi Y \to X \otimes_A Y$ exists which fulfils the following condition: for any balanced map $f : X \Pi Y \to G$ a unique mormorphism $\bar{f} : X \otimes_A Y \to G$ of abelian groups exists such that $\bar{f}k = f$. Derive the following construction of $X \otimes_A Y$. Consider the abelian group $\mathbf{Z}^{(X \Pi Y)}$ and let $\{e_{(x,y)}\}_{(x,y) \in X \Pi Y}$ be a basis of this group. Denote by H the subgroup generated by the systems of elements:

$$e_{(x+x', y)} - e_{(x, y)} - e_{(x', y)}$$
$$e_{(x, y+y')} - e_{(x, y)} - e_{(x, y')}$$
$$e_{(xa, y)} - e_{(x, ay)}$$

for all x, $x' \in X$, y, $y' \in Y$ and $a \in A$.

Then show that $X \otimes_A Y$ is canonically isomorphic to the quotient group $\mathbf{Z}^{(X \Pi Y)}/H$.

3. Let $\mathscr{C}$ be a Grothendieck category, U a generator of $\mathscr{C}$ and

$$A = \mathrm{Hom}_{\mathscr{C}}(U, U).$$

Then for any subobject U' in U it is clear that $\mathrm{Hom}_{\mathscr{C}}(U, U')$ is identified to a right ideal in A, and we shall denote it by $\mathfrak{r}_{U'}$. Let K be a subset in A and let us denote by $K(U)$ the image of the morphism

$$p : U^{(K)} \longrightarrow U$$

for which $pu_k = k$, for any $k \in K$, u_k being the structural morphism. Prove the following assertions.

(a) For any subobject U' in U one has that $\mathfrak{r}_{U'}(U) = U'$;

(b) For any two right ideals $\mathfrak{r}_1$, $\mathfrak{r}_2$ in A, one has the following equalities:

$$(\mathfrak{r}_1 + \mathfrak{r}_2)(U) = \mathfrak{r}_1(U) + \mathfrak{r}_2(U)$$
$$(\mathfrak{r}_1 \cap \mathfrak{r}_2)(U) = \mathfrak{r}_1(U) \cap \mathfrak{r}_2(U).$$

4. (Harada, [86, Proposition 3]). Let $\mathscr{C}$ be a Grothendieck category, U—a generator of $\mathscr{C}$ and A, T, S as in Theorem 7.9. Then prove that the following statements are equivalent:

(a) T is an equivalence of categories;

(b) With the notations of the previous exercise, the assignment $U' \rightsquigarrow \mathfrak{r}_{U'}$ is in fact a bijection between right ideals of A and subobjects of U;

(c) For any maximal right ideal $\mathfrak{r}$ in A, one has that $T(A/\mathfrak{r}) \neq 0$;

(d) U is projective and small in $\mathscr{C}$.

5. Let $\mathscr{C}$ be a Grothendieck category and U a projective generator of $\mathscr{C}$. Here we shall make use of the notations of Theorem 7.9. If X' is a finitely generated submodule of the module $S(X)$, prove that there exists a subobject X'' of X such that $S(X'') = X'$.

6. Let A be a ring and $t : \mathbf{Z} \to A$ the unique ring morphism. Then show that the functors t_* and $A^0 \otimes_A ?$ are isomorphic.

7. Let A be a ring, X—an A^0-module and Y—an A-module. Next, consider the functors:

$$X \otimes_A ? : \mathrm{Mod}\, A \to \mathscr{Ab}$$
$$? \otimes_A Y : \mathrm{Mod}\, A^0 \to \mathscr{Ab}.$$

Then show that the abelian groups $(X \otimes_A ?)(Y) = X \otimes_A Y$ and

$$(? \otimes_A Y)(X) = X \otimes_A Y$$

are canonically isomorphic. This is our manner of identifying these groups.

Let $f : X' \to X$ and $g : Y' \to Y$ be two morphisms in Mod A^0 and respectively in Mod A. Then prove the existence of the following commutative diagram:

$$\begin{array}{ccc} X' \otimes_A Y' & \xrightarrow{X' \otimes g} & X' \otimes_A Y \\ {\scriptstyle f \otimes Y'} \downarrow & & \downarrow {\scriptstyle f \otimes Y} \\ X \otimes_A Y' & \xrightarrow{X \otimes g} & X \otimes_A Y \end{array}$$

which is canonically constructed.

8. Let A be a ring, X—an A^0-module and Y—an A-module. Assume that $\{x_i\}_{i \in I}$ and $\{y_j\}_{j \in J}$ are systems of generators of X and Y, respectively. Then show that the elements $\{x_i \otimes y_j\}_{(i,j) \in I \otimes J}$ form a system of generators of $X \otimes_A Y$.

9. Let $\mathscr{C}$ be a preadditive category. We define the *centre* of $\mathscr{C}$ as the ring of all functorial morphisms $u : \mathrm{Id}\mathscr{C} \to \mathrm{Id}\mathscr{C}$. Prove that the centre of $\mathscr{C}$ is a commutative ring. If $\mathscr{C}$ is a Grothendieck category and U a generator of $\mathscr{C}$, then the centre of $\mathscr{C}$ and the centre of the ring $\mathrm{Hom}_{\mathscr{C}}(U, U)$ are isomorphic. (Hint: If one makes use of the Theorem 7.9 then this exercise may be reduced to the case of Mod A^0 for a suitable ring A).

References

Bourbaki [21]; Cartan–Eilenberg [33]; Freyd [57], [58]; Gabriel [59]; Gabriel–Popescu [60]; Harada [86]; Mitchell [133]; Morita [134]; Năstăsescu–Nitá [140]; Northcott [152]; Pareigis [162]; Tachikawa [218]; Watts [238].

3.8 Flat modules

Let A, B be two rings and U an (A^0, B)-bimodule. We shall say that U is *flat* if the functor $U \otimes_A ? : \mathrm{Mod}\, A \to \mathrm{Mod}\, B$ is exact. If $S : \mathrm{Mod}\, B \to \mathscr{Ab}$ is the underlying functor then U is flat if and only if $S(U \otimes_A ?)$ is exact. Now, if we consider U to be an arbitrary A^0-module, then the functor

$$U \otimes_A ? : \mathrm{Mod}\, A \to \mathscr{Ab}$$

is canonically isomorphic to the functor $S(U \otimes_A ?)$. Therefore, in order to study flat modules, it will be sufficient to consider a ring A and an A^0-module U.

PROPOSITION 8.1. *Let A be a ring. Then the following assertions are true:*

(1) *If $\{U_i\}_{i\in I}$ is a set of A^0-modules then $\coprod_{i\in I} U_i$ is flat if and only if U_i is flat for all $i \in I$.*

(2) *If $\{U_i, f_{ij}\}_{i\in I}$ is a direct system of flat A^0-modules then the module $\varinjlim_i U_i$ is also flat.*

(3) *Any projective A^0-module is flat.*

Proof. Let $f : X' \to X$ be a monomorphism in the category Mod A. Then $U_i \otimes f$ is a monomorphism for all $i \in I$ and the morphism $\coprod_{i\in I} (U_i \otimes f)$ is canonically isomorphic to the morphism $(\coprod_{i\in I} U_i) \otimes f$. Consequently, the last morphism is a monomorphism.

Conversely, let us suppose that $\coprod_{i\in I} U_i$ is flat and that u_i are the structural morphisms. Then, for any monomorphism $f : X' \to X$ of Mod A, we have the following commutative diagram:

$$\begin{array}{ccc} (\coprod_{i\in I} U_i) \otimes_A X' & \xrightarrow{(\coprod_{i\in I} U_i)\otimes f} & (\coprod_{i\in I} U_i) \otimes_A X \\ {\scriptstyle u_i\otimes X'}\uparrow & & \uparrow{\scriptstyle u_i\otimes X} \\ U_i \otimes_A X' & \xrightarrow{U_i\otimes f} & U_i \otimes_A X \end{array}$$

with the vertical morphisms being sections and $(\coprod_{i\in I} U_i) \otimes f$ being a monomorphism. This implies that $U_i \otimes f$ is a monomorphism.

The second assertion is a consequence of the condition Ab 5 which is naturally valid in $\mathscr{Ab}$.

The third assertion is a consequence of Exercise 6 of 3.7 and of assertion (1). ■

The next lemma is a basic result in the theory of flat modules.

LEMMA 8.2. (Bourbaki, [26, Chapter 1]). *Let X be an A^0-module, Y an A-module, $\{y_i\}_{i\in I}$ a set of generators of Y and $\{x_i\}_{i\in I}$ a set of finite support of X (that is, $x_i = 0$ for all but a finite number of i). The following assertion are equivalent:*

(1) $\sum_i x_i \otimes y_i = 0$;

(2) *There exists a finite set J, a set $\{t_j\}_{j\in J}$ of elements of X and a set $\{a_{ji}\}_{(i,j)\in I\times J}$ of elements of A, such that:*

(a) *The elements a_{ji} are 0 for all but a finite number of (i, j);*

(b) $\sum_{i\in I} a_{ji} y_i = 0$ *for all $j \in J$;*

(c) $x_i = \sum_{j\in J} t_j a_{ji}$ *for all $i \in I$.*

Proof. (1) $\Rightarrow$ (2). Let us consider the exact sequence of modules:

$$0 \longrightarrow R \xrightarrow{v} A^{(I)} \xrightarrow{p} Y \longrightarrow 0$$

where p is defined such that $p(e_i) = y_i$ for all $i \in I$. Here $\{e_i\}_i$ is the canonical basis of $A^{(I)}$ and v is the kernel of p. Moreover, we have the exact sequence:

$$X \otimes_A R \xrightarrow{X \otimes v} X \otimes_A A^{(I)} \xrightarrow{X \otimes p} X \otimes_A Y \longrightarrow 0$$

Let t be the isomorphism from $X^{(I)}$ into $X \otimes_A A^{(I)}$ defined such that $t\big((x_i)_i\big) = \sum_{i \in I} x_i \otimes e_i$. Then we must have:

$$0 = \textstyle\sum_{i \in I} x_i \otimes y_i = \sum_{i \in I} X \otimes p\,(x_i \otimes e_i) = (X \otimes p)\,(\sum_{i \in I} x_i \otimes e_i).$$

In other words $\sum_i x_i \otimes e_i \in \ker(X \otimes p) = \operatorname{im}(X \otimes v)$. But the right exactness of the tensor product implies that we can find a finite set $\{t_j\}_{j \in J}$ of X, and a set of elements $\{r_j\}_{j \in J}$ of R, such that:

$$(X \otimes v)\,(\textstyle\sum_{j \in J} t_j \otimes r_j) = \sum_{j \in J} t_j \otimes v\,(r_j) = \sum_{i \in I} x_i \otimes e_i.$$

On the other hand, for any j we may write

$$v(r_j) = \textstyle\sum_i a_{ji}\, e_i.$$

Obviously, for any j we have:

$$pv(r_j) = p(\textstyle\sum_{i \in I} a_{ji}\, e_i) = \sum_{i \in I} a_{ji}\, p(e_i) = \sum_{i \in I} a_{ji}\, y_i = 0.$$

Furthermore, the relation becomes:

$$\textstyle\sum_{j \in J} t_j \otimes \sum a_{ji}\, e_i = \sum_{i \in I} x_i \otimes e_i$$

or equivalently:

$$\textstyle\sum_{i \in I} (\sum_{j \in J} t_j \otimes a_{ji}) \otimes e_i = \sum_{i \in I} x_i \otimes e_i.$$

A glance of the definition of the isomorphism t suggests that

$$x_i = \textstyle\sum_{j \in J} t_j\, a_{ji}$$

and this holds for all i.

A few simple computations enable one to prove that (2) $\Rightarrow$ (1). ■

THEOREM 8.3. *Let X be an A^0-module. The following assertions are equivalent:*

(1) *X is flat;*

(2) *For any left-ideal $\mathfrak{a}$ of A the morphism $X \otimes j$ is a monomorphism, if $j : \mathfrak{a} \to A$ is the canonical inclusion;*

(3) *Assuming that $\mathfrak{a}$ is finitely generated,* (2) *still holds;*

(4) *If* $x_1, \ldots, x_n$ *are elements of* X *and if* $a_1, \ldots, a_n$ *are elements of* A *such that* $\sum_{i=1}^n x_i a_i = 0$, *then there exist elements* $t_1, \ldots, t_m$ *in* X *and elements*

$$\{a_{ji}\}_{\substack{1 \leqslant j \leqslant m \\ 1 \leqslant i \leqslant n}}$$

in A *which are subject to the following conditions*:

(a) $\sum_{i=1}^{n} a_{ji} a_i = 0$ *for all* $1 \leqslant j \leqslant m$;

(b) $x_i = \sum_{j=1}^{m} t_j a_{ji}$ *for all* $1 \leqslant i \leqslant n$.

Proof. According to Proposition 5.8 we have that (2) $\Leftrightarrow$ (3).

(3) $\Rightarrow$ (4): Let $x_1, \ldots, x_n$ be elements of X and $a_1, \ldots, a_n$ be some elements of A such that

$$\sum_{i=1}^{n} x_i a_i = 0.$$

Also let $\mathfrak{a}$ be the ideal generated by $a_1, \ldots, a_n$ and $j: \mathfrak{a} \to A$ be the natural inclusion. Then $X \otimes j: X \otimes_A \mathfrak{a} \to X \otimes_A A = X$ is a monomorphism. Still we have

$$(X \otimes j)\left(\sum_{i=1}^{n} x_i \otimes a_i\right) = \sum_{i=1}^{n} x_i a_i = 0$$

and thus

$$\sum_{i=1}^{n} x_i \otimes a_i = 0.$$

Now the result is obtained by making use of Lemma 8.2.

(4) $\Rightarrow$ (3). Let $\mathfrak{a}$ be a left ideal which is finitely generated by the elements $a_1, \ldots, a_n$. Also, let $j: \mathfrak{a} \to A$ be the natural inclusion and $y \in X \otimes_A \mathfrak{a}$ an element such that $(X \otimes j)(y) = 0$. We may write

$$y = \sum_{i=1}^{n} x_i \otimes a_i,$$

for some $x_i \in X$ and thus

$$(X \otimes j)(y) = \sum_{i=1}^{n} x_i \otimes j(a_i) = 0.$$

According to (4) we have that $y = 0$ that is, $X \otimes j$ is a monomorphism.

The implication (1) $\Rightarrow$ (2) is obvious.

Now, we have only to prove that (2) ⇒ (1). We shall make use of Theorem 2.8 taking $\mathcal{A}b$ as a category with sufficiently many injectives. Let Q be an injective object in $\mathcal{A}b$, and let $j: \mathfrak{a} \to A$ a monomorphism. Consider the commutative diagram:

$$\begin{array}{ccc} \mathrm{Hom}_A(A, \mathrm{Hom}_{\mathbf{Z}}(X, Q)) & \xrightarrow{u} & \mathrm{Hom}_A(\mathfrak{a}, \mathrm{Hom}_{\mathbf{Z}}(X, Q)) \\ \big\downarrow\!\wr & & \big\downarrow\!\wr \\ \mathrm{Hom}_{\mathbf{Z}}(X \otimes_A A, Q) & \xrightarrow{v} & \mathrm{Hom}_{\mathbf{Z}}(X \otimes_A \mathfrak{a}, Q) \end{array}$$

where the vertical morphisms are functorial isomorphisms and where the horizontal morphisms are induced by j and $X \otimes j$. According to (2) we have that v is an epimorphism, since Q is injective. Then, u is also an epimorphism, that is, $\mathrm{Hom}_{\mathbf{Z}}(X, Q)$ is injective by Lemma 3.1. Consequently, $X \otimes_A ?$ is exact, according to Theorem 2.8. ∎

COROLLARY 8.4. *An abelian group is flat if and only if it has no non-zero torsion elements.*

Proof. Let X be a (non-null) flat abelian group, and $x \in X$ such that $nx = 0$ for some integer $n \neq 0$. According to Theorem 8.3, there exist certain elements $x_1, \ldots x_m$ in X and $n_1, \ldots, n_m$ in $\mathbf{Z}$ such that $n_j n = 0$ for all $1 \leqslant j \leqslant m$ and

$$x = \sum_{j=1}^{m} n_j x_j.$$

Obviously we also have that $n_j = 0$ for any j and thus $x = 0$.

Conversely, let X be an abelian group without torsion, $j: n\mathbf{Z} \to \mathbf{Z}$ is a monomorphism and $x \in X$ an element such that $(X \otimes j)(x \otimes m) = mx = 0$ for an element $m \in n\mathbf{Z}$. Then, according to the hypothesis we necessarily have that $x = 0$ or that $m = 0$. ∎

Now we shall characterize the rings A such that any A^0-module is flat. In order to do that we shall make use of the following result.

LEMMA 8.5. *Let a be an element of the ring A. The following assertions are equivalent:*

(1) $a \in aAa$;

(2) *The right ideal aA is a direct summand of A*;

(3) *A/aA is a flat A^0-module*;

(4) *For each left-ideal $\mathfrak{b}$ of A we have:*

$$(aA) \cap \mathfrak{b} = a\mathfrak{b}$$

(i.e. the set of all products ax with $x \in \mathfrak{b}$).

Proof. (1) ⇒ (2). Let x be an element such that $a = axa$. Then, $ax = (ax)^2$ that is, $ax = e$ is idempotent. Obviously, $Aa = Ae$, and thus Aa is a direct summand in A according to Section 4, Exercise 1.6.

(2) ⇒ (3). Is obviously obtained on the basis of Proposition 8.1.

(3) ⇒ (4). This implication results from the equality $A\mathfrak{b} = \mathfrak{b}$ and from Exercise 4.

(4) ⇒ (1). Particularly, $(aA) \cap (Aa) = aAa$. ■

THEOREM 8.6. *For a ring A the following assertions are equivalent:*

(1) *Any element of A satisfies the conditions of the above lemma;*

(2) *Any finitely generated right ideal of A is a direct summand;*

(3) *Any finitely generated left ideal of A is a direct summand;*

(4) *Any A-module is flat;*

(5) *Any A^0-module is flat.*

Proof. (1) ⇒ (2). It suffices to consider a right ideal $aA + bA$ which is generated by two elements. Now, $aA = eA$, where $e^2 = e$ and $bA \subseteq ebA + (1-e)bA$. Therefore, $aA + bA = eA + (1-e)bA = eA + fA$ with $f^2 = f$, $fA = (1-e)bA$ and $ef = 0$. Let us take $g = f(1-e)$. Then we have $gf = f(1-e)f = f(f - ef) = f^2 = f$,

$$g^2 = gf(1-e) = f(1-e) = g.$$

Finally, we get $eg = 0 = ge$, that is, $(e + g)^2 = e + g$.

Now $g \in fA$ and $f \in gA$, hence $fA = gA$. Therefore, $aA + bA = eA + gA$. We claim that this is the same as $(e + g)A$. Indeed, $ex + gx' = (e + g)(ex + gx')$, hence $eA + gA \subseteq (e + g)A$ and the converse assertion is obvious. Thus, $aA + bA = (e + g)A$ as required.

In the same way, we may derive that any finitely generated left ideal of A is a direct summand, that is (1) ⇒ (3) and also we have obviously, (2) ⇒ (1) and (3) ⇒ (1).

Implications (2) ⇒ (5) and (3) ⇒ (4) are direct consequences of the Theorem 8.3.

(5) ⇒ (1) according to Lemma 8.5. ■

A ring A which is subject to the conditions of the above theorem is called a *von Neumann regular ring*, or simply, *a regular ring*.

Obviously, any division ring is a regular ring (see 3.4, Exercise 9).

COROLLARY 8.7. *Let X be an object of $\mathscr{C}$ such that for any endomorphism u of X,* im u *is a direct summand of X. Then* $\mathrm{Hom}_{\mathscr{C}}(X, X)$ *is a regular ring.*

Proof. Let $p: X \to \operatorname{im} u$ be the canonical epimorphism and $t: \operatorname{im} u \to X$ the natural inclusion such that $u = tp$. Also let s be a section of p and v a retraction of t. Then $u = usvu$. ∎

Exercises

1. (Lazard, [115]). Let X be an A^0-module. The following assertions are equivalent:

(a) X is flat;

(b) If, I, K are finite sets, $x_i \in X$, $a_{ik} \in A$ for all $i \in I$, $k \in K$ and

$$\textstyle\sum_{i \in I} x_i a_{ik} = 0$$

for all $k \in K$, then, there exists a finite set J and some elements $t_j \in X$, $s_{ji} \in A$ for all $j \in J$, $i \in I$, such that $\sum_{i \in I} s_{ji} a_{ik} = 0$ and $x_i = \sum_{j \in J} t_j s_{ji}$ for all $j \in J$, $k \in K$, $i \in I$.

Also the condition (a) is equivalent to the following condition:

(b′) If $\{x_i\}_{i \in I}$ is a set of generators of X, K is finite set, $a_{ik} \in A$ for all $i \in I$, $k \in K$ and if $\sum_{i \in I} x_i a_{ik} = 0$ for $k \in K$, then there exist elements $s_{ji} \in A$ for $i, j \in I$ such that $x_i = \sum_{j \in I} x_j s_{ji}$ and $\sum_{i \in I} s_{ji} a_{ik} = 0$ for all $i, j \in I$, $k \in K$.

2. Let X be an A^0-module and $\mathfrak{a}$ a left ideal of A. We shall denote by $X\mathfrak{a}$ the subgroup of X generated by the elements xa for all $x \in X$ and $a \in \mathfrak{a}$. If X is flat, then for any two left ideals $\mathfrak{a}$, $\mathfrak{a}'$ of A we have: $X(\mathfrak{a} \cap \mathfrak{a}') = X\mathfrak{a} \cap X\mathfrak{a}'$.

3. Let X be an A^0-module. The following assertions are equivalent:

(a) X is flat;

(b) For every exact sequence in Mod A^0:

$$0 \to G \xrightarrow{v} H \to X \to 0$$

and for any A-module Y, we have that the morphism $v \otimes Y$ is a monomorphism;

(c) There exists the above exact sequence with H flat such that for any left ideal $\mathfrak{a}$ of A the morphism $v \otimes (A/\mathfrak{a})$ is a monomorphism;

(d) The same conditions as in (c) together with the assumption that $\mathfrak{a}$ is finitely generated.

4. Let X be an A^0-module and X' a submodule of X. Assume that X/X' is flat. Then for any left ideal $\mathfrak{a}$ of A we have that $X'\mathfrak{a} = X' \cap (X\mathfrak{a})$.

Conversely, suppose that X is flat and for any finitely generated left ideal $\mathfrak{a}$ of A we have that $X'\mathfrak{a} = X' \cap (X\mathfrak{a})$. Then X/X' is flat.

5. Let A be a regular ring. Then, the ring A_n of $n \times n$ matrices over A is also regular.

6. Let $0 \to R \to L \to E \to 0$ an exact sequence of A^0-modules where L is free and $\{e_i\}_i$ a basis of L. The following assertions are equivalent:

(a) E is flat;

(b) If for any $x \in R$, denote by $\mathfrak{a}_x$ the left ideal of A generated by the components of x for the basis $\{e_i\}_i$, then $x \in \mathfrak{a}_x R$.

(c) For any $x \in R$ there exists a morphism $u_x : L \to R$ such that $u_x(x) = x$.

(d) For any finite sequence $\{x_i\}_{1 \leqslant i \leqslant n}$ of elements of R there exists a morphism $u : L \to R$ such that $u(x_i) = x_i$ for any $1 \leqslant i \leqslant n$.

7. (Lazard [117]). Let M be an A^0-module. The following assertions are equivalent:

(a) M is flat;

(b) For any finitely presented A^0-module P and any morphism $u : P \to M$, there exists a finitely generated free A^0-module L and the morphisms $v : P \to L$, $w : L \to M$ such that $wv = u$.

(c) There exist a directed set I and a direct system $\{L_i\}_{i \in I}$ of free modules such that

$$M = \varinjlim_i L_i.$$

8. Let $\mathscr{C}$ be a Grothendieck category. An exact sequence

$$0 \longrightarrow X' \xrightarrow{\ i\ } X \xrightarrow{\ f\ } X'' \longrightarrow 0$$

of $\mathscr{C}$ is called *pure* if for any finitely presented object P, the morphism $h^P(f) : \mathrm{Hom}_{\mathscr{C}}(P, X) \to \mathrm{Hom}_{\mathscr{C}}(P, X'')$ is an epimorphism. If $\mathscr{C} = \mathrm{Mod}\, A$, where A is a ring, then a sequence is pure if and only if for any A^0-module Y, the morphism $Y \otimes i$ is a monomorphism.

9. An object M of a Grothendieck category $\mathscr{C}$ is *flat* if any exact sequence $0 \to K \to L \to M \to 0$ is pure. Prove that every projective object is flat and every inductive direct limit of flat objects is also flat. Moreover every finitely presented flat object is projective. Finally, prove that the following assertions are equivalent:

(a) M is flat;

(b) There exists a pure exact sequence $0 \to K \to P \to M \to 0$ with projective P;

(c) Every morphism $F \to M$, where F is finitely presented, may be factored through a projective object.

10. (Stenström [211]). Let $\mathscr{C}$ be a Grothendieck category with a family $\mathscr{F}$ of finitely presented generators. The following statements are equivalent:

(a) All objects are flat;

(b) All short exact sequences are pure;

(c) All finitely presented objects are projective;

(d) Every $\mathscr{F}$-finitely generated subobject of any finitely presented object Y is a direct summand;

(d′) Every object of $\mathscr{F}$ is projective and (d) is true for any $Y \in \mathscr{F}$;

(e) For any finitely presented object F, the ring $\mathrm{Hom}_{\mathscr{C}}(F, F)$ is regular;

(f) If F and G are finitely presented, then for every morphism $f : F \to G$ there exists a morphism $g : G \to F$ such that $f = fgf$;

(f′) If F and G are in $\mathscr{F}$, then they are projective and the condition (f) is true for them.

11. (Jensen, [98]). Let A be a domain and M an A^0-module such that for any $x \in M$, the morphism $u_x : A \to M$ associated with x is a monomorphism. Then M is flat if and only if $M(\mathfrak{a} \cap \mathfrak{b}) = M\mathfrak{a} \cap M\mathfrak{b}$ for all left ideals $\mathfrak{a}$ and $\mathfrak{b}$ in A.

References

Bănică–Popescu [13]; Bourbaki [21], [26]; Cartan–Eilenberg [33]; Endo [51]; Govorov [78]; Jensen [98]; Lambek [113]; Lazard [115]; [117], Northcott [152]; Stenström [209], [210], [211].

3.9 Some remarks on projective modules

LEMMA 9.1. *Suppose that X is an A^0-module. Then the following two conditions are equivalent:*

(a) *X is projective;*

(b) *If $\{x_i\}_{i \in I}$ is a set of generators of X, then there exist some elements $s_{ij} \in A$, for $i, j \in I$ such that*

$$x_i = \textstyle\sum_{j \in I} x_j s_{ji} \tag{14}$$

and if for a set K, we have

$$\textstyle\sum_{i \in I} x_i a_{ik} = 0 \tag{15}$$

for all $k \in K$ then we get

$$\textstyle\sum_{i \in I} s_{ji} a_{ik} = 0 \tag{16}$$

for all $j \in I$, $k \in K$.

Proof. Assume that X is a projective module generated by the elements x_i, $i \in I$. Then there exists a free module $X \amalg X'$ with a set $\{e_i = x_i + x_i'\}_{i \in I}$ of free generators, for some $x_i' \in X'$. Since $x_i = \sum_{j \in J} e_j s_{ji}$ for some $s_{ji} \in A$, then (14) holds.

If the equalities (15) are satisfied, then we have the following relation for any $k \in K$:

$$\textstyle\sum_{j \in I} e_j (\sum_{i \in I} s_{ji} a_{ik}) = \sum_{i \in I} x_i a_{ik} = 0.$$

Now, since e_j, $j \in I$ are free generators, relation (16) holds.

Conversely, assume that the condition (b) is satisfied and let F be a free module with basis $\{e_i\}_{i \in I}$. Then, there exists an epimorphism $f : F \to X$ such that $f(e_i) = x_i$. Let us define a homomorphism

$$g : F \to F \quad \text{by} \quad g(e_i) = \textstyle\sum_{j \in I} e_i s_{ji}.$$

If ker f is generated by the elements:

$$n_k = \textstyle\sum_{i \in I} e_i a_{ik} \quad \text{for} \quad k \in K,$$

then according to (b) we have that $g(n_k) = 0$. Thus, the morphism g induces a morphism $\bar{g} : X \to F$ such that $\bar{g}(x_i) = g(e_i) = \sum_{j \in I} e_j s_{ji}$. We derive from relation (14) that $f\bar{g} = 1_X$ which shows that X is a projective module. ∎

COROLLARY 9.2. *If an A^0-module X is flat and finitely presented, then X is projective.*

Proof. We shall make use of Exercise 1 from Section 3.8 and the above lemma. ∎

LEMMA 9.3. *Let X be an A^0-projective module which is generated by the elements x_i, $i \in I$ and let the elements s_{ji}, $i, j \in I$ be subject to condition* (b) *of Lemma* 9.1. *If I' is a subset of I, such that $s_{ji} = 0$ for $i \in I'$, $j \in I'$, then the submodule X' which is generated by the elements $x_{i'}$, $i' \in I'$ is a direct summand of X.*

Proof. It will be sufficient to show that the module $X'' = X/X'$ is projective. This module is generated by the elements $x_i'' = p(x_i)$ where $p : X \to X''$ is the canonical epimorphism. According to our assumption, we have:

$$x_i'' = \textstyle\sum_{j \in I} x_j'' s_{ji} = \sum_{j \notin I'} x_j'' s_{ji}$$

because $x_j'' = 0$ for all $j \in I'$. Let us assume that the equalities

$$\textstyle\sum_{i \in I'} x_i'' a_{ik} = 0$$

hold for any $k \in K$. Then there exist some elements $a_{ik} \in A$ for $i \in I$, $k \in K$, such that:

$$\sum_{i \in I} x_i a_{ik} = 0.$$

Consequently, $\sum_{i \in I} s_{ji} a_{ik} = 0$, for all $j \in I$, $k \in K$. It follows that for $j \in I'$, $k \in K$,

$$\sum_{i \in I'} s_{ji} a_{ik} = \sum_{i \in I} s_{ji} a_{ik} = 0$$

and X'' is projective according to Lemma 9.1. ∎

THEOREM 9.4. (Kaplansky [102]). *Any projective A^0-module is a direct sum of countably generated projective modules.*

Proof. Let X be a projective A^0-module which is generated by the elements x_i, $i \in I$, and also let us consider the elements $s_{ji} \in A$, $i, j \in I$, as in Lemma 9.1. It will be sufficient to prove that X is the sum of a well-ordered increasing sequence of submodules X_α which have the following properties:

(i) each X_α is a direct sum of countably generated projective modules;

(ii) if α is a limit ordinal number then $X = \sum_{\beta < \alpha} X_\beta$;

(iii) for any ordinal α, $X_{\alpha+1}/X_\alpha$ is a countably generated projective module.

Let $X_0 = 0$. If X_α is constructed, then we can construct $X_{\alpha+1}$ as follows. Choose any element x_i which does not belong to X_α. Then, define the sets $J_0 \subseteq J_1 \subseteq \ldots \subseteq J_n \subseteq \ldots$. Let $J_0 = \{x_i\}$ and if $J_n = \{x_{i_1}, \ldots, x_{i_n}\}$ then J_{n+1} is the union of J_n and of those elements x_j which appear with non-zero coefficients in the representation of the elements $x_{i_1}, \ldots, x_{i_n}$ by x_i's. Let $X_{\alpha+1}$ be a submodule of X which is generated by X_α and the set

$$J = \bigcup_{n=0}^{\infty} J_n.$$

Obviously, $X_{\alpha+1}/X_\alpha$ is countably generated and according to Lemma 9.3, $X_{\alpha+1}$ and $X_{\alpha+1}/X_\alpha$ are projective. If α is a limit ordinal then we have

$$X_\alpha \simeq \coprod_{\beta < \alpha} X_{\beta+1}/X_\beta$$

and thus, for any α, conditions (i)—(iii) are satisfied. ∎

Now, we shall begin a study of a class of rings for which all projective modules are free.

A ring A is called a *local ring* if it is subject to the following condition: the elements of A which do not have a left inverse form a left ideal $\mathfrak{a}$.

PROPOSITION 9.5. *$\mathfrak{a}$ is a two-sided ideal and contains all proper left and right ideals of A. The elements of $\mathfrak{a}$ have neither a left or a right inverse, while the elements which are not in $\mathfrak{a}$ have a two-sided inverse (that is, they are invertible). The quotient ring $A/\mathfrak{a}$ is a division ring.*

Proof. If $\mathfrak{a}'$ is a proper left ideal then no element of $\mathfrak{a}'$ has a left inverse. Thus, $\mathfrak{a}' \subseteq \mathfrak{a}$. Next, we have to show that no element of $\mathfrak{a}$ has a right inverse. Indeed, suppose that $xy = 1$ for some $x \in \mathfrak{a}$. Then $(1 - yx)y = 0$ and since $yx \in \mathfrak{a}$ it follows that $1 - yx$ is not in $\mathfrak{a}$ so that $1 - yx$ has a left inverse y'. Then, $y = y'(1 - yx)y = 0$ which is a contradiction.

For each $y \in A$, $\mathfrak{a}y$ is a left ideal and since $xy \neq 1$ for $x \in \mathfrak{a}$, $\mathfrak{a}y$ is a proper left ideal. Thus $\mathfrak{a}y \subseteq \mathfrak{a}$ so that $\mathfrak{a}$ is a right ideal.

Now, assume that $y \notin \mathfrak{a}$ and let y' be a left inverse of y. Then, $y'y = 1$ and since $\mathfrak{a}$ is a right ideal, it follows that $y' \notin \mathfrak{a}$. Thus, y' has a left inverse, y''. Then we have that $y = y''$, which shows that y is invertible. Since $\mathfrak{a}$ consists of all elements which have not a right inverse, it follows as above that $\mathfrak{a}$ contains all proper right ideals.

The conclusion that $A/\mathfrak{a}$ is a division ring follows from the above established facts. ∎

THEOREM 9.6. (Kaplansky [102]). *Any projective module over a local ring is free.*

Proof. According to Exercise 3 and Theorem 9.4 it will suffice to prove that any element of a projective module P over a local ring, can be embedded in a free summand of P.

Take $F = P \amalg Q$ where F is free. Let x be an element of P whose insertion into a free direct summand of P is required. Select a basis u_i of F such that the expression of x in terms of that basis has the smallest possible number of non-zero components. Say that $x = a_1 u_1 + \dots a_n u_n$ is the expression in question. We point out that none of the a's can be a right linear combination of the remaining ones. For suppose that $a_n = a_1 b_1 + \dots + a_{n-1} b_{n-1}$. If we replace u_i by $u_i + b_i u_n$ for $i = 1, \dots, n-1$ and leave all other elements of the basis unchanged, we get a new basis for F. Thus, $x = a_1(u_1 + b_1 u_n) + \dots + a_{n-1}(u_{n-1} + b_{n-1} u_n)$ is a shorter expression for x, and this is a contradiction.

Let $u_i = y_i + z_i$ be the decomposition if u_i into P and Q components. Necessarily, we have:

$$a_1 y_1 + \dots + a_n y_n = a_1 u_1 + \dots + a_n u_n \tag{17}$$

because any of those elements is the component in P of x. Thus, we put

$$y_i = \sum_{j=1}^{n} c_{ij} u_j + t_i, \qquad i = 1, \dots n \tag{18}$$

where t_i is a linear combination of the basis elements other than $u_1, \ldots u_n$. Then, combining this equation with (17) and equating the coefficients of u_j, we get

$$a_1 c_{1j} + \ldots + a_n c_{nj} = a_j. \tag{19}$$

It follows from (19) that the elements c_{ij} with $i \neq j$ must be noninvertible in A; otherwise some a_i would be a right linear combination of the others, contrary to what was shown above. On the other hand, the elements c_{jj} are invertible, otherwise $1 - c_{jj}$ would be invertible and thus a_j would be a right linear combination of the other and again we have a contradiction to what was shown above. A glance at relations (18) shows that we may write

$$u_i = \sum_{j=1}^{n} b_{ij} y_j + v_i$$

with v_i being a linear combination of the elements of the basis which are distinct from $u_1, \ldots, u_n$. Thus the elements $y_1, \ldots, y_n$ form a basis for F, when augmented by the u's other than $u_1, \ldots, u_n$. If we write S for the submodule spanned by $y_1, \ldots, y_n$, then we have that $x \in S \subseteq P$ and S is a free direct summand of F. Hence, S is a free direct summand of P. ∎

A ring is said to be *left* (*respectively*, *right*) *semi-hereditary* if every finitely generated left (respectively, right) ideal is projective. In the commutative case one omits of course the distinction between left and right.

PROPOSITION 9.7. (Cartan–Eilenberg [33]). *If A is left semi-hereditary then every finitely generated submodule of a free A-module is the direct sum of a finite number of modules, each of them being isomorphic to a finitely generated ideal of A (and obviously, it is projective).*

Proof. Let $\{e_i\}_i$ be a basis for the free module F and let X be a finitely generated submodule of F. Then, X must be contained in a submodule of F generated by a finite number of the elements e_i. Thus we may assume that F has a finite basis $e_1, \ldots, e_n$.

We proceed by induction with respect to n. Let Y be the submodule of those elements of X which can be expressed using $e_1, \ldots, e_{n-1}$. Then each $x \in X$ can be written uniquely as $x = ae_n + y$, $a \in A$, $y \in Y$ (note that $Y = 0$ if $n = 1$). The mapping $x \to a$ maps X onto a left ideal $\mathfrak{a}$ of A, and the kernel of this morphism is Y. There results an exact sequence:

$$0 \to Y \to X \xrightarrow{p} \mathfrak{a} \to 0.$$

It follows that the ideal $\mathfrak{a}$ is finitely generated, and therefore is a projective module. Then p is a retraction and the above exact sequence is split, consequently X is isomorphic with the direct sum of $\mathfrak{a}$ and Y. This implies that Y is finitely generated and therefore by the inductive assumption, satisfies the conclusion of the proposition. ∎

COROLLARY 9.8. *Let A be a ring. The following assertions are equivalent:*

(1) *A is left semi-hereditary;*

(2) *Each finitely generated submodule of a projective A-module is projective.*

Proof. The implication (1) ⇒ (2) follows from Proposition 9.7 and the facts that each projective module is a submodule of a free module and that the direct sum of projective modules is projective. The implication (2) ⇒ (1) is obvious since A itself is free and thus projective. ∎

We shall prove that for left semi-hereditary rings, Proposition 9.7 is true for any projective module. We shall utilize the following result:

LEMMA 9.9. *Any projective module over a left semi-hereditary ring is a direct sum of finitely generated modules.*

Proof. Let A be a left semi-hereditary ring and P a projective A-module. First, we shall prove that any finite subset X of P is contained in a finitely generated direct summand of P. Indeed, P is a direct summand of a free module, i.e. $P \amalg P' = F$ is free. Denote $f : F \to F$ be the composition:

$$F \xrightarrow{p} P' \xrightarrow{u} F$$

where p is the projection onto the factor P' and u is the natural inclusion.

Now let $B = \{e_i\}_{i\in I}$ be a basis of F and $\{e_j\}_{1\leqslant j\leqslant n}$ a finite subset of B such that X is contained in F', the submodule of F generated by e_j's. Denote $f_j = p_j f v$, where $p_j: F \to A$ are canonical projections and $v: F' \to F$ the natural inclusion. Thus, im f_1 is a finitely generated left ideal of A, hence it is projective, i.e. ker f_1 is a direct summand of F'. Also $f_2(\ker f_1)$ is a projective left ideal of A, hence $\ker f_1 \cap \ker f_2$ is a direct summand of ker f_1, and obviously a direct summand of F'. Furthermore, by induction we can check that $F'' = \bigcap_{1\leqslant j\leqslant n} \ker f_j$ is a direct summand of F', which contains X (because $X \subseteq \ker f_j$ for every $1 \leqslant j \leqslant n$).

A few computations prove that F' is a finitely generated direct summand of F included in $\ker f = P$, i.e. a direct summand of P.

Furthermore, according to Theorem 9.4 we can assume that P is countably generated by $e_1, e_2 \dots e_n \dots$ say. Suppose that we have already defined a direct decomposition $P = P_1 \amalg P_2 \amalg \dots \amalg P_n \amalg P_n'$, where $P_1, \dots P_n$ are finitely generated and their sum contains $e_1 \dots, e_n$. Now, let Q be a finitely generated direct summand of P which contains $e_1, \dots, e_{n+1}$ and a generating set for $P_1 \ldots P_n$. Thus $P_1 \amalg \ldots \amalg P_n$ is contained in Q, and a direct summand of Q, since it is a direct summand of P, i.e. $Q = P_1 \amalg \ldots \amalg P_n \amalg P_{n+1}$. Then $P = P_1 \amalg P_2 \ldots \amalg P_{n+1} \amalg Q'$. By induction, we get a direct summand $P_1 \amalg P_2 \ldots$ containing $e_1, e_2, \ldots, e_n \ldots$, so that P is a direct sum of finitely generated modules. ■

The following result gives the structure of projective modules over left semi-hereditary rings.

THEOREM 9.10. *Any projective module over a left semi-hereditary ring A is a direct sum of finitely generated left ideals of A.*

The proof is a direct consequence of Lemma 9.9 and of Proposition 9.7. ■

Now we are able to show that any projective abelian group is free.

COROLLARY 9.11. *Let A be an integral domain in which all finitely generated ideals are principal. Then any projective A-module is free. In particular, any projective abelian group is free.* ■

LEMMA 9.12. *Any regular ring is left and right semi-hereditary.*

Proof. Indeed, according to the Theorem 8.6 (implication $(1) \Rightarrow (2)$) we deduce that any left (right) finitely generated ideal of a regular ring is a direct summand, that is, it is projective. ■

COROLLARY 9.13. *Any projective module over a regular ring is a direct sum of modules, which are isomorphic to principal left ideals.*

The proof follows by Lemma 9.12 and Theorem 9.10. ■

Exercises

1. In order that an A-module X be projective, it is necessary and sufficient that there exist a family $\{x_i\}_{i \in I}$ of elements of X and a family $\{f_i\}_{i \in I}$ of morphisms $f_i : X \to A$ such that for all $x \in X$, we have:

$$x = \sum_{i \in I} f_i(x)\, x_i$$

where $f_i(x)$ is zero for all but a finite number of indices i. (Hint: Use Lemma 9.1. The dependence between elements s_{ji} and morphisms $f_j : X \to A$ is given by $f_j(x_i) = s_{ji}$).

2. A ring A is local if and only if for any two non-invertible elements x, y of A, it follows that $x + y$ is also non-invertible.

3. Let A be a ring, X a countably generated A-module. Assume that any direct summand X' of X has the following property: any element of X' can be embedded in a free direct summand of X'. Then X is free.

4. If $\mathfrak{a}$ is an ideal in a ring A, we say that $\mathfrak{a}$ is *left T-nilpotent* ("T" for transfinite) if, given any sequence $\{a_i\}_i$ of elements in $\mathfrak{a}$, there exists an n such that $a_1 a_2 \ldots a_n = 0$. (Right T-nilpotence requires instead that

$$a_n a_{n-1} \ldots a_1 = 0).$$

Prove that any flat A-module is free if and only if A is a local ring whose radical is left T-nilpotent. This result is due to Govorov, [78].

5. (Bass, [16]). The following are equivalent for a ring A:

(a) A finitely generated projective submodule of a projective A-module is always a direct summand;

(b) The left annihilator of a finitely generated proper right ideal is always non-zero;

(c) If P is a finitely generated projective right A-module and P' is a finitely generated proper submodule, then $\mathrm{Hom}_A(P/P', A) \neq 0$.

References

Bass [16]; Bourbaki [21]; Cartan–Eilenberg [33]; Cohn [39]; Govorov [78]; Kaplansky [102]; Kuroš [110]; Lazard [115], [117]; Simson [206].

Note. The presentation of 3.9 follows Kaplansky [102], Simson [206] and Cohn [39].

3.10 Injective envelopes

Throughout this section all categories considered will be abelian.

Lemma 10.1. *Let $u : X' \to X$ be a morphism. The following assertions are equivalent:*

(1) *For any subobject Y of X, $u^{-1}(Y) = 0$ implies $Y = 0$;*

(2) *If $f : X \to Z$ is a morphism such that $\ker(fu) = \ker u$, then f is a monomorphism.*

Proof. (2) ⇒ (1). Let Y be a subobject of X such that $u^{-1}(Y) = 0$. If $Y \neq 0$, then the canonical morphism $p : X \to X/Y$ is not a monomorphism and $\ker(pu) = \ker u$. This is a contradiction.

The implication (1) ⇒ (2) is obvious. ∎

We define an *essential extension* of an object X' to be a monomorphism $u : X' \to X$ which is subject to the conditions of the above lemma. A subobject X' of X is called an *essential subobject* if the canonical inclusion $X' \to X$ is an essential extension of X'. An inclusion $X' \subseteq X$ in the category Mod A is an essential extension if and only if for each $x \in X$, $x \neq 0$, there exists an $a \in A$ such that $0 \neq ax \in X'$.

Sometimes, an essential extension $u : X' \to X$ is called an *essential monomorphism*. Also an essential subobject is called *large*.

COROLLARY 10.2. *If $X_1' \to X_1$ and $X_2' \to X_2$ are essential extensions, then the sum inclusion $X_1' \amalg X_2' \to X_1 \amalg X_2$ is an essential extension.*

Proof. Let Y be a subobject of $X_1 \amalg X_2$. We must prove that if $Y \neq 0$, then $Y \cap (X_1' \amalg X_2') \neq 0$. Let p_1, p_2 be the canonical projections of $X_1 \amalg X_2$ onto X_1 and X_2. It follows from the hypothesis that either $p_1(Y) \neq 0$ or $p_2(Y) \neq 0$. Assume that $p_1(Y) \neq 0$. Since $p_1(Y) \subseteq X_1$, it follows from the fact that $X_1' \to X_1$ is an essential extension, that $p_1(Y) \cap X_1' \neq 0$. Thus, we have $Y_1 = Y \cap p_1^{-1}(X_1') \neq 0$.

Two cases are possible: either $p_2(Y_1) \neq 0$ or $Y_1 \subseteq p_2^{-1}(X_2')$. In the first case it follows that $p_2(Y_1) \cap X_2' \neq 0$ and therefore, $Y_1 \cap p_2^{-1}(X_2') \neq 0$. But $Y_1 \cap p_2^{-1}(X_2') \subseteq X_1' \amalg X_2'$. In the second case, it follows that

$$Y_1 \subseteq X_1' \amalg X_2'. \blacksquare$$

COROLLARY 10.3. *Let $u : X \to Y$, $v : Y \to Z$ be monomorphisms. Then vu is an essential monomorphism if and only if u and v are essential monomorphisms.*

Proof. Let us assume that u and v are essential monomorphisms, and let $f : Z \to T$ be a morphism such that fvu is a monomorphism. Then, fv is a monomorphism. Hence, f is a monomorphism, according to Lemma 10.1.

Conversely, assume that vu is an essential monomorphism, and let

$$f : Z \to T$$

be a morphism such that fv is a monomorphism. Then fvu is a monomorphism. Hence f is a monomorphism. In other words, v is an essential monomorphism.

Furthermore, let Y' be a subobject of Y such that $u_M^{-1}(Y') = 0$ and $h : Z \to Z/v(Y')$ the canonical epimorphism. Then hvu is a monomorphism and thus, h is a monomorphism. Hence, $v(Y') = 0$ or equivalently $Y' = 0$. Finally, we have that u is an essential monomorphism. ∎

We call u a *proper extension* if u is not an isomorphism.

LEMMA 10.4. *An object X is injective (respectively, projective) if and only if any monomorphism $f : X \to Y$ (respectively, any epimorphism $g : Y \to X$) is a section (respectively, retraction).*

Proof. Assume that any monomorphism $f : X \to Y$ is a section. We consider the diagram

$$\begin{array}{ccccc} 0 & \longrightarrow & X' & \xrightarrow{\ i\ } & Z \\ & & {\scriptstyle f}\downarrow & & \downarrow{\scriptstyle f'} \\ & & X & \xrightarrow{\ i'\ } & X \coprod_{Z'} Z \end{array}$$

with i being a monomorphism.

Then, i' is also a monomorphism and consequently a morphism

$$u : X \coprod_{Z'} Z \to X$$

exists such that $ui' = 1_X$. Then we get $uf'\,i = ui'\,f = f$, that is, X is injective.∎

Let X' be a subobject of an object X. A subobject X'' of X is called a *complement* of X' if $X' \cap X'' = 0$ and if $X' + X''$ is an essential subobject of X. Then we shall say that X is *complemented.* Obviously, X'' is also complemented and X' is a complement of X''.

LEMMA 10.5. *Let $\mathscr{C}$ be a locally small* Ab 5*-category. Then any subobject X' of an object X is complemented. Moreover, any complement of X' is contained in a maximal one.*

Proof. Let M be the set of subobjects of X which trivially intersect X'. Under the natural ordering of subobjects if $\{X_i\}_i$ is a linearly ordered subset of M, then according to Ab 5 we have that $(\sum_i X_i) \cap X' = \sum_i(X_i \cap X') = 0$. Thus, M is inductive and according to Zorn's lemma we can find a maximal member X'' of M. Then X'' is a complement of X'. In fact, if Y is a nonzero subobject of X such that $(X' + X'') \cap Y = 0$, then $X' \cap (X'' + Y) = 0$ and $X'' + Y$ is greater than Y. But this is a contradiction. The last part of the proof follows in the canonical way. ∎

LEMMA 10.6. *Let $\mathscr{C}$ be a category which is subject to the following condition. If X' is a subobject of X, then X' is essential, or X' is complemented and any* *complement of X' is contained in a maximal one. Then, an object Q of $\mathscr{C}$ is injective if and only if Q admits no proper essential extension.*

Proof. Let $u: Q \to X$ be a monomorphism. If u is not essential, then a maximal complement X' of $u(Q)$ exists and pu is an essential monomorphism with

$$p: X \to X/X'$$

being the canonical epimorphism. Indeed, if $\bar{Y}$ is a non-null subobject of X/X' such that $pu(Q) \cap \bar{Y} = 0$, then $Y = p^{-1}(\bar{Y})$ is such that $u(Q) \cap Y = 0$ and obviously, we have that $Y \supset X'$. This is a contradiction. Thus pu is essential, that is, it is an isomorphism. According to Chapter 2, Theorem 7.2, it follows that u is a section. The proof results from Lemma 10.4. ∎

An *injective envelope* of an object X is an essential extension $X \to Q$ with Q being injective.

PROPOSITION 10.7. *Let $u: X \to Q$ and $u': X \to Q'$ be two injective envelopes of X. Then there is an isomorphism $f: Q \to Q'$ (not necessarily unique) such that $fu = u'$.*

Proof. Let $f: Q \to Q'$ be such that $fu = u'$ (here we made use of the injectivity of Q'). Then, f is an essential monomorphism (according to Corollary 10.3). Hence, it is an isomorphism because as Q is injective. ∎

LEMMA 10.8. *Let $\mathscr{C}$ be a category as in* Lemma 10.6. *If X is a subobject of an injective object Q, then X has an injective envelope.*

Proof. Let Y be a maximal complement of X in Q and $\bar{Q}$ a maximal complement of Y, which contains X. Then X is an essential subobject of $\bar{Q}$ and $\bar{Q}$ is injective. Let us note first that if X' is a subobject of $\bar{Q}$ such that $X' \neq 0$ and $X' \cap X = 0$, then $X' \cap Y = 0$ and $X' + Y$ is a complement of X which is greater than Y. But this is a contradiction. Hence, X is essential in $\bar{Q}$.

Furthermore, let $f: \bar{Q} \to T$ be an essential monomorphism. Then according to the injectivity of Q there exists a morphism $g: T \to Q$ such that gf is a monomorphism. Now, according to Lemma 10.1, we derive that g is a monomorphism, and that $g(T)$ is a subobject of Q which contains $\bar{Q}$ with $\bar{Q}$ being essential in $g(T)$. Thus, we have that $g(T) \cap Y = 0$. Now, because of the maximality of $\bar{Q}$ as a complement of Y, we have that $g(T) = \bar{Q}$; that is, f is an isomorphism. Then $\bar{Q}$ is injective according to Lemma 10.6. ∎

A category $\mathscr{C}$ is said to *have injective envelopes* if every object of $\mathscr{C}$ has an injective envelope.

COROLLARY 10.9. *If* $\mathscr{C}$ *is an arbitrary preadditive small category, then* Mod $\mathscr{C}$ *is a category with injective envelopes.*

The proof results from Lemmas 10.5 and 10.8. ■

The following theorem is a basic result in the theory of abelian categories and in homological algebra. This theorem will be proved on the basis of Theorem 7.9.

THEOREM 10.10. *An* Ab 5-*category* $\mathscr{C}$ *with generators* (*that is a Grothendieck category*) *is in fact a category with injective envelopes.*

Proof. We shall use the notations which were introduced in the proof of Theorem 7.9. Let X be a non-zero object of $\mathscr{C}$ and let $f : S(X) \to Q$ be an essential monomorphism, with Q being injective (see Corollary 10.9). If $u : \mathrm{Id}\,\mathrm{Mod}\,A^0 \to ST$ is an adjunction arrow, then we get the following commutative diagram:

$$\begin{array}{ccc} S(X) & \xrightarrow{\ f\ } & Q \\ {\scriptstyle u_{S(x)}}\downarrow & & \downarrow{\scriptstyle u_Q} \\ STS(X) & \xrightarrow{\ ST(f)\ } & ST(Q) \end{array}$$

where $u_{S(X)}$ is an isomorphism and $S\,T(f)$ is a monomorphism. Then, we derive from Lemma 10.1 that u_Q is a monomorphism. For instance, let us assume that $ST(f)$ is essential. Then u_Q is also an essential monomorphism. Hence it is an isomorphism. Furthermore, let $r : X' \to X$ be a monomorphism in $\mathscr{C}$ and $g : X' \to T(Q)$ a morphism. Then, there exists a morphism $h : S(X) \to ST(Q)$, such that $hS(r) = S(g)$. According to Theorem 7.9, S is full and faithful, and thus a unique morphism $t : X \to T(Q)$ exists such that $S(t) = h$. Hence, $tr = g$. Consequently, $T(Q)$ is injective.

Now we have only to prove that $ST(f)$ is essential. We shall prove first that $T(f)$ is essential. In fact, let Y be a non-null subobject of $T(Q)$ such that $Y \cap X = 0$. (We identify X with $TS(X)$ and X with the image of $T(f)$). Then, $S(Y) \cap S(X) = 0$, and $S(Y) \neq 0$. Furthermore,

$$S(Y) \cap u_Q(Q) = 0.$$

Hence we have the following split exact sequence

$$0 \longrightarrow Q \xrightarrow{u_Q} ST(Q) \longrightarrow Q' \longrightarrow 0$$

and $S(Y) \subseteq Q'$. But $T(u_Q)$ is an isomorphism and thus, $T(Q') = 0$. Consequently, $TS(Y) = Y = 0$. This is a contradiction. The result is that $T(f)$

is essential. In order to complete the proof, we have only to make specific use of the following general lemma, where the same notations as in Theorem 7.9 are used.

LEMMA 10.11. *Let $\mathscr{C}$ be a Grothendieck category. If $t : X' \to X$ is an essential extension then $S(t)$ is an essential extension too.*

Proof. Suppose that $g \in \mathrm{Hom}_{\mathscr{C}}(U, X) = S(X)$ is a non-zero element. We must find an $a \in A = \mathrm{Hom}_{\mathscr{C}}(U, U)$ such that $0 \neq ga \in \mathrm{Hom}_{\mathscr{C}}(U, X') = S(X')$, or more correctly, such that $\mathrm{im}(ga) \subseteq \mathrm{im}\, t$. Let us consider the following diagram

$$\begin{array}{ccccc} V & \xrightarrow{k} & X' \cap X'' & \xrightarrow{w} & X' \\ {\scriptstyle h}\downarrow & & \downarrow & & \downarrow{\scriptstyle t} \\ U & \xrightarrow{q} & X'' & \xrightarrow{v} & X \end{array}$$

where the row at the bottom is the factorization of g through its image and where each square is cartesian. Since $g \neq 0$, we have that $X'' \neq 0$. Now, since t is essential, we find that $X' \cap X'' \neq 0$. Since q is an epimorphism, k will also be an epimorphism and consequently, $k \neq 0$. Therefore, there exists a morphism $s : U \to V$ such that $ks \neq 0$. Since t and w are monomorphisms then $twks \neq 0$. Therefore, $vqhs = ghs \neq 0$ and we may take $a = hs$. ∎

NOTE 1. For an object X of an abelian category $\mathscr{C}$, we shall sometimes denote by $E(X)$ an injective envelope of X (of course, if it exists). Moreover X will be considered as a subobject of $E(X)$.

Exercises

1. An epimorphism $f : X \to X''$ is said to be *essential* if for any morphism $g : Y \to X$ such that fg is an epimorphism, g is an epimorphism. Let A be a local ring with maximal ideal $\mathfrak{m}$. Then the canonical epimorphism

$$f : A \to A/\mathfrak{m}$$

is an essential epimorphism in Mod A. (Hint: Use the Lemma of Nakayama, Section 3.5, Exercise 11).

2. In the category $\mathscr{Ab}$ the additive group $\mathbf{Q}$ of rational numbers is an injective envelope of each of its non-null subgroups. Also the group $\mathbf{Q}_p$ which is defined in Section 3, Example 2, is an injective envelope of each of its non-null subgroups.

3. Prove that the converse of Lemma 10.11 is also true. Namely, show that if $S(t)$ is an essential extension, then t is essential too.

4. Prove that any Grothendieck category $\mathscr{C}$ has sufficiently many injectives, and consequently, it is a category with injective envelopes. In doing this, make specific use of a procedure due to Cartan and Eilenberg in [33]. Then obtain a new proof of the Theorem 7.9 by showing that the functor S preserves injective objects and using Theorem 2.8.

5. Let $\mathscr{C}$ be a Grothendieck category, U a generator of $\mathscr{C}$, and $T : \mathscr{C} \to \mathscr{C}'$ a covariant functor and S a right adjoint of T. We assume that $\mathscr{C}'$ has sufficiently many injectives. Then T is exact if and only if for any monomorphism $f : X \to U$, $T(f)$ is also a monomorphism. (Hint: Use Lemma 3.1, and Theorems 2.8 and 10.10).

6. A non-null object X is called *simple* if any proper subobject of X is null. An A-module X is simple if and only if $X = A/\mathfrak{m}$ with $\mathfrak{m}$ being a left maximal ideal of A. An injective A-module V is a cogenerator of Mod A if and only if V contains any simple A-module. Particularly the injective envelope of the module $\Pi A/\mathfrak{m}$ with $\mathfrak{m}$ running through the set of all left maximal ideals of A, is a minimal cogenerator. In other words, it is contained in any injective cogenerator.

7. Let A be an integral domain. An A-module X is called *divisible* if for every non-zero element $a \in A$ and every $x \in X$, there is an element $y \in X$ such that $ay = x$. Then an injective A-module is divisible. Conversely, a divisible A-module without torsion is injective. Show that the field of quotients of A (see Chapter 4, Section 4.1) is an injective envelope of A.

References

Baer [12]; Bucur–Deleanu [31]; Cartan–Eilenberg [33]; Eckmann–Schopf [48]; Gabriel [59]; Gabriel–Popescu [60]; Grothendieck [79]; MacLane [120]; Mitchell [133]; Pareigis [162]; Tisseron [225]; Walker–Walker [235].

3.11 Semisimple rings

LEMMA 11.1. *Let A be a ring and X an A-module. The following assertions are equivalent*:

(1) *X is a sum of simple submodules*;

(2) *X is a direct sum of simple modules*;

(3) *Any submodule of X is a direct summand.*

Proof. (1) $\Rightarrow$ (3). Let X' be a proper submodule of X. Denote by M the set of non-zero submodules X'' of X such that $X' \cap X'' = 0$. The set M is non-void. Indeed, there is a simple submodule Y of X such that $Y \nsubseteq X'$; (otherwise X' contains all simple submodules of X, that is, it is not proper). Then $Y \cap X' = 0$. Clearly M, ordered by inclusion, is an inductive set and by Zorn's Lemma it has a maximal element X''. If $X' + X'' \neq X$, then as above, there exists also a simple submodule Y of X, with $Y \nsubseteq X' + X''$. Then $Y \cap (X' + X'') = 0$ and thus $X' \cap (Y + X'') = 0$. But it is clear that $Y + X'' \in M$ and X'' are strictly included in $X'' + Y$, in contradiction with the maximality of X''.

(3) $\Rightarrow$ (2). Let $x \in X$ be a non-zero element; denote by M the set of all submodules of X which do not contain x. Obviously M, ordered by inclusion, is an inductive set and so, by Zorn's Lemma it contains a maximal element X'. Let X'' be a supplement of X'. If X'' is not simple, then it contains a proper submodule Y and $X' + Y$ is greater than X'. Next, let Y' be a supplement of $X' + Y$. Then $(X' + Y) \cap (X' + Y') = X'$ and thus the maximality of X' is violated. In conclusion, necessarily, X'' is simple, that is X contains simple submodules.

Now denote by T the set of all submodules X' of X such that X' is a direct sum of simple modules. In an obvious manner, we deduce that T has a maximal element X'. If X' is not equal to X, then as above, we deduce that there exists a simple submodule Y of X such that $X' \cap Y = 0$. Then the maximality of X' is also violated.

The implication (1) $\Rightarrow$ (2) is obvious. ∎

An A-module X as in the above lemma is called *semisimple*.

COROLLARY 11.2. *Let K be a division ring. Then any K-module is semisimple.* ∎

If A is a ring and X an A-module, then X becomes an $\mathrm{End}_A(X)$-module in the following manner. If $f \in \mathrm{End}_A(X)$ and $x \in X$, then $fx = f(x)$. Furthermore we denote by $\mathrm{Cent}_A(X)$, the ring $\mathrm{End}_{\mathrm{End}_A(X)}(X)$. Let $a \in A$; denote by $\phi(a) : X \to X$ the morphism of abelian groups defined by multiplication with a, that is $\phi(a)(x) = ax$ for all $x \in X$. Clearly, $\phi(a)$ is in fact an element of $\mathrm{Cent}_A(X)$, and so the assignment $a \leadsto \phi(a)$ defines a ring morphism

$$\phi : A \to \mathrm{Cent}_A(X).$$

The ring $\mathrm{Cent}_A(X)$ is called the *bicentralizer* of X and the morphism ϕ (or ϕ_X if there is danger of confusion) is called the *centralizing morphism*. The kernel of the centralizing morphism is denoted by $\mathrm{Ann}(X)$ and it is called the *annihilator* of X. An element $a \in A$ is in $\mathrm{Ann}(X)$ if and only if

$ax = 0$ for any $x \in X$, that is, $\mathrm{Ann}(X) = \bigcap_{x \in X} \mathrm{Ann}(x)$. If $\mathrm{Ann}(X) = 0$ then X is called a *faithful* module.

LEMMA 11.3. *Let A be a ring. Then for any natural number n, the centralizing morphism*

$$\phi : A \to \mathrm{Cent}_A(A^n)$$

is an isomorphism.

Proof. Let $a \in A$ and $x = (x_1, \ldots, x\ ) \in A^n$; then $\phi(a)x = ax = (ax_1, \ldots, ax_n)$. Thus $\phi(a) = 0$ implies $a(1, 0, \ldots, 0) = (a, 0, \ldots, 0) = 0$, that is $a = 0$. Hence ϕ is a monomorphism.

Let $f \in \mathrm{Cent}_A(A^n)$ and $x = (x_1, \ldots, x_n) \in A^n$. Thus $f(x) = f(x_1 e_1 + \ldots + x_n e_n) = f(x_1 e_1) + \ldots + f(x_n e_n)$, where $\{e_i\}_{1 \leqslant i \leqslant n}$ is the canonical basis of A^n. But $x_i e_i$ can be written as $m_i e_i$, where m_i is the $n \times n$ matrix (a_{ij}) with $a_{ii} = x_i$ and the other elements zero. Thus

$$f(x) = f\left(\sum_{i=1}^{n} m_i e_i\right) = \sum_{i=i}^{n} m_i f(e_i).$$

Let us put $f(e_i) = (t_{i1,\ldots,} t_{in}) \in A^n$. Then

$$\sum_{i=1}^{n} m_i f(e_i) = \sum_{i=1}^{n} m_i(t_{i1}, \ldots, t_{in}) = (t_{11} x_1, \ldots, t_{nn} x_n).$$

In particular we have $f(e_i) = t_{ii} e_i$. Now we shall prove that $t_{11} = \ldots = t_{nn}$. For that let m_{ij} be an element of $\mathrm{End}_A(A^n) = A_n$ such that $m_{ij}(e_i) = e_j$. Thus $f(m_{ij}(e_i)) = m_{ij} f(e_i) = m_{ij}(t_{ii} e_i) = t_{ii} m_{ij}(e_i) = t_{ii} e_j$. Also we have $f(m_{ij} e_i) = f(e_j) = t_{jj} e_j$, so that $t_{ii} = t_{jj}$. Hence $f = \phi(a)$ for a suitable element $a \in A$, or equivalently ϕ is an isomorphism. ∎

THEOREM 11.4. *Let A be a ring. Then the categories* Mod A *and* Mod A_n *are equivalent for each natural number n.*

The proof is a direct consequence of Corollary 7.7 and Lemma 11.3. ∎

THEOREM 11.5. (Wedderburn). *Let A be a ring. The following assertions are equivalent.*

(1) *Any A-module is semisimple and any two simple A-modules are isomorphic.*

(2) *Any A^0-module is semisimple and any two simple A^0-modules are isomorphic.*

(3) *A is isomorppic with the ring K_n for a suitable division ring.*

Proof. The implications (3) ⇒ (1) and (3) ⇒ (2) are a direct consequence of Theorem 11.4 and Corollary 11.2.

(1) ⇒ (3). Let M be a simple module of Mod A. Then A is isomorphic with a direct sum $\coprod_{i \in I} \mathfrak{a}_i$ of left ideals $\mathfrak{a}_i$, each being isomorphic with M and I is a finite set. Let $K = \mathrm{End}_A(M)$. Clearly, M is a K-module and by Theorem 7.2 we have the functors:

$$S_M = \mathrm{Hom}_A(M, ?) : \mathrm{Mod}\, A \to \mathrm{Mod}\, K^0$$

$$T_M = ? \otimes_K M : \mathrm{Mod}\, K^0 \to \mathrm{Mod}\, A$$

such that S_M is right adjoint of T_M. By Theorem 7.9, we see that S_M is full and faithful, since M is a generator of Mod A. Furthermore, $T_M(K^0) = M$; in addition, the natural morphism

$$T_M(K^0, K^0) : \mathrm{Hom}_{K^0}(K^0, K^0) \to \mathrm{Hom}_A(M, M)$$

is an isomorphism. By a slight computation we check that T_M is also a full and faithful functor, that is, T_M and S_M define an equivalence of categories. Finally, by Corollary 7.7 we see that A is canonically isomorphic with $\mathrm{End}_{K^0}(S_M(A))$, that is, it is a full matrix ring over a division ring.

The implication (2) ⇒ (3) is checked in the same way. ■

A ring A as in the previous Theorem is called a *simple ring*. It is clear that any division ring is a simple ring.

THEOREM 11.6. (Artin–Wedderburn). *Let A be a ring. The following assertions are equivalent*:

(1) *Any A-module is injective*;

(1^0) *Any A^0-module is injective*;

(2) *Any A-module is projective*;

(2^0) *Any A^0-module is projective*;

(3) *Any A-module is semisimple*;

(3^0) *Any A^0-module is semisimple*;

(4) *A is a semisimple module in* Mod A;

(4^0) *A is a semisimple module in* Mod A^0;

(5) *A is a direct product (in the category of rings) of finitely many simple rings.*

Proof. The equivalences (1) ⇔ (2) ⇔ (3) ⇔ (4) and (1^0) ⇔ (2^0) ⇔ (3^0) ⇔ (4^0) are easy consequences of Lemma 11.1 and their proofs are left to the reader.

(4) ⇒ (5). It is clear that $A = \coprod_{i\in I} \mathfrak{a}_i$ where $\mathfrak{a}_i$ are simple left ideals, and I is a finite set (see Lemma 11.1). Let M be a simple A-module. Then M is isomorphic with a direct summand of A (see Lemma 11.1), and so it can be considered to be a left ideal. Then M is isomorphic with a left ideal $\mathfrak{a}_i$ for a suitable $i \in I$. Indeed, we make the choice of i such that a non-null element x of M has a non-null component x_i in the ideal $\mathfrak{a}_i$. Then the mapping

$$ax \rightsquigarrow ax_i \qquad a \in A$$

establishes an isomorphism of M onto $\mathfrak{a}_i$. In conclusion, in the category Mod A there exist only a finite number $M_1, \ldots, M_r$ of non-isomorphic simple modules. Denote by $\mathfrak{b}_j, j = 1, \ldots, r$ the sum of all simple left ideals of A isomorphic with M_j. Then

$$A \simeq \coprod_{j=1}^{r} \mathfrak{b}_j.$$

If $f \in \mathrm{End}_A(A)$, then $f(\mathfrak{b}_j) \subseteq \mathfrak{b}_j$ and so the restriction of f to $\mathfrak{b}_j$ is a morphism $f_j : \mathfrak{b}_j \to \mathfrak{b}_j$. Let $A_j = \mathrm{End}_A(\mathfrak{b}_j)$. We see that the mapping $f \rightsquigarrow (f_j)_{1 \leqslant j \leqslant r}$ is an isomorphism of A^0 onto the direct product

$$\prod_{j=1}^{r} A_j.$$

Now, if $\mathfrak{a}, \mathfrak{a}'$ are two non-isomorphic simple left ideals of a ring A, then $\mathfrak{a}\mathfrak{a}' = 0$ or $\mathfrak{a}\mathfrak{a}' = \mathfrak{a}'$, and if $\mathfrak{a} \simeq A/\mathfrak{m}$ where $\mathfrak{m}$ is a maximal left ideal, then, $\mathfrak{m} = \mathrm{Ann}(y)$ for a suitable non-zero element y of $\mathfrak{a}'$, so that $\mathfrak{a}' \simeq A/\mathfrak{m}$, a contradiction.

Thus, we check that for any $1 \leqslant j, k \leqslant r, j \neq k$ one has $\mathfrak{b}_j M_k = 0$. Hence M_j becomes canonically a simple A_j-module, so that A_j is a simple ring (see also Exercise 2). Finally, we use Theorem 11.5 and Exercise 2.

The implication (5) ⇒ (4) follows from Theorem 11.5 and Exercise 2. The other implications are obvious. ∎

A ring as in the above theorem is called a *semisimple ring*.

Exercises

1. (Schur's Lemma). An A-module M is simple if and only if $M = Ax$ for any non-zero element x of M. Also for a simple A-module M, the ring $\mathrm{End}_A(M)$ is a division ring.

2. Let $A_1, \ldots, A_n$ be rings and

$$A = \prod_{i=1}^{n} A_i$$

(the direct product in the category of rings). Then Mod A is canonically equivalent with

$$\prod_{i=1}^{n} \text{Mod } A_i.$$

Particularly A is semisimple if and only if all A_i are semisimple.

3. A ring A is semisimple if and only if any simple A-module is projective.

4. Let A be a ring. There exists a set $\{M_i\}_{i\in I}$ of simple non-isomorphic modules, such that any simple module M is isomorphic with M_i for a suitable i. Let $K_i = \text{End}_A(M_i)$. Also let $\text{So}^i(\text{Mod } A)$ be the full subcategory of semisimple modules X, such that any simple module contained in X is isomorphic with M_i (that is X is M_i-*isotypic*). Then the categories $\text{Mod } K_i{}^0$ and $\text{So}^i(\text{Mod } A)$ are equivalent. Moreover if we denote by $\text{So}(\text{Mod } A)$ the full subcategory of all semisimple modules, then $\text{So}(\text{Mod } A)$ is equivalent with the product $\prod_{i\in I} \text{So}^i(\text{Mod } A)$. (In this manner we may show that any two decompositions of a semisimple module in a direct sum of simple modules are necessarily isomorphic. [See Chapter 5, Section 5.1]).

5. Let K be a division ring and A a matrix ring over K. Then the centres of A and K are isomorphic. Particularly, if K is commutative, then K is a subring of A.

6. Let A be a simple ring. Then each finitely generated A-module X is a generator of Mod A and $\text{End}_A(X)$ is also a simple ring.

7. An A-module X is semisimple if and only if X has no proper essential submodules.

8. (Sandomierski–Kasch). For any A-module X, denote by $\text{So}(X)$ and call *socle* of X the sum of all simple submodules of X. Show that the intersection of all essential submodules of X is the socle of X.

9. (Brauer). If $\mathfrak{a}$ is a simple left ideal of a ring A, then either $\mathfrak{a}^2 = 0$ or $\mathfrak{a} = Ae$ where $e^2 = e \in \mathfrak{a}$.

10. Show that A is semisimple if and only if no maximal left ideal is essential. Compare with Exercises 3 and 7.

11. Show that the ring of all 2×2 matrices over an infinite field has an infinite number of distinct simple left ideals.

12. If K and K' are division rings and $K_n \simeq K'_{n'}$ show that $K \simeq K'$ and $n = n'$.

13. (Jacobson, [96]). Let A be a ring, X a semisimple A-module,

$$x_1, \ldots, x_n \in X \quad \text{and} \quad f \in \mathrm{Cent}_A(X).$$

Then an element $a \in A$ may be found such that $ax_i = f(x_i)$ for any $i = 1, \ldots, n$.

14. Let K be a division ring and X a K-module. Then the centralizing morphism

$$\phi : K \to \mathrm{Cent}_K(X)$$

is an isomorphism.

References

Bourbaki [23]; Cartan–Eilenberg [33]; Jacobson [96]; Lambek [113]; Morita [134]; Northcott [152]; Pareigis [162]; Rieffel [185].

4. Localization

The concept of localization stems from various problems, one of them being localization at a prime ideal in a commutative ring. Another localization is given in topology through the study of sheaves of abelian groups over a topological space. These two notions have a common underlying idea. In fact, let A be a commutative ring, S a multiplicative system, A_S the associated ring of fractions (see Section 4.18) and $\phi : A \to A_S$ the canonical morphism. Then the restriction functor $\phi_* : \text{Mod } A_S \to \text{Mod } A$ is full and faithful and has an exact left adjoint, namely the tensor product with A_S. On the other hand, let E be a topological space, $\mathscr{P}(E, \mathscr{A}b)$ the category of presheaves of abelian groups over E and $\mathscr{F}(E, \mathscr{A}b)$ the category of sheaves over E. If $I : \mathscr{F}(E, \mathscr{A}b) \to \mathscr{P}(E, \mathscr{A}b)$ is the inclusion functor, then I is full and faithful and has an exact left adjoint: the functor "associated sheaf".

The above examples (as well as many others) lead to the consideration of the following situation: let $\mathscr{C}$ be an abelian category; we shall call a localization of $\mathscr{C}$ a pair $(\mathscr{C}', T)$ where $\mathscr{C}'$ is an abelian category and

$$T : \mathscr{C} \to \mathscr{C}'$$

an exact functor which has a full and faithful right adjoint. All known examples of abelian localization reduce to this concept. It is known that the localizations of abelian categories are completely determined by certain subcategories, which are called localizing subcategories.

Note that the notion of localization can be defined for any category: the localization of a category $\mathscr{C}$ will be a pair $(\mathscr{C}', T)$ with $\mathscr{C}'$ a category and $T : \mathscr{C} \to \mathscr{C}'$ a functor commuting with finite projective limits (of course with those that do exist!) and having a full and faithful right adjoint. As in the case of abelian categories, the description of the localizations of an arbitrary category is carried out only in particular cases (for example for the categories of set-valued functors over a small category using Grothendieck's topologies; see [65]).

The arbitrary localizations are examples of categories of fractions, and the abelian localizations are important examples of additive fractions.

We remark that the additive fractions appeared first (the ordinary fractions, rings of fractions, etc) and the arbitrary fractions appeared later (the monoids of fraction). Homotopy theory deals with important examples of categories of fractions (see [62]).

The reader will observe that there are significant differences between the additive and general fractions.

In the first section of this chapter we define the notion of additive fraction and give conditions ensuring the existence of such fractions. An example of a category of additive fractions is given by the spectral category associated with an abelian category (Section 3.2). Another example, which has great importance for what follows is given in Section 4.3: namely the quotient category of an abelian category with respect to a dense subcategory. The idea of the quotient category of an abelian category by a dense subcategory was suggested by Serre's work on the homology groups of certain topological spaces modulo a certain dense subcategory of $\mathscr{Ab}$. In [79], Grothendieck indicates the idea for the construction of the quotient category and Gabriel, in his thesis [59], develops the technique of the quotient categories and gives many interesting applications.

Among the dense subcategories $\mathscr{A}$ of an abelian category $\mathscr{C}$, the subcategory for which the canonic functor $T : \mathscr{C} \to \mathscr{C}/\mathscr{A}$ has a right adjoint plays an important role. Gabriel called such a subcategory "localizing" and we maintain his terminology which has been largely adopted.

In Sections 4.4, 4.5 and 4.6 we expose some results about localized subcategories. The seventh section contains an important example of localization (which was one of the starting points in the localization theory, as already remarked) proving that the category of sheaves over a topological space with values in certain abelian categories, is a quotient category of the category of presheaves with respect to a suitable localizing subcategory.

Papers dealing with some more or less direct aspects of localization have appeared in the past few years. Thus, Dickson [43] defined torsion subcategories of an abelian category. Torsion subcategories are more general than the localizing subcategories but one obtains important results about torsion theories when torsion subcategories are also localizing. Maranda [121] constructed a technique which is only formally different from the construction due to Gabriel.

The basic tool in Maranda's theory is the notion of the radical. A similar technique was developed also by Goldman [76] but he obtained deeper results. In the eighth section we establish a close connection between torsion theories, radicals and localizing subcategories.

In [59] Gabriel proved that localizing subcategories in the category of modules over a ring A are in one-to-one correspondence with some systems of left ideals called idempotent and topologizing systems. In papers [144] and [213] idempotent topologizing systems are called "additive topologies". Indeed, they represent the additive variant of the topologies of Grothendieck (see [65]).

Now, we think that it is more suggestive that the idempotent topologizing systems be called "localizing systems". In the ninth section the localizing subcategories and the localizing systems are studied. In the tenth section some special aspects of localization in categories of modules are exhibited. Using the localization technique, in the eleventh section we prove that for each abelian small category $\mathscr{C}$ there exists an exact full and faithful functor $\bar{h} : \mathscr{C} \to \operatorname{Lex} \mathscr{C}$, where $\operatorname{Lex} \mathscr{C}$ is the Grothendieck category of left exact functors from $\mathscr{C}$ into $\mathscr{Ab}$. These results have been proved for the first time by Gabriel in [59].

Also in this section the following important result of Mitchell is proved: for any small abelian category $\mathscr{C}$ there exists a ring A and an exact full and faithful functor $S : \mathscr{C} \to \operatorname{Mod} A$.

This result is useful because it reduces the study of the objects and morphisms of an abelian category to the study of modules and their morphisms.

In Section 4.12 we study the localization ring of a ring with respect to a localizing system. As important examples of localization rings we notice the complete ring of quotients introduced by Utumi [232] (see also [113]) and the regular ring of quotients defined by Johnson [99]. These notions are discussed in Section 4.13.

Using the technique of localization we give in Section 4.14 a new proof to Theorem 7.9 of Chapter 3 (this is, in fact, the author's original proof, [60]). We take the opportunity to define the notion of canonical localizing system which will help in making some observations about Grothendieck categories.

The notion of flat epimorphism of rings (see Section 4.16) is very useful for computation of quotient categories of a category of modules with respect to some localizing systems and also in other applications. Among flat epimorphisms we distinguish the quasi-orders, whose study occupies Section 4.17. Here it is shown that for any ring A there exists a maximal epimorphic flat extension, denoted by $M(A)$. In general, $M(A)$ is a subring of the complete ring of quotients and in Section 4.17, we also give some sufficient conditions for these two rings to coincide. Furthermore, we also give a characterization of those rings A that are quasi-orders in semisimple rings.

In Section 4.18 we consider rings of fractions. Now the rings of fractions appear as a particular case of localization rings. Among the ring of fractions one distinguishes the so-called "left orders". One shows that for any ring A there exists a ring $O(A)$ such that A is an order in $O(A)$ and maximal with this property (see Theorem 19.2). As an application of the localization theory following Gabriel [59] and Stenström [213] we give a proof of a Theorem of Goldie [73] (see Theorem 19.5).

Prime ideals play a very important part in the study of commutative rings. The notion of prime ideal has been extended to arbitrary rings, so we have two-sided prime ideals and two-sided completely prime ideals, which are essentially generalizations of the notion of prime ideal (in commutative rings), but not the analogues of this notion.

The analogous notion of prime ideal in the non-commutative case seems to be the notion of prime localizing system (see Section 4.20) introduced by Goldman [76]. Another definition for the concept of prime for arbitrary rings was given by Popescu [171] and this notion is equivalent to that given by Goldman.

In Section 4.20 we review some definitions and results about left primes of an arbitrary ring.

The last section (4.21) deals with the study of bilocalizing subcategories and the proof of a result due to Roos [190] which gives a characterization of the Ab 6 and Ab 4* Grothendieck categories. The proof of this Theorem has been communicated to me by Remi Goblot and I use this opportunity to thank him for it.

4.1 Categories of fractions. Calculus of fractions

Let $\mathscr{C}$ be a category and let Σ be a class of morphisms of $\mathscr{C}$. We shall say that the couple $(T, \mathscr{C}_\Sigma)$ where $\mathscr{C}_\Sigma$ is a category and $T: \mathscr{C} \to \mathscr{C}_\Sigma$ a covariant functor is a *category of fractions* of $\mathscr{C}$ relative to Σ if for any $s \in \Sigma$, $T(s)$ is an isomorphism in $\mathscr{C}_\Sigma$ and if $T': \mathscr{C} \to \mathscr{C}'$ is a covariant functor such that for any morphism s in Σ, $T'(s)$ is an isomorphism, then a unique functor $\bar{T}': \mathscr{C}_\Sigma \to \mathscr{C}'$ exists such that the following diagram of categories and functors is commutative:

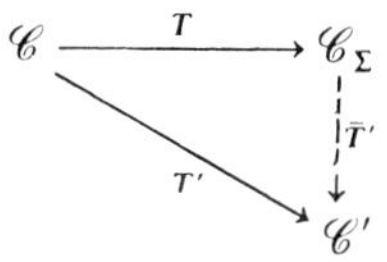

Sometimes, instead of saying that $(T, \mathscr{C}_\Sigma)$ is a category of fractions of $\mathscr{C}$ relative to Σ, we shall simply say that $\mathscr{C}_\Sigma$ is a category of fractions of $\mathscr{C}$ with respect to Σ (the functor T not being explicitly mentioned). In the general theory of categories it is shown that under sufficiently general conditions, the category $\mathscr{C}_\Sigma$ of fractions exists (see [62] and [7]).

In subsequent sections we shall be concerned with the additive category of fractions. Precisely, let $\mathscr{C}$ be a preadditive category and let Σ be a class of morphisms of $\mathscr{C}$. A couple $(T, \mathscr{C}_\Sigma)$ is called a *category of additive fractions* of $\mathscr{C}$ relative to Σ if T is an additive covariant functor which is subject to the above conditions, only for additive functors.

Generally, the categories of additive fractions are essentially different from the general categories of fractions (see for example the rings of fractions).

We remark that the ring of fractions of a ring relative to a suitable system of elements was the first example of a category of additive fractions and the ideas introduced here are fundamental in the theory of additive fractions.

Let $\mathscr{C}$ be a category and Σ a system of morphisms in $\mathscr{C}$. We shall say that Σ is a *multiplicative system* if for any two composable morphisms s, s' of Σ the composition ss' is also in Σ and all identity morphisms are in Σ. We shall say that Σ is *right* (*respectively*, *left*)-*permutable* if any angle (respectively co-angle):

$$\begin{array}{ccc} & & Y' \\ & & \downarrow \\ X & \xrightarrow{s} & Y \end{array} \qquad \begin{array}{ccc} X & \longrightarrow & X' \\ {\scriptstyle s}\downarrow & & \\ Y & & \end{array}$$

with $s \in \Sigma$ can be embedded in the commutative square:

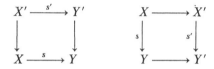

$$\begin{array}{ccc} X' & \xrightarrow{s'} & Y' \\ \downarrow & & \downarrow \\ X & \xrightarrow{s} & Y \end{array} \qquad \begin{array}{ccc} X & \longrightarrow & X' \\ {\scriptstyle s}\downarrow & & \downarrow{\scriptstyle s'} \\ Y & \longrightarrow & Y' \end{array}$$

such that s' is also in Σ. We shall say that Σ is *right* (*respectively*, *left*)-*simplifiable* if for any couple of morphisms $f, g : X \to Y$ for which a $s : Y \to Y'$ exists (respectively $s : X' \to X$) such that $s \in \Sigma$ and $sf = sg$ (respectively $fs = gs$) there exists an $s' : Z \to X$ (respectively $s' : Y \to Z$) with $s' \in \Sigma$ and $fs' = gs'$ (respectively $s'f = s'g$).

Furthermore, we shall say that Σ is *right* (*respectively*, *left*)-*calculable* if Σ is multiplicative and if it is also right (respectively, left)-permutable and simplifiable. Finally, we shall say that Σ is *bicalculable* if Σ is simultaneously right and left-calculable.

Obviously, the notions of right and left permutable, simplifiable and calculable are respectively dual.

EXAMPLES 1. Let A be a commutative ring. Then any multiplicative subset of A is a bicalculable system.

2. Let us denote by Σ the class of all morphisms f in $\mathscr{Ab}$ for which ker f and coker f are torsion groups. Then Σ is a bicalculable system.

Now, we shall make some observations which will be useful later on. A category $\mathscr{I}$ is called *quasi-direct* if the following conditions are fulfilled:

(1) Any co-angle:

$$\begin{array}{ccc} & \nearrow & i' \\ i & & \\ & \searrow & i'' \end{array}$$

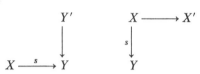

of $\mathscr{I}$ can be embedded in a commutative square:

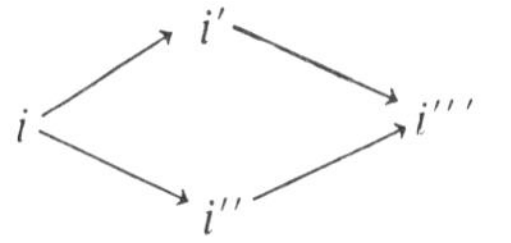

(2) For any couple $u, v : i \to j$ of morphisms in $\mathscr{I}$, there exists a morphism $w : j \to k$ such that $wu = wv$.

A full subcategory $\mathscr{I}'$ of an arbitrary category $\mathscr{I}$ is called *cofinal* if for any object i of $\mathscr{I}$ there exists a morphism $u : i \to i'$ with $i' \in \mathrm{Ob}\,\mathscr{I}'$.

LEMMA 1.1. *Let $\mathscr{I}$ be a quasi-direct category, $\mathscr{I}'$ a cofinal small subcategory of $\mathscr{I}$ and $H : \mathscr{I}' \to \mathscr{I}$ the inclusion functor. Then, a covariant functor $T : \mathscr{I} \to \mathscr{C}$ has an inductive limit if and only if TH has an inductive limit. Moreover,* $\varinjlim T \equiv \varinjlim (TH)$.

Proof. Let $X = \varinjlim(TH)$ and $u_{i'} : T(i') \to X$ the structural morphisms; if i is an object of $\mathscr{I}$, then a morphism $v : i \to i'$ exists with $i' \in \mathrm{Ob}\,\mathscr{I}'$. Let us put $u_i = u_{i'}T(v)$. We must prove that u_i does not depend on i'. In order to do this, consider $v' : i \to i''$, with $i'' \in \mathrm{Ob}\,\mathscr{I}'$. Then, morphisms $t : i' \to j$ and $t' : i'' \to j$ exist such that $tv = t'v'$. Let $v'' : j \to i'''$ be a morphism with $i''' \in \mathrm{Ob}\,\mathscr{I}'$. Then: $u_{i'''}T(v'')\,T(t)\,T(v) = u_{i'''}T(v'')\,T(t')\,T(v') = u_{i'''}(TH)(v''t)\,T(v) = u_{i'}T(v) = u_{i'''}(TH)\,(v''t')\,T(v') = u_{i''}T(v')$, and so u_i is independent of i' and v. The proof follows from this. ∎

A category $\mathscr{I}$ is said to be *connected* if for any two objects i, j there exists a finite set $i_1, \dots i_n$ of objects and morphisms

$$u_1 : i \to i_1, \quad u_2 : i_2 \to i_1, \quad u_3 : i_2 \to i_3 \dots, \quad u_{n-1} : i_{n-1} \to i_n, \quad u_n : j \to i_n.$$

This means that two objects may be connected by a path containing a finite set of oriented edges.

LEMMA 1.2. *Let $\mathscr{I}$ be a quasi-direct connected category and let $T : \mathscr{I} \to$ Mod A be a covariant functor where A is a ring, such that $\varinjlim T$ exists (for example, when $\mathscr{I}$ has a small cofinal subcategory). Also let us denote by $S :$ Mod $A \to \mathscr{S}et$ the underlying functor. Then, $S(\varinjlim T) = \varinjlim(ST)$, that is, the underlying functor commutes with quasi-direct and connected inductive limits.*

The proof is left to the reader. ∎

Let Σ be a system of morphisms in $\mathscr{C}$. For any object X of $\mathscr{C}$ we shall denote by Σ/X (respectively by X/Σ) the full subcategory of $\mathscr{C}/X$ (respecspectively of $X/\mathscr{C}$) which is generated by all objects (Y, s) (respectively (s, Y)) with $s \in \Sigma$.

LEMMA 1.3. *Let Σ be a left calculable system of morphisms of $\mathscr{C}$. Then for any object X of $\mathscr{C}$, the category X/Σ is quasi-direct and connected.*

Proof. Let us consider in X/Σ the co-angle:

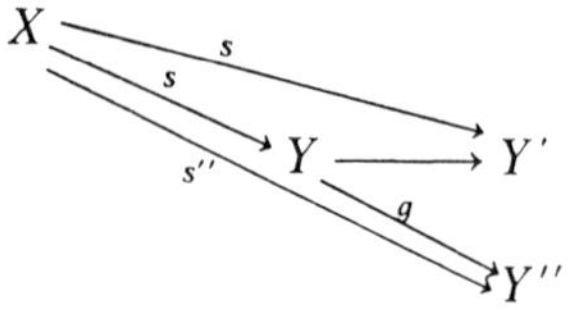

Then there exist morphisms $u: Y' \to Y'''$, $v: Y'' \to Y'''$ such that $u \in \Sigma$ and $us' = vs''$. Then $ufs = vgs$, so that a morphism $s''': Y''' \to Z$, $s''' \in \Sigma$ exists such that $s''' uf = s''' vg$. Hence $(s'''us', Z)$ is an object of X/Σ, that is, the above coangle has been embedded in a square which is commutative in X/Σ.

Now, let $f, g : (s, Y) \to (s', Y')$ be a couple of morphisms in X/Σ; then, one has that $fs = gs$ and according to the definition of a left-calculable system, a morphism $s'' : Y' \to Y''$ exists with $s'' \in \Sigma$ and $s'' f = s''g$. Obviously, $(s'' s', Y'')$ is an object in X/Σ and $s'' : (s', Y') \to (s''s', Y'')$ is a morphism such that $s''f = s''g$.

Obviously X/Σ is connected. ■

THEOREM 1.4. *Let $\mathscr{C}$ be a preadditive category and Σ a system of morphisms in $\mathscr{C}$ such that*:

(1) *Σ is left-calculable*;

(2) *For any object X of $\mathscr{C}$ the category X/Σ has a small cofinal subcategory denoted by $F(X)$.*

In these conditions, the category of additive fractions of $\mathscr{C}$ relative to Σ exists.

Proof. The objects of $\mathscr{C}_\Sigma$ are the same as the objects of $\mathscr{C}$. An object X of $\mathscr{C}$ when considered in $\mathscr{C}_\Sigma$ will be denoted by $\overline{X}$. Let X, Y be two objects in $\mathscr{C}$. We shall denote by $t_X{}^Y : Y/\Sigma \to \mathscr{Ab}$ the functor which is defined by taking $t_X{}^Y(s, Z) = \mathrm{Hom}_\mathscr{C}(X, Z)$ for any object (s, Z) of Y/Σ. If

$$f : (s, Z) \to (s', Z')$$

is a morphism in Y/Σ, then one has that $t_X{}^Y(f) = h^X(f)$. Then we define:

$$\mathrm{Hom}_{\mathscr{C}_\Sigma}(\overline{X}, \overline{Y}) = \varinjlim t_X{}^Y \tag{1}$$

(the existence of this inductive limit is ensured by Lemma 1.1). According to Lemma 1.2 one gets the following construction of the group (1). Let us consider the direct sum

$$M = \coprod_{(s,\,Z) \in \operatorname{Ob} F(Y)} t_X{}^Y(s, Z)$$

in the category $\mathcal{S}et$. We introduce in M the following equivalence relation: the elements $f \in t_X{}^Y(s, Z)$ and $g \in t_X{}^Y(s', Z')$ are equivalent if and only if there exists an object (s'', Z'') in $F(Y)$ and morphisms $u : (s, Z) \to (s'', Z'')$, $v : (s', Z') \to (s'', Z'')$ such that $t_X{}^Y(u)(f) = t_X{}^Y(v)(g)$, that is, such that $uf = vg$. In other words, there should exist the following commutative diagram:

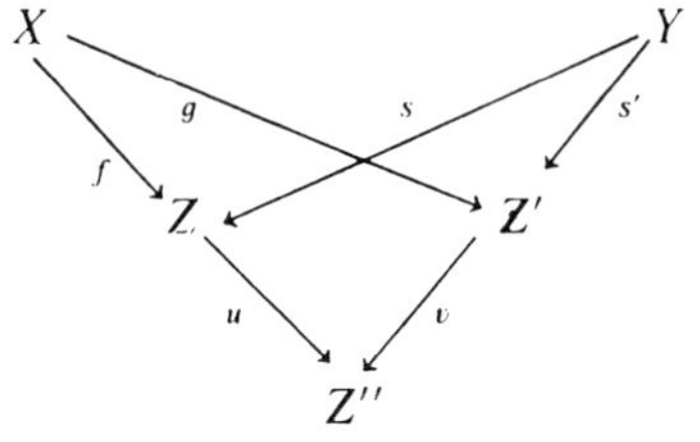

Obviously, $us = vs' \in \Sigma$.

Let $\alpha_{(s,Z)} : t_X{}^Y(s, Z) \to \operatorname{Hom}_{\mathscr{C}_\Sigma}(\overline{X}, \overline{Y})$ be the structural morphisms and let us put $\alpha_{(s,Z)}(f) = (s/f)$ for any $f \in \operatorname{Hom}_{\mathscr{C}}(X, Z) = t_X{}^Y(s, Z)$.

Furthermore, we must define the composition of these morphisms in $\mathscr{C}_\Sigma$. Let $(s/f) : \overline{X} \to \overline{Y}$, $(t/g) : \overline{Y} \to \overline{Z}$ be two morphisms in $\mathscr{C}_\Sigma$. Then one gets the following diagram in $\mathscr{C}$:

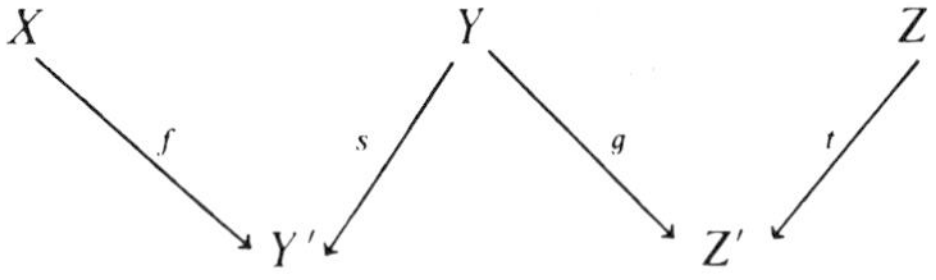

Since Σ is left-calculable, there exist morphisms $s' : Z' \to H, g' : Y' \to H$ such that $s' g = g' s$ and $s' \in \Sigma$. We shall define:

$$(t/g)\,(s/f) = (s'\,t/g'f). \tag{3}$$

It is not difficult to see that with the equivalence relation (2) introduced above one also has a well-defined composition law, that is, it does not depend on the representative elements used.

According to Lemmas 1.1 and 1.2, one can easily derive that $\operatorname{Hom}_{\mathscr{C}_\Sigma}(\overline{X}, \overline{Y})$ has a canonical structure of an abelian group and the composition of morphisms is biadditive: that is $\mathscr{C}_\Sigma$ is a preadditive category. The definition of the functor $T : \mathscr{C} \to \mathscr{C}_\Sigma$ is now obvious: for any object X in $\mathscr{C}$, let $T(X) = \overline{X}$, and for any morphism $f \in \mathrm{Hom}_{\mathscr{C}}(X, Y)$, let $T(f) = (1_Y/f) = \alpha_{(1_Y, Y)}(f)$. From the definition of the abelian group structure of $\mathrm{Hom}_{\mathscr{C}_\Sigma}(\bar{X}, \bar{Y})$ it follows immediately that T is additive.

The proof will be complete if we prove that $(T, \mathscr{C}_\Sigma)$ is a category of additive fractions of $\mathscr{C}$ with respect to Σ. In order to do this, let $s : X \to Y$ be a morphism of Σ. Also let us consider the morphism $(s/1_Y) \in \mathrm{Hom}_{\mathscr{C}_\Sigma}(\bar{Y}, \bar{X})$. Then according to (3), one has that $(1_Y/s)(s/1_Y) = (s/s)$. We derive that $(s/s) = 1_{\bar{X}}$. In the same way, one can derive that $(s/1_Y)(1_Y/s) = 1_{\bar{Y}}$. Hence the morphism $(s/1_Y)$ is the inverse of $(1_Y/s) = T(s)$, that is, $T(s)$ is an isomorphism in $\mathscr{C}_\Sigma$. Finally, let $T' : \mathscr{C} \to \mathscr{C}'$ be a covariant additive functor such that for every morphism $s \in \Sigma$, $T'(s)$ is an isomorphism. We shall define the functor $\bar{T}' : \mathscr{C}_\Sigma \to \mathscr{C}'$ as follows. For any object $\bar{X}$ write $\bar{T}'(\bar{X}) = T'(X)$. Also, for any morphism (s/f) we take $\bar{T}'(s/f) = (T'(s))^{-1} T'(f)$, with $(T'(s))^{-1}$ being the inverse of $T'(s)$. Simple computations show that $\bar{T}'$ is well defined and that $\bar{T}' T = T'$. The uniqueness of $\bar{T}'$ is now obvious. ∎

NOTES 1. The morphisms of $\mathscr{C}_\Sigma$ (constructed in the previous theorem) are usually called *left additive fractions* because the system Σ was previously assumed to be left-calculable. Because of this fact, the denominators, that is, the morphisms of Σ are written in the first place of left-fractions.

2. Naturally, this theorem has a dual which is obtained by replacing in the theorem the left-calculable systems by right-calculable ones. In this case, a morphism of $\mathscr{C}_\Sigma$ is written as (f/s) with $s \in \Sigma$. This means that the denominators are written on the right.

3. Remark that sometimes it will be more convenient to write the morphism (s/f) as $(s/1_{Y'})(1_{Y'}/f)$ with $s : Y \to Y'$ and $f : X \to Y'$, $s \in \Sigma$. But then $(s/1_{Y'}) = T(s)^{-1}$, and so $(s/f) = T(s)^{-1} T(f)$. Consequently the composition, of two morphisms of $\mathscr{C}_\Sigma$ is carried out in the following way:

$$(T(s)^{-1} T(f))(T(s'))^{-1} T(f') = T(s'' s')^{-1} T(f'' f')$$

with $f : X \to Y'$, $s : Y \to Y'$, $f' : Y \to Z'$, $s' : Z \to Z'$ being given morphisms with $s'' : Z' \to H$, $f'' : Y' \to H$ such that $f'' s = s'' f'$ and with s, s', s'' in Σ.

The following results will complete Theorem 1.4. Let us assume in the following that conditions of Theorem 1.4 are always fulfilled.

COROLLARY 1.5. *If $\mathscr{C}$ is an additive category, then $\mathscr{C}_\Sigma$ is also additive.*

Proof. Let X, Y be two objects in $\mathscr{C}$ and let $Z = X \amalg Y$. Also let $u : X \to Z$, $v : Y \to Z$ be the structural morphisms and p, q respectively the associated projections such that conditions (1) of Chapter 2 are fulfilled. Then, it is not difficult to see that $T(Z)$ is a direct sum of $T(X)$ and $T(Y)$, $T(u)$ and $T(v)$ are structural morphisms (see Chapter 2, Theorem 1.1). ∎

Corollary 1.6. *If $\mathscr{C}$ is a category with cokernels, then $\mathscr{C}_\Sigma$ also has cokernels, and T is right exact. Dually, if Σ is right-calculable, and if $\mathscr{C}$ is a category with kernels, then $\mathscr{C}_\Sigma$ also has kernels and T is left exact.*

Proof. Let $f : X \to Y$ be a morphism in $\mathscr{C}$ and $p : Y \to Y'$ a cokernel of f. Then, we have $T(pf) = T(p)\,T(f) = 0$. If $(s/g) : \overline{Y} \to \overline{Z}$ is a morphism in $\mathscr{C}_\Sigma$ such that $(s/g)\,T(f) = 0$, then we get in $\mathscr{C}$ the diagram:

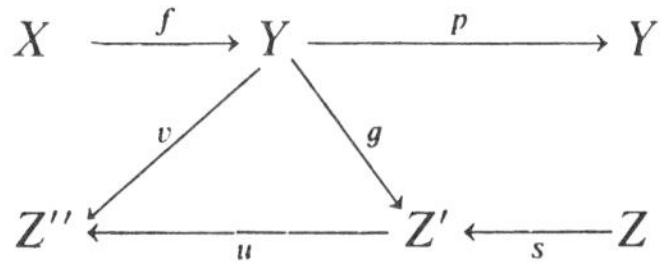

where $ug = v$, and u is a morphism in Σ, such that $(us/vf) = (s/g)\,T(f)$. Then according to (2), we may assume that $vf = 0$. This implies that a unique morphism $g' : Y' \to Z''$ exists such that $g'p = v$. In this case, according to (2) we get $(us/g')\,T(p) = (us/g')(1_{Y'}/p) = (us/g'p) = (us/v) = (us/ug) = (s/g)$. Hence T is right exact.

Now let us assume that $\mathscr{C}$ is a category with cokernels, and that

$$(s/f) : \overline{X} \to \overline{Y}$$

is a morphism in $\mathscr{C}_\Sigma$. Then, $s : Y \to Y'$, $f : X \to Y'$ are morphisms in $\mathscr{C}$, and $s \in \Sigma$. Furthermore, let (p, Y'') be a cokernel of f. Thus we obtain the following diagram:

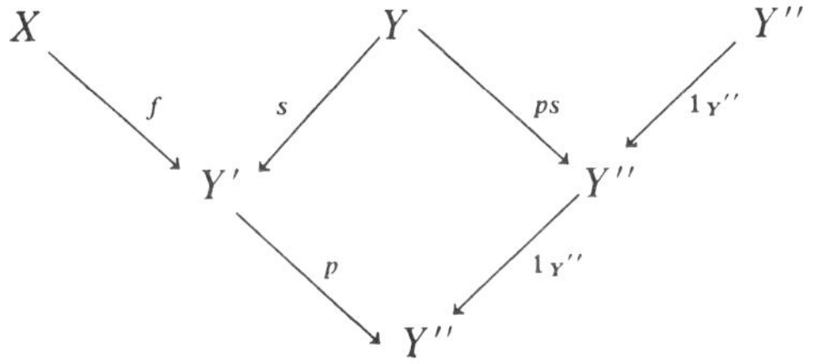

which commutes in $\mathscr{C}$. This implies that $(1_{Y''}/ps)(s/f) = (1_{Y''}/pf) = 0$. Let $(s'/g) : \overline{Y} \to \overline{Z}$ be a morphism, such that $(s'/g)(s/f) = 0$. Then we also have the following diagram:

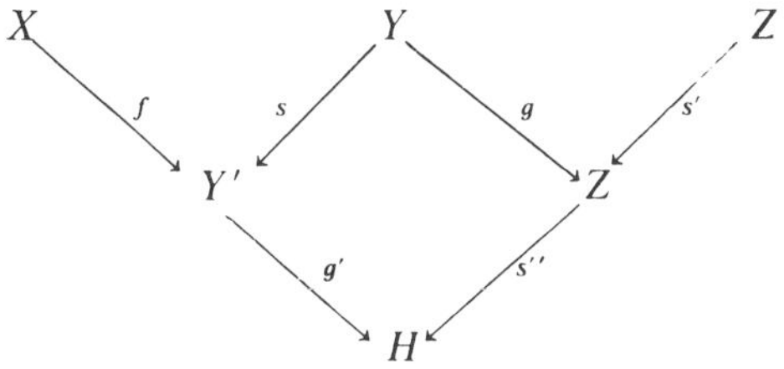

which commutes in $\mathscr{C}$ and from which we derive the composition $(s'/g)(s/f)$. According to the hypothesis, we have that $0 = (s'/g)(s/f) = (s''s'/g'f)$. Hence, according to (2) we may assume that $g'f = 0$. Then one can find a unique morphism $h: Y'' \to H$ such that $hp = g'$. This implies that $(s''s'/h)(1_{Y''}/ps) = (s''s'/hps) = (s''s'/g's)$. But $g's = s''g$ and consequently $(s''s'/g's) = (s''s'/s''g) = (s'/g)$. Finally, we can conclude that $((1_{Y''}/ps), \overline{Y}'')$ is a cokernel of (s/f). ∎

COROLLARY 1.7. *Assume that Σ is a bicalculable system, $\mathscr{C}$ is abelian and that for any X of $\mathscr{C}$, any category X/Σ and $(\Sigma/X)^0$ contains a small co-final subcategory. In these conditions, $\mathscr{C}_\Sigma$ is an abelian category and T is exact.*

Proof. According to Theorem 1.4 and its dual we have that there are in fact two constructions for $\mathscr{C}_\Sigma$. The first makes use of the fact that Σ is left-calculable and the second uses the fact that Σ is right-calculable. Both categories thus constructed are subject to the same universality condition and consequently, they can be identified with each other.

Then, on the basis of Corollary 1.6, we can derive that $\mathscr{C}_\Sigma$ is a preabelian category, and that T commutes with kernels and cokernels. In order to prove that $\mathscr{C}_\Sigma$ is abelian, it remains only to show that the parallel to a morphism from its canonical factorization is an isomorphism. Let

$$(s/f) \in \mathrm{Hom}_{\mathscr{C}_\Sigma}(\overline{X}, \overline{Y})$$

with $f: X \to Y'$, $s: Y \to Y'$, $s \in \Sigma$ being morphisms of $\mathscr{C}$. Also, let us consider the following commutative diagram of $\mathscr{C}$ which gives the canonical decomposition of f in $\mathscr{C}$:

$$\begin{array}{ccc} X & \xrightarrow{f} & Y' \\ {\scriptstyle p}\downarrow & & \uparrow{\scriptstyle j} \\ X' & \xrightarrow{\bar{f}} & Y'' \end{array}$$

$\bar{f}$ being, of course, the parallel of f. Then, we get another commutative diagram in $\mathscr{C}_\Sigma$:

$$\begin{array}{ccc} \overline{X} & \xrightarrow{T(f)} & \overline{Y}' \\ {\scriptstyle T(p)}\downarrow & & \uparrow{\scriptstyle T(j)} \\ \overline{X}' & \xrightarrow{T(\bar{f})} & \overline{Y}'' \end{array} \qquad (4)$$

where $T(p)$ is an epimorphism and $T(j)$ a monomorphism (T is left and right exact). According to Section 2.2 Chapter 2, we derive that the diagram (4) gives a canonical factorization in $\mathscr{C}_\Sigma$ of $T(f)$.

Then one has that:

$$(s/f) = T(s)^{-1}\,T(f) = T(s)^{-1}\,T(j)\,T(\bar{f})\,T(p), \tag{5}$$

where $T(s)^{-1}\,T(j)$ is a monomorphism. Obviously, (5) is a canonical factorization of (s/f) in $\mathscr{C}_\Sigma$ with $T(\bar{f})$ being the parallel of (s/f). Finally, it results that $\mathscr{C}_\Sigma$ is abelian. ∎

COROLLARY 1.8. *The hypothesis is the same as in Theorem* 1.4. *For a morphism $f : X \to Y$ of $\mathscr{C}$, the following assertions are equivalent*:

(1) $T(f) = 0$;

(2) *There exists a morphism $s : Y \to Z$, $s \in \Sigma$ such that $sf = 0$.*

Proof. (1) ⇒ (2). We have that $T(f) = (1_Y/f) = 0$. Then from the definition of $\mathrm{Hom}_{\mathscr{C}_\Sigma}(\bar{X}, \bar{Y})$, we may construct in $\mathscr{C}$ a commutative diagram

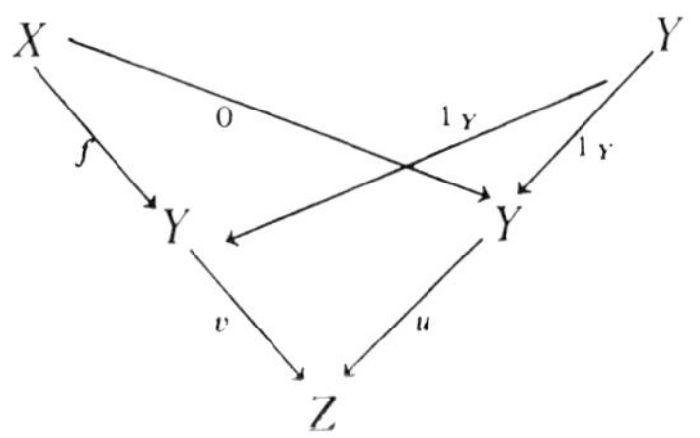

with $u, v \in \Sigma$ and where $vf = u0 = 0$ [in fact, this diagram is an expression of the equivalence of $(1_Y, f)$ with $(1_Y, 0)$, from the construction of the group $\mathrm{Hom}_{\mathscr{C}_\Sigma}(\bar{X}, \bar{Y})$ (see the proof of Theorem 1.4)].

(2) ⇒ (1). In this case, the above diagram exists for $u = v = s$ and thus one has that $T(f) = 0$. ∎

Let A be a ring. In the sequel by a "*ring of fractions*" we mean a ring of additive fractions conforming to the classical terminology.

Exercises

1. Let $T : \mathscr{C} \to \mathscr{C}'$ be a covariant functor and S a full and faithful right adjoint of T. Denote by Σ the class of all morphisms s in $\mathscr{C}$ such that $T(s)$ is an isomorphism. Prove that $(T, \mathscr{C}')$ is a category of fractions of $\mathscr{C}$ relative to Σ.

2. The hypothesis is the same as in Theorem 1.4. For a morphism

$$f : X \to Y,$$

show that the following assertions are equivalent:

(a) $T(f)$ is invertible;

(b) A commutative diagram:

$$\begin{array}{ccc} X & \xrightarrow{f} & Y \\ {\scriptstyle s}\downarrow & \overset{g}{\swarrow} & \downarrow{\scriptstyle s'} \\ X' & \xrightarrow{f'} & Y' \end{array}$$

exists in $\mathscr{C}$, with $s, s' \in \Sigma$.

3. Let $\mathscr{C}$ be a preabelian category, which is subject to the following conditions. If f is an epimorphism (respectively, g is a monomorphism) in the cartesian (respectively, cocartesian) square of $\mathscr{C}$:

$$\begin{array}{ccc} X & \xrightarrow{g} & Y \\ \downarrow & & \downarrow \\ Z & \xrightarrow{f} & T \end{array}$$

then g(respectively, f) is also an epimorphism (respectively, a monomorphism). Moreover, assume that $\mathscr{C}$ is locally small and colocally small. Prove that the system Σ of all bimorphisms of $\mathscr{C}$ is a bicalculable system. Also show that the category $\mathscr{C}_\Sigma$ exists and is abelian. In addition T is faithful.

4. Let A be a domain. Prove that the set S of non-zero elements of A is a multiplicative set. A is called a *left* (respectively, *right*) *Ore domain* if the set S is a left (respectively, right) calculable system. Prove that the ring A_S is a division ring. Moreover, a domain A is a left (respectively, right) Ore domain if and only if the intersection of any non-zero left (respectively, right)-ideals is a non-zero left (respectively, right) ideal.

References

Almkvist [7]; Bănică–Popescu [14], [15]; Cohn [39]; Elizarov [49]; Gabriel–Oberst [61]; Gabriel–Zisman [62]; Gabriel–Ulmer [63]; Giraud [65]; Ore [158]; Ouzilou [160]; Popescu [167]; Raikov [179]; Roos [199].

4.2 The spectral category of an abelian category

A very important example of a category of additive fractions is the spectral category associated to an abelian category.

Let $\mathscr{C}$ be an abelian category and E the system of all its essential monomorphisms. According to Corollary 10.3 of Chapter 3, E is a multiplicative system and all isomorphisms of $\mathscr{C}$ are contained in E.

LEMMA 2.1. *If in the cartesian square of* $\mathscr{C}$:

$$\begin{array}{ccc} X' & \xrightarrow{s'} & Y' \\ {\scriptstyle g}\downarrow & & \downarrow{\scriptstyle f} \\ X & \xrightarrow{s} & Y \end{array}$$

$s \in E$ *then also* $s' \in E$.

Proof. In fact, if $h: Y' \to Z$ is a morphism such that hs' is a monomorphism, then $(\ker h) \cap s'(X') = 0$ and thus, $f(\ker h) \cap s(X) = 0$, that is,

$$f(\ker h) = 0.$$

If $j: \ker h \to Y'$ is the canonical inclusion, then $fj = 0$. Consequently, we can find a unique morphism $t: \ker h \to X'$ with $s't = j$, and thus, $hs't = hj = 0$. Finally, $j = 0$ because hs' is a monomorphism. Hence, h is a monomorphism. ∎

COROLLARY 2.2. *The system E is right calculable.* ∎

Generally, we can find the conditions which are to be imposed on $\mathscr{C}$ such that the category $\mathscr{C}_E$ will be calculable (for example, $\mathscr{C}$ has to be locally small), but in the sequel we shall analyse only two important cases.

PROPOSITION 2.3. *For an abelian category* $\mathscr{C}$, *the following assertions are equivalent*:

(1) *Any object of* $\mathscr{C}$ *is injective*;

(2) *Any object of* $\mathscr{C}$ *is projective*;

(3) *Any monomorphism of* $\mathscr{C}$ *is a section*;

(4) *Any epimorphism of* $\mathscr{C}$ *is a retraction.* ∎

A category $\mathscr{C}$ defined as in Proposition 2.3 is called a *spectral* category.

COROLLARY 2.4. *Let A be a ring. Then the category* Mod A *is spectral if and only if A is semisimple.*

The proof follows by Chapter 3, Section 3.11. ∎

THEOREM 2.5. *Let* $\mathscr{C}$ *be a locally small abelian category which is subject to the following condition*: *if* X' *is any subobject of an object* X, *then* X' *is essential or* X' *has a complement. In these conditions, the category* $\mathscr{C}_E$ *is a spectral category.*

Proof. According to Corollary 2.2, E is right calculable and obviously, for any object X of $\mathscr{C}$, the category $(E/X)^0$ has a cofinal small subcategory. Let $P:\mathscr{C}\to\mathscr{C}_E$ be the canonical functor. According to Corollary 1.5 and Corollary 1.6, we have that $\mathscr{C}_E$ is an additive category with kernels. Now, let $(f/s):P(X)\to P(Y)$ be a morphism of $\mathscr{C}_E$ (according to Theorem 1.4, for any object $\overline{X}$ of $\mathscr{C}_E$, there exists a unique object X in $\mathscr{C}$ such that $P(X)=\overline{X}$), where $f:X'\to Y$, $s:X'\to X$ and $s\in E$. Then, we can construct in $\mathscr{C}$ the following diagram:

$$\begin{array}{ccccc} & & \ker f \amalg X_1' & \xrightarrow{f'} & f(X_1')\amalg Y' \\ & & \Big\downarrow{\scriptstyle i} & & \Big\downarrow{\scriptstyle j} \\ X & \xleftarrow{s} & X' & \xrightarrow{f} & Y \end{array}$$

in which X_1' is a complement of $\ker f$ (initially we assumed that (f/s) is not null, that is, $\ker f$ is not essential), Y' is a complement of $f(X_1')$, (or $Y'=0$), i,j are the sum morphisms, $f'=vf''ip$ with $p:\ker f\amalg X_1'\to X_1'$, $v:f(X_1')\to f(X_1')\amalg Y'$ being structural morphisms, and with

$$f'':X_1'\to f(X_1')$$

being canonically induced by f. Moreover, i and j are essential monomorphisms and thus, we have that $(fi/si)=(jf'/si)$. Consequently (f/s) is canonically isomorphic to (jf'/si). But one has the exact sequence of $\mathscr{C}$:

$$0\longrightarrow \ker f\xrightarrow{r}\ker f\amalg X_1'\xrightarrow{f'} f(X_1')\amalg Y'\xrightarrow{q} Y'\longrightarrow 0$$

r being the structural inclusion and q being the structural projection. But this sequence is composed of two short split exact sequences, and P is additive; then one obtains the following exact sequence of $\mathscr{C}_E$:

$$0\longrightarrow P(\ker f)\xrightarrow{P(r)} P(\ker f)\amalg P(X_1')\xrightarrow{P(f')} P(f(X_1'))$$
$$\amalg P(Y')\xrightarrow{P(q)} P(Y')\longrightarrow 0$$

which is also composed of two short split exact sequences. It is easily seen that $\mathscr{C}_E$ is a category in which any monomorphism is a section and any epimorphism is a retraction. Again, it may be easily checked that $\mathscr{C}_E$ is an abelian category, and naturally it is a spectral category too. ∎

NOTE. Generally, the functor P is not right exact as we shall see later in Chapter 5, Section 5.6.

THEOREM 2.6. *Let $\mathscr{C}$ be a locally small* Ab 5-*category. Then $\mathscr{C}_E$ is an* Ab 5-*category. Moreover, if $\mathscr{C}$ has a set of generators, then $\mathscr{C}_E$ also has a set of generators.*

Proof. According to Theorem 2.5, $\mathscr{C}_E$ is a spectral category and the canonical functor $P:\mathscr{C}\to\mathscr{C}_E$ is left exact. We shall prove first that $\mathscr{C}_E$ has direct sums. In order to do this, let $\{P(X_i)\}_{i\in I}$ be a set of objects in $\mathscr{C}_E$, $X=\coprod_{i\in I}X_i$ and also let $u_i:X_i\to X$ be the structural morphisms. We shall prove that $P(X)$ and $P(u_i)$ define a direct sum of the above named set of objects of $\mathscr{C}_E$. Suppose that for any i a morphism $(f_i/s_i):P(X_i)\to P(Y)$, is given with $f_i:X_i'\to Y$ and $s_i:X_i'\to X_i$, $s_i\in E$. Let $s:X'=\coprod_{i\in I}X_i'\to X$ be the sum morphisms. Then s is an essential monomorphism. In fact, if X'' is a nonzero subobject of X, then $X''=\sum_F(X''\cap\coprod_{j\in F}X_j)$, with F running through all finite subsets of I. But then

$$s^{-1}(X'')=\textstyle\sum_F s^{-1}(X''\cap\coprod_{j\in F}X_j)=\sum_F s^{-1}(X'')\cap\coprod_{j\in F}X_j'$$

and for a suitable F, we also have $s^{-1}(X'')\cap\coprod_{j\in F}X_j'\neq 0$, according to Corollary 10.2 of Chapter 3.

Now, let $f:X'\to Y$ be the morphism induced by the morphisms f_i. Thus, we get in $\mathscr{C}_E: P(f)P(s)^{-1}P(u_i)=P(f)P(u_i')P(s_i)^{-1}=P(fu_i')P(s_i)^{-1}$ $=P(f_i)P(s_i)^{-1}=(f_i/s_i)$, that is, $(f/s)P(u_i)=(f_i/s_i)$, $u_i':X_i'\to X'$ are structural morphisms. The uniqueness of (f/s) is obvious. Moreover, we have thus proved that the functor P commutes with direct sums. Now we have only to prove that $\mathscr{C}_E$ is an Ab 5-category. We shall apply Theorem 8.6 from Chapter 2. For that, we consider a set $\{P(X_i)\}_{i\in I}$ of objects in $\mathscr{C}_E$ and $P(X)$ a subobject of $\coprod_{i\in I}P(X_i)=P(\coprod_{i\in I}X_i)$. Let $(f/s):P(X)\to P(\coprod_{i\in I}X_i)$ be the inclusion morphism, with $s:X'\to X$, $s\in E$ and $f:X'\to\coprod_{i\in I}X_i$. According to Theorem 8.6 of Chapter 2, we have that

$$f(X')=\textstyle\sum_F(f(X')\cap\coprod_{j\in F}X_j),$$

with F running through all finite subsets of I.

Let us assume for instance, that P commutes with sums of direct sets of subobjects, that is, if $\{X_k\}_k$ is a direct set of subobjects of X, then

$$P(\textstyle\sum_k X_k)=\sum_k P(X_k).$$

Then we have:

$$P(X)=P(f(X'))=P(\textstyle\sum_F f(X')\cap\coprod_{j\in F}X_j)=\sum_F P(f(X')\cap\coprod_{j\in F}X_j)$$
$$=\textstyle\sum_F P(X)\cap P(\coprod_{j\in F}X_j)=\sum_F(P(X)\cap\coprod_{j\in F}P(X_j)).$$

Finally, we have to prove that P commutes with sums of direct sets of subobjects. In doing this, we shall make use of the following lemma:

LEMMA 2.7. *Let $\{X_j\}_j$ be a direct set of subobjects of X. Also suppose that for any j, an essential subobject X_j' of X_j is given, such that $\{X_j'\}_j$ is a direct set of subobjects. Then $\sum_j X_j'$ is also an essential subobject of $\sum_j X_j$.*

The proof is not difficult and is left to the reader. ∎

Furthermore, let $\{X_i\}_i$ be a direct and complete set of subobjects of X. Then, $\{P(X_i)\}_i$ is obviously a direct set of subobjects of $P(X)$. To complete the proof, it will be sufficient to show that $\sum_i P(X_i) = P(X)$. Let us assume the contrary, that is, suppose that $P(Y)$ is a supplement of $\sum_i P(X_i)$. Then, for any i, one has that $P(X_i) \cap P(Y) = 0$. Let $(v/s): P(Y) \to P(X)$ be the canonical inclusion. This implies that $v: Y' \to X$ is a monomorphism and that $s: Y' \to Y$ is an essential monomorphism. Suppose that $P(Y) \neq 0$ and that Y' is a subobject (not necessarily null) of X. Then, $Y' = \sum_i(X_i \cap Y')$ and $Y' \cap X_i \neq 0$ for an index i. But then, we should have that

$$P(X_i \cap Y') = P(X_i) \cap P(Y') = P(X_i) \cap P(Y) \neq 0,$$

and this contradicts the conditions of our Theorem. Consequently, we must have that $\sum_i P(X_i) = P(X)$.

Finally, let $\{U_i\}_i$ be a set of generators of $\mathscr{C}$ and let $\{V_j\}_j$ be the set of all non-null quotient objects of U_i, for any i. Then, $\{P(V_j)\}_j$ is a set of generators of $\mathscr{C}_E$. Indeed, if $(f/s): P(X) \to P(Y)$ is a monomorphism which is not an isomorphism with $f: X' \to Y$ and with $s: X' \to X$, $s \in E$, then f is not essential and an index j exists such that $V_j \subseteq Y$ and $V_j \cap \operatorname{im} f = 0$. Obviously, the canonical inclusion $P(V_j) \to P(Y)$ cannot be factorized through $P(X)$. With this, the proof of the Theorem is complete. ∎

THEOREM 2.8. *Let $\mathscr{C}$ be an* Ab 5-*spectral category and U a generator of $\mathscr{C}$. Then, the ring $A = \operatorname{End}_{\mathscr{C}}(U)$ is a regular ring and injective as an A^0-module. Moreover, the functor*

$$S: \mathscr{C} \to \operatorname{Mod} A^0$$

defined in the Theorem 7.9, *Chapter* 3, *defines an equivalence between $\mathscr{C}$ and the full subcategory of* Mod A^0 *consisting of all objects Y which are direct summands of an object of the form A^I, I being a set.*

Proof. By Exercise 1 we see that A is regular. Also by Theorem 7.9 and 2.8 of Chapter 3 we deduce that $S(U) = A$ is an injective A^0-module. Now, let X be an object of $\mathscr{C}$. Then X is a direct summand of U^I for a suitable set I, and thus $S(X)$ is a direct summand of $S(U^I) = A^I$. Conversely, let Y be a direct summand of A^I, for a set I, T a left adjoint of S and $u: \operatorname{IdMod} A^0 \to S\,T$ an arrow of adjunction. Then we have in Mod A^0 the commutative diagram:

$$\begin{array}{ccc} A^I & \xrightarrow{u_{A^I}} & S\,T(A^I) \\ \uparrow{\scriptstyle i} & & \uparrow{\scriptstyle ST(i)} \\ X & \xrightarrow{u_X} & S\,T(X) \end{array}$$

i being the canonical inclusion. We see that $u_{A\iota}$ is an isomorphism, and $ST(i)$ is a section. A slight computation shows that u_X is also an isomorphism.

The Theorem follows from Theorem 7.9 of Chapter 3. ∎

Exercises

1. Let $\mathscr{C}$ be a spectral category. Then, for any object X of $\mathscr{C}$ the ring $\mathrm{End}_{\mathscr{C}}(X)$ is regular.

2. Let $\mathscr{C}$ be a locally small abelian category and Q an injective object of $\mathscr{C}$. Denote by $\mathfrak{a}$ the ideal of $\mathrm{End}_{\mathscr{C}}(Q)$ consisting of all endomorphisms $f : Q \to Q$ such that $\ker f$ is essential. Then we have a canonical isomorphism of rings:

$$\mathrm{End}_{\mathscr{C}}(Q) \quad \mathfrak{a} \simeq \mathrm{End}_{\mathscr{C}_E}(P(Q)).$$

3. Let $\mathscr{C}$ be a locally small abelian category with injective envelopes. For any two objects X, Y of $\mathscr{C}$, the following assertions are equivalent:

(a) $P(X)$ is isomorphic with $P(Y)$;

(b) There is an object Z of $\mathscr{C}$ with essential monomorphisms $u : Z \to X$ and $v : Z \to Y$;

(c) X and Y have isomorphic injective envelopes.

References

Gabriel–Oberst [61]; Gabriel–Zisman [62]; Johnson [99]; Năstăsescu–Popescu [142]; Popescu [168]; Roos [192], [193], [194], [195], [196], [197], [199]; Tisseron [226], [227].

4.3 The quotient category of an abelian category relative to a dense subcategory

A full subcategory $\mathscr{A}$ of an abelian category $\mathscr{C}$ is called *dense* if for any exact sequence from $\mathscr{C}$:

$$0 \to X' \to X \to X'' \to 0$$

X is in $\mathscr{A}$ if and only if both X' and X'' are in $\mathscr{A}$.

COROLLARY 3.1. *Let $\mathscr{A}$ be a dense subcategory in $\mathscr{C}$. The following assertions are true:*

(1) *If $X \in \mathrm{Ob}\mathscr{A}$ and $X \simeq X'$, then $X' \in \mathrm{Ob}\mathscr{A}$;*

(2) *If $X, Y \in \mathrm{Ob}\mathscr{A}$, then $X \amalg Y \in \mathrm{Ob}\mathscr{A}$;*

(3) *If $X \in \mathrm{Ob}\mathscr{A}$, then any subobject or quotient object of X is also in $\mathscr{A}$.* ∎

Let $\mathscr{A}$ be a dense subcategory of $\mathscr{C}$. We shall denote by $\sum_{\mathscr{A}}$ or simply by Σ (if no confusion is possible) the system of all morphisms s of $\mathscr{C}$ such that $\ker s$ and $\operatorname{coker} s$ are in $\mathscr{A}$.

PROPOSITION 3.2. *The system Σ is a bicalculable system of morphisms.*

Proof. Using duality, it will suffice to prove that Σ is right calculable.

Let us notice first that Σ is a multiplicative system. Indeed, if $s: X \to Y$ and $s': Y \to Z$ are morphisms in Σ then from the canonically induced exact sequence

$$0 \to \ker s \to \ker(s's) \to (\operatorname{im} s) \cap \ker s' \to 0,$$

one can derive that $\ker(s's) \in \operatorname{Ob}\mathscr{A}$. Dually, one can also derive that $\operatorname{coker}(s's) \in \operatorname{Ob}\mathscr{A}$. These imply that $s's \in \Sigma$.

Now, consider the angle: $X \xrightarrow{s} Y \xleftarrow{f} Z$ with $s \in \Sigma$. Then the following cartesian square:

$$\begin{array}{ccc} Z \prod_Y X & \xrightarrow{f'} & X \\ {\scriptstyle s'}\downarrow & & \downarrow{\scriptstyle s} \\ Z & \xrightarrow{f} & Y \end{array}$$

can be constructed with $\ker s' \simeq \ker s$, and with $\operatorname{coker} s'$ being a canonical subobject of $\operatorname{coker} s$ (see Section 2.5 of Chapter 2). Obviously, one has that $s' \in \Sigma$.

Furthermore, let

$$X \xrightarrow{f} Y \xrightarrow{s} Z$$

be morphisms of $\mathscr{C}$, with $s \in \Sigma$ and such that $sf = 0$. Then $\operatorname{im} f \subseteq \ker s$ and from the exact sequence:

$$0 \longrightarrow \ker f \xrightarrow{u} X \xrightarrow{p} \operatorname{im} f \longrightarrow 0$$

one derives that $u \in \Sigma$. Obviously we now have that $fu = 0$. ∎

THEOREM 3.3. *Let $\mathscr{A}$ be a dense subcategory of a locally small abelian category $\mathscr{C}$ and let $\Sigma = \Sigma_{\mathscr{A}}$. Then the category $\mathscr{C}_\Sigma$ of additive fractions of $\mathscr{C}$ relative to Σ exists.*

Proof. As before we consider the objects of $\mathscr{C}_\Sigma$ to be the same as the objects of $\mathscr{C}$. An object X of $\mathscr{C}$ when considered in $\mathscr{C}_\Sigma$ will be denoted by $\overline{X}$. Let X, Y be two objects of $\mathscr{C}$. Let us denote by $\mathscr{L}(X, Y)$ the following directed set. The elements of $\mathscr{L}(X, Y)$ are couples (X', Y') with X' being a subobject in X such that $X/X' \in \mathrm{Ob}\mathscr{A}$ and with Y' being a subobject of Y in $\mathscr{A}$. Take $(X', Y') \leqslant (X_1', Y_1')$ if and only if $X_1' \subseteq X'$ and $Y' \subseteq Y_1'$. If (X', Y') and (X_1', Y_1') are two elements in $\mathscr{L}(X, Y)$, then $(X', Y') \leqslant (X' \cap X_1', Y' + Y_1')$, $(X_1', Y_1') \leqslant (X' \cap X_1', Y' + Y_1')$ and obviously

$$(X' \cap X_1', Y' + Y_1') \in \mathscr{L}(X, Y).$$

Now let $q(X, Y) : \mathscr{L}(X, Y) \to \mathscr{Ab}$ be the functor which is defined by the equality $q(X, Y)(X', Y') = \mathrm{Hom}_{\mathscr{C}}(X', Y/Y')$. If $(X', Y') \leqslant (X_1', Y_1')$, then the canonical morphisms $u : X_1' \to X'$, $v : Y/Y' \to Y/Y_1'$ define a morphism of groups $\mathrm{Hom}_{\mathscr{C}}(X', Y/Y') \to \mathrm{Hom}_{\mathscr{C}}(X_1', Y/Y_1')$ which associates the morphism $vfu : X_1' \to Y/Y_1'$ to the morphism $f : X' \to Y/Y'$. Now we introduce a new definition:

$$\mathrm{Hom}_{\mathscr{C}_\Sigma}(\bar{X}, \bar{Y}) = \varinjlim q(X, Y). \tag{6}$$

Obviously, (6) has canonically a structure of an abelian group. We define the composition of morphisms as follows: If $\alpha \in \mathrm{Hom}_{\mathscr{C}_\Sigma}(\bar{X}, \bar{Y})$ and $\beta \in \mathrm{Hom}_{\mathscr{C}}(\bar{Y}, \bar{Z})$ then α and β are represented by the morphisms $f : X' \to Y/Y'$ and $g : Y'' \to Z/Z'$ respectively with $(X', Y') \in \mathscr{L}(X, Y)$ and with

$$(Y'', Z') \in \mathscr{L}(Y, Z)$$

correspondingly. Let $p : Y \to Y/Y'$ and $q : Z \to Z/Z'$ be canonical epimorphisms. Take $X'' = f^{-1}(p(Y''))$ and $Z'' = q^{-1}(g(Y' \cap Y''))$. Then $(X'', Z'') \in \mathscr{L}(X, Z)$. In other words, we have now to prove that X/X'' and Z'' are in $\mathscr{A}$. Indeed, consider the exact sequence:

$$0 \to X'/X'' \to X/X'' \to X/X' \to 0$$

and remark that X'/X'' is a canonical subobject of

$$(Y/Y')/p(Y'') \simeq Y/(Y' + Y''),$$

that is, it is an object of $\mathscr{A}$. Thus, X/X'' is an object of $\mathscr{A}$. Furthermore, $Y' \cap Y''$ is an object in $\mathscr{A}$. Hence, $g(Y' \cap Y'')$ is in $\mathscr{A}$ as a quotient object of $Y' \cap Y''$. Then, from the canonically constructed exact sequence

$$0 \to Z' \to Z'' \to g(Y' \cap Y'') \to 0$$

we derive that Z'' is an object in $\mathscr{A}$.

Let f' be the composition of morphisms: $X'' \to p(Y'') \simeq Y''/Y'' \cap Y'$ with the first morphism being induced by f. Also, let $g' : Y''/Y'' \cap Y' \to Z/Z''$ be the canonical morphism induced by g. Then, we shall define $\beta\alpha$ as the image of $g'f'$ in the group $\mathrm{Hom}_{\mathscr{C}_\Sigma}(\bar{X}, \bar{Z})$. Through a simple computation we derive that the definition of the composition law does not depend on representative elements and is biadditive. The assignments $X \rightsquigarrow \bar{X}$ and $f \rightsquigarrow \bar{f}$ (with $\bar{f}$ being canonically defined) define a covariant additive functor $T : \mathscr{C} \to \mathscr{C}_\Sigma$.

Now, let $s : X \to Y$ be a morphism of Σ. From the following commutative diagram which defines the canonical factorization of s:

$$\begin{array}{ccc} X & \xrightarrow{\ s\ } & Y \\ {\scriptstyle p}\downarrow & & \uparrow{\scriptstyle j} \\ \operatorname{coim} s & \xrightarrow{\ \bar{s}\ } & \operatorname{im} s \end{array}$$

we derive that the image of $(\bar{s})^{-1}$ in the group $\operatorname{Hom}_{\mathscr{C}_\Sigma}(\bar{Y}, \bar{X})$ is an inverse of $T(s)$. That is, for any element s in Σ, $T(s)$ is an isomorphism in $\mathscr{C}_\Sigma$. Furthermore, we notice that any morphism $\alpha : \bar{X} \to \bar{Y}$ in $\mathscr{C}_\Sigma$ can be written as $T(s)^{-1}\, T(f)\, T(s')^{-1}$ with s, s' being in Σ.

In order to complete the proof, it will be sufficient to show that for any covariant functor $T' : \mathscr{C} \to \mathscr{C}'$ such that $T'(s)$ is an isomorphism for any element in Σ, a unique functor $\bar{T}' : \mathscr{C}_\Sigma \to \mathscr{C}'$ exists such that $\bar{T}'\, T = T'$. This functor $\bar{T}'$ is defined by the equality $\bar{T}'\,(X) = T'\,(X)$ and for any morphism

$$\alpha = T(s)^{-1}\, T(f)\, T(s')^{-1},$$

we put $\bar{T}'(\alpha) = T'(s)^{-1}\, T'(f)\, T'(s')^{-1}$. It is not difficult to see that if $T(s)^{-1}\, T(f)\, T(s')^{-1} = T(s_1)^{-1}\, T(f_1)\, T(s_1')^{-1}$, then also,

$$T'(s)^{-1}\, T'(f)\, T'(s')^{-1} = T'(s_1)^{-1}\, T'(f_1)\, T'(s_1')^{-1}.$$

Now the uniqueness of $\bar{T}'$ is obvious.

An alternative proof. Let Σ' be the system of all monomorphisms in Σ. Then, Σ' is a bicalculable system and for any object X of $\mathscr{C}$ the category Σ'/X has a coinitial subcategory which consists of all subobjects X' of X so that X/X' is in $\mathscr{A}$. Then, we can construct the category $\mathscr{C}_{\Sigma'}$ as in Theorem 1.4, and consequently the canonical functor $T' : \mathscr{C} \to \mathscr{C}_{\Sigma'}$ is left exact. Let $\bar{\Sigma}''$ be the system of all morphisms α in $\mathscr{C}_{\Sigma'}$ which can be inserted in the commutative diagram:

$$\begin{array}{ccc} T'(X) & \xrightarrow{\ T'(s)\ } & T'(X'') \\ \wr\!\| & & \|\!\wr \\ T'(X_1) & \xrightarrow{\ \alpha\ } & T'(X_1'') \end{array}$$

where $s : X \to X''$ is an epimorphism in Σ and where the vertical morphisms are isomorphisms. Also, let us denote by Σ'' the class of all epimorphisms of Σ. Now we shall prove that $\bar{\Sigma}''$ is a left-calculable system in $\mathscr{C}_{\Sigma'}$. Obviously we may suppose that any morphism in $\bar{\Sigma}''$ has the form $T'(s'')$ with $s'' \in \Sigma''$.

Further, let:

$$\begin{array}{ccc} T'(X) & \xrightarrow{T'(s'')} & T'(Y) \\ {\scriptstyle (f/s')}\downarrow & & \\ T'(Z) & & \end{array}$$

be a co-angle in $\mathscr{C}_{\Sigma'}$ with $f: X' \to Z$, $s': X' \to X$, and with $s' \in \Sigma'$. Then $s''s' \in \Sigma$ and the canonical morphisms $u: X' \to \operatorname{im}(s''s')$ and $v: \operatorname{im}(s''s') \to Y$ are in Σ'' and Σ' respectively. Thus, we obtain the following commutative diagram of $\mathscr{C}$:

$$\begin{array}{ccc} X & \xrightarrow{s''} & Y \\ {\scriptstyle s'}\uparrow & & \uparrow{\scriptstyle v} \\ X' & \xrightarrow{u} & \operatorname{im}(s''s') \\ {\scriptstyle f}\downarrow & & \downarrow{\scriptstyle f'} \\ Z & \xrightarrow{u''} & Z \coprod_{X'} (\operatorname{im} s''s') \end{array}$$

in which $u'' \in \Sigma''$. Then $T'(s')$ and $T'(v)$ are isomorphisms, and thus

$$\begin{aligned} T'(u'')\,T'(f)\,T'(s')^{-1} &= T'(f')\,T'(u)\,T'(s')^{-1} = T'(f')\,T'(v)^{-1}\,T'(s'') \\ &= (f'/v)\,T'(s''). \end{aligned}$$

These relations show that $\bar{\Sigma}''$ is left permutable.

Now, we must prove that $\bar{\Sigma}''$ is left simplifiable. In order to do this, we shall prove first that $T'(s'')$ is an epimorphism in $\mathscr{C}_{\Sigma'}$ for any $s'' \in \Sigma''$. Indeed, if $s'': X \to Y$ and $(f/s'): T'(Y) \to T'(Z)$ are morphisms such that

$$(f/s')\,T'(s'') = 0,$$

then we may assume that s' is an identity and that $(f/s') = T'(f)$. This implies that $T'(fs'') = 0$. According to the dual of Corollary 1.8 a morphism $u: X' \to X$ of Σ' exists such that $fs''u = 0$. Since the canonical morphism $v: \operatorname{im}(s''u) \to Y$ is in Σ', we have $fv = 0$, that is if we apply once more the dual of Corollary 1.8 then we get $T'(f) = 0$. From the definition of the system $\bar{\Sigma}''$ we derive that for any object $T'(X)$ of $\mathscr{C}_{\Sigma'}$, the category $T'(X)/\bar{\Sigma}''$ has a cofinal small subcategory. Also, according to Theorem 1.4, we can construct the category $(\mathscr{C}_{\Sigma'})_{\bar{\Sigma}''}$. Let $T'': \mathscr{C}_{\Sigma'} \to (\mathscr{C}_{\Sigma'})_{\bar{\Sigma}''}$ be the canonical functor and let $T = T''\,T'$. Then $((\mathscr{C}_{\Sigma'})_{\bar{\Sigma}''}, T)$ is a category of additive fractions of $\mathscr{C}$ relatively to Σ.

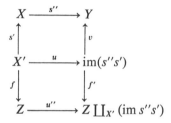

In order to prove this, let $s \in \Sigma$. Then $s = s' s''$, where $s'' \in \Sigma''$ and $s' \in \Sigma'$. Then it is clear that $T(s)$ is an isomorphism. Furthermore, let $H : \mathscr{C} \to \mathscr{C}'$ be a covariant additive functor so that the morphism $H(s)$ is an isomorphism in $\mathscr{C}'$ for any morphism s in Σ. But $\Sigma' \subseteq \Sigma$ and thus a unique functor

$$\bar{H}' : \mathscr{C}_{\Sigma'} \to \mathscr{C}'$$

exists such that $\bar{H}' T' = H$. Now, if $s'' \in \Sigma''$, one has that $H(s'')$ is an isomorphism in $\mathscr{C}'$ and consequently, $\bar{H}'(T'(s'')) = H(s'')$ is also an isomorphism. Again, a unique functor $\bar{H} : (\mathscr{C}_{\Sigma'})_{\bar{\Sigma}''} \to \mathscr{C}'$ may be found such that $\bar{H} T'' = \bar{H}'$. Finally, we see that $\bar{H} T = \bar{H} T'' T' = (\bar{H} T'') T' = \bar{H}' T' = H$. It is clear that $\bar{H}$ is the unique functor which has the property that $\bar{H} T = H$. ∎

NOTE. We could construct first the category $\mathscr{C}_{\Sigma''}$ and then the category $(\mathscr{C}_{\Sigma''})_{\bar{\Sigma}'}$. In the end, we would have obtained the same thing, that is $(\mathscr{C}_{\Sigma''})_{\bar{\Sigma}'}$ is equivalent with $(\mathscr{C}_{\Sigma'})_{\bar{\Sigma}''}$.

Let us consider $\mathscr{C}$ and $\mathscr{A}$ as in Theorem 3.3. Then the category $\mathscr{C}_{\Sigma_{\mathscr{A}}}$ will usually be denoted by $\mathscr{C}/\mathscr{A}$ and it will be called the *quotient category* of $\mathscr{C}$ relative to $\mathscr{A}$. In the sequel $\mathscr{C}$ and $\mathscr{A}$ will be always considered as in Theorem 3.3. However, we shall still preserve the notations introduced above.

LEMMA 3.4. *Let $f : X \to Y$ be a morphism in $\mathscr{C}$. Then $T(f) = 0$ if and only if the image of f is an object of $\mathscr{A}$.*

Proof. Let us suppose first that $\operatorname{im} f \in \operatorname{Ob}\mathscr{A}$. Then the canonical morphism $u : \ker f \to X$ is in Σ' and thus, $T'(f) = 0$ according to the dual of Corollary 1.8.

Conversely, assume now that $T(f) = 0$. Then, $T''(T'(f)) = 0$. Thus, a morphism $s'' \in \Sigma''$ can be found so that $T'(s'')\, T'(f) = 0$ (see Corollary 1.8). Consequently, a morphism $s' \in \Sigma'$ exists such that $s'' f s' = 0$. Consider now the following canonically constructed exact sequence of $\mathscr{C}$:

$$0 \to f^{-1}(\ker s'')/\ker f \to X/\ker f \to X/f^{-1}(\ker s'') \to 0$$

where $f^{-1}(\ker s'')/\ker f \simeq (\operatorname{im} f) \cap \ker s''$, hence an object in $\mathscr{A}$ and $X/f^{-1}(\ker s'')$ is a quotient object of $X/\operatorname{im} s'$, also an object of $\mathscr{A}$. Finally one obtains that $\operatorname{im} f \simeq X/\ker f$ is an object in $\mathscr{A}$. ∎

LEMMA 3.5. *Let $f : X \to Y$ be a morphism in $\mathscr{C}$. Then, $T(f)$ is a monomorphism, (respectively an epimorphism), if and only if* $\ker f \in \operatorname{Ob}\mathscr{A}$ *(respectively* $\operatorname{coker} f \in \operatorname{Ob} \mathscr{A}$).

Proof. Because of the duality (see the above note), it will be sufficient to prove that $T(f)$ is a monomorphism if and only if $\ker f \in \operatorname{Ob}\mathscr{A}$.

Indeed, let us suppose that $\ker f \in \mathrm{Ob}\mathscr{A}$ and that α is a morphism in $\mathscr{C}_\Sigma$ so that $T(f)\,\alpha = 0$. We may assume that $\alpha = T''(\beta)$ with β being a morphism of $\mathscr{C}_{\Sigma'}$ and moreover with $\beta = T(g), g : Z \to X$ being morphism in $\mathscr{C}$. Thus, $T(fg) = 0$, and according to the above lemma, one has that

$$\mathrm{im}(fg) \in \mathrm{Ob}\mathscr{A}.$$

The ends of the canonically constructed exact sequence

$$0 \to \ker(fg)/\ker g \to Z/\ker g \to Z/\ker(fg) \to 0$$

are in $\mathscr{A}$,$\big(\ker(fg)/\ker g \simeq (\mathrm{im}\, g) \cap (\ker f)$ and $Z/\ker(fg) \simeq \mathrm{im}(fg)\big)$, therefore $Z/\ker g \simeq \mathrm{im}\, g \in \mathrm{Ob}\mathscr{A}$. Then from the Corollary 3.4 one also has that $T(g) = 0$.

Conversely, let us assume that $T(f)$ is a monomorphism and that

$$u : \ker f \to X$$

is a kernel of f. Then $T(f)\,T(u) = 0$ and consequently, $T(u) = 0$. According to Lemma 3.4, this implies that $\mathrm{im}\, u \in \mathrm{Ob}\mathscr{A}$. ∎

LEMMA 3.6. *The category $\mathscr{C}/\mathscr{A}$ is preabelian. Moreover, if $f : X \to Y$ is a morphism in $\mathscr{C}$ and if (X', i) (respectively (p, Y')) is a kernel (respectively, cokernel) of f, then $\big(T(X'), T(i)\big)$, (respectively $\big(T(p), T(Y')\big)$ is a kernel (respectively, cokernel) of $T(f)$.*

Proof. According to the dual of Corollary 1 6, $\big(T'(X'), T'(i)\big)$ is a kernel of $T'(f)$. If $\alpha : T(Z) \to T(X)$ is a morphism of $\mathscr{C}_\Sigma$, such that $T(f)\,\alpha = 0$ one can reduce it canonically to the case when $\alpha = T(g)$. Then, $T(fg) = 0$ and thus a morphism $s' : Z' \to Z$, $s' \in \Sigma'$ exists such that $fgs' = 0$. This implies that a morphism $h : Z' \to X'$ can be found so that $ih = gs'$. Hence

$$T(i)\,T(h)\,T(s')^{-1} = T(g).$$

Now, Corollary 3.5 directly implies that $T(i)$ is a monomorphism. ∎

LEMMA 3.7. *Let $f : X \to Y$ be a morphism in $\mathscr{C}$. Then, the morphism $T(f)$ is an isomorphism if and only if* $\ker f$ *and* $\operatorname{coker} f$ *are in $\mathscr{A}$.*

Proof. If $\ker f$ and $\operatorname{coker} f$ are in $\mathscr{A}$, then $f \in \Sigma$ and so $T(f)$ is an isomorphism. According to Lemma 3.5 the converse assertion is obvious. ∎

A simple glance at the above lemmas enables us to prove the following result.

THEOREM 3.8. *The category $\mathscr{C}/\mathscr{A}$ is abelian and the canonical functor*

$$T : \mathscr{C} \to \mathscr{C}/\mathscr{A}$$

is exact. ∎

When studying the category $\mathscr{C}/\mathscr{A}$ the following lemma will be very useful.

LEMMA 3.9. *Any morphism of $\mathscr{C}/\mathscr{A}$ can by written (and this form is not necessarily unique) as:*

$$T(s'')^{-1}\, T(f)\, T(s')^{-1}$$

with $s' \in \Sigma'$, $s'' \in \Sigma''$.

The proof is a simple computation which involves the construction of $\mathscr{C}/\mathscr{A}$ given in the second proof of Theorem 3.3. ∎

COROLLARY 3.10. *Let*

$$0 \longrightarrow \bar{X}' \xrightarrow{\ u\ } \bar{X} \xrightarrow{\ v\ } \bar{X}'' \longrightarrow 0$$

be an exact sequence in $\mathscr{C}/\mathscr{A}$. Then, an exact sequence

$$0 \longrightarrow Y' \xrightarrow{\ i\ } Y \xrightarrow{\ p\ } Y'' \longrightarrow 0$$

exists in $\mathscr{C}$, so that we have the commutative diagram in $\mathscr{C}/\mathscr{A}$

$$\begin{array}{ccccccccc} 0 & \longrightarrow & \bar{X}' & \xrightarrow{\ u\ } & \bar{X} & \xrightarrow{\ v\ } & \bar{X}'' & \longrightarrow & 0 \\ & & \wr\!\downarrow & & \wr\!\downarrow & & \wr\!\downarrow & & \\ 0 & \longrightarrow & T(Y') & \xrightarrow{T(i)} & T(Y) & \xrightarrow{T(p)} & T(Y'') & \longrightarrow & 0 \end{array}$$

where all vertical arrows are isomorphisms.

Proof. According to Lemma 3.9, the morphism u is of the form

$$T(s'')^{-1}\, T(f)\, T(s')^{-1}$$

with $s' : R \to X'$, $f : R \to H$, $s'' : X \to H$, $T(X') = \bar{X}'$ and $T(X) = \bar{X}$. But u is a monomorphism and so thus, from Lemma 3.5, one has that

$$\ker f \in \mathrm{Ob}\mathscr{A}.$$

Also v is of the form $T(s_1'')^{-1}T(f_1)T(s_1')^{-1}$ with

$$s'_1 : R_1 \to X, \qquad f_1 : R_1 \to H_1, \qquad s''_1 : X'' \to H_1$$

and with $T(X'') = \bar{X}''$. Again, from Lemma 3.5, one has that coker $f_1 \in \mathrm{Ob}\mathscr{A}$. Let $Y = R_1$, $Y'' = \mathrm{im}\, f_1$ and let $p : Y \to Y''$ be the canonical epimorphism.

Also, let $Y' = \ker p$ and $i: Y' \to Y$ the natural inclusion. The natural inclusion $h: Y'' \to H_1$ is a morphism of Σ and thus one obtains the following diagram in $\mathscr{C}/\mathscr{A}$:

$$\begin{array}{ccccccccc}
0 \longrightarrow & \overline{X}' & \xrightarrow{T(s'')^{-1}T(f)T(s')^{-1}} & \overline{X} & \xrightarrow{T(s''_1)^{-1}T(f_1)T(s'_1)^{-1}} & \overline{X}'' & \longrightarrow 0 \\
& \downarrow{\scriptstyle \alpha} & & \downarrow{\scriptstyle T(s'_1)^{-1}} & & \downarrow{\scriptstyle T(h)^{-1}T(s''_1)} & \\
0 \longrightarrow & T(Y') & \xrightarrow{T(i)} & T(Y) & \xrightarrow{T(p)} & T(Y'') & \longrightarrow 0
\end{array}$$

where the last square is commutative and where the vertical arrows are isomorphisms. Finally, an isomorphism $\alpha: \overline{X}' \to T(Y')$ exists such that the first square is also commutative. ■

COROLLARY 3.11. *Let $H: \mathscr{C} \to \mathscr{C}'$ be an exact functor so that $H(X) = 0$ for any object X in $\mathscr{A}$. Then, a unique functor $\overline{H}: \mathscr{C}/\mathscr{A} \to \mathscr{C}'$ exists with $\overline{H}\,T = H$.*

The proof is a direct consequence of Lemma 3.7. ■

COROLLARY 3.12. *Let $G: \mathscr{C}/\mathscr{A} \to \mathscr{C}'$ be a covariant functor. Then G is exact if and only if $G\,T$ is exact. In particular, the functor $\overline{H}$ of the above corollary is exact.*

The proof is a consequence of Corollary 3.11. ■

Exercises

1. Let A be an integral domain and let $\mathscr{T}ors\ \mathscr{A}$ be the full subcategory of $\operatorname{Mod} A$ consisting of all modules X such that $\operatorname{Ann}(x) \neq 0$ for each $x \in X$. Prove that $\mathscr{T}ors\ \mathscr{A}$ is a dense subcategory.

2. Show that any intersection of dense subcategories is again a dense subcategory.

3. Let $\mathscr{A}$ be a dense subcategory of $\mathscr{C}$ and let Σ' and Σ'' the monomorphisms and epimorphisms in $\Sigma_{\mathscr{A}}$ respectively. Then Σ' and Σ'' are bicalculable systems.

4. Let $\mathscr{A}$ be a dense subcategory of $\mathscr{C}$. Give sufficient conditions (more general than in Theorem 3.3) for the existence of the category $\mathscr{C}/\mathscr{A}$. (In this case $\mathscr{A}$ is called a "quotifiable" subcategory). Then the category $\mathscr{C}/\mathscr{A}$ is abelian and the canonical functor $T: \mathscr{C} \to \mathscr{C}/\mathscr{A}$ is exact.

5. Let $H: \mathscr{C} \to \mathscr{C}'$ be a covariant exact functor. We denote by $\ker H$ the full subcategory of $\mathscr{C}$ which consists of all objects X for which $H(X) = 0$. Then $\ker H$ is a dense subcategory. Furthermore, let us assume that $\ker H$

is quotifiable. Then H can be written as $\bar{H}\,T$, where $T : \mathscr{C} \to \mathscr{C}/\ker H$ is canonically defined and $\bar{H}$ is faithful and exact.

6. Let $\mathscr{A}$ be a quotifiable subcategory of $\mathscr{C}$ and $T : \mathscr{C} \to \mathscr{C}/\mathscr{A}$ the canonical functor. If $\mathscr{A}'$ is a dense subcategory of $\mathscr{C}/\mathscr{A}$, then prove that the full subcategory $T^{-1}(\mathscr{A}')$ of $\mathscr{C}$ which consists of all objects X so that $T(X) \in \mathrm{Ob}\mathscr{A}'$ is also dense, and the assignment $\mathscr{A}' \rightsquigarrow T^{-1}(\mathscr{A}')$ is a bijection between the class of all dense subcategories of $\mathscr{C}/\mathscr{A}$ and all dense subcategories of $\mathscr{C}$ which contain $\mathscr{A}$. Moreover $\mathscr{A}'$ is quotifiable if and only if $T^{-1}(\mathscr{A}')$ is quotifiable and thus $(\mathscr{C}/\mathscr{A})/\mathscr{A}'$ is equivalent with $\mathscr{C}/T^{-1}(\mathscr{A}')$.

7. Let $\mathscr{A}$ and $\mathscr{A}'$ be two dense subcategories in $\mathscr{C}$ such that $\mathscr{A}$ is quotifiable. Then $\mathscr{A} \cap \mathscr{A}'$ is a quotifiable subcategory in $\mathscr{A}'$ and $\mathscr{A}'/\mathscr{A}' \cap \mathscr{A}$ is equivalent to a dense subcategory of $\mathscr{C}/\mathscr{A}$. If $\mathscr{A}'$ is also quotifiable then $\mathscr{A}'/\mathscr{A}' \cap \mathscr{A}$ is quotifiable in $\mathscr{C}/\mathscr{A}$.

References

Gabriel [59]; Gabriel–Zisman [62]; Grothendieck [79]; Ouzilou [160]; Popescu [167]; Roos [199]; Walker–Walker [236].

4.4 The section functor

The category $\mathscr{C}$ will always be considered locally small and abelian. Also $\mathscr{A}$ will be a dense subcategory of $\mathscr{C}$ and $T : \mathscr{C} \to \mathscr{C}/\mathscr{A}$ will be the canonical functor. The other notations of the previous paragraph are all preserved here.

Now let us assume that the functor T has a right adjoint which is denoted by S so that we have the functorial isomorphism:

$$\mathrm{Hom}_{\mathscr{C}}(X, S(Y)) \simeq \mathrm{Hom}_{\mathscr{C}/\mathscr{A}}(T(X), Y).$$

Our aim is to study the functor S and naturally, the category $\mathscr{C}/\mathscr{A}$.

In this case the subcategory $\mathscr{A}$ is called *localizing* and the functor S is called a *section functor.*

LEMMA 4.1. *If M is an object of $\mathscr{C}$ then the following assertions are equivalent:*

(1) *For any morphism $s : X \to Y$, $s \in \Sigma = \Sigma_{\mathscr{A}}$, the canonical morphism $h_M(s) : \mathrm{Hom}_{\mathscr{C}}(Y, M) \to \mathrm{Hom}_{\mathscr{C}}(X, M)$ is an isomorphism;*

(2) *If $f : X \to M$ is a monomorphism and $X \in \mathrm{Ob}\mathscr{A}$ then $f = 0$; moreover any monomorphism $s : M \to Y$, with $s \in \Sigma$ is a section;*

(3) *For any object X in $\mathscr{C}$ the following morphism of abelian groups:*

$$T(X, M) : \mathrm{Hom}_{\mathscr{C}}(X, M) \to \mathrm{Hom}_{\mathscr{C}/\mathscr{A}}(T(X), T(M))$$

is an isomorphism.

Proof. (1) $\Rightarrow$ (2). Let (p, X') be a cokernel of f; then $\ker p = (X, f)$, that is $p \in \Sigma$, and thus a unique morphism $t: X' \to M$ exists such that $tp = 1_M$. Therefore p is a monomorphism and we get $f = 0$.

Also, if in the exact sequence:

$$0 \to M \to X \to X' \to 0$$

X' is an object of $\mathscr{A}$, then a unique morphism $t: X \to M$ exists such that $ts = 1_M$.

(2) $\Rightarrow$ (1). Let $f: X \to M$ be a morphism and $i: \ker s \to X$ the canonical inclusion. Then $fi = 0$ and hence there exists a unique morphism

$$t: \operatorname{coim} s \to M$$

such that $tp = f$, where $p: X \to \operatorname{coim} s$ canonically. Let $j: \operatorname{coim} s \to Y$ be the inclusion, then clearly $j \in \Sigma$. Consider in $\mathscr{C}$ the following diagram:

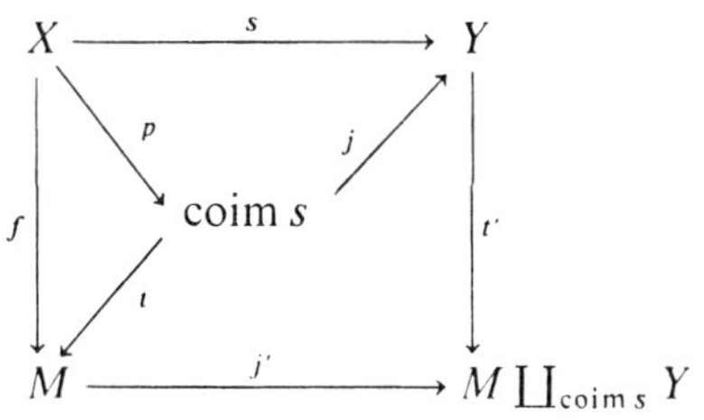

in which the bottom quadrilateral is cocartesian, j' is a monomorphism and $\operatorname{coker} j' \in \operatorname{Ob}\mathscr{A}$. Then there exists a morphism $h: M \coprod_{\operatorname{coim} s} Y \to M$ such that $hj' = 1_M$. But then $ht's = ht'jp = hj'tp = tp = f$. The uniqueness of ht' is obvious.

(2) $\Rightarrow$ (3). Let $f: X \to M$ be a morphism such that $T(f) = 0$. Then, from Lemma 3.4, it results that $\operatorname{im} f \in \operatorname{Ob}\mathscr{A}$ and thus $f = 0$, according to (2). In other words, $T(X, M)$ is a monomorphism of groups. Now let $u: T(X) \to T(M)$ be a morphism in $\mathscr{C}/\mathscr{A}$. From Lemma 3.9 one gets

$$u = T(s'')^{-1}\, T(f)\, T(s')^{-1}$$

with $s': P \to X$, $s'': M \to Q$, $f: P \to Q$, $s' \in \Sigma'$ and $s'' \in \Sigma''$. But one also has that $\ker s'' \in \operatorname{Ob}\mathscr{A}$ and consequently s'' is an isomorphism. Also we have that $\operatorname{coker} s' \in \operatorname{Ob}\mathscr{A}$ and that a unique morphism $t: X \to M$ exists, with $ts' = (s'')^{-1} f$. Furthermore, $f = s''ts'$ and thus $T(f) = T(s'')\, T(t)\, T(s')$. Hence, one obtains that $u = T(s'')^{-1}\, T(f)\, T(s')^{-1} = T(t)$. Finally, it follows that $T(X, M)$ is an epimorphism of groups, that is, it is an isomorphism.

(3) ⇒ (1). Consider the morphism $s: X \to Y$, $s \in \Sigma$. Then we have the following diagram of abelian groups:

$$\begin{array}{ccc}
\mathrm{Hom}_{\mathscr{C}}(X, M) & \xrightarrow{T(X,M)} & \mathrm{Hom}_{\mathscr{C}/\mathscr{A}}(T(X), T(M)) \\
\downarrow{\scriptstyle h_M(s)} & & \downarrow{\scriptstyle h_{T(M)}(T(s))} \\
\mathrm{Hom}_{\mathscr{C}}(Y, M) & \xrightarrow{T(Y,M)} & \mathrm{Hom}_{\mathscr{C}/\mathscr{A}}(T(Y), T(M))
\end{array}$$

the horizontal morphisms and the morphism $h_{T(M)}(T(s))$ being isomorphisms Then it follows that $h_M(s)$ is also an isomorphism. ∎

An object M in $\mathscr{C}$ which is subject to the conditions of the above lemma, is called *$\mathscr{A}$-closed* or simply, *closed* if there is no danger of confusion.

COROLLARY 4.2. *Let $\mathscr{C}$, $\mathscr{A}$, T, S, be as at the beginning of this section. For any object Z in $\mathscr{C}/\mathscr{A}$ the object $S(Z)$ is closed.*

Proof. Consider the following diagram of abelian groups:

$$\begin{array}{ccc}
\mathrm{Hom}_{\mathscr{C}}(X, S(Z)) & \xrightarrow{\phi(X,Z)} & \mathrm{Hom}_{\mathscr{C}/\mathscr{A}}(T(X), Z) \\
\uparrow{\scriptstyle h_{S(Z)}(s)} & & \uparrow{\scriptstyle h_Z(T(s))} \\
\mathrm{Hom}_{\mathscr{C}}(Y, S(Z)) & \xrightarrow{\phi(Y,Z)} & \mathrm{Hom}_{\mathscr{C}/\mathscr{A}}(T(Y), Z)
\end{array}$$

where $s: X \to Y$ is a morphism in Σ, and the horizontal morphisms are the adjunction isomorphisms. Clearly, $h_{S(Z)}(s)$ is an isomorphism, because $T(s)$ is an isomorphism in $\mathscr{C}/\mathscr{A}$. In order to complete the proof we make specific use of Lemma 4.1, case (1). ∎

PROPOSITION 4.3. *Let $\mathscr{C}$, $\mathscr{A}$, T, S be as at the beginning of this section. Also let $u: \mathrm{Id}\mathscr{C} \to ST$ and $v: TS \to \mathrm{Id}(\mathscr{C}/\mathscr{A})$ be the arrows of adjunction. The following assertions are true:*

(1) *The functorial morphism v is an isomorphism;*

(2) *For any object X in $\mathscr{C}$ the morphism u_X is in Σ, that is,* $\ker u_X$ *and* $\operatorname{coker} u_X$ *are in $\mathscr{A}$.*

Proof. (1) Let us consider the following commutative diagram of abelian groups:

$$\begin{array}{ccc}
\mathrm{Hom}_{\mathscr{C}}(X, S(Y)) & \xrightarrow{\phi(X,Y)} & \mathrm{Hom}_{\mathscr{C}/\mathscr{A}}(T(X), Y) \\
& \searrow{\scriptstyle T(X,S(Y))} \quad \nearrow{\scriptstyle h^{T(X)}(v_Y)} & \\
& \mathrm{Hom}_{\mathscr{C}/\mathscr{A}}(T(X), (TS)(Y)) &
\end{array}$$

with $Y \in \mathrm{Ob}(\mathscr{C}/\mathscr{A})$. According to Corollary 4.2, $T(X, S(Y))$ is an isomorphism and thus $h^{T(X)}(v_Y)$ is an isomorphism. Since any object of $\mathscr{C}/\mathscr{A}$ has the form $T(X)$ with $X \in \mathrm{Ob}\mathscr{C}$, the first assertion is proved.

(2) Remark that $T(u_X)$ is an inverse of $v_{T(X)}$, that is, according to Lemma 3.5, $\ker u_X$ and $\operatorname{coker} u_X$ are in $\mathscr{A}$. ∎

From the construction of the functor T (see Theorem 3.3) it follows that we may assume that for any object X in $\mathscr{C}/\mathscr{A}$, one has: $TS(X) = X$ (that is, $TS = \mathrm{Id}(\mathscr{C}/\mathscr{A})$). Moreover, we suppose that v is the identity of TS.

COROLLARY 4.4. *If M is an object of $\mathscr{C}$, the following assertions are equivalent*:

(1) *M is closed*;

(2) *u_M is an isomorphism.*

Proof. (1) ⇒ (2). Let M be closed. Then, according to Lemma 4.1, (2), any subobject of M in $\mathscr{A}$ is null and thus $\ker u_M = 0$ because of Proposition 4.3. Furthermore, we see that u_M is a section and thus, $\operatorname{coker} u_M$ is isomorphic to a subobject of $ST(M)$. Now, from Lemma 4.1, (2) we obtain $\operatorname{coker} u_M = 0$. Therefore, u_M is an isomorphism. The implication (2) ⇒ (1) follows from Corollary 4.2. ∎

We derive from the previous results that the section functor is full and faithful (Chapter 1, Section 1.5). Obviously, according to Corollary 3.12, S is exact if and only if ST is exact. The functor ST is sometimes called the *localization functor*.

The following result provides us with sufficiently general conditions for a dense subcategory to be a localizing subcategory.

THEOREM 4.5. *Let $\mathscr{A}$ be a dense subcategory in $\mathscr{C}$. The following assertions are equivalent*:

(1) *$\mathscr{A}$ is a localizing subcategory*;

(2) *For any object X in $\mathscr{C}$, the set of subobjects of X in $\mathscr{A}$ has the greatest subobject. In addition, if X has no non-zero subobject in $\mathscr{A}$, there exists a monomorphism from X into a closed object.*

Proof. (1) ⇒ (2). Let $s: X \to X''$ be an epimorphism in Σ, that is $\ker s \in \mathrm{Ob}\mathscr{A}$. Then, $T(s)$ is an isomorphism in $\mathscr{C}/\mathscr{A}$ and consequently we have the following commutative diagram in $\mathscr{C}$:

$$\begin{array}{ccc} X & \xrightarrow{u_X} & (ST)(X) \\ {\scriptstyle s}\downarrow & & \downarrow{\scriptstyle ST(s)} \\ X'' & \xrightarrow{u_{X''}} & (ST)(X'') \end{array}$$

since $ST(s)$ is an isomorphism, $\ker(u_{X''}\,s) = \ker u_X$. Otherwise, $\ker s \subseteq \ker u_X$. Hence, according to Proposition 4.3 it is now clear that $\ker u_x$ is the greatest subobject of X in $\mathscr{A}$.

Obviously, if $\ker u_X = 0$, u_X is a monomorphism and $ST\ (X)$ is closed, according to Corollary 4.2.

(2) $\Rightarrow$ (1). Let X' be the greatest subobject of X in $\mathscr{A}$. Then X/X' has no non-zero subobject in $\mathscr{A}$ and thus, there exists a monomorphism $i: X/X' \to R$ with R closed. Let $p: R \to \operatorname{coker} i$ be the canonical epimorphism, H the smallest subobject of $\operatorname{coker} i$ in $\mathscr{A}$ and $\overline{X} = p^{-1}(H)$. Then, $\overline{X}$ is closed.

Let us notice first that $\overline{X}$ has no non-zero subobject in $\mathscr{A}$. Furthermore, consider the following diagram:

$$\begin{array}{ccccccccc} 0 & \longrightarrow & \overline{X} & \xrightarrow{u} & Y & \xrightarrow{r} & Y' & \longrightarrow & 0 \\ & & {\scriptstyle v}\downarrow & & & & & & \\ & & R & & & & & & \end{array}$$

whose row is exact, $Y' \in \operatorname{Ob}\mathscr{A}$ and with v being the canonical inclusion. It results from Lemma 4.1 that a unique morphism $t: Y \to R$ exists such that $tu = v$. Now, we shall prove that $t(Y) \subseteq v(\overline{X})$. Consequently, t will define a morphism $t': Y \to \overline{X}$ so that $t'u = 1_{\overline{X}}$. In order to do this, let us remark that $t(Y) \supseteq v(\overline{X})$ and $t(Y)/v(\overline{X}) \simeq Y/\big(u(\overline{X}) + \ker t\big) \simeq Y'/r(\ker t) \in \operatorname{Ob}\mathscr{A}$. Obviously, $t(Y)/v(\overline{X}) \subseteq R/v(\overline{X})$ and from the definition of $\overline{X}$, we derive that $R/v(\overline{X})$ has no non-zero subobject in $\mathscr{A}$. Therefore, $t(Y)/v(\overline{X}) = 0$, that is, $t(Y) = v(\overline{X})$.

We shall define the functor S as follows. For any object X in $\mathscr{C}$, we choose a couple $\big(u_X, H(X)\big)$, with $H(X)$ being closed, and $\ker u_X$, $\operatorname{coker} u_X$ being objects of $\mathscr{A}$. The existence of such a couple was proved above. It is not difficult to see that two such couples are canonically isomorphic.

For any $X \in \operatorname{Ob}\mathscr{C}$, one finds that $T(u_X)$ is an isomorphism in $\mathscr{C}/\mathscr{A}$. To finish, it will be sufficient to prove that for any object Y in $\mathscr{C}/\mathscr{A}$ the functor $h_Y T: \mathscr{C} \to \mathscr{Ab}$ is representable (Chapter 1, Section 5). In order to do this, let Y' be an object of $\mathscr{C}$ so that $T(Y') = Y$, and $u_X: Y' \to H(Y')$ defined as above. Since $T(u_{Y'})$ is an isomorphism, the functors $h_Y T$ and $h_{T(H(Y'))}\,T$ are isomorphic and thus it suffices to prove that the last functor is representable. But, according to Lemma 4.1, for any object X in $\mathscr{C}$ the canonical morphism $T\big(X, H(Y')\big): \operatorname{Hom}_{\mathscr{C}}\big(X, H(Y')\big) \to \operatorname{Hom}_{\mathscr{C}/\mathscr{A}}\big(T(X), T\big(H(Y')\big)\big)$ is an isomorphism and this establishes an isomorphism between $h_{H(Y')}$ and

$$h_{T(H(Y'))}\,T. \blacksquare$$

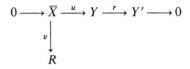

LEMMA 4.6. *Let $\mathscr{A}$ be a localizing subcategory and X an object in $\mathscr{C}$. We shall denote by $X_{\mathscr{A}}$ the largest subobject of X in $\mathscr{A}$. If $u : X \to Y$ is a morphism in $\mathscr{C}$, then $u(X_{\mathscr{A}}) \subseteq Y_{(\mathscr{A})}$ and u defines a canonical morphism*

$$u' : X/X_{\mathscr{A}} \to Y/Y_{\mathscr{A}}.$$

Then:

(1) *$T(u)$ is a monomorphism if and only if u' is a monomorphism;*

(2) *For any object X in $\mathscr{C}$ the image of u_X which is canonically isomorphic to $X/X_{\mathscr{A}}$, is an essential subobject in $(ST)(X)$;*

(3) *$T(u)$ is an essential monomorphism if and only if u' is an essential monomorphism.*

Proof (1). Let us assume that $T(u)$ is a monomorphism and that

$$i : Z \to X/X_{\mathscr{A}}$$

is a kernel of u'. Then, $T(i) = 0$ and thus $\operatorname{im} i \in \operatorname{Ob}\mathscr{A}$ (see Lemma 3.4), that is, $i = 0$. The converse assertion is obvious.

(2) If Y is a non-zero subobject in $(ST)(X)$ so that $Y \cap (\operatorname{im} u_X) = 0$, then Y is isomorphic to a subobject of $\operatorname{coker} u_X$. But then we get a contradiction with Corollary 4.2 and Proposition 4.3.

(3) Let us assume first that u' is an essential monomorphism. From the commutative diagram:

$$\begin{array}{ccc} X/X_{\mathscr{A}} & \xrightarrow{\ h\ } & (ST)(X) \\ {\scriptstyle u'}\downarrow & & \downarrow{\scriptstyle (ST)(u)} \\ Y/Y_{\mathscr{A}} & \xrightarrow{\ g\ } & (ST)(Y) \end{array}$$

with h and g being canonically induced by u_X and u_Y respectively, we derive that $(ST)(u)$ is an essential monomorphism. Now let Z be a non-zero subobject in $T(Y)$, such that $Z \cap (\operatorname{im} T(u)) = 0$. Then, $S(Z) \neq 0$ and

$$S(Z) \cap S(\operatorname{im} T(u)) = 0.$$

But $S(\operatorname{im} T(u)) = \operatorname{im}(ST)(u)$, and this contradicts the hypothesis.

Conversely, if $T(u)$ is essential, then in the same manner we derive that $(ST)(u)$ is essential, and finally, it follows that u' is essential. ■

COROLLARY 4.7. *The section functor S preserves injectives and essential extensions.* ■

NOTE. Compare this result with Lemma 10.11 of Chapter 3.

The following result is in fact, a direct consequence of Corollary 5.5 in Chapter 1, and of the above results. Here we give a direct proof.

LEMMA 4.8. *Let $\mathscr{A}$ be a localizing subcategory of $\mathscr{C}$. If $\{U_i\}_i$ is a set of generators of $\mathscr{C}$, then $\{T(U_i)\}_i$ is a set of generators of $\mathscr{C}/\mathscr{A}$.*

Proof. Let $f : X \to Y$ be a non-zero morphism in $\mathscr{C}/\mathscr{A}$. Then $S(f) \neq 0$ and since S is faithful, there exists an index i_0 and a morphism $g : U_{i_0} \to S(X)$ such that $S(f)\ g \neq 0$. Then, $T(S(f))\,T(g) \neq 0$, because im g is a non-zero subobject in $S(X)$ and $S(X)$ has no non-zero subobjects in $\mathscr{A}$ (see Corollary 4.2 and Lemma 3.4). ∎

The following result is characteristic for localizing subcategories.

THEOREM 4.9. *Let $T : \mathscr{C} \to \mathscr{C}'$ be a covariant exact functor and S a full and faithful right adjoint of T. Then,* ker T *is a localizing subcategory in $\mathscr{C}$ and S establishes an equivalence between $\mathscr{C}'$ and $\mathscr{C}/$*ker T.

Proof. Let Σ be the class of morphisms s of $\mathscr{C}$, so that ker s and coker s are in ker T. Clearly, Σ is a bicalculable system because ker T is dense, and thus, there exists a unique functor $H : \mathscr{C}_\Sigma \to \mathscr{C}'$ such that $HT' = T$, with $T' : \mathscr{C} \to \mathscr{C}_\Sigma$ being the canonical functor. Now, we shall prove that H is an equivalence of categories. Indeed, if $X' \in \mathrm{Ob}\mathscr{C}'$, then $X' \simeq (TS)(X')$ and thus $X' \simeq (HT'S)(X')$.

Let $u = T'(s'')^{-1}\,T'(f)\,T'(s')^{-1}$ be a morphism of $\mathscr{C}_\Sigma$, with $s'' \in \Sigma''$, $s' \in \Sigma'$, and f a morphism of $\mathscr{C}$. Then we have $H(u) = T(s'')^{-1}\,T(f)\,T(s')^{-1}$. If $H(u) = 0$, then $T(f) = 0$ and consequently im $f \in$ ker T, that is $u = 0$, because of Lemma 3.4. Finally, if $u' : X' \to Y'$ is a morphism of $\mathscr{C}'$, we may assume that $u' = (TS)(u')$ and consequently $u' = (HT')(S(u'))$. Therefore H is full and faithful and any object of $\mathscr{C}'$ is isomorphic to an object of the form $H(X)$ with X being in $\mathscr{C}_\Sigma$. It results from Theorem 5.3 of Chapter 1 that H is an equivalence of categories. ∎

COROLLARY 4.10. *Let $\mathscr{C}$ be a Grothendieck category. Then there exists a ring A and a localizing subcategory $\mathscr{A}$ in* Mod A^0 *such that $\mathscr{C}$ is equivalent to* Mod $A^0/\mathscr{A}$.

The proof follows from Theorem 7.9 of Chapter 3 and the above Theorem 4.9. ∎

NOTE. The concept of localization is defined by an abelian category $\mathscr{C}$ and a localizing subcategory of $\mathscr{C}$. From these data we are able to construct the category $\mathscr{C}/\mathscr{A}$ and to make some useful observations concerning many problems. In the next paragraphs, we shall give many examples of this nature which will illustrate the importance of this method.

Exercises

1. Let $\mathscr{A}$ be a localizing subcategory of an abelian category $\mathscr{C}$. An object X in $\mathscr{C}$ is called $\mathscr{A}$-torsionfree (or briefly *torsionfree*, if there is no danger of confusion) if $X_{\mathscr{A}} = 0$. The following assertions are true:

(a) Any subobject of a torsionfree object is torsionfree.

(b) A projective limit of torsionfree objects is also torsionfree.

(c) If in the exact sequence

$$0 \to X' \to X \to X'' \to 0$$

X' and X'' are torsionfree, then X is torsionfree.

(d) An essential extension of a torsionfree object is always torsionfree.

(e) An object Y is in $\mathscr{A}$ if and only if $\mathrm{Hom}_{\mathscr{C}}(Y, X) = 0$ for all torsionfree objects in $\mathscr{C}$.

2. Let $\mathscr{A}$ and $\mathscr{A}'$ be two localizing subcategories of $\mathscr{C}$. Then $\mathscr{A} = \mathscr{A}'$ if and only if any $\mathscr{A}$-torsionfree object is also $\mathscr{A}'$-torsionfree and vice versa.

References

Gabriel [59]; Grothendieck [79]; Ouzilou [160]; Popescu [167]; Roos [199]; Walker–Walker [236].

4.5 Localization in categories with injective envelopes

LEMMA 5.1. *Let $\mathscr{A}$ be a dense subcategory of $\mathscr{C}$.*

(1) *If Q is an injective and $\mathscr{A}$-torsionfree object in $\mathscr{C}$, then Q is closed.*

(2) *If $\mathscr{A}$ is localizing, $i: X \to Q$ is an injective envelope of X and if X is $\mathscr{A}$-torsionfree, then $T(Q)$ is also an injective envelope of $T(X)$ in $\mathscr{C}/\mathscr{A}$.*

Proof. (1) The first assertion results from Lemma 4.1.

(2) We derive from Lemma 4.6 that $T(i)$ is an essential morphism. To complete the proof, it will be sufficient to show that $T(Q)$ is injective. Indeed, if $u: T(Q) \to Y$ is a monomorphism in $\mathscr{C}/\mathscr{A}$, then a morphism

$v: S(Y) \to Q$ exists such that $vS(u)\, u_Q = 1_Q$. But $TS(u) = u$ and

$$T(u_Q) = 1_{T(Q)}$$

and consequently $T(v)\, u = 1_{T(Q)}$. Now, we apply Lemma 10.1 of Chapter 3. ∎

PROPOSITION 5.2. *Let $\mathscr{C}$ be a category with injective envelopes and let $\mathscr{A}$ be a dense subcategory of $\mathscr{C}$. The following assertions are equivalent*:

(1) *$\mathscr{A}$ is a localizing subcategory*;

(2) *For any object X, the set of all subobjects of X in $\mathscr{A}$ has a maximal subobject.*

Proof. The implication (1) ⇒ (2) results from Theorem 4.5.

(2) ⇒ (1). Let X be an object in $\mathscr{C}$ and let X' be the largest subobject of X in $\mathscr{A}$. Then X/X' is $\mathscr{A}$-torsionfree and any injective envelope of X/X' is $\mathscr{A}$-closed because of Lemma 5.1. Again we shall make use of Theorem 4.5. ∎

PROPOSITION 5.3. *Let $\mathscr{A}$ be a localizing subcategory of a category $\mathscr{C}$ with injective envelopes. Then, the following assertions are true*:

(1) *$\mathscr{C}/\mathscr{A}$ is a category with injective envelopes*;

(2) *Any injective object in $\mathscr{C}/\mathscr{A}$ is of the form $T(H)$ with H being an injective and torsion-free object in $\mathscr{C}$.*

(3) *Any injective object Q of $\mathscr{C}$ is a direct sum $Q_1 \amalg Q_2$ where Q_1 is $\mathscr{A}$-closed and Q_2 the injective envelope of an object in $\mathscr{A}$. The objects Q_1 and Q_2 are determined up to isomorphism.*

Proof. (1) Let $X \in \operatorname{Ob} \mathscr{C}/\mathscr{A}$ and let $i: S(X) \to Q$ be an injective envelope of $S(X)$. Then Q is $\mathscr{A}$-torsionfree and consequently $T(Q)$ is injective because $T(i)$ is an essential extension and from the above, $X = TS(X)$.

(2) If Q is an injective object of $\mathscr{C}/\mathscr{A}$, then $Q \simeq TS(Q)$, and $S(Q)$ is also injective.

(3) Let Q be injective in $\mathscr{C}$ and let X be the largest subobject of Q in $\mathscr{A}$. Also, let Q_1 be the injective envelope of X. Then $Q = Q_1 \amalg Q_2$ and Q_2 is closed. The uniqueness of Q_1 and Q_2 is obvious. ∎

COROLLARY 5.4. *Assume that $\mathscr{A}$ contains the injective envelope of each of its objects in $\mathscr{C}$. Then, for any injective object Q in $\mathscr{C}$, the object $T(Q)$ is also injective in $\mathscr{C}/\mathscr{A}$.*

Proof. According to Proposition 5.3, if Q is an object in $\mathscr{C}$, then $Q = Q_1 \amalg Q_2$ and $Q_1 \in \operatorname{Ob} \mathscr{A}$. Consequently $T(Q) \simeq T(Q_2)$. ∎

PROPOSITION 5.5. *Let $\mathscr{A}$ be a localizing subcategory of a category $\mathscr{C}$ with injective envelopes. The following assertions are equivalent:*

(1) *The localizing functor (or, equivalently, the section functor) is exact;*

(2) *If in the exact sequence of $\mathscr{C}$:*

$$0 \to X' \to X \to X'' \to 0$$

X' and X are closed, then X'' is also closed;

(3) *If in the exact sequence of $\mathscr{C}$:*

$$0 \to X' \to Q \to X'' \to 0$$

Q is injective, Q is $\mathscr{A}$-torsionfree and X' is closed, then X'' is closed.

Proof. (1) ⇒ (2). Consider the following commutative diagram in $\mathscr{C}$, with the exact rows:

$$\begin{array}{ccccccc}
0 \to X' & \longrightarrow & X & \longrightarrow & X'' & \longrightarrow & 0 \\
\downarrow{\scriptstyle u_{X'}} & & \downarrow{\scriptstyle u_X} & & \downarrow{\scriptstyle u_{X''}} & & \\
0 \to (ST)(X') & \longrightarrow & (ST)(X) & \longrightarrow & (ST)(X'') & \longrightarrow & 0
\end{array}$$

where the first two vertical morphisms are isomorphisms. It is then clear that $u_{X''}$ is also an isomorphism. Now we make specific use of Corollary 4.4.

(2) ⇒ (3). It follows from Lemma 5.1 that Q is closed.

(3) ⇒ (1). Consider in $\mathscr{C}/\mathscr{A}$ the following commutative diagram:

$$\begin{array}{ccccccccc}
 & & & & 0 & & 0 & & \\
 & & & & \downarrow & & \downarrow & & \\
0 & \longrightarrow & X' & \longrightarrow & X & \xrightarrow{f} & X'' & \longrightarrow & 0 \\
 & & \| & & \downarrow{\scriptstyle u} & & \downarrow{\scriptstyle v} & & \\
0 & \longrightarrow & X' & \longrightarrow & Q & \xrightarrow{p} & Y & \longrightarrow & 0 \\
 & & & & \downarrow & & \downarrow & & \\
 & & & & Z & = & Z & & \\
 & & & & \downarrow & & \downarrow & & \\
 & & & & 0 & & 0 & &
\end{array}$$

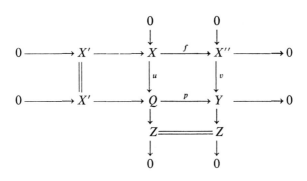

with the rows and columns being exact and Q being injective. If we operate with S on the above diagram, then we get in $\mathscr{C}$ the following diagram:

$$\begin{array}{ccccccccc}
0 & \longrightarrow & S(X') & \longrightarrow & S(X) & \xrightarrow{S(f)} & S(X'') & & \\
 & & \Vert & & \downarrow{\scriptstyle S(u)} & & \downarrow{\scriptstyle S(v)} & & \\
 & & S(X') & \longrightarrow & S(Q) & \xrightarrow{S(p)} & S(Y) & \longrightarrow & 0 \\
 & & & & \downarrow{\scriptstyle S(q)} & & \downarrow & & \\
 & & & & S(Z) & = & S(Z) & & \\
 & & & & \downarrow & & \downarrow & & \\
 & & & & 0 & & 0 & &
\end{array}$$

with $S(p)$ and $S(q)$ being epimorphisms. Indeed, if $R = \operatorname{im} S(p)$, then R is closed and coker $S(p) \in \mathrm{Ob}\mathscr{A}$. Consequently the canonical inclusion $R \to S(Y)$ is a section. But $S(Y)$ is also closed, that is, $S(p)$ is an epimorphism. Then it follows that $S(f)$ is an epimorphism (see Chapter 2, Section 2.5). ∎

Let $R = \{Q_i\}_i$ be a class of objects of $\mathscr{C}$. We shall say that the *R-dominant dimension* of an object X is larger than n, and we shall write R dom.dim.$X \geqslant n$ if an exact sequence

$$0 \to X \to Q_1 \to Q_2 \to \dots \to Q_n$$

exists so that each Q_i is in the class R.

PROPOSITION 5.6. *Let $\mathscr{C}$ be a category with injective envelopes and let $\mathscr{A}$ be a localizing subcategory of $\mathscr{C}$. Let us denote by R the class of all $\mathscr{A}$-closed objects and by R' the class of all $\mathscr{A}$-closed injective objects. For any object X in $\mathscr{C}$ the following assertions are equivalent:*

(1) *X is closed;*

(2) *R dom.dim.$X \geqslant 2$;*

(3) *R' dom.dim.$X \geqslant 2$.*

Proof. (1) ⇒ (2). Let $i: X \to Q$ be an essential monomorphism with Q being injective; then Q is also closed (see Lemma 5.1). From the following canonically constructed commutative diagram:

$$\begin{array}{ccccccccc}
0 & \longrightarrow & X & \xrightarrow{i} & Q & \longrightarrow & Q/X & \longrightarrow & 0 \\
 & & \Vert{\scriptstyle u_X} & & \Vert{\scriptstyle u_Q} & & \downarrow{\scriptstyle u_{Q/X}} & & \\
0 & \longrightarrow & ST(X) & \longrightarrow & ST(Q) & \longrightarrow & ST(Q/X) & &
\end{array}$$

we derive that $u_{Q/X}$ is a monomorphism (Chapter 2, Section 2.5). But then the injective envelope of Q/X is closed, that is, R dom.dim.$X \geqslant 2$. Again it is not difficult to check that (1) ⇒ (3).

(3) ⇒ (1). Let us consider the commutative diagram:

$$\begin{array}{ccccccc} 0 \longrightarrow & X & \longrightarrow & Q_1 & \longrightarrow & Q_2 \\ & \downarrow{\scriptstyle u_X} & & \downarrow{\scriptstyle u_{Q_1}} & & \downarrow{\scriptstyle u_{Q_2}} \\ 0 \longrightarrow & ST(X) & \longrightarrow & ST(Q_1) & \longrightarrow & ST(Q_2) \end{array}$$

of $\mathscr{C}$ with exact rows and with Q_1, Q_2 being in R'. Obviously u_{Q_1} and u_{Q_2} are isomorphisms, and consequently u_X is an isomorphism too. ■

NOTE. The concept of dominant dimension was first formulated by Kato [104] and Morita [136]. Later, we shall obtain a deeper insight into this concept.

Exercises

1. Let $\mathscr{C}$ be a locally small Ab 3-category and $\mathscr{M}$ a full subcategory of $\mathscr{C}$ satisfying the following axioms:

(M_1) A projective limit of objects in $\mathscr{M}$, calculated in $\mathscr{C}$ is also an object in $\mathscr{M}$;

(M_2) If $X' \to X$ is a monomorphism in $\mathscr{C}$ and $X \in \text{Ob}\mathscr{M}$ then $X' \in \text{Ob}\mathscr{M}$;

(M_3) For each $X \in \text{Ob}\mathscr{M}$ there exists a monomorphism $X \to Q$ in $\mathscr{M}$ such that Q is injective considered as an object in $\mathscr{C}$.

We shall call $\mathscr{M}$ a *monosubcategory* of $\mathscr{C}$ if it satisfies the above conditions M_1, M_2 and M_3.

Denote by $\mathscr{A}$ the full subcategory of $\mathscr{C}$ consisting of all objects X such that $\text{Hom}_{\mathscr{C}}(X, Y) = 0$ for any object Y of $\mathscr{M}$. Then $\mathscr{A}$ is a localizing subcategory. Moreover an object Y in $\mathscr{C}$ is $\mathscr{A}$-torsionfree if and only if Y is isomorphic with an object in $\mathscr{M}$.

2. Let $\mathscr{C}$ be an Ab 3* locally small category and $\mathscr{M}$ a full subcategory verifying the above conditions M_1 and M_2. Then the inclusion functor $\mathscr{M} \to \mathscr{C}$ has a left adjoint.

References

Bănică–Popescu [14]; Bucur–Deleanu [31]; Dickson [42]; Gabriel [59]; Goldman [76]; Gruson [81]; Haque [83]; Hudry [94]; Kato [104]; Lambek [114]; Maranda [121]; Mitchell [133]; Morita [136]; Popescu [167]; Roos [199] Tachikawa [218], [219]; Tisseron [227]; Walker–Walker [236].

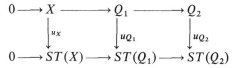

4.6 Localization in Ab 3-categories

PROPOSITION 6.1. *Let $\mathscr{A}$ be a localizing subcategory of $\mathscr{C}$. Then:*

(1) *If $\mathscr{C}$ is an* Ab 3 *(respectively,* Ab 3**)-category, then $\mathscr{C}/\mathscr{A}$ is also an* Ab 3 *(respectively,* Ab 3**)-category;*

(2) *If $\mathscr{C}$ is an* Ab 4*-category, then $\mathscr{C}/\mathscr{A}$ is also an* Ab 4*-category;*

(3) *If $\mathscr{C}$ is an* Ab 5*-category then $\mathscr{C}/\mathscr{A}$ is an* Ab 5*-category.*

Proof. The first point results from Section 1.5 of Chapter 1. The remaining statements result from the exactness of the functor $T:\mathscr{C}\to\mathscr{C}/\mathscr{A}$ and from the definition of conditions Ab 4 and Ab 5 (see Chapter 2, Section 2.8). ■

COROLLARY 6.2. *If $\mathscr{C}$ is a Grothendieck category, then $\mathscr{C}/\mathscr{A}$ is also a Grothendieck category.*

The proof follows from the above proposition and from Lemma 4.8. ■

PROPOSITION 6.3. *Let $\mathscr{C}$ be an* Ab 3*-category with injective envelopes, and $\mathscr{A}$ a dense subcategory of $\mathscr{C}$. The following assertions are equivalent:*

(1) *$\mathscr{A}$ is a localizing subcategory;*

(2) *If $\{X_i\}_{i\in I}$ is a set of objects in $\mathscr{A}$, then $\coprod_{i\in I} X_i$ (the direct sum being carried out in $\mathscr{C}$) is also an object in $\mathscr{A}$.*

Proof. The implication (1) ⇒ (2) results from the commutativity of

$$T:\mathscr{C}\to\mathscr{C}/\mathscr{A}$$

with inductive limits.

(2) ⇒ (1). According to Theorem 4.5 and Lemma 5.1 it will be sufficient to prove that any object X has a largest subobject in $\mathscr{A}$. Indeed, if $\{X_i\}_i$ is the set of subobjects of X in $\mathscr{A}$, then $\sum_i X_i$ is a quotient object of the direct sum $\coprod_i X_i$, that is $\sum_i X_i$ is an object of $\mathscr{A}$. ■

COROLLARY 6.4. *Let $\mathscr{C}$ be an* Ab 3*-category with injective envelopes and $\mathscr{A}$ a full subcategory of $\mathscr{C}$. The following assertions are equivalent:*

(1) *There exists a class $\{Q_i\}_i$ of injective objects in $\mathscr{C}$, such that $X\in \mathrm{Ob}\mathscr{A}$ if and only if $\mathrm{Hom}_{\mathscr{C}}(X, Q_i)=0$ for any i;*

(2) *$\mathscr{A}$ is a localizing subcategory.*

Proof. (1) ⇒ (2). Let

$$0\to X'\to X\to X''\to 0$$

be an exact sequence in $\mathscr{C}$, with $X \in \mathrm{Ob}\mathscr{A}$. Then for any i we have the exact sequence of abelian groups:

$$0 \to \mathrm{Hom}_{\mathscr{C}}(X'', Q_i) \to \mathrm{Hom}_{\mathscr{C}}(X, Q_i) \to \mathrm{Hom}_{\mathscr{C}}(X', Q_i) \to 0$$

All terms of this exact sequence must be zero since the group in the middle is null. Conversely, if X' and X'' are in $\mathscr{A}$, then the ends of the last exact sequence are null. Consequently the middle group is also null. Therefore $\mathscr{A}$ is a dense subcategory. Furthermore, if $\{X_j\}_j$ is a set of objects in $\mathscr{A}$, then $\mathrm{Hom}_{\mathscr{C}}(\coprod_j X_j, Q_i) = \prod_j \mathrm{Hom}_{\mathscr{C}}(X_j, Q_i) = 0$ for any i. According to Proposition 6.3 it follows that $\mathscr{A}$ is localizing.

(2) $\Rightarrow$ (1). Let $\{E_i\}_i$ be the class of all injective objects in $\mathscr{C}/\mathscr{A}$. Then, for any object X in $\mathscr{C}$, $X \in \mathrm{Ob}\mathscr{A}$ if and only if $T(X) = 0$, that is, if and only if $\mathrm{Hom}_{\mathscr{C}}(X, S(Y)) = 0$ for any object Y in $\mathscr{C}/\mathscr{A}$. But if E is the injective envelope of Y, then $\mathrm{Hom}_{\mathscr{C}}(X, S(Y)) = 0$ if and only if $\mathrm{Hom}_{\mathscr{C}}(X, S(E)) = 0$ (according to Lemma 4.6). Finally, it will be sufficient to select the class $\{S(E_i)\}_i$ of injective objects of $\mathscr{C}$. ■

Assume that $\mathscr{C}$ is an Ab 3 locally small category with injective envelopes. For any object X in $\mathscr{C}$ we denote by $\mathscr{A}(X)$ the localizing subcategory of $\mathscr{C}$, which consists of all objects Y such that $\mathrm{Hom}_{\mathscr{C}}(Y, Q) = 0$ where Q is the injective envelope of X. If X' is an essential subobject of X or if X_1 is an essential extension of X, then one obviously has $\mathscr{A}(X') = \mathscr{A}(X) = \mathscr{A}(X_1)$. Assume that $X \neq 0$. Then one gets the following lemma.

LEMMA 6.5.(1) *A non-zero object Y is in $\mathscr{A}(X)$ if and only if any non-zero quotient object Y'' of Y has no non-null subobject isomorphic to a subobject of X;*

(2) *The subcategory $\mathscr{A}(X)$ is the largest localizing subcategory so that Q—the injective envelope of X—is $\mathscr{A}(X)$-closed (that is, if $\mathscr{A}$ is any subcategory for which Q is $\mathscr{A}$-closed, then $\mathscr{A} \subseteq \mathscr{A}(X)$).*

Proof. (1) Let $Y \in \mathrm{Ob}\mathscr{A}(X)$, then $\mathrm{Hom}_{\mathscr{C}}(Y, Q) = 0$. Let Y'' be a non-null quotient object of Y, Y_1 a non-null subobject and $f : Y_1 \to X$ a monomorphism; then there exists a non-null morphism $f_1 : Y'' \to Q$. Consequently $Y'' \notin \mathrm{Ob}\mathscr{A}(X)$ and this contradicts the definition of a localizing subcategory. Conversely, if Y is a non-null object and if $Y \notin \mathrm{Ob}\mathscr{A}$, then a non-null morphism $f : Y \to Q$, exists and $f(Y) \cap X \neq 0$, that is a non-null quotient of Y has a non-null subobject which is isomorphic to a subobject of X.

(2) Obviously, Q has no non-null subobjects in $\mathscr{A}(X)$ and thus, it follows from Lemma 5.1 that Q is closed. Now if $\mathscr{A}$ is a localizing subcategory such that Q is $\mathscr{A}$-closed, then for any object Y in $\mathscr{A}$ we have that $\mathrm{Hom}_{\mathscr{C}}(Y, Q) = 0$. Therefore, $\mathscr{A} \subseteq \mathscr{A}(X)$. ■

Let $\mathscr{A}$ be a localizing subcategory of $\mathscr{C}$; if an object X of $\mathscr{C}$ exists such that $\mathscr{A} = \mathscr{A}(X)$, we call X a *defining object* of $\mathscr{A}$. As a matter of fact this situation is elsewhere described by saying that X *cogenerates* $\mathscr{A}$.

PROPOSITION 6.6. *A localizing subcategory $\mathscr{A}$ has a defining object if and only if the category $\mathscr{C}/\mathscr{A}$ has a cogenerator.*

In particular, if $\mathscr{C}$ has a set of generators, any localizing subcategory has a defining object.

Proof. Let $\mathscr{A} = \mathscr{A}(Q)$; we may assume that Q is injective. Then $T(Q)$ is an injective cogenerator in $\mathscr{C}/\mathscr{A}$. Indeed, if Y is an object of $\mathscr{C}/\mathscr{A}$, then $Y = T(Y')$ for a suitable Y' in $\mathscr{C}$, and so

$$\mathrm{Hom}_{\mathscr{C}/\mathscr{A}}(Y, T(Q)) \simeq \mathrm{Hom}_{\mathscr{C}}(Y', ST(Q)),$$

that is $Y = 0$ if and only if $Y' \in \mathrm{Ob}\mathscr{A}(Q)$.

Conversely, if V is an injective cogenerator of $\mathscr{C}/\mathscr{A}$, then $S(V)$ is an injective defining object of $\mathscr{A}$.

In order to complete the proof, it will be sufficient to note that the category $\mathscr{C}/\mathscr{A}$ has an injective cogenerator whereas $\mathscr{C}$ has a set of generators (see Lemma 4.8 and Chapter 3, Lemma 7.12). ∎

COROLLARY 6.7. *Let Q be an injective defining object of $\mathscr{A}$. Assume that for any set I, the object Q^I exists. Also, let R be the class of all objects Q^I with I being any set. Then, an object X is closed, if and only if*

$$R\text{-dom.dim.}\,X \geqslant 2.$$

Proof. Obviously, for any set I, the object Q^I is closed, and then it follows from Proposition 5.6 that if R-dom.dim.$X \geqslant 2$ then X is closed. Conversely, if X is closed, then it follows from Proposition 6.6 that we have in $\mathscr{C}/\mathscr{A}$ the following injective resolution:

$$0 \to T(X) \to T(Q)^I \to T(Q)^J$$

If we operate now with the functor S on this resolution, then we derive that R-dom.dim.$X \geqslant 2$. ∎

NOTE. Corollary 6.7 is in essence similar to a result of Morita [136, Theorem 5.4].

LEMMA 6.8. *Let $\{Q_i\}_i$ be a set of injective objects of the category $\mathscr{C}$. Assume that the object $\prod_i Q_i = Q$ exists in $\mathscr{C}$. Then, one has that $\mathscr{A}(Q) = \bigcap_i \mathscr{A}(Q_i)$.*

The proof is not difficult and is left to the reader. ∎

COROLLARY 6.9. *If $\mathscr{C}$ is an Ab 3*-category and if $\mathscr{C}/\mathscr{A}$ has a set of cogenerators, then $\mathscr{A}$ has a defining object.* ∎

PROPOSITION 6.10. *Let X be an object in a locally small* Ab 5*-category, with injective envelope. Denote by $\mathscr{L}(X)$ the intersection of localizing subcategories which contain X.*

A non-zero object Y is in $\mathscr{L}(X)$ if and only if any non-zero quotient object of Y contains a non-zero quotient object of a subobject of X.

Proof. Let $\mathscr{A}$ be the full subcategory of $\mathscr{C}$ which consists of zero and all objects Y that are subject to conditions of the above proposition. We shall prove that $\mathscr{A}$ is a localizing subcategory and that $X \in \mathrm{Ob}\mathscr{A}$. In order to do this let $Y \in \mathrm{Ob}\mathscr{A}$, Y' a non-null subobject of Y and Y_1' a quotient of $Y.'$ Then, Y_1' may be considered as a subobject in a quotient Y_1 of Y. Assume that $Y_1' \neq 0$ and let Z be a complement of Y_1'. Then Y_1/Z is a quotient of Y and Y_1' is essential in Y_1/Z. If X'' is a non-null quotient of a subobject of X which is contained in Y_1/Z, then $X'' \cap Y_1' \neq 0$, and thus Y_1' contains a non-null quotient of a suitable subobject of X. Hence, any subobject of an object in $\mathscr{A}$ is also in $\mathscr{A}$. Obviously, any quotient object of an object in $\mathscr{A}$ is necessarily in $\mathscr{A}$.

We claim that $\mathscr{A}$ is dense. In order to prove this, let Y be an object in $\mathscr{C}$ and Y' a subobject of Y such that Y' and Y/Y' are in $\mathscr{A}$. Also, let Y_1' be a subobject of Y. If $Y_1' \supseteq Y'$, then Y/Y_1' is a quotient of Y/Y' and is an object of $\mathscr{A}$. If $Y' \nsubseteq Y_1'$, then $Y'/Y' \cap Y_1'$ is a non-null subobject of Y/Y_1' and from the hypothesis a non-null quotient object of a subobject of X is contained in $Y'/Y' \cap Y_1'$ and also in Y/Y_1'. In any case we see that Y is in $\mathscr{A}$, that is, $\mathscr{A}$ is dense, as claimed.

We leave it to the reader to prove that any direct sum of objects of $\mathscr{A}$ is also an object of $\mathscr{A}$.

In order to complete the proof, it is sufficient to show that any localizing subcategory $\mathscr{B}$ of $\mathscr{C}$ which contains X, also necessarily contains $\mathscr{A}$. Indeed, if $Y \in \mathrm{Ob}\mathscr{A}$, let $Y_{\mathscr{B}}$ be the largest subobject of Y in $\mathscr{B}$.

If $Y_{\mathscr{B}} \neq Y$, then $Y/Y_{\mathscr{B}} \neq 0$ and from the hypothesis we know that it contains a non-null subobject of $\mathscr{B}$ contrary to the definition of $Y_{\mathscr{B}}$. Hence one necessarily has that $\mathscr{A} \subseteq \mathscr{B}$. ∎

We shall say that a family $\{X_i\}_i$ of objects of a category $\mathscr{C}$ *generates* a localizing subcategory $\mathscr{A}$ of $\mathscr{C}$, if $\mathscr{A}$ is the smallest localizing subcategory containing all objects X_i. In this way, the subcategory described above $\mathscr{L}(X)$ is generated by X.

Now we shall give an important example of localization.

LEMMA 6.11. *Let $\mathscr{C}$ be a Grothendieck category and Σ a multiplicative set of elements of $Z(\mathscr{C})$, the centre of $\mathscr{C}$. Denote by $\mathscr{A}_\Sigma = \mathscr{A}$ the full subcategory of $\mathscr{C}$ which consists of all objects X such that $X = \sup_{s\in\Sigma}(\ker s_X)$. Then $\mathscr{A}$ is a localizing subcategory of $\mathscr{C}$.*

We recall that each element s of Σ is a functorial morphism $s : \mathrm{Id}\,\mathscr{C} \to \mathrm{Id}\,\mathscr{C}$, whose component for the object X is denoted by s_X.

Proof. We shall prove that $\mathscr{A}$ is a dense subcategory. In order to do this we remark that for any object X the set $\{\ker s_X\}_{s\in\Sigma}$ of subobjects of X is directed (in fact $\ker(ss')_X$ contains both $\ker(s_X)$ and $\ker(s'_X)$). Now it is clear that any subobject of an object of $\mathscr{A}$ is also in $\mathscr{A}$.

Furthermore, if $p : X \to X''$ is an epimorphism and $X \in \mathrm{Ob}\,\mathscr{A}$ then

$$\ker(s_{X''}) \supseteq p(\ker(s_X)), \text{ that is, } X'' = p(X) = p(\sup_{s\in\Sigma}\ker(s_X)) = \sup_{s\in\Sigma} p(\ker(s_X)) = \sup_{s\in\Sigma}\ker(s_{X''}).$$

Hence X'' is an object in $\mathscr{A}$. Suppose that in the exact sequence:

$$0 \longrightarrow X' \xrightarrow{\ i\ } X \xrightarrow{\ p\ } X'' \longrightarrow 0$$

the outside terms are in $\mathscr{A}$. We may assume that $X' = \sup_{s\in\Sigma}(\ker s_X)$. If $X' \neq X$, that is, $X'' \neq 0$, there exists an element $s \in \Sigma$ such that $\ker(s_{X''}) \neq 0$. Let U be a generator of $\mathscr{C}$. Then a non-zero morphism $f : U \to X''$ can be found so that $s_{X''}f = 0$. Then in the commutative diagram:

$$\begin{array}{ccc} U \prod_{X''} X & \xrightarrow{\ p'\ } & U \\ {\scriptstyle f'}\big\downarrow & & \big\downarrow{\scriptstyle f} \\ X & \xrightarrow{\ p\ } & X'' \end{array}$$

the morphism f' is non-zero, and $pf' = fp' \neq 0$, since p' is an epimorphism. But $ps_X f' = s_{X''} pf' = s_{X''} fp' = 0$. In conclusion $s_X(\mathrm{im}\, f') \subseteq X'$ or equivalently $\mathrm{im}\, f' \subseteq \sup_{\bar{s}\in\Sigma}\ker(\bar{s}_X s_X) = \sup_{\bar{s}\in\Sigma}\ker((\bar{s}s)_X) = X'$. In other words $\mathrm{im}\, f' \subseteq \ker p$, contrary to what was supposed above.

In order to complete the proof, let us notice that any Grothendieck category has injective envelopes (see Chapter 3, Section 3.10) and thus it will be sufficient to prove that any direct sum of objects in $\mathscr{A}$ is also an object in $\mathscr{A}$. Indeed, let $\{X_i\}_{i\in I}$ be a set of objects in $\mathscr{A}$, and $X = \coprod_{i\in I} X_i$; then $X = \sup_F X_F$, with F running through all finite subsets of I, and

$$X_F = \coprod_{j\in F} X_j.$$

Obviously, one has that $X_F \in \mathrm{Ob}\mathscr{A}$ for any F, and thus $X \in \mathrm{Ob}\mathscr{A}$. ∎

PROPOSITION 6.12. *All conditions and notations are as above. The following assertions are equivalent:*

(1) *The object M is $\mathscr{A}$-closed;*

(2) s_M *is an automorphism of M for any $s \in \Sigma$.*

Proof. (1) $\Rightarrow$ (2). Obviously for any $s \in \Sigma$, $\ker(s_M) \in \mathrm{Ob}\mathscr{A}$, and consequently s_M is a monomorphism according to Lemma 4.1. But $\operatorname{coker}(s_M) \in \mathrm{Ob}\mathscr{A}$. Making use of Lemma 6.1 we see that s_M is a section. Hence $\operatorname{coker}(s_M)$ is isomorphic with a subobject of M and it is necessarily null. Finally, s_M is an automorphism.

(2) $\Rightarrow$ (1). It is clear that M is $\mathscr{A}$-torsionfree. Let $M_\Sigma = ST\,(M)$ and let $u_M\colon M \to M_\Sigma$ be the canonical inclusion. Then we can construct the commutative diagram:

$$\begin{array}{ccccccccc}
0 & \longrightarrow & M & \xrightarrow{u_M} & M_\Sigma & \xrightarrow{p} & M' & \longrightarrow & 0 \\
 & & \downarrow{\scriptstyle s_M} & & \downarrow{\scriptstyle s_{M_\Sigma}} & & \downarrow{\scriptstyle s_{M'}} & & \\
0 & \longrightarrow & M & \xrightarrow{u_M} & M_\Sigma & \xrightarrow{p} & M' & \longrightarrow & 0
\end{array}$$

with exact rows in which s_M and s_{M_Σ} are isomorphisms and with M_Σ being closed. Obviously, $s_{M'}$ is also an isomorphism and $M' \in \mathrm{Ob}\mathscr{A}$. Thus, one necessarily has that $M' = 0$, that is, u_M is an isomorphism. ∎

COROLLARY 6.13. *All conditions are as above. Then the localization functor is exact.*

Proof. Use Propositions 5.5 and 6.12. ∎

Exercises

1. Let A be a simple ring. Then Mod A has only two localizing subcategories. Describe all localizing subcategories of Mod A where A is a semisimple ring.

2. Let p be a prime number. Denote by $\mathscr{F}(p)$ the full subcategory of $\mathscr{Ab}$ consisting of all objects X such that for any $x \in X$ there exists a natural number n with $p^n x = 0$. $\mathscr{F}\,(p)$ is the smallest localizing subcategory containing $\mathbf{Z}_p$.

References

Dickson [42]; Gabriel [59]; Goldman [76]; Hacque [83]; Hudry [94]; Lambek [114]; Maranda [121]; Mitchell [133]; Morita [136]; Popescu [167]; Roos [199]; Tisseron [227]; Walker–Walker [236].

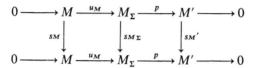

4.7 Sheaves over a topological space

Throughout this paragraph, $\mathscr{C}$ will denote a locally small Ab 5 and Ab 3*-category with sufficiently many injectives.

Let E be a topological space, and let $D(E)$ denote the family of open sets in E. We consider $D(E)$ as an ordered set by defining $U \leqslant V$ if and only if $V \subseteq U$. The functor category $[D(E), \mathscr{C}]$ is called *category of presheaves* in $\mathscr{C}$ over E and is denoted by $\mathscr{P}(E, \mathscr{C})$. Thus a presheaf X assigns to each open set U in E an object $X(U)$ in $\mathscr{C}$, and to each inclusion $V \subseteq U$ of open sets a morphism $X_{UV}: X(U) \to X(V)$ such that if $W \subseteq V \subseteq U$ then

$$X_{VW}\, X_{UV} = X_{UW} \quad \text{and} \quad X_{UU} = 1_{X(U)}$$

for all open sets U in X. A morphism $t: X \to Y$ of presheaves is a family $t_U: X(U) \to Y(U)$ such that for each inclusion of open sets $V \subseteq U$ we have a commutative diagram:

$$\begin{array}{ccc} X(U) & \xrightarrow{t_U} & Y(U) \\ \downarrow{\scriptstyle X_{UV}} & & \downarrow{\scriptstyle Y_{UV}} \\ X(V) & \xrightarrow{t_V} & Y(V) \end{array}$$

By Chapter 3, Theorem 4.1, $\mathscr{P}(E, \mathscr{C})$ is an Ab 5 and Ab 3*-category. Also it is easy to check that this category is locally small (we utilize essentially the fact that $D(E)$ is small).

For each $x \in E$, the open sets in E which contain x form a directed subset of $D(E)$ and thus, a presheaf X over E determines a direct system in $\mathscr{C}$. We write

$$X_x = \varinjlim_{x \in U} X(U)$$

and we call X_x the *stalk* of X over x. The structural morphism from $X(U)$ to the inductive limit is denoted by $X_U{}^x$. If $t: X \to Y$ is a morphism of presheaves, then we have an induced morphism of stalks $t_x: X_x \to Y_x$ in $\mathscr{C}$ and in this way the process of taking the stalks becomes functorial. We let $P_x: \mathscr{P}(E, \mathscr{C}) \to \mathscr{C}$ be the functor defined by $P_x(X) = X_x$ and $P_x(t) = t_x$. According to the construction of inductive limits in $\mathscr{P}(E, \mathscr{C})$ (Chapter 3, Theorem 4.1) we deduce that P_x commutes with inductive limits; also P_x is an exact functor. P_x is called the "*point functor*" associated with x.

Given a family $H = \{H_x\}_{x \in E}$ of objects in $\mathscr{C}$, we form a presheaf $R(H)$ as follows. For an open set U we define

$$R(H)(U) = \textstyle\prod_{x \in U} H_x$$

and for $V \subseteq U$ we have the morphism $\prod_{x \in U} H_x \to \prod_{x \in V} H_x$ which composed with the xth projection from the codomain, gives the xth projection from the domain. The rules for a presheaf are trivially satisfied.

If X is an object in $\mathscr{P}(E, \mathscr{C})$, then we denote by RX the presheaf $R(H)$ with $H = \{X_x\}_{x \in E}$. Then a canonical morphism $s^X : X \to RX$ is defined in the following manner. For any open set U, $s_U{}^X : X(U) \to \prod_{x \in U} X_x$ is the unique morphism such that $k_x s_U{}^X = X_U{}^x$, where $k_x : \prod_{x \in U} X_x \to X_x$ is the xth projection.

Now we denote by $\mathscr{N}$ the full subcategory of $\mathscr{P}(E, \mathscr{C})$ consisting of all objects X, such that $X_x = 0$ for all $x \in E$, that is $\mathrm{P}_x(X) = 0$ for all $x \in E$. We see that $\mathscr{N}$ is a dense subcategory, since all the functors P_x are exact.

An object in $\mathscr{N}$ is called a *negligible presheaf*. Furthermore, $\mathscr{N}$ is closed with respect to direct sums. We shall see that $\mathscr{N}$ is in fact a localizing subcategory. First we observe that any object X in $\mathscr{P}(E, \mathscr{C})$ has a largest subobject $X_{\mathscr{N}}$ in $\mathscr{N}$ and $X/X_{\mathscr{N}} = \overline{X}$ has no non-null subobjects in $\mathscr{N}$.

LEMMA 7.1. *For any object X in* $\mathscr{P}(E, \mathscr{C})$, $(RX)_{\mathscr{N}} = 0$, *that is RX has no non-null subobjects in* $\mathscr{N}$. *In addition,* $X_{\mathscr{N}}$ *is the kernel of* $s^X : X \to RX$.

Proof. Let Y be a non-zero subobject in $\mathscr{N}$ of RX. Then there exists an open set U and a point $x \in U$, such that $k_x\big(Y(U)\big) \neq 0$, $k_x : (RX)(U) \to X_x$ being the xth projection. We see that there exists a unique morphism

$$r_x : (RX)_x \to X_x$$

such that $r_x (RX)_U{}^x = k_x$ for any open set U and $x \in U$. If $j: Y \to RX$ is the canonical inclusion, then $r_x j_x Y_U{}^x = k_x j_U \neq 0$ for a suitable point x, contrary to our assumptions.

From the exact sequence:

$$0 \longrightarrow X_{\mathscr{N}} \longrightarrow X \xrightarrow{\ p\ } \overline{X} \longrightarrow 0$$

we deduce that for any $x \in E$, p_x is an isomorphism and these isomorphisms define an isomorphism $\bar{p} : RX \to R\overline{X}$ such that $\bar{p}s^X = s^{\overline{X}} p$. The proof will be complete if we prove that $s^{\overline{X}}$ is a monomorphism. First, notice that the morphism $s_x{}^{\overline{X}} : \overline{X}_x \to (R\overline{X})_x$ is a section, or equivalently $r_x s_x{}^{\overline{X}} = 1_{\overline{X}_x}$ by the definition of the morphisms r_x. Now if $\ker s^{\overline{X}} \neq 0$, then there exists a point $x \in E$ such that $(\ker s^{\overline{X}})_x = \ker s_x{}^{\overline{X}} \neq 0$, contrary to what was said above. ■

Let $H = \{H_x\}_{x \in E}$ be a set of objects in $\mathscr{C}$, and X an object in $\mathscr{P}(E, \mathscr{C})$; for any morphism $\bar{t} : X \to R(H)$ we denote by $\alpha(t)$ the set of morphisms $\{\bar{t}_x\}_{x \in E}$, where t_x: $X_x \to H_x$ is the composition $r_x\ t_x$, r_x: $R(H)_x \to H_x$ being defined canonically as above.

LEMMA 7.2. *The assignment* $t \rightsquigarrow \alpha(t)$ *defines a bijection*

$$\alpha\colon \mathrm{Hom}_{\mathscr{P}(E, \mathscr{C})}\, X, \big(R(H)\big) \to \textstyle\prod_{x \in E} \mathrm{Hom}_{\mathscr{C}}(X_x, H_x)$$

which is functorial in X.

Proof. Let $f = \{f_x\}_{x\in E}$, $f_x : X_x \to H_x$ be a set of morphisms in $\mathscr{C}$. Denote by $\beta(f) : X \to R(H)$ the morphism defined as follows. For any open set U, $\beta(f)_U : X(U) \to \prod_{x\in U} H_x = R(H)(U)$ is the unique morphism which when composed with the xth projection of the direct product gives $f_x X_U{}^x$.

It is easy to check that α and β are inverses of each other. The functoriality in X is obvious. ∎

COROLLARY 7.3. *If for every $x \in E$, the object H_x is injective in $\mathscr{C}$, then $R(H)$ is injective in $\mathscr{P}(E, \mathscr{C})$.*

Proof. If $X' \xrightarrow{t} X$ is a monomorphism in $\mathscr{P}(E, \mathscr{C})$, then we must show that the induced morphism of groups:

$$\mathrm{Hom}_{\mathscr{P}(E,\mathscr{C})}(X, R(H)) \to \mathrm{Hom}_{\mathscr{P}(E,\mathscr{C})}(X', R(H)) \qquad (6)$$

is an epimorphism. But for any $x \in E$, the morphism $t_x : X_x' \to X_x$ is a monomorphism, hence since H_x is injective in $\mathscr{C}$, the morphism

$$\mathrm{Hom}_{\mathscr{C}}(X_x, H_x) \to \mathrm{Hom}_{\mathscr{C}}(X_x', H_x)$$

induced by t_x is an epimorphism. Therefore, the morphism (6) is an epimorphism, by Lemma 7.2 ($\mathscr{A}\mathscr{b}$ is an Ab 4*-category). ∎

Now we are able to prove the main result.

THEOREM 7.4. *The subcategory $\mathscr{N}$ of all negligible presheaves is a localizing subcategory of $\mathscr{P}(E, \mathscr{C})$.*

Proof. Let X be an object in $\mathscr{P}(E, \mathscr{C})$ for which $X_{\mathscr{N}} = 0$. Also for any $x \in E$ let Q_x be an injective envelope of X_x, $f_x : X_x \to Q_x$ the inclusion morphism, and write $Q = \{Q_x\}_{x\in E}$. Then the morphisms $\{f_x\}_{x\in E}$ define a monomorphism $f : RX \to R(Q)$ and thus fs^X is a monomorphism to X into $R(Q)$, by Lemma 7.1. By Corollary 7.3, $R(Q)$ is an injective object in $\mathscr{P}(E, \mathscr{C})$ and so, to finish it will be sufficient to prove that $R(Q)_{\mathscr{N}} = 0$ (then we can apply Lemma 5.1 and Theorem 4.5). To do this, let Y be a non-null subobject of $R(Q)$ in $\mathscr{N}$. Then for a suitable open set U and $x \in U$, we deduce that $k_x(Y(U)) \neq 0$, where $k_x : R(Q)(U) \to Q_x$ is the xth projection. Thus, $k_x(Y(U)) \cap X_x \neq 0$. But by hypothesis, if we consider X as a subobject of $R(Q)$, we have that $Y \cap X = 0$ and by Ab 5, in $\mathscr{C}$, we deduce that $Y_x \cap X_x = 0$, contrary to what was said above. ∎

We shall denote by $\mathscr{F}(E, \mathscr{C})$ the category $(E, \mathscr{C})/\mathscr{N}$, by F: $\mathscr{P}(E, \mathscr{C}) \to \mathscr{F}(E, \mathscr{C})$ the canonical functor and by I a right adjoint of F. An object in $\mathscr{F}(E, \mathscr{C})$ is called a *sheaf* in $\mathscr{C}$ over E.

Corollary 7.5. *The category $\mathscr{F}(E, \mathscr{C})$ has injective envelopes.*

The proof results by the proof of the above theorem and by Corollary 7.3. ■

Now we shall give an alternative definition of sheaves for some categories $\mathscr{C}$.

Let $\{U_i\}_{i\in I}$ be an open cover for an open set U (i.e. each U_i is open and $\bigcup_{i\in I} U_i = U$). Let $U_{ik} = U_i \cap U_k$, $i, k \in I$ and for a presheaf X let

$$p_i : \prod_i X(U_i) \to X(U_i)$$

the structural projection. Then we have a unique morphism

$$v : X(U) \to \prod_i X(U_i)$$

such that $p_i v = X_{UU_i}$ for all $i \in I$. Also we have the morphisms:

$$f, g : \prod_i X(U_i) \rightrightarrows \prod_{i,k} X(U_i \cap U_k) \tag{7}$$

defined as follows. Denote by $p_{ik} : \prod_{i,k} X(U_{ik}) \to X(U_{ik})$ the structural projection; then $p_{ik} f = X_{U_i U_{ik}} p_i$, and g is defined such that $p_{ik} g = X_{U_k U_{ik}} p_k$. Clearly $fv = gv$.

Now we consider the following property:

(f): For every open set U and every open cover of U the morphism v is the kernel of $f - g$.

A presheaf X will be called a *monopresheaf* if it satisfies the weaker condition that v is always a monomorphism.

Theorem 7.6. *Let $\mathscr{C}$ be a locally small* Ab 6 *and* Ab 3*-*category with sufficiently many injectives. Then, an object X in $\mathscr{P}(E, \mathscr{C})$ is a sheaf if and only if it has the property* (f).

Proof. Denote by $\mathscr{F}'(E, \mathscr{C})$ the full subcategory of $\mathscr{P}(E, \mathscr{C})$ generated by all objects X which have the property (f) and by $I' : \mathscr{F}'(E, \mathscr{C}) \to \mathscr{P}(E, \mathscr{C})$ the inclusion functor. We shall show that I' has an left exact adjoint denoted by F' and $\ker F' = \mathscr{N}$.

We shall divide the proof into several steps.

(a) Let $H = \{H_x\}_{x\in E}$ be a set of objects in $\mathscr{C}$. Then $R(H)$ has the property (f).

Indeed, relative to an open cover $U = \bigcup_{i\in I} U_i$ consider the diagram

$$\begin{array}{c} B \\ \downarrow u \\ \prod_{x\in U} H_x \xrightarrow{v} \prod_{i\in I} \left(\prod_{x\in U_i} H_x\right) \underset{g}{\overset{f}{\rightrightarrows}} \prod_{(j,k)\in I\times I} \left(\prod_{x\in U_{jk}} H_x\right) \end{array}$$

where v, f and g are defined by (7) and u is such that $fu = gu$. We wish to find a morphism $h: B \to R(H)(U) = \prod_{x \in U} H_x$ such that $vh = u$. For $y \in U$, let $i_y \in I$ such that $y \in U_{i_y}$. Let $h_y : B \to H_y$ be the composition

$$B \to \prod_{i \in I} (\prod_{x \in U_i} H_x) \to \prod_{x \in U_{i_y}} H_x \to H_y.$$

It follows from the equality $fu = gu$ that h_y is independent of the choice of i_y, and from this it is easy to verify that the morphism h induced by the family $\{h_y\}_{y \in U}$ is such that $vh = u$. It is also easy to see that v is a monomorphism. Hence v is a kernel of $f - g$ and so $R(H)$ has the property (f).

(b) Let X be a monopresheaf. Then the morphism $s^X : X \to RX$ is a monomorphism (that is, for any open set U, $s_U{}^X$ is a monomorphism).

Indeed, suppose that we have a morphism $f : B \to X(U)$ such that

$$s_U{}^X f = 0.$$

Then for each $x \in U$ we have the commutative diagram:

$$\begin{array}{ccccc} B & \xrightarrow{f} & X(U) & \xrightarrow{s_U{}^X} & \prod_{x \in U} X_x \\ & & {\scriptstyle X_U{}^x} \downarrow & & \downarrow {\scriptstyle k_x} \\ & & X_x & =\!= & X_x \end{array}$$

and so $X_U{}^x f = 0$. But $\mathscr{C}$ is an Ab 6 category and thus by Section 2.8, Exercise 6 we may write:

$$B = \sum_{V \subseteq U} \ker(X_{UV} f).$$

Furthermore for any $x \in U$, let $T(x)$ be the set of all open sets V such that $V \subseteq U$ and $x \in V$. $\mathscr{C}$ being an Ab 6-category we have:

$$B = \bigcap_{x \in U} (\sum_{V \in T(x)} \ker(X_{UV} f)) = \sum_{(V(x)) \in \prod_{x \in U} T(x)} (\bigcap_{x \in U} \ker(X_{UV(x)} f))$$

Hence to show that $f = 0$, it suffices to show that for each

$$(V(x)) \in \prod_{x \in U} T(x),$$

the restriction of f to $\bigcap_{x \in U} \ker(X_{UV(x)} f)$ is null. Now, for each $(V(x))_{x \in U}$ we see that the sets $\{V(x)\}_{x \in U}$ give an open cover for U and the composition $\bigcap_{x \in U} \ker(X_{UV(x)} f) \to B \xrightarrow{f} X(U) \to \prod_{x \in U} X(V(x))$ is necessarily null. Therefore the conclusion follows since X is a monopresheaf.

(c) A presheaf X is $\mathscr{N}$-torsionfree if and only if it is a monopresheaf.

Indeed, let X be $\mathscr{N}$-torsionfree, that is $X_{\mathscr{N}} = 0$, $\{U_i\}_i$ an open cover of an open set U, and $v: X(U) \to \prod_{i \in I} X(U_i)$ the morphism defined in (7).

Let Y be the kernel of v. Denote also by $\tilde{Y}$ the presheaf defined as follows. For any open set V, we write $\tilde{Y}(V) = X_{U, U \cap V}(Y)$; the restriction morphisms are defined canonically. If $Y \neq 0$ then $\tilde{Y}$ is $\mathcal{N}$ torsionfree and there exists a point $x \in U$ such that $\tilde{Y}_x \neq 0$, or equivalently $X_U{}^x(Y) \neq 0$. Let $j \in I$ such that $x \in U_j$. Then we have the commutative diagram:

$$\begin{array}{ccccc} Y = \tilde{Y}(U) & \xrightarrow{t} & X(U) & \xrightarrow{v} & \prod_{i \in I} X(U_i) \\ & & \downarrow{\scriptstyle X_U{}^x} & & \downarrow{\scriptstyle p_j} \\ & & X_x & \xleftarrow{X_{U_j}{}^x} & X(U_j) \end{array}$$

where t is the inclusion. But it is clear that $X_{U_j}{}^x p_j vt = 0$, contrary to what was shown above.

The converse result follows by (b) and Lemma 7.1.

(d) Consider a morphism of presheaves $t: X \to Y$ where Y has the property (f), that is an object in $\mathscr{F}'(E, \mathscr{C})$. Form the diagram:

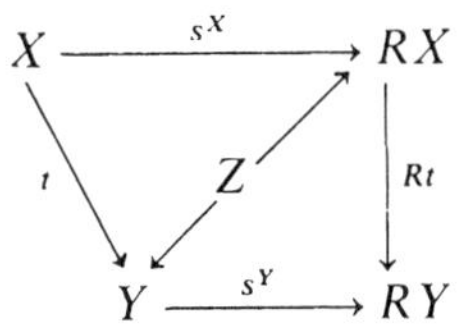

where Rt is canonically induced by t, the inner square of this diagram is cartesian and the outer square is commutative since s^X is functorial in X. By the definition of a cartesian square, we obtain a morphism $X \to Z$ keeping the diagram commutative. Now a slight computation proves that Z also has the property (f) [see (a)]. Since s^Y is a monomorphism, by (b), so is $Z \to RX$. Let $F'(X)$ be the intersection of all subobjects of RX which has the property (f) through which s^X factors. Again a slight computation shows that $F'(X)$ has the property (f) and t factors through $F'(X)$. Furthermore, if there were two distinct such factorizations, then the equalizer of the two factorizations would provide us with a proper subobject of $F'(X)$ having the property (f) which s^X factors, contradicting the definition of $F'(X)$. Hence the factorization is unique and so the assignment $X \rightsquigarrow F'(X)$ defines a left adjoint of the functor I'. Moreover, if $r^X: X \to F'(X)$ is the canonical morphism induced by s^X, then the morphisms $\{r^X\}_X$ define an arrow of adjunction of F' with I'. By Lemma 7.1 we deduce that $\ker r^X = X_{\mathcal{N}}$.

(e) The functor F': $\mathscr{P}(E, \mathscr{C}) \to \mathscr{F}'(E, \mathscr{C})$ is exact.

It will be sufficient to show that F' is left exact. Indeed if $X' \xrightarrow{t} X$ is a monomorphism in $\mathscr{P}(E, \mathscr{C})$, then for any $x \in E$, the morphism $t_x: X_x' \to X_x$ is also a monomorphism and so the morphism $Rt: RX' \to RX$ induced by

t is monomorphism too. But a small computation shows that we may write the following commutative diagram in $\mathscr{P}(E, \mathscr{C})$:

$$\begin{array}{ccc} X' & \xrightarrow{F'(t)} & X \\ {\scriptstyle r^{X'}}\downarrow & & \downarrow{\scriptstyle r^{X}} \\ F'(X') & \xrightarrow{t} & F'(X) \\ {\scriptstyle f'}\downarrow & & \downarrow{\scriptstyle f} \\ RX' & \xrightarrow{Rt} & RX \end{array}$$

where f' and f are canonical inclusions. In conclusion we deduce that $F'(t)$ is a monomorphism in $\mathscr{P}(E, \mathscr{C})$ that $F'(t)$ is a monomorphism in $\mathscr{P}(E, \mathscr{C})$ and also in $\mathscr{F}'(E, \mathscr{C})$, the inclusion functor being full and faithful.

(f) $\mathscr{N} = \ker F'$.

Let $X \in \operatorname{Ob} \mathscr{N}$ and X'' the image of X in $F'(X)$; then by (c) we deduce that X'' is $\mathscr{N}$-torsionfree and so necessarily $X'' = 0$ [again by (c)]. In conclusion $\mathscr{N} \subseteq \ker F'$.

Conversely, if $X \in \ker F'$ and $X_{\mathscr{N}} \neq X$ then $X/X_{\mathscr{N}}$ is a non-zero monopresheaf by (c) and so $F'(X) \neq 0$ contradicting the assumption.

Finally $\mathscr{N} = \ker F'$ and by Theorem 4.9 we deduce that the categories $\mathscr{F}(E, \mathscr{C})$ and $\mathscr{F}'(E, \mathscr{C})$ are equivalent. ∎

Let $\mathscr{C}$ be as in Theorem 7.6.

COROLLARY 7.7. *Let X be an object in $\mathscr{P}(E, \mathscr{C})$. Then, for any $x \in E$, the morphism $r_x^X : X_x \to F'(X)_x$ is an isomorphism.* ∎

COROLLARY 7.8. *A sequence $0 \to X' \xrightarrow{f} X \xrightarrow{g} X'' \to 0$ of sheaves is exact in $\mathscr{F}(E, \mathscr{C})$ if and only if for any point $x \in E$ the sequence*

$$0 \longrightarrow X_x' \xrightarrow{f_x} X_x \xrightarrow{g_x} X_x'' \longrightarrow 0$$

is exact in $\mathscr{C}$.

The proof is left to the reader. ∎

Exercises

1. Let $f : E \to E'$ be a continuous map of topological spaces (i.e. f^{-1} is a functor of DE' into DE). Then a restriction functor $f_*{}^0 : \mathscr{P}(E, \mathscr{C}) \to \mathscr{P}(E', \mathscr{C})$ is defined such that for any presheaf $X : DE \to \mathscr{C}$ we have $f_*{}^0(X) = Xf^{-1}$. Assume that $\mathscr{C}$ is as in Theorem 7.6. Then the functor $f_*{}^0$ has a left exact

adjoint denoted by f^0. If $\mathscr{C}$ is as in Theorem 7.6, then $f_*{}^0$ defines a functor $f_* : \mathscr{F}(E, \mathscr{C}) \to \mathscr{F}(E', \mathscr{C})$ and $f^* = Ff^0$ is a left exact adjoint of f_*. Furthermore, if $g : E' \to E''$ is another continuous map, then $(gf)^* = f^* g^*$. Also for any $X \in \mathrm{Ob}\mathscr{P}(E', \mathscr{C})$ and $x \in E$ we have $(f^0 X)_x \xrightarrow{\sim} X_{f(x)}$ and this isomorphism is natural with respect to morphisms in $\mathscr{P}(E', \mathscr{C})$.

2. Let $\mathscr{C}$ be as in Theorem 7.6, and $0 \to X' \to X \to X'' \to 0$ an exact sequence of presheaves over E. Then

(i) If X' has the property (f) and X is a monopresheaf, then X'' is a monopresheaf.

(ii) If X'' is a monopresheaf and X has the property (f), then X' also has the property (f).

3. Let $\mathscr{C}$ be an Ab 6 and Ab 3*-category. Suppose that $t : X \to Y$ is a morphism in $\mathscr{P}(E, \mathscr{C})$ where X has the property (f) and Y is a monopresheaf. Then t is an isomorphism if and only if t_x is an isomorphism for all $x \in E$.

4. Let $\mathscr{C}$ be as in Theorem 7.6, and U an open set in E. Denote by $R_U : \mathscr{P}(E, \mathscr{C}) \to \mathscr{P}(U, \mathscr{C})$ the restriction functor defined as follows. For any open set $V \subseteq U$, $R_U(X)(V) = X(V)$, for every $X \in \mathrm{Ob}\,\mathscr{P}(E, \mathscr{C})$. Then R_U is an exact functor and it defines a functor $\bar{R}_U : \mathscr{F}(E, \mathscr{C}) \to \mathscr{F}(U, \mathscr{C})$ such that $\bar{R}_U F = F_U R_U$, where $F_U : \mathscr{P}(U, \mathscr{C}) \to \mathscr{F}(U, \mathscr{C})$ is the canonical functor. Prove that $\bar{R}_U$ is exact and has a left adjoint.

References

Bourbaki [24]; Bucur–Deleanu [31]; Freyd [57]; Gabriel [59]; Giraud [65]; Godement [66]; Gray [77]; Grothendieck [79]; Heller–Rowe [89]; Mitchell [133]; Pareigis [162]; Popescu [167].

4.8 Torsion theories

We shall assume throughout this paragraph that $\mathscr{C}$ is a locally small Ab 3-category.

A *torsion theory* for the category $\mathscr{C}$ consists of a couple $(\mathscr{T}, \mathscr{F})$ of full subcategories of $\mathscr{C}$ satisfying the following axioms:

(i) $\mathscr{T} \cap \mathscr{F} = 0$;

(ii) if X is an object of $\mathscr{T}$, then any quotient object of X is also in $\mathscr{T}$;

(iii) if Y is an object in $\mathscr{F}$, then any subobject of Y is also in $\mathscr{F}$;

(iv) for each object X of $\mathscr{C}$ there is an exact sequence:

$$0 \to X' \to X \to X'' \to 0$$

with $X' \in \mathrm{Ob}\mathscr{T}$, and $X'' \in \mathrm{Ob}\mathscr{F}$.

NOTE. A subcategory $\mathscr{A}$ of $\mathscr{C}$ is called *hereditary* (respectively *cohereditary*) if it verifies the above condition (iii) [respectively (ii)]. Finally a subcategory $\mathscr{A}$ is called *stable for extensions* if for any exact sequence

$$0 \to X' \to X \to X'' \to 0$$

with X' and X'' in $\mathscr{A}$, we deduce that $X \in \mathrm{Ob}\mathscr{A}$.

For example a dense subcategory is at the same time hereditary, cohereditary and stable for extensions, and conversely.

Axiom (iv) will be referred to as the extension axiom. Axioms (i)–(iii) may be replaced by the orthogonality axiom:

(v) $\mathrm{Hom}_{\mathscr{C}}(X, Y) = 0$ for each $X \in \mathrm{Ob}\mathscr{T}$ and $Y \in \mathrm{Ob}\mathscr{F}$,

in the presence of (iv) and the additional assumption that $\mathscr{T}$ and $\mathscr{F}$ contain an object together with all its isomorphic copies in $\mathscr{C}$.

If for the full subcategory $\mathscr{T}$ there exists a full subcategory $\mathscr{F}$ such that $\mathscr{T}$ and $\mathscr{F}$ together satisfy (i)–(iv) we then say $\mathscr{T}$ is a *torsion subcategory* or T-subcategory. Similarly we define a *torsion-free* subcategory or F-subcategory.

EXAMPLES. 1. Let $\mathscr{D}$ be the full subcategory of $\mathscr{Ab}$ consisting of all divisible groups and $\mathscr{R}$ the full subcategory consisting of all reduced groups (an abelian group is *reduced* if the largest divisible group contained in it is null). Then the couple $(\mathscr{D}, \mathscr{R})$ is a torsion theory.

2. Let $\mathscr{A}$ be a hereditary torsion subcategory of $\mathscr{C}$ (i.e. a dense subcategory of $\mathscr{C}$ closed under direct sums, that is the direct sum in $\mathscr{C}$ of any set of objects in $\mathscr{A}$ is also in $\mathscr{A}$). Then any object X in $\mathscr{C}$ has a largest subobject in $\mathscr{A}$ denoted by $X_{\mathscr{A}}$ and the quotient $X/X_{\mathscr{A}}$ has no non-null subobjects in $\mathscr{A}$. Denote by $\mathscr{A}^*$ (the orthogonal of $\mathscr{A}$) the full subcategory of $\mathscr{C}$ consisting of all objects Y such that $Y_{\mathscr{A}} = 0$. Clearly, the couple $(\mathscr{A}, \mathscr{A}^*)$ is a torsion theory for $\mathscr{C}$. Most interesting torsion theories are of this kind. If $\mathscr{A}$ is a localizing subcategory, then $\mathscr{A}$ consists exactly of $\mathscr{A}$-torsionfree objects.

COROLLARY 8.1. *Let $(\mathscr{T}, \mathscr{F})$ be a torsion theory. Then an object X is in $\mathscr{T}$ if and only if* $\mathrm{Hom}_{\mathscr{C}}(X, Y) = 0$ *for any $Y \in \mathrm{Ob}\mathscr{F}$. Dually, an object Y is in $\mathscr{F}$ if and only if* $\mathrm{Hom}_{\mathscr{C}}(X, Y) = 0$ *for any $X \in \mathrm{Ob}\mathscr{T}$.*

The proof is a direct consequence of the axioms (i)–(iv). ■

Generally, let $\mathcal{T}$ and $\mathcal{F}$ be two subcategories of $\mathscr{C}$. We shall say that $\mathcal{T}$ is *complete* with respect to the orthogonality relation (v) if for any object X, such that $\mathrm{Hom}_{\mathscr{C}}(X, Y) = 0$ for all objects Y in $\mathcal{F}$ we deduce that $X \in \mathrm{Ob}\mathcal{T}$. The above corollary states that $\mathcal{T}$ and $\mathcal{F}$ are each complete with respect to the orthogonality relation (v).

COROLLARY 8.2. *Let $(\mathcal{T}, \mathcal{F})$ be a torsion theory for $\mathscr{C}$. Then $\mathcal{T}$ is closed under direct sums and $\mathcal{F}$ under direct products (when they exist in $\mathscr{C}$). In addition, any object X of $\mathscr{C}$ has a greatest subobject in $\mathcal{T}$ denoted by $X_{\mathcal{T}}$. $\mathcal{T}$ and $\mathcal{F}$ are each stable under extensions.*

The proof is obvious by Corollary 8.1. ■

PROPOSITION 8.3. *Let $\mathcal{T}$ and $\mathcal{F}$ be two full subcategories of $\mathscr{C}$ each complete with respect to the orthogonality relation* (v). *Then the couple $(\mathcal{T}, \mathcal{F})$ is a torsion theory for $\mathscr{C}$.*

Proof. It suffices to verify axiom (iv). To this end, we first show that $\mathcal{T}$ is closed under direct sums and stable under extensions. Let $\{X_i\}_i$ be a set of objects of $\mathcal{T}$. If Y is an arbitrary member of $\mathcal{F}$, we have:

$$\mathrm{Hom}_{\mathscr{C}}(\textstyle\coprod_i X_i, Y) \simeq \prod_i \mathrm{Hom}_{\mathscr{C}}(X_i, Y) = 0$$

so that $\coprod_i X_i \in \mathrm{Ob}\mathcal{T}$ by the completeness of $\mathcal{T}$. To see that $\mathcal{T}$ is stable under extensions, consider the exact sequence:

$$0 \to X_1 \to X \to X_2 \to 0$$

with $X_1, X_2 \in \mathrm{Ob}\mathcal{T}$. For any $Y \in \mathrm{Ob}\mathcal{F}$ we have the exact sequence of abelian groups:

$$0 \to 0 = \mathrm{Hom}_{\mathscr{C}}(X_2, Y) \to \mathrm{Hom}_{\mathscr{C}}(X, Y) \to \mathrm{Hom}_{\mathscr{C}}(X_1, Y) = 0$$

from which it follows that $X \in \mathrm{Ob}\mathcal{T}$.

Now it is easy to see that any object X in $\mathscr{C}$ has a largest subobject in $\mathcal{T}$: the sum of all subobjects of X in $\mathcal{T}$; this subobject is denoted by $X_{\mathcal{T}}$. It remains to show that $X/X_{\mathcal{T}} \in \mathrm{Ob}\mathcal{F}$. Indeed, if X' is the greatest subobject in $\mathcal{T}$ of $X/X_{\mathcal{T}}$, then we have the exact sequence:

$$0 \to X_{\mathcal{T}} \to H \to X' \to 0$$

H being a suitable subobject of X. If $X' \neq 0$ then H is an object of $\mathcal{T}$, greater than $X_{\mathcal{T}}$. The contradiction is obvious. Hence $X/X_{\mathcal{T}}$ has no non-null subobjects in $\mathscr{T}$, and so $\mathrm{Hom}_{\mathscr{C}}(X', X/X_{\mathscr{T}}) = 0$ for any $X' \in \mathrm{Ob}\mathscr{T}$. In conclusion $X/X_{\mathscr{T}} \in \mathrm{Ob}\mathscr{F}$. ∎

Finally, we obtain the following characterization of torsion theories which will be useful for computations.

THEOREM 8.4. *The full subcategory $\mathscr{T}$ of $\mathscr{C}$ is a torsion subcategory if and only if $\mathscr{T}$ is cohereditary, closed under direct sums and stable under extensions. Dually, the full subcategory $\mathscr{F}$ is a torsion free subcategory of $\mathscr{C}$ if and only if $\mathscr{F}$ is hereditary, closed under direct products (when they exist in $\mathscr{C}$) and stable under extensions.*

The proof follows from the above results. ∎

COROLLARY 8.5. *Let $(\mathscr{T}, \mathscr{F})$ be a torsion theory for $\mathscr{C}$. Then $\mathscr{T}$ is a hereditary torsion subcategory and $\mathscr{F} = \mathscr{T}^*$ if and only if $\mathscr{T}$ is hereditary.* ∎

A *preradical* of $\mathscr{C}$ is defined as a subfunctor of the identity functor of $\mathscr{C}$. Thus a preradical $\mathfrak{r}$ is given if for any object X a subobject $\mathfrak{r}(X)$ of X is defined such that if $f: X \to Y$ there exists a morphism $\mathfrak{r}(f): \mathfrak{r}(X) \to \mathfrak{r}(Y)$ such that the following diagram is commutative:

$$\begin{array}{ccc} \mathfrak{r}(X) & \xrightarrow{i_X} & X \\ {\scriptstyle \mathfrak{r}(f)}\downarrow & & \downarrow{\scriptstyle f} \\ \mathfrak{r}(Y) & \xrightarrow{i_Y} & Y \end{array}$$

where i_X and i_Y are the canonical inclusions.

If a preradical $\mathfrak{r}$ is such that for any object X we have $\mathfrak{r}(X/\mathfrak{r}(X)) = 0$ then $\mathfrak{r}$ is termed a *radical*. A preradical $\mathfrak{r}$ is *hereditary* if $\mathfrak{r}$ is left exact, equivalently, if $X' \subseteq X$ implies $\mathfrak{r}(X') = X' \cap \mathfrak{r}(X)$. Finally, a preradical $\mathfrak{r}$ is *idempotent* if $\mathfrak{r}^2 = \mathfrak{r}$, that is $\mathfrak{r}(\mathfrak{r}(X)) = \mathfrak{r}(X)$ for any object X. One easily verifies that a hereditary radical is idempotent. However the converse is not true. For example, the radical $X \rightsquigarrow X_{\mathscr{D}}$ of the category of abelian groups assigning to each group its maximum divisible subgroup is idempotent, but is not hereditary.

COROLLARY 8.6. *Let $(\mathscr{T}, \mathscr{F})$ be a torsion theory for $\mathscr{C}$. Then the correspondence $X \rightsquigarrow X_{\mathscr{T}}$ defines an idempotent radical.*

The proof is easy and left to the reader. ∎

Next we prove a converse result.

PROPOSITION 8.7. *Let* $\mathfrak{r}$ *be a preradical of* $\mathscr{C}$. *Then* $\mathfrak{r}$ *is an idempotent radical if and only if the full subcategory* $\mathscr{T}$ *of objects fixed by* $\mathfrak{r}$ *is a* T*-subcategory. Equivalently,* $\mathfrak{r}$ *is an idempotent radical if and only if the full subcategory* $\mathscr{F}$ *consisting of all objects* Y *such that* $\mathfrak{r}(Y) = 0$ *is an* F*-subcategory.*

Proof. Assume $\mathfrak{r}$ is an idempotent radical. Let $\mathscr{T}$ be the full subcategory of $\mathscr{C}$ consisting of objects X such that $\mathfrak{r}(X) = X$, and $\mathscr{F}$ the full subcategory of objects Y with $\mathfrak{r}(Y) = 0$. We verify the axioms for a torsion theory. First, (i) is clear, (ii) is just Exercise 4 and (iii) is immediate. Now the exact sequence:

$$0 \to \mathfrak{r}(X) \to X \to X/\mathfrak{r}(X) \to 0$$

satisfies the axiom (iv) in view of the hypothesis on $\mathfrak{r}$.

Conversely, define $\mathscr{T}$ and $\mathscr{F}$ as above and suppose $\mathscr{T}$ is a T-subcategory. Denoting by $X_{\mathscr{T}}$ the maximum subobject of X lying in $\mathscr{T}$, the result will follow by Corollary 8.6 if we show that for any object X, $\mathfrak{r}(X) = X_{\mathscr{T}}$. But, given X, $\mathfrak{r}(X) \subseteq X_{\mathscr{T}}$, as $X_{\mathscr{T}}$ is the maximum subobject fixed by $\mathfrak{r}$. On the other hand, $X_{\mathscr{T}} \subseteq X$ shows $X_{\mathscr{T}} = \mathfrak{r}(X_{\mathscr{T}}) \subseteq \mathfrak{r}(X)$, since $\mathfrak{r}$ is a subfunctor of the identity functor. ∎

We have noted that $\mathscr{T}$ is closed under subobjects if and only if the functor $X \rightsquigarrow X_{\mathscr{T}}$ is left exact, that is, a hereditary radical. We have the following characterization of this situation when $\mathscr{C}$ has injective envelopes.

THEOREM 8.8. *Let* $\mathscr{C}$ *have injective envelopes and let* $(\mathscr{T}, \mathscr{F})$ *be a torsion theory for* $\mathscr{C}$. *Then* $\mathscr{T}$ *is hereditary (that is, a localizing subcategory) if and only if* $\mathscr{F}$ *is closed under taking injective envelopes.*

The proof is simple and left to the reader. ∎

In the following we shall make some useful observations on subcategories of an Ab 3-category.

A full subcategory of $\mathscr{C}$ is called *prelocalizing* if it is hereditary, cohereditary and closed under direct sums. Obviously, any hereditary torsion subcategory, particularly any localizing subcategory is prelocalizing. Later we shall see examples of prelocalizing subcategories which are not localizing; when $\mathscr{P}$ is a prelocalizing subcategory, then any object X has a largest subobject in $\mathscr{P}$, denoted by $X_{\mathscr{P}}$. $X_{\mathscr{P}}$ is of course, the sum of all subobjects of X in $\mathscr{P}$.

PROPOSITION 8.9. *There exists a bijection between the class of all prelocalizing subcategories of* $\mathscr{C}$ *and the class of all hereditary idempotent preradicals of* $\mathscr{C}$.

Proof. Let $\mathscr{P}$ be a prelocalizing subcategory; denote by $\mathfrak{r}_{\mathscr{P}}$ the preradical which assigns to each object X the subobject $\mathfrak{r}_{\mathscr{P}}(X) = X_{\mathscr{P}}$. If $f : X \to Y$

is a morphism, then $f(X_{\mathscr{P}})$ is a quotient of $X_{\mathscr{P}}$ and so an object in $\mathscr{P}$, that is, $f(X_{\mathscr{P}}) \subseteq Y_{\mathscr{P}}$. Obviously $\mathfrak{r}^2_{\mathscr{P}}(X) = \mathfrak{r}_{\mathscr{P}}(\mathfrak{r}_{\mathscr{P}}(X)) = (X_{\mathscr{P}})_{\mathscr{P}} = X_{\mathscr{P}} = \mathfrak{r}_{\mathscr{P}}(X)$. Clearly $\mathfrak{r}_{\mathscr{P}}$ is a hereditary preradical.

Conversely, let $\mathfrak{r}$ be an idempotent hereditary preradical; denote by $\mathscr{P}_{\mathfrak{r}}$ the full subcategory of $\mathscr{C}$ consisting of all objects X such that $\mathfrak{r}(X) = X$. We see immediately that $\mathscr{P}_{\mathfrak{r}}$ is hereditary and cohereditary.

Now let $I : \mathscr{P}_{\mathfrak{r}} \to \mathscr{C}$ be the inclusion functor; then the assignment $X \rightsquigarrow \mathfrak{r}(X)$ gives a right adjoint H for I, and thus I commutes with inductive limits. But for any set $\{X_i\}_i$ of objects in $\mathscr{P}_{\mathfrak{r}}$, we see that $\mathfrak{r}(\coprod_i X_i)$ is a direct sum of this set of objects in $\mathscr{P}_{\mathfrak{r}}$ (here $\coprod_i X_i$ is the direct sum in $\mathscr{C}$). Therefore, $I(\mathfrak{r}(\coprod_i X_i))$ is necessarily equal to $\coprod_i X_i$, that is, $\coprod_i X_i$ is also an object in $\mathscr{P}_{\mathfrak{r}}$.

To complete the proof, we leave it to the reader to prove that $\mathfrak{r}_{\mathscr{P}_{\mathfrak{r}}} = \mathfrak{r}$ and $\mathscr{P}_{\mathfrak{r}_{\mathscr{P}}} = \mathscr{P}$. ∎

PROPOSITION 8.10. *A prelocalizing subcategory $\mathscr{P}$ is a hereditary torsion subcategory if and only if for each object X we have $(X/X_{\mathscr{P}})_{\mathscr{P}} = 0$ that is, if and only if the associated preradical $\mathfrak{r}_{\mathscr{P}}$ is a radical.*

Proof. If $\mathscr{P}$ is hereditary torsion and $(X/X_{\mathscr{P}})_{\mathscr{P}} = X_1$ is non-zero, then we may write the exact sequence:

$$0 \to X_{\mathscr{P}} \to H \to X_1 \to 0$$

where H is a suitable subobject of X and by hypothesis an object of $\mathscr{P}$, greater than $X_{\mathscr{P}}$, contrary to our assumptions.

Conversely, assume that $\mathscr{P}$ is a prelocalizing subcategory, such that $(X/X_{\mathscr{P}})_{\mathscr{P}} = 0$ for each object X, and

$$0 \to X' \to X \to X'' \to 0$$

is an exact sequence with X' and X'' in $\mathscr{P}$. If X is not in $\mathscr{P}$, then $X/X_{\mathscr{P}}$ is isomorphic with a quotient of X'', that is, an object in $\mathscr{P}$. The contradiction is obvious. ∎

If $\mathscr{P}$ is a prelocalizing subcategory, then we denote by $\overline{\mathscr{P}}$ the intersection of all hereditary torsion subcategories containing $\mathscr{P}$. Obviously, $\overline{\mathscr{P}}$ is the smallest hereditary torsion subcategory which contains $\mathscr{P}$. Generally, we may not be able to construct the subcategory $\overline{\mathscr{P}}$, but under additional conditions a reasonable construction of $\overline{\mathscr{P}}$ can be given. Sometimes $\overline{\mathscr{P}}$ is called the *closure* of $\mathscr{P}$.

Let $\mathscr{P}$ and $\mathscr{P}'$ be two prelocalizing subcategories. Denote by $\mathscr{P} \circ \mathscr{P}'$ and call the *composition* of $\mathscr{P}$ with $\mathscr{P}'$, the full subcategory of $\mathscr{C}$ formed by all objects X such that $X/X_{\mathscr{P}} \in \operatorname{Ob}\mathscr{P}'$.

LEMMA 8.11. *The subcategory $\mathscr{P} \circ \mathscr{P}'$ is prelocalizing.*

Proof. If $X \in \mathrm{Ob}(\mathscr{P} \circ \mathscr{P}')$ and X' is a subobject in X, then $X'_{\mathscr{P}} = X_{\mathscr{P}} \cap X'$ and so $X'/X'_{\mathscr{P}}$ is a subobject of $X/X_{\mathscr{P}}$, that is, an object in $\mathscr{P}'$. Therefore X' is also in $\mathscr{P} \circ \mathscr{P}'$. On the other hand, if X'' is a quotient object of X, then $X''/X''_{\mathscr{P}}$ is a quotient object of $X/X_{\mathscr{P}}$, hence an object in $\mathscr{P}'$. Again we conclude that X'' is in $\mathscr{P} \circ \mathscr{P}'$.

Finally, let $\{X_i\}_i$ be a set of objects in $\mathscr{P} \circ \mathscr{P}'$, then for any i,

$$X_i/(X_i)_{\mathscr{P}} \in \mathrm{Ob}\mathscr{P}'.$$

But the structural morphisms $u_i : X_i \to \coprod_i X_i$, define a canonical morphism $t: \coprod_i (X_i)_{\mathscr{P}} \to (\coprod_i X_i)_{\mathscr{P}}$ such that $rt = h$ (where $r: (\coprod_i X_i)_{\mathscr{P}} \to \coprod_i X_i$ is the canonical inclusion and $h: \coprod_i (X_i)_{\mathscr{P}} \to \coprod_i X_i$ is the sum morphism induced by the canonical inclusions $(X_i)_{\mathscr{P}} \to X_i$) and so $(\coprod_i X_i)/(\coprod_i (X_i))_{\mathscr{P}}$ is a quotient object of $\coprod_i (X_i/(X_i)_{\mathscr{P}})$; in other words an object of $\mathscr{P}'$. In conclusion $\coprod_i X_i$ is also an object of $\mathscr{P} \circ \mathscr{P}'$. ∎

COROLLARY 8.12. *A prelocalizing subcategory $\mathscr{P}$ is hereditary torsion subcategory if and only if $\mathscr{P}^2 = \mathscr{P}$.*

Proof. If $\mathscr{P}$ is hereditary torsion subcategory, then an object X is in $\mathscr{P} \circ \mathscr{P}$ if and only if $X/X_{\mathscr{P}} \in \mathrm{Ob}\mathscr{P}$, that is if and only if $X/X_{\mathscr{P}} = 0$. We conclude that $\mathscr{P} \circ \mathscr{P} \subseteq \mathscr{P}$. The reverse inclusion always holds (Exercise 10). Conversely, if $\mathscr{P} \circ \mathscr{P} = \mathscr{P}$ and X is an object, then $X/X_{\mathscr{P}}$ has no non-zero subobjects in $\mathscr{P}$. Indeed by the exact sequence:

$$0 \to X_{\mathscr{P}} \to H \to (X/X_{\mathscr{P}})_{\mathscr{P}} \to 0$$

where H is a suitable subobject of X, we derive that $H \in \mathrm{Ob}\mathscr{P}$, hence necessarily $X_{\mathscr{P}} = H$. ∎

NOTE 2. Let $\mathscr{P}$, $\mathscr{P}'$ be two prelocalizing subcategories of $\mathscr{C}$ and $\mathfrak{r}$, $\mathfrak{r}'$ the associated preradicals (see Proposition 8.9). Denote by $\mathfrak{r} \circ \mathfrak{r}'$ the preradical associated with $\mathscr{P} \circ \mathscr{P}'$. It is easy to see, that for any object X of $\mathscr{C}$, one has:

$$(\mathfrak{r} \circ \mathfrak{r}')(X) = p^{-1}\big(\mathfrak{r}'(X/\mathfrak{r}(X)\big)$$

where $p: X \to X/\mathfrak{r}(X)$ is the canonical epimorphism.

This raises the question: under what conditions can a prelocalizing subcategory be embedded in a hereditary torsion subcategory? This question can be solved in the following important case.

Let $\mathscr{C}$ be a locally small Ab 5-category, $\mathscr{P}$ a prelocalizing subcategory of $\mathscr{C}$ and $\mathfrak{r}$ the associated preradical. For any ordinal α denote by $\mathfrak{r}_\alpha$ the preradical of $\mathscr{C}$ defined as follows:

$$\mathfrak{r}_1 = \mathfrak{r}$$
$$\mathfrak{r}_{\alpha+1} = \mathfrak{r}_\alpha \circ \mathfrak{r}$$

If α is a limit ordinal, then for any object X one has

$$\mathfrak{r}_\alpha(X) = \textstyle\sum_{\beta<\alpha} \mathfrak{r}_\beta(X).$$

We leave to the reader to verify that $\mathfrak{r}_\alpha$ is hereditary idempotent preradical for every α.

Lemma 8.13. *Let $\mathscr{C}$ be a locally small* Ab 5-*category and $\mathfrak{r}$ a hereditary idempotent preradical of $\mathscr{C}$. Assume that an ordinal α exists such that $\mathfrak{r}_\alpha = \mathfrak{r}_{\alpha'}$, for any $\alpha' \geqslant \alpha$. Then $\mathfrak{r}_\alpha$ is a hereditary idempotent radical.*

Proof. By Corollary 8.12 it is enough to prove that $\mathfrak{r}_\alpha \circ \mathfrak{r}_\alpha = \mathfrak{r}_\alpha$. First we notice that $\mathfrak{r}_{\alpha+1} = \mathfrak{r}_\alpha \circ \mathfrak{r} = \mathfrak{r}_\alpha$. Let β be an ordinal such that $\mathfrak{r}_\alpha \circ \mathfrak{r}_\gamma = \mathfrak{r}_\alpha$ for any $\gamma < \beta$, in other words, for every $X \in \operatorname{Ob}\mathscr{C}$, one has

$$\mathfrak{r}_\gamma\big(X/\mathfrak{r}_\alpha(X)\big) = 0.$$

If $\beta = \gamma + 1$, then $\mathfrak{r}_\alpha \circ \mathfrak{r}_\beta = \mathfrak{r}_\alpha \circ \mathfrak{r}_{\gamma+1} = (\mathfrak{r}_\alpha \circ \mathfrak{r}_\gamma) \circ \mathfrak{r} = \mathfrak{r}_\alpha \circ \mathfrak{r} = \mathfrak{r}$ (see Exercise 10). If β is a limit ordinal, then for every object X of $\mathscr{C}$ have

$$(\mathfrak{r}_\alpha \circ \mathfrak{r}_\beta)\,(X) = p^{-1}\big(\textstyle\sum_{\gamma<\beta} \mathfrak{r}_\gamma(X/\mathfrak{r}_\alpha(X))\big) = \mathfrak{r}_\alpha(X)$$

where $p : X \to X/\mathfrak{r}_\alpha(X)$ is the canonical epimorphism.

Therefore, for any ordinal β, we have $\mathfrak{r}_\alpha \circ \mathfrak{r}_\beta = \mathfrak{r}_\alpha$, particularly

$$\mathfrak{r}_\alpha \circ \mathfrak{r}_\alpha = \mathfrak{r}_\alpha,$$

i.e. $\mathfrak{r}_\alpha$ is an idempotent hereditary radical, or equivalently, the associated prelocalizing subcategory is hereditary torsion subcategory. ∎

Corollary 8.14. *Let $\mathscr{C}$ be a locally small* Ab 5-*category. Assume that the family of all prelocalizing subcategory of $\mathscr{C}$ is a set. Then any prelocalizing subcategory of $\mathscr{C}$ can be embedded in a smallest hereditary torsion subcategory.* ∎

The following result can be derived from Exercise 11 but we give a direct proof.

Proposition 8.15. *Let U be a generator of $\mathscr{C}$, $\mathscr{P}$ a prelocalizing subcategory and let $c(\mathscr{P})$ be the set of quotients objects of U lying in $\mathscr{P}$. Then, $c(\mathscr{P})$ determines $\mathscr{P}$ in the following sense. If $\mathscr{P}$ and $\mathscr{P}'$ are two prelocalizing subcategories of $\mathscr{C}$, then $c(\mathscr{P}) = c(\mathscr{P}')$ if and only if $\mathscr{P} = \mathscr{P}'$. Particularly, $\mathscr{C}$ has a set of prelocalizing subcategories.*

Proof. Assume that $c(\mathscr{P}) = c(\mathscr{P}')$ and $X \in \mathrm{Ob}\mathscr{P}$. If the inclusion $X_{\mathscr{P}'} \to X$ is not an isomorphism, then a non-zero quotient U' of U exists such that $U' \subseteq X$, and U' is not included in $X_{\mathscr{P}'}$. By hypothesis, U' is also an object in $\mathscr{P}'$, and so $X_{\mathscr{P}'} + U'$ is an object in $\mathscr{P}'$, greater than $X_{\mathscr{P}'}$. The contradiction is obvious. ∎

Exercises

1. Suppose that $\mathscr{T}$ and $\mathscr{F}$ are subcategories of $\mathscr{C}$ which satisfy (V) and each of them are closed under isomorphic images. Then a necessary and sufficient condition that $\mathscr{T}$ and $\mathscr{F}$ be complete with respect to (V) is that the extension axiom holds.

2. An element a of an abelian group X is of *infinite height* if for any natural number n the equation $nx = a$ has a solution in X. If X^1 denotes the subgroup of the elements of infinite height of the abelian group X, the functor $X \rightsquigarrow X^1$ is a radical but not idempotent (examine the Prüfer group [110]).

3. If $\mathfrak{r}$ is a preradical of $\mathscr{C}$ and if $X' \subseteq \mathfrak{r}(X)$ then $\mathfrak{r}(X/X') = \mathfrak{r}(X)/X'$.

4. The subcategory of objects fixed by a radical $\mathfrak{r}$ (an object X is fixed if $\mathfrak{r}(X) = X$) is cohereditary.

5. Let $\mathscr{C}$ be an Ab 5-category and let $(\mathscr{T}, \mathscr{F})$ be a torsion theory for $\mathscr{C}$. Then $\mathscr{T}$ is hereditary if and only if $\mathscr{F}$ is closed under essential extensions.

6. Given a class $\mathscr{A}$ of objects of $\mathscr{C}$, we define the operators L and R as follows:

$L(\mathscr{A})$ is the full subcategory of $\mathscr{C}$ consisting of all objects X such that $\mathrm{Hom}_{\mathscr{C}}(X, Y) = 0$ for all $Y \in \mathscr{A}$.

$R(\mathscr{A})$ is the full subcategory of $\mathscr{C}$ consisting of all objects Y such that $\mathrm{Hom}_{\mathscr{C}}(X, Y) = 0$ for all $X \in \mathscr{A}$.

Then we have the following relations:

(i) $\mathscr{A} \cap L(\mathscr{A}) = 0$; $\mathscr{A} \cap R(\mathscr{A}) = 0$;

(ii) $\mathscr{A} \subseteq L(R(\mathscr{A}))$; $\mathscr{A} \subseteq R(L(\mathscr{A}))$;

(iii) If $\mathscr{A} \subseteq \mathscr{A}'$ then $L(\mathscr{A}) \supseteq L(\mathscr{A}')$ and $R(\mathscr{A}) \supseteq R(\mathscr{A}')$;

(iv) $L(R(L(\mathscr{A}))) = L(\mathscr{A})$; $R(L(R(\mathscr{A}))) = R(\mathscr{A})$;

(v) The operators $T = LR$ and $F = RL$ are idempotent;

(vi) There is an order reversing bijection between images of T and those of F, wherein $T(\mathscr{A})$ corresponds to $R(T(\mathscr{A})) = F(R(\mathscr{A}))$, and $F(\mathscr{A})$ corresponds to $LF(\mathscr{A}) = TL(\mathscr{A})$.

7. The following statements are equivalent for the pair $(\mathcal{T}, \mathcal{F})$ of full subcategories of $\mathcal{C}$.

(a) $(\mathcal{T}, \mathcal{F})$ is a torsion theory of $\mathcal{C}$;

(b) $T(\mathcal{T}) = \mathcal{T}$, with $R(\mathcal{T}) = \mathcal{F}$;

(c) $F(\mathcal{F}) = \mathcal{F}$, with $L(\mathcal{F}) = \mathcal{T}$;

(d) $R(\mathcal{T}) = \mathcal{F}$ and $L(\mathcal{F}) = \mathcal{T}$;

8. If $\{\mathcal{A}_i\}_i$ are classes of objects of $\mathcal{C}$, then

$$L\left(\bigcup_i \mathcal{A}_i\right) = L\left(\bigcup_i RL(\mathcal{A}_i)\right) = \bigcap_i L(\mathcal{A}_i).$$

The dual result holds, interchanging L and R. Moreover if $T(\mathcal{A}_i) = \mathcal{A}_i$ for any i, then $T\left(\bigcup_i \mathcal{A}_i\right) = \bigcup_i \mathcal{A}_i$.

9. Let $T(\mathcal{A}_i) = \mathcal{A}_i$ for any i. Then $F\left(R\left(\bigcup_i \mathcal{A}_i\right)\right) = R\left(\bigcup_i \mathcal{A}_i\right)$, and $F\left(\bigcap_i R(\mathcal{A}_i)\right) = \bigcap_i R(\mathcal{A}_i)$.

10. If $\mathcal{O}$ is the *null subcategory* of $\mathcal{C}$ i.e. the full subcategory generated by the zero object (obviously prelocalizing), then $\mathcal{O} \circ \mathcal{P} = \mathcal{P} \circ \mathcal{O} = \mathcal{P}$ for any prelocalizing subcategory and $\mathcal{P} \circ \mathcal{C} = \mathcal{C} \circ \mathcal{P} = \mathcal{C}$. The composition of prelocalizing subcategories is associative. In addition if $\mathcal{P} \subseteq \mathcal{P}'$ then

$$\mathcal{P} \circ \mathcal{P}'' \subseteq \mathcal{P}' \circ \mathcal{P}'' \quad \text{and} \quad \mathcal{P}'' \circ \mathcal{P} \subseteq \mathcal{P}'' \circ \mathcal{P}'$$

and also $\mathcal{P} \subseteq \mathcal{P} \circ \mathcal{P}$ for any prelocalizing subcategories $\mathcal{P}$, $\mathcal{P}'$, $\mathcal{P}''$ of $\mathcal{C}$.

11. Let U be a generator of $\mathcal{C}$. A set F of subobjects of U is called a *prelocalizing system of subobjects*, if the following conditions are fulfilled:

(a) If $U' \in F$ and $U' \subseteq U'' \subseteq U$, then $U'' \in F$;

(b) If U', $U'' \in F$, then $U' \cap U'' \in F$;

(c) If $g : U' \to U$ is a morphism, and $U' \in F$ then $g^{-1}(U)' \in F$.

Denote by $\mathcal{P}(F)$ the full subcategory of $\mathcal{C}$ generated by all objects X such that the kernel of any morphism $f : U \to X$ is an element of F. Then $\mathcal{P}(F)$ is a prelocalizing subcategory and the map $F \rightsquigarrow \mathcal{P}(F)$ is a bijection between the set of all prelocalizing systems of subobjects of U and the set of all prelocalizing subcategories of $\mathcal{C}$.

12. Let B be an over-ring of a ring A which is flat as an A^0-module. For any A-module X define $s_X : X \to B \otimes_A X$ by $s_X(x) = 1 \otimes x$. Let $\mathcal{T}$ the full subcategory of Mod A consisting of all modules X such that $s_X = 0$ and $\mathcal{F}$ the full subcategory generated by all objects Y for which s_Y is a monomorphism. Then $(\mathcal{T}, \mathcal{F})$ is a hereditary torsion theory.

13. If $(\mathcal{T}, \mathcal{F})$ is a torsion theory for Mod A such that every finitely generated torsion free A-module is projective, then $X_{\mathcal{T}}$ is a direct summand of X for every finitely generated A-module X.

References

Alin [4], [5]; Alin–Dickson [6]; Dickson [42]; Gabriel [59]; Goldman [76]; Hacque [83]; Hudry [94]; Kuroš [110]; Lambek [114]; Maranda [121]; Mitchell [133]; Popescu [167]; Roos [199]; Stenström [213]; Tisseron [227]; Turnidge [228].

4.9 Localizing subcategories in categories of modules

Let $\mathscr{C}$ be a small preadditive category. We shall make some observations about the prelocalizing and localizing subcategories in $\mathrm{Mod}\,\mathscr{C}$.

If $\mathfrak{a}$ is a left ideal of an object X of $\mathscr{C}$ and $g : X \to Y$ is a morphism in $\mathscr{C}$, we shall denote by $(\mathfrak{a} : g)$ the left ideal of Y consisting of all couples (f, Z) such that $(fg, Z) \in \mathfrak{a}$.

A *left prelocalizing system* P on $\mathscr{C}$ is given if for each object X of $\mathscr{C}$ a set $P(X)$ of left ideals on X is prescribed, such that the following conditions are fulfilled.

(1) If $\mathfrak{a} \in P(X)$ and $\mathfrak{a}'$ is a left ideal on X such that $\mathfrak{a} \subseteq \mathfrak{a}'$, then $\mathfrak{a}' \in P(X)$.

(2) If $\mathfrak{a}, \mathfrak{a}' \in P(X)$, then $\mathfrak{a} \cap \mathfrak{a}' \in P(X)$.

(3) If $\mathfrak{a} \in P(X)$, then for any morphism $g : X \to Y$ it follows that $(\mathfrak{a} : g) \in P(Y)$.

A left prelocalizing system is called *localizing* if in addition the following axiom is true:

(4) If $\mathfrak{a} \in P(X)$ and $\mathfrak{a}'$ is a left ideal on X such that for any $(f, Y) \in \mathfrak{a}$ it follows that $(\mathfrak{a}' : f) \in P(Y)$, then $\mathfrak{a}' \in P(X)$.

Usually, instead of "left prelocalizing (respectively, left localizing) system" we write simply, "prelocalizing (respectively, localizing) system" if there is no danger of confusion.

Let $F \in \mathrm{Ob}\,\mathrm{Mod}\,\mathscr{C}$, $X \in \mathrm{Ob}\,\mathscr{C}$ and $x \in F(X)$; as always, we denote by $u_x : X \to F$ the morphism induced by x. Also by $\mathrm{Ann}(x)$ we shall denote the kernel of u_x. Moreover, $\mathrm{Ann}(x)$ is a left ideal on X, consisting of all couples (f, Y) such that $F(f)(x) = 0$.

Now, let P be a prelocalizing system on $\mathscr{C}$. Let us denote by $s(P)$ the full subcategory of $\mathrm{Mod}\,\mathscr{C}$ consisting of all the objects F such that for any $X \in \mathrm{Ob}\,\mathscr{C}$ and any $x \in F(X)$, we deduce that $\mathrm{Ann}(x) \in P(X)$. Then, $s(P)$ is a prelocalizing subcategory. Indeed, if $F \in \mathrm{Ob}\,s(P)$ and $F' \subseteq F$, then obviously $F' \in \mathrm{Ob}\,s(P)$. In addition, if $x \in F(X)$ and $\bar{x}$ is the image of x in the group $(F/F')(X) = F(X)/F'(X)$, then $\mathrm{Ann}(\bar{x}) \supseteq \mathrm{Ann}(x)$ and it follows from the above condition (1) that $F/F' \in \mathrm{Ob}\,s(P)$. Finally, if $\{F_i\}_i$ is a set of objects in $s(P)$ and $x \in (\coprod_i F_i)(X) = \coprod_i F_i(X)$, then $x = (x_i)_i$ with $x_i \in F_i(X)$ and

$x_i \neq 0$ for only a finite number of i, say for $x_{i_1}, \dots x_{i_n}$. Then

$$\operatorname{Ann}(x) = \bigcap_{j=1}^{n} \operatorname{Ann} x_{i_j}$$

and so from (2) we deduce that $\coprod_i F_i \in \operatorname{Ob} s(P)$. Therefore, $s(P)$ is a prelocalizing subcategory. Moreover, we have the following fundamental result.

THEOREM 9.1. (Gabriel, [59, Chapter 5]). *With the above notation, the assignment*

$$P \rightsquigarrow s(P)$$

is a bijection from the set of prelocalizing systems of $\mathscr{C}$ to the set of all the prelocalizing subcategories of Mod $\mathscr{C}$. *Moreover, P is a localizing system if and only if $s(P)$ is a localizing subcategory.*

Proof. Let $\mathscr{F}$ be a prelocalizing subcategory of Mod $\mathscr{C}$; then for any $X \in \operatorname{Ob}\mathscr{C}$ we denote by $i(\mathscr{F})(X)$ the set of the all left ideals $\mathfrak{a}$ on X such that $X/\mathfrak{a} \in \operatorname{Ob}\mathscr{F}$. By a slight computation, we can derive that the above axioms (1)–(3) are fulfilled, hence a prelocalizing system, denoted by $i(\mathscr{F})$, can be defined. Then an object F in Mod $\mathscr{C}$ belongs to the subcategory $si(\mathscr{F})$ if and only if, for any $X \in \operatorname{Ob}\mathscr{C}$ and any $x \in F(X)$ we deduce that $\operatorname{Ann}(x) \in i(\mathscr{F})$. But then F is a quotient object of a direct sum of objects in $\mathscr{F}$ hence belongs to $\mathscr{F}$. Finally, we have $\mathscr{F} \supseteq si(\mathscr{F})$. The reverse inclusion is obvious.

Also, if P is a prelocalizing system in $\mathscr{C}$, then a left ideal $\mathfrak{a}$ on X belongs to $is(P)(X)$ if and only if $X/\mathfrak{a} \in \operatorname{Ob} s(P)$. But then $\mathfrak{a} = \operatorname{Ann}(\bar{1}_x)$, where $\bar{1}_x$ is the class of 1_x in $(X/\mathfrak{a})(X)$; hence $\mathfrak{a} \in P(X)$. Therefore we have $is(P) \subseteq P$. The reverse inclusion is immediate. Thus the first part of the theorem is proved.

Furthermore, let us assume that P is a localizing system. We must show that $s(P)$ is a localizing subcategory, i.e. is stable under extensions. To that end we consider in Mod $\mathscr{C}$ the exact sequence:

$$0 \longrightarrow F' \longrightarrow F \xrightarrow{p} F'' \longrightarrow 0$$

such that F', F'' belongs to $s(P)$. Now, let $X \in \operatorname{Ob}\mathscr{C}$, $x \in F(X)$ and $\bar{x}$ the image of x in $F''(X)$. Then, $\mathfrak{a} = \operatorname{Ann}(\bar{x}) \in P(X)$ and for any $(g, Y) \in \mathfrak{a}$, we deduce that $(\operatorname{Ann}(x) : g) \in P(Y)$. Indeed, $F''(g)(\bar{x}) = p_Y(F(g)(x)) = 0$ and then $F(g)(x) \in F'(Y)$. But it is clear that $\operatorname{Ann} F(g)(x) = (\operatorname{Ann}(x) : g)$. Then, from (4), we deduce that $F \in \operatorname{Ob} s(P)$.

Conversely, if $s(P)$ is a localizing subcategory, then P is a localizing system. Indeed, let $\mathfrak{a} \in P(X)$ and let $\mathfrak{a}'$ be a left ideal on X such that for any $(g, Y) \in \mathfrak{a}$; then we deduce that $(\mathfrak{a}' : g) \in P(Y)$. We may assume that $\mathfrak{a}' \subseteq \mathfrak{a}$. Then we have the following exact sequence in Mod $\mathscr{C}$:

$$0 \longrightarrow \mathfrak{a}/\mathfrak{a}' \longrightarrow X/\mathfrak{a}' \longrightarrow X/\mathfrak{a} \longrightarrow 0.$$

By hypothesis one has that $\mathfrak{a}/\mathfrak{a}'$ and $X/\mathfrak{a}$ are in $s(P)$ and so $X/\mathfrak{a}'$ is also in $s(P)$. In conclusion P is a localizing system. ∎

Usually we shall denote by dis($\mathscr{C}$) the "discrete" localizing system whose localizing subcategory is Mod $\mathscr{C}$. This system is considered to be an "improper" system.

We denote by $P_0(\mathscr{C})$ (respectively, $L_0(\mathscr{C})$) the set of the all prelocalizing (respectively, localizing) systems on $\mathscr{C}$. If P is a prelocalizing system on $\mathscr{C}$, we denote by $\mathscr{P}$ the associated prelocalizing subcategory and by $P(\mathscr{C})$ (respectively, $L(\mathscr{C})$) the set of all the "proper" prelocalizing (respectively, localizing) systems.

Let $P \in P(\mathscr{C})$; for any ordinal α, we denote by P^α the prelocalizing system on $\mathscr{C}$ defined as follows:

$P^1 = P$.

If $\alpha = \beta + 1$, that is, α has a predecessor β, then P^α can be defined as follows. If $X \in \mathrm{Ob}\mathscr{C}$, then $P^\alpha(X)$ is formed by all the left ideals $\mathfrak{a}$ on X for which there exists $\mathfrak{a}' \in P^\beta(X)$ such that for any $(g, Y) \in \mathfrak{a}'$ we can deduce that $(\mathfrak{a} : g) \in P^\beta(Y)$.

If α is a limit ordinal, then $P^\alpha = \bigcup_{\beta<\alpha} P^\beta$.

LEMMA 9.2. *A prelocalizing system P is localizing if and only if $P^2 = P$.* ∎

COROLLARY 9.3. *Let $P \in P(\mathscr{C})$ and let us write $\bar{P} = \bigcup_\alpha P^\alpha$. Then, $\bar{P}$ is the smallest localizing system containing P.*

Proof. Let $\mathfrak{a} \in \bar{P}(X)$ and let $\mathfrak{a}'$ be a left ideal on X such that $(\mathfrak{a}' : g) \in \bar{P}(Y)$, for any $(g, Y) \in \mathfrak{a}$. Then, $\mathfrak{a} \in P^\alpha$, for any α and so $\mathfrak{a}' \in P^{\alpha+1}$. Hence, $\mathfrak{a}' \in \bar{P}$. Moreover, if G is a localizing system containing P, we deduce that $P^\alpha \subseteq G$, for any α; hence $\bar{P} \subseteq G$. ∎

$\bar{P}$ is called the *closure* of P.

Now we consider a ring A. A left prelocalizing (respectively, localizing) system on A is a set of left ideals of A subject to the above axiom (1)–(3) [respectively, (1)–(4)].

If X is an A-module, we denote by $F(X)$ the localizing system formed by left ideals $\mathfrak{a}$ such that $\mathrm{Hom}(A/\mathfrak{a}, E(X)) = 0$, where $E(X)$ is the injective envelope of X.

PROPOSITION 9.4. *Let X be an A-module. Then for a left ideal $\mathfrak{a}$ of A, the following assertions are equivalent*:

(1) $\mathfrak{a} \in F(X)$.

(2) For any non-zero element $x \in X$ and for any element $a \in A$, there exists an element $b \in A$ such that $bx \neq 0$ and $ba \in \mathfrak{a}$.

Proof. (1) ⇒ (2). Let x be a non-zero element in X and a an element in A such that for any element $b \in A$ satisfying $ba \in \mathfrak{a}$, it follows that $bx = 0$. Then we have $(\mathfrak{a} : a) \in F(X)$ and by hypothesis $(\mathfrak{a} : a) \subseteq \mathrm{Ann}(x)$; but

$$\mathrm{Ann}(x) \in F(X) \quad \text{and} \quad Ax \in F(X).$$

This is a contradiction.

(2) ⇒ (1). If $\mathfrak{a} \notin F(X)$, then there exists a non-null morphism $f : A/\mathfrak{a} \to E(X)$ Let $a \in A$ be an element such that $\bar{a}$, the image of a in $A/\mathfrak{a}$ satisfies $f(\bar{a}) \in X$ and $f(\bar{a}) \neq 0$. Then, an element $b \in A$ can be found such that $ba \in \mathfrak{a}$ and $bf(\bar{a}) \neq 0$. But then $f(b\bar{a}) = 0 = bf(\bar{a}) \neq 0$. The contradiction is obvious. ∎

Let $\mathfrak{a}$ be a left ideal of A and let us write $F_{\mathfrak{a}} = F(A/\mathfrak{a})$.

COROLLARY 9.5. *A left ideal* $\mathfrak{b}$ *is in* $F_{\mathfrak{a}}$ *if and only if, for any two elements* $x_1, x_2 \in A$ *such that* $x_1 \notin \mathfrak{a}$ *there exists an element* $x \in A$ *satisfying* $xx_1 \notin \mathfrak{a}$ *and* $xx_2 \in \mathfrak{b}$.

Moreover, if A is commutative, an ideal $\mathfrak{b}$ *is in* $F_{\mathfrak{a}}$ *if and only if for any element* $x \in A$ *satisfying* $\mathfrak{b}x \subseteq \mathfrak{a}$ *it follows that* $x \in \mathfrak{a}$.

Proof. The first part is a direct consequence of Proposition 9.4. Now let A be commutative, $\mathfrak{b} \in F_{\mathfrak{a}}$ and $x \in A$ satisfying $\mathfrak{b}x \subseteq \mathfrak{a}$. If $x \notin \mathfrak{a}$, then there exists an element such that $yx \notin \mathfrak{a}$ and $y \in \mathfrak{b}$. This is a contradiction. Conversely, let $\mathfrak{b}$ be an ideal as in the assumptions. If $\mathfrak{b} \notin F_{\mathfrak{a}}$, then there exist elements $x_1, x_2 \in A$ such that $x_1 \notin \mathfrak{a}$ and $(\mathfrak{b} : x_2) \subseteq (\mathfrak{a} : x_1)$. But $\mathfrak{b} \subseteq (\mathfrak{b} : x_2)$, so that $\mathfrak{b}x_1 \subseteq \mathfrak{a}$. Again, we have a contradiction. ∎

NOTE. We denote by $\mathfrak{o}$ the null ideal of A. Then as in [113], an element of $F_{\mathfrak{o}}$ is called *left-dense* or simply *dense* if there is no confusion. The localizing system consisting of dense ideals is denoted by D.

Let $P \in P(\mathscr{C})$; for any $\mathscr{C}$-module X, we denote by $X_{\mathscr{P}}$, the greatest submodule of X in $\mathscr{P}$. By Proposition 8.10 we deduce that P is localizing if and only if $(X/X_{\mathscr{P}})_{\mathscr{P}} = 0$, for any $\mathscr{C}$-module X. Particularly, if A is a ring, then $A_{\mathscr{P}}$ is usually denoted by $\mathfrak{r}_P$. An element $x \in A$ is in $\mathfrak{r}_P$ if and only if $\mathrm{Ann}(x) \in P$. We see that $\mathfrak{r}_P$ is in fact, an ideal of A. If $F, F' \in L(A)$, and $F \leqslant F'$ then $\mathfrak{r}_F \subseteq \mathfrak{r}_{F'}$. If $\mathfrak{r}_F = \mathfrak{r}_{F'}$ we say that F and F' are *associated*.

COROLLARY 9.6. *Let* $F \in L(A)$. *Then* $F_{\mathfrak{r}_F}$ *is the greatest element in the set of localizing systems associated with* F.

Proof. It is enough to prove that $\mathfrak{r}_F = \mathfrak{r}_{F_{\mathfrak{r}_F}}$ and $F \subseteq F_{\mathfrak{r}_F}$. Indeed, if $x \in \mathfrak{r}_{F_{\mathfrak{r}_F}}$ and $x \notin \mathfrak{r}_F$, then there exists an ideal $\mathfrak{a} \in F_{\mathfrak{r}_F}$ such that $\mathfrak{a}x = 0$. On the other hand, by Corollary 9.5, there exists an element $y \in A$ such that $yx \notin \mathfrak{r}_F$ and

$y \in \mathfrak{a}$. This is a contradiction. Furthermore, if Q is the injective envelope of $A/\mathfrak{r}_F$ in Mod A, then from Lemma 5.1 we obtain that Q is $\mathscr{F}$-closed and so for any element $\mathfrak{a} \in F$ we have $\operatorname{Hom}_A(A/\mathfrak{a}, Q) = 0$, that is $F \subseteq F_{\mathfrak{r}_F}$. ∎

Particularly, we deduce that $D = F_{\mathfrak{o}}$ is the greatest element in the set of localizing systems F with $\mathfrak{r}_F = \mathfrak{o}$.

For any ring A, we shall denote by E the set of all the essential left ideals of A. Usually instead of "essential submodule" we shall say "large submodule" in accordance with customary terminology (see for example [213]). E is called the *Goldie prelocalizing system.*

PROPOSITION 9.7. *For any ring A the following assertions are true:*

(a) *E is a prelocalizing system on A;*

(b) *If $\mathfrak{r}_E = 0$, then E is localizing;*

(c) *Always $D \leqslant E$; equality holds if and only if $\mathfrak{r}_E = 0$;*

(d) *E^2 is a localizing system, namely the closure of E.*

Proof. (a) If $\mathfrak{a} \in E$ and $\mathfrak{a} \subseteq \mathfrak{b}$ then clearly $\mathfrak{b}$ is large. Also if $\mathfrak{a}, \mathfrak{b} \in E$, then $\mathfrak{a} \cap \mathfrak{b} \in E$. Finally, let $\mathfrak{a} \in E$, $x, y \in A$ with $y \neq 0$. If $yx \neq 0$, then there exists an element $z \in A$, such that $0 \neq zyx \in \mathfrak{a}$, that is $0 \neq zy \in (\mathfrak{a} : x)$. Finally we have $(\mathfrak{a} : x) \in E$; hence E is a prelocalizing system.

(b) Let us assume that $\mathfrak{r}_E = 0$, $\mathfrak{a} \in E$ and $\mathfrak{b}$ is a left ideal such that

$$(\mathfrak{b} : x) \in E \quad \text{for any} \quad x \in \mathfrak{a}.$$

We must prove that $\mathfrak{b} \in E$. Thus let y be a non-zero element of A. By hypothesis, there exists an element t such that $0 \neq ty \in \mathfrak{a}$. Then we have

$$(\mathfrak{b} : ty) \in E \quad \text{and so} \quad (\mathfrak{b} : ty)\, ty \neq 0.$$

Hence there exists an element $z \in (\mathfrak{b} : ty)$, such that $0 \neq zty \in \mathfrak{b}$.

(c) Let $\mathfrak{a} \in D$ and let $x \neq 0$. Then, by Corollary 9.5, there exists an element y such that $yx \neq 0$ and $yx \in \mathfrak{a}$. Therefore $\mathfrak{a}$ is large. Furthermore, we see that always $\mathfrak{r}_D = 0$ and so if $E = D$ then $\mathfrak{r}_E = \mathfrak{r}_D = 0$. Conversely, if $\mathfrak{r}_E = 0$, then from Corollary 9.6 and (b) we deduce that $E = D$.

(d) We shall prove that for any module X we have $X_{\mathscr{E}^2} = X_{\mathscr{E}^3}$ and thus $E^2 = E^3$, hence from Lemma 9.2 we can deduce that E^2 is a localizing system. Firstly we notice that X' is a submodule of X such that if

$$X' \cap X_{\mathscr{E}} = 0,$$

then $X' \cap X_{\mathscr{E}^2} = X' \cap X_{\mathscr{E}^3} = 0$. Indeed, if $x \in X' \cap X_{\mathscr{E}^2}$, then there exists $\mathfrak{a} \in E$ satisfying $\mathfrak{a}x \in X_{\mathscr{E}} \cap X' = 0$. Then, we have $x \in X_{\mathscr{E}}$ and so

$$x \in X_{\mathscr{E}} \cap X' = 0.$$

Also, if $x \in X' \cap X_{\mathscr{E}^3}$ then $\mathfrak{a}x \subseteq X_{\mathscr{E}^2}$ for an element $\mathfrak{a} \in E$, and thus

$$\mathfrak{a}x \subseteq X_{\mathscr{E}^2} \cap X' = 0,$$

hence $\mathfrak{a}x = 0$. Again, we can deduce that $x \in X_{\mathscr{E}}$ and finally that $x = 0$.

Now, let $x \in X_{\mathscr{E}^3}$ and let us write $\mathfrak{a} = \{a \in A \mid ax \in X_{\mathscr{E}}\}$. Then $\mathfrak{a}$ is a large left ideal of A. Indeed, if $\mathfrak{a}'$ is another left ideal such that $\mathfrak{a} \cap \mathfrak{a}' = 0$ then $(\mathfrak{a}' x) \cap X_{\mathscr{E}} = 0$, and thus $(\mathfrak{a}' x) \cap X_{\mathscr{E}^3} = 0$. Then we have $\mathfrak{a}' x = 0$ and necessarily $\mathfrak{a}' = 0$. Thus we deduce that $x \in X_{\mathscr{E}^2}$ and so $X_{\mathscr{E}^3} \subseteq X_{\mathscr{E}^2}$. The reverse inclusion is always valid. ∎

NOTE. (1) For any A-module X, the submodule $X_{\mathscr{E}}$ is called the *singular submodule* of X; it is denoted by some authors by $Z(X)$ or $J(X)$. We shall use the latter notation. Also the submodule $X_{\bar{\mathscr{E}}} = X_{\mathscr{E}^2}$ is denoted by $\bar{J}(X)$ and is called the *closure of the singular submodule.*

Let $u : A \to B$ be a morphism of rings and let $P \in P(A)$. We denote by $u(P)$ the set of left ideals $\mathfrak{b}$ of B such that for any $y \in B$ it follows that $u^{-1}((\mathfrak{b} : y)) \in P$. Then $u(P)$ is an element of $P(B)$. Indeed, if $\mathfrak{b} \in u(P)$ and $\mathfrak{b} \subseteq \mathfrak{b}'$, then clearly $\mathfrak{b}' \in u(P)$. Furthermore, if $\mathfrak{b}, \mathfrak{b}' \in u(P)$, then $u^{-1}((\mathfrak{b} \cap \mathfrak{b}' : y)) = u^{-1}((\mathfrak{b} : y) \cap (\mathfrak{b}' : y)) = u^{-1}((\mathfrak{b} : y)) \cap u^{-1}((\mathfrak{b} : y))$ for any $y \in B$, i.e.

$$\mathfrak{b} \cap \mathfrak{b}' \in u(P).$$

Also, if $\mathfrak{b} \in u(P)$ and $y \in B$, then for any $z \in B$ we have

$$u^{-1}((\mathfrak{b} : y) : z)) = u^{-1}((\mathfrak{b} : zy)) \in P.$$

In other words we can claim that $u(P) \in P(B)$.

On the other hand, if $H \in P(B)$ we denote by $u^{-1}(H)$ the set of left ideals $\mathfrak{a}$ of A such that $Bu((\mathfrak{a} : x)) \in H$ for any $x \in A$. We claim that $u^{-1}(H)$ is an element of $P(A)$. Firstly we notice that if $\mathfrak{a} \in u^{-1}(H)$ and $\mathfrak{a} \subseteq \mathfrak{a}'$, then $\mathfrak{a}' \in u^{-1}(H)$. In addition, it is simple to prove that if $\mathfrak{a}$, $\mathfrak{a}' \in u^{-1}(H)$ then $\mathfrak{a} \cap \mathfrak{a}' \in u^{-1}(H)$ and for any $x \in A$ we have $(\mathfrak{a} : x) \in u^{-1}(H)$. Therefore, we obtain $u^{-1}(H) \in P(A)$.

PROPOSITION 9.8. *Let $u : A \to B$ be a ring morphism. Then the assignments:*

$$P \rightsquigarrow u(P) \quad \text{and} \quad H \rightsquigarrow u^{-1}(H)$$

are order preserving maps of $P(A)$ into $P(B)$ and respectively of $P(B)$ into $P(A)$ such that:

(1) *For any $P \in P(A)$ we have $P \supseteq u^{-1}(u(P))$ and $u(P) = u(u^{-1}(u(P)))$.*

(2) *For any* $H \in P(B)$ *we have* $H \subseteq u(u^{-1}(H))$ *and*

$$u^{-1}(H) = u^{-1}(u(u^{-1}(H))).$$

(3) *If* $F \in L(A)$ *then* $u(F) \in L(B)$. *Also if* $G \in L(B)$, *then* $u^{-1}(G) \in L(A)$.

Proof. The assertions (1) and (2) are left to the reader. Now we shall prove the assertions (3). Let $F \in L(A)$, $\mathfrak{b} \in u(F)$ and $\mathfrak{b}'$ a left ideal of B such that $(\mathfrak{b}' : y) \in u(F)$, for any $y \in \mathfrak{b}$. We must prove that $u^{-1}((\mathfrak{b}' : z)) \in F$ for any $z \in B$. In order to see that $u^{-1}((\mathfrak{b}' : z)) \in F$, let $x \in u^{-1}((\mathfrak{b} : z))$; then we have:

$$(u^{-1}((\mathfrak{b}' : z)) : x) = u^{-1}(((\mathfrak{b}' : z) : u(x))) = u^{-1}((\mathfrak{b}' : u(x)z)).$$

But $u(x)z \in \mathfrak{b}$ and so $u^{-1}((\mathfrak{b}' : u(x)z)) \in F$. Finally we see that $\mathfrak{b}' \in u(F)$.

Again we invite the reader to prove that for any $G \in L(B)$, we have that $u^{-1}(G) \in L(A)$. ∎

We denote by $u(\mathscr{P})$ the prelocalizing subcategory associated with $u(P)$. Also $u^{-1}(\mathscr{H})$ is the prelocalizing subcategory associated with $u^{-1}(H)$.

COROLLARY 9.9. *Let* $u : A \to B$ *be a ring morphism and let* $P \in P(A)$, $H \in P(B)$. *A B-module* Y *is in* $u(\mathscr{P})$ *if and only if* $u_*(Y) \in \mathrm{Ob}\mathscr{P}$. *Also, an A-module* X *is in* $u^{-1}(\mathscr{H})$ *if and only if for any* $x \in X$, *we deduce that* $B \otimes_A (Ax) \in \mathrm{Ob}\mathscr{H}$. ∎

Again let $\mathscr{C}$ be a small preadditive category and S a multiplicative system of morphisms. For any $X \in \mathrm{Ob}\mathscr{C}$ we denote by $F_S(X)$ the set of left ideals on X defined as follows. A left ideal $\mathfrak{a}$ on X is in $F_S(X)$ if and only if for any morphism $f : X \to Y$, there exists $s \in S$ such that $sf \in \mathfrak{a}$.

PROPOSITION 9.10. *The assignment* $X \rightsquigarrow F_S(X)$ *defines a localizing system on* $\mathscr{C}$, *denoted by* F_S.

Proof. Let $\mathfrak{a}$, $\mathfrak{a}' \in F_S(X)$ and let $f : X \to Y$. Then there exists an element $s \in S$ such that $sf \in \mathfrak{a}$. Also an element $s' \in S$ can be found, such that $s' sf \in \mathfrak{a}'$. We see that $(s's)f \in \mathfrak{a} \cap \mathfrak{a}'$. If $\mathfrak{a} \in F_S(X)$ and $f : X \to Y$ is a morphism, then for any morphism $g : Y \to Z$ an element $s \in S$ can be chosen such that $sgf \in \mathfrak{a}$, that is, $sg \in (\mathfrak{a} : f)$. Finally, let $\mathfrak{a}$ be an element of $F_S(X)$ and $\mathfrak{a}'$ a left ideal of X such that for any morphism $f : X \to Y$, $(f, Y) \in \mathfrak{a}$, we deduce that $(\mathfrak{a}' : f) \in F_S(Y)$. If $g : X \to Z$ is a morphism of $\mathscr{C}$, then there exists an element $s \in S$, $s : Z \to T$ such that $sg \in \mathfrak{a}$. But then we have $(\mathfrak{a}' : sg) \in F_S(T)$ and so an element $s' \in S$ can be found such that $s'sg \in \mathfrak{a}'$. In conclusion F_S is an element of $L(\mathscr{C})$. ∎

As above we deduce that an object H of $\mathrm{Mod}\,\mathscr{C}$ is in $\mathscr{F}_S$ if and only if for any $X \in \mathrm{Ob}\mathscr{C}$, $x \in H(X)$ and morphism $f : X \to Y$ of $\mathscr{C}$, there exists a morphism $s \in S$ such that $H(sf)(x) = 0$.

PROPOSITION 9.11. *Let A be a commutative ring and $\mathfrak{a}$ an ideal. The following assertions are equivalent*:

(1) *The difference set $A-\mathfrak{a}$ is multiplicative*;

(2) *If $\mathfrak{b}$ is an ideal such that $\mathfrak{b} \notin F_\mathfrak{a}$ then $\mathfrak{b} \subseteq \mathfrak{a}$.*

Proof. (1) $\Rightarrow$ (2). Let $\mathfrak{b} \notin F_\mathfrak{a}$; if $\mathfrak{b} \nsubseteq \mathfrak{a}$ then there exists an element $s \in A-\mathfrak{a}$ such that $s \in \mathfrak{b}$. Now if $\mathfrak{b}x \subseteq \mathfrak{a}$, then we have $sx \in \mathfrak{a}$ and so $x \in \mathfrak{a}$, i.e. $\mathfrak{b} \in F_\mathfrak{a}$ from Corollary 9.5. But then we have a contradiction.

(2) $\Rightarrow$ (1). Let $s \in A-\mathfrak{a}$; then $As \in F_\mathfrak{a}$. If $s' \in A-\mathfrak{a}$ then we must have that $ss' \notin \mathfrak{a}$. ∎

Let A be a commutative ring. A proper ideal $\mathfrak{a}$ as in the above proposition is called *prime*.

COROLLARY 9.12. *For any prime ideal $\mathfrak{a}$ of a commutative ring we have $F_\mathfrak{a} = F_{A-\mathfrak{a}}$.* ∎

Finally we make some observations on the uniqueness of the injective envelope of certain modules. Namely, in Chapter 3, Section 3.10 we have seen that the injective envelope of an object X in a suitable category is unique in the following sense: if $X \xrightarrow{i} E$, $X \xrightarrow{i'} E'$ are two injective envelopes of X, then an isomorphism $f: E \to E'$ can be found such that $fi = i'$. Generally, we are not able to assert the uniqueness of f. However, under some additional hypotheses we can prove the uniqueness of f.

LEMMA 9.13. *Let X be a non-zero A-module such that $J(X) = 0$. Then $E \subseteq F(X)$.*

Proof. Indeed, let $\mathfrak{a}$ be a large left ideal of A, x a non-zero element of X and $a \in A$. Then $(\mathfrak{a} : a)$ is also a large left ideal of A and by hypothesis $(\mathfrak{a} : a)x \neq 0$, i.e. an element $b \in A$ can be chosen such that $ba \in \mathfrak{a}$ and $bx \neq 0$. Then $\mathfrak{a} \in F(X)$ according to Proposition 9.4. ∎

THEOREM 9.14. *Let X be a* non-singular *A-module* (*i.e.* $J(X) = 0$). *If*

$$i: X \to E, \qquad i': X \to E'$$

are two injective envelopes of X, then a unique isomorphism $f: E \to E'$ can be found such that $fi = i'$.

Proof. By Lemma 9.4, and Exercise 16 it follows that coker i and coker i' are objects in $\mathscr{F}(X)$. Then by Lemma 5.1, one has that E and E' are $\mathscr{F}(X)$-

closed. Now, if $f, f' : E \to E'$ are isomorphisms such that $(f - f')\, i = 0$, then necessarily $f = f'$. ∎

Exercises

1. Let $P, H \in P(\mathscr{C})$. We denote by $P \circ H$ the prelocalizing system associated with the composition of associated prelocalizing subcategories. Find a description of $P \circ H$. In this way the set $P(\mathscr{C})$ becomes canonically an ordered monoid with a zero element.

2. Let $\{F_i\}_i$ a set of elements belonging to $L(A)$. Then $\mathfrak{r}_{\cap_i F_i} = \bigcap_i \mathfrak{r}_{F_i}$. Therefore, the subset of $L(A)$ consisting of the elements associated with an element F, is a complete lattice under inclusion.

3. Let F be a localizing system on A. If $\mathfrak{a}$, $\mathfrak{a}'$ are two ideals of F then the product $\mathfrak{a}\,\mathfrak{a}'$ is also in F.

4. Let $\mathscr{C}$ be a small preadditive category and S a multiplicative system on $\mathscr{C}$. The following assertions are equivalent:

(a) For any $s \in S$, $s : X \to Y$ it follows that the principal left ideal generated by s is in $F_S(X)$.

(b) S is a left permutable system.

5. Let A be a commutative ring and $\mathfrak{a}$ an ideal of A. We denote by $S_{\mathfrak{a}}$ the set of elements s such that for any $x \in A$ with $sx \in \mathfrak{a}$, we deduce that $x \in \mathfrak{a}$. Give a description of all the ideals $\mathfrak{a}$ such that $F_{\mathfrak{a}} = F_{S_{\mathfrak{a}}}$. Any prime ideal is such an ideal.

6. Let $u : A \to B$ be a ring morphism. If $\mathfrak{b}$ is a left ideal of B then

$$Fu^{-1}(\mathfrak{b}) \supseteq u^{-1}(F_{\mathfrak{b}}).$$

The equality is valid if for example u is surjective. If A and B are commutative and $\mathfrak{b}$ is prime, then $u^{-1}(\mathfrak{b})$ is also prime and thus $Fu^{-1}(\mathfrak{b}) = u^{-1}(F_{\mathfrak{b}})$.

7. An element $P \in P(A)$ is called "*finitely generated*" if P has a coinitial set F' of finitely generated left ideals. We denote by $P'(A)$ the subset of $P(A)$ consisting of all the finitely generated prelocalizing systems and we put $L'(A) = L(A) \cap P'(A)$; assuming that A is commutative, prove that

$$L'(A) \equiv P'(A).$$

An element $F \in L'(A)$ is maximal if and only if $F = F_{\mathfrak{a}}$ where $\mathfrak{a}$ is a minimal prime ideal of A. Also an element F belonging to $L'(A)$ is of the form $\bigcap_{\mathfrak{a}} F_{\mathfrak{a}}$ where $\mathfrak{a}$ runs through all the prime ideals $\mathfrak{a}$ which are not in F.

8. A set M of ideals in a commutative ring A, such that $\mathfrak{a}, \mathfrak{b} \in M$ implies $\mathfrak{a}\mathfrak{b} \in M$, and $\mathfrak{a} \in M, \mathfrak{b} \supseteq \mathfrak{a}$, implies $\mathfrak{b} \in M$ is a *multiplicative system of ideals.* Every multiplicative system of ideals (in a commutative ring) having a cofinal system of finitely generated ideals is an element of $L'(A)$.

9. Let R be a non-empty set of prime ideals of a commutative ring A and let F_R be the set of all ideals $\mathfrak{a}$ such that $V(\mathfrak{a}) \subseteq R[V(\mathfrak{a})$ is the set of all prime ideals containing $\mathfrak{a}]$. F_R is a localizing system. If $\mathfrak{a}$ is a finitely generated ideal, then $F_{V(\mathfrak{a})}$ is the set of all ideals containing a power of $\mathfrak{a}$; moreover, if $\mathfrak{a} \in F$, then $F_{V(\mathfrak{a})} \subseteq F$, i.e. $F_{V(\mathfrak{a})}$ is the smallest localizing system containing $\mathfrak{a}$.

10. Let A be a commutative ring and $F \in L(A)$. If $\mathfrak{p}$ is an ideal maximal in the set of ideals not in F, then $\mathfrak{p}$ is a prime ideal. If $F \in L'(A)$ and $\mathfrak{a} \notin F$, there exists an ideal $\mathfrak{p} \supseteq \mathfrak{a}$ and maximal in the set of ideals not in F. In particular, if S is a multiplicative set and $\mathfrak{a}$ an ideal such that $\mathfrak{a} \cap S = \varnothing$, then an ideal $\mathfrak{p} \supseteq \mathfrak{a}$ can be found such that $\mathfrak{p} \cap S = \varnothing$ and maximal with such properties. Clearly $\mathfrak{p}$ is a prime ideal.

The result of Exercises 7–10 are given in [144].

11. Let A be a commutative ring. Then the Goldie prelocalizing system E of large ideals is localizing if and only if A is reduced (that is, has no nilpotent elements).

12. Let A be a commutative ring and $F \in L(A)$. Denote by F^* the set of all ideals of $A[X]$, containing an ideal of the form $A[X]\,\mathfrak{a}$, with $\mathfrak{a} \in F$. Then F^* is a localizing system of $A[X]$.

13. A two-sided ideal $\mathfrak{a}$ of A is dense (as a left ideal) if and only if the relation $\mathfrak{a}x = 0$ implies $x = 0$, for any $x \in A$.

14. Let $A \xrightarrow{u} B$ be a surjective ring morphism. Then for any $F \in L(A)$ we have $F = u^{-1}(u(F))$.

15. Let $\mathscr{P}$ be a prelocalizing subcategory of $\mathscr{C}$, $\overline{\mathscr{P}}$ its closure and $X \in \operatorname{Ob} \mathscr{C}$. Then $X_{\mathscr{P}} = 0$ if and only if $X_{\overline{\mathscr{P}}} = 0$.

16. If X' is an essential submodule of X, then for any $x \in X$, the left ideal $(X' : x)$ is also essential in A.

17. Give a direct proof of the Theorem 9.14 (using Lemma 9.13).

References

Bourbaki [26]; Bucur–Deleanu [31]; Gabriel [59]; Goldie [74]; Goldman [76]; Gruson [81]; Hacque [83]; Hacque–Maury [85]; Harada [87]; Hudry [93]; [94], Kasu [103]; Lambek [113]; [114], Leclerc [118]; Maranda [121];

Morita [136], [137]; Năstăsescu–Popescu [144]; Ouzilou [160]; Popescu [167]; Popescu–Spircu [175], [176]; Popescu–Spulber [177]; Roos [199]; Stenström [212], [213]; Tisseron [226], [227].

4.10 Localization in categories of modules

Throughout $\mathscr{C}$ is a small preadditive category. Let F be a localizing system of $\mathscr{C}$, $\mathscr{F}$ the associated localizing subcategory, $T : \operatorname{Mod}\mathscr{C} \to \operatorname{Mod}\mathscr{C}/\mathscr{F}$ the canonical functor and S a right adjoint of T. Also we denote by

$$u : \operatorname{Id}\operatorname{Mod}\mathscr{C} \to ST = L$$

an arrow of adjunction; as a quasi-inverse of u we may take the identity of $\operatorname{Id}(\operatorname{Mod}\mathscr{C}/\mathscr{F})$. We always preserve this notation.

LEMMA 10.1. *For a non-zero object H of* $\operatorname{Mod}\mathscr{C}$ *the following assertions are equivalent:*

(1) *H is $\mathscr{F}$-closed (or simply "closed");*

(2) *For any object X of $\mathscr{C}$, any ideal $\mathfrak{a} \in F(X)$ and any morphism $f : \mathfrak{a} \to H$ there exists a unique morphism $\bar{f} : X \to H$, prolonging f.*

Proof. The implication (1) ⇒ (2) follows from Lemma 4.1 and Theorem 9.1.

(2) ⇒ (1). We shall prove that $u_H : H \to ST(H)$ is an isomorphism. First we remark that H is $\mathscr{F}$-torsionfree. Indeed, if $y \in H(Y)$ is such that $\operatorname{Ann}(y) \in F(Y)$, for an object Y of $\mathscr{C}$, then the morphism $u_y : Y \to H$, associated with y, has null restriction to $\operatorname{Ann}(y)$ and so, $u_y = 0$ that is, $y = 0$.

Now we consider the following exact sequence:

$$0 \longrightarrow H \xrightarrow{u_H} ST(H) \xrightarrow{p} H'' \longrightarrow 0$$

in which H'' is in $\mathscr{F}$. Assuming that $p \neq 0$, then there exists an object $X \in \operatorname{Ob}\mathscr{C}$ and $x \in \big(ST(H)\big)(X)$ such $p_X(x) \neq 0$ and obviously, we have $pu_x \neq 0$. By hypothesis, $\ker(pu_x) = \mathfrak{a} \in F(X)$. But $ST(H)$ is $\mathscr{F}$-torsionfree and consequently we deduce that $u_x(\mathfrak{a}) \neq 0$. According to Lemma 4.7 we deduce that u_H is an essential morphism hence $u_x(\mathfrak{a}) \cap (\operatorname{im} u_H) = H' \neq 0$. Furthermore, $u_x(\mathfrak{a})/H'$ is an object of $\mathscr{F}$ and so $u_x^{-1}(H') = \mathfrak{a}' \in F(X)$. Let us denote by $j : H' \to H$ the canonical morphism induced by u_H and by $f : \mathfrak{a}' \to H'$ the morphism induced by u_x. Then jf is a morphism of $\mathfrak{a}'$ into H, and by hypothesis there exists a unique morphism $\bar{f} : X \to H$ which prolongs jf. But then we have $\mathfrak{a}' \subseteq \ker(u_x - u_H\bar{f})$, that is $u_x = u_H\bar{f}$. The contradiction is obvious. ∎

For any object H of Mod $\mathscr{C}$, we denote by $\overline{H}$ the object $H/H_{\mathscr{F}}$ and by $p^H : H \to \overline{H}$ the canonical epimorphism. Also, for any object X of $\mathscr{C}$, we denote by $H_F(X)$ the abelian group

$$H_F(X) = \varinjlim_{\mathfrak{a} \in F(X)} \operatorname{Hom}_{\mathscr{C}}(\mathfrak{a}, \overline{H}).$$

(The inductive limit is filtered by the inclusion of the elements of $F(X)$ in a canonical way). For any $\mathfrak{a} \in F(X)$ we denote by $t_{\mathfrak{a}}{}^H : \operatorname{Hom}_{\mathscr{C}}(\mathfrak{a}, \overline{H}) \to H_F(X)$ the structural morphism. Sometimes for any $f \in \operatorname{Hom}_{\mathscr{C}}(\mathfrak{a}, \overline{H})$ we shall write:

$$t_{\mathfrak{a}}{}^H(f) = (\mathfrak{a}, f).$$

If $f \in \operatorname{Hom}_{\mathscr{C}}(\mathfrak{a}, \overline{H})$, $f' \in \operatorname{Hom}_{\mathscr{C}}(\mathfrak{a}', \overline{H})$ then we have $(\mathfrak{a}, f) = (\mathfrak{a}', f')$ if and only if there exists an element $\mathfrak{a}'' \in F(X)$ such that the restrictions of f and f' to $\mathfrak{a}''$ are identical. Clearly $\overline{H}(X) \simeq \operatorname{Hom}_{\mathscr{C}}(X, \overline{H})$ and thus $t_X{}^H$ can be considered as a morphism of abelian groups of $\overline{H}(X)$ to $H_F(X)$. Also we write $v_X = t_X{}^H p_X{}^H$.

Lemma 10.2. *The assignment $X \rightsquigarrow H_F(X)$ defines an object of* Mod $\mathscr{C}$. *Moreover the morphisms $\{v_X\}_{X \in \mathrm{Ob}\mathscr{C}}$ define a morphism of $\mathscr{C}$-modules, denoted by v^H.*

Proof. Let $f : X \to Y$ be a morphism of $\mathscr{C}$. Then the assignment $\mathfrak{a} \rightsquigarrow (\mathfrak{a} : f)$ defines a map $f^* : F(X) \to F(Y)$. If we consider $F(X)$ as a category (the morphisms are inclusions) then f^* is in fact a covariant functor. For any $\mathfrak{a} \in F(X)$ we denote by $f_{\mathfrak{a}} : \operatorname{Hom}_{\mathscr{C}}(\mathfrak{a}, \overline{H}) \to \operatorname{Hom}_{\mathscr{C}}((\mathfrak{a} : f), \overline{H})$ the morphism of abelian groups defined as follows. Let $g : \mathfrak{a} \to \overline{H}$ be a morphism; then $f_{\mathfrak{a}}(g) : (\mathfrak{a} : f) \to \overline{H}$ has the components $(f_{\mathfrak{a}}(g))_Z : (\mathfrak{a} : f)(Z) \to \overline{H}(Z)$, defined by the rule: if $h \in (\mathfrak{a} : f)(Z)$ then $h : Y \to Z$ is a morphism in $\mathscr{C}$ such that $hf \in \mathfrak{a}$; thus, $(f_{\mathfrak{a}}(g))_Z(h) = g_Z(hf)$. By the functoriality of f^* we deduce that the morphisms $(f_{\mathfrak{a}})$, $\mathfrak{a} \in F(X)$ define a morphism of direct systems, denoted by $\bar{f}^*$, of the system $\{\operatorname{Hom}_{\mathscr{C}}(\mathfrak{a}, \overline{H})\}_{\mathfrak{a} \in F(X)}$ to the system $\{\operatorname{Hom}_{\mathscr{C}}(\mathfrak{a}', \overline{H})\}_{\mathfrak{a}' \in F(Y)}$. Finally $\bar{f}^*$ defines a morphism of abelian groups $H_F(f) : H_F(X) \to H_F(Y)$. We invite the reader to prove that for any other morphism $g : Y \to Z$ of $\mathscr{C}$, we have $H_F(gf) = H_F(g) H_F(f)$, that is, the assignment $X \rightsquigarrow H_F(X)$ is an object of Mod $\mathscr{C}$.

To finish it is enough to prove that for any morphism $f : X \to Y$ of $\mathscr{C}$ the following diagram of abelian groups is commutative:

$$\begin{array}{ccc} H(X) & \xrightarrow{v_X} & H_F(X) \\ {\scriptstyle H(f)}\downarrow & & \downarrow{\scriptstyle H_F(f)} \\ H(Y) & \xrightarrow{v_Y} & H_F(Y) \end{array}$$

But this follows on observing that the construction of $H_F(f)$ is natural. The details are left to the reader. ∎

Now let $f: H \to G$ be a morphism in $\operatorname{Mod}\mathscr{C}$; then $f(H_{\mathscr{F}}) \subseteq G_{\mathscr{F}}$ and thus there exists a unique morphism $\bar{f}: \bar{H} \to \bar{G}$ such that $\bar{f}p^H = p^G f$. For any $X \in \operatorname{Ob}\mathscr{C}$ and $\mathfrak{a} \in F(X)$ we denote by $f_X{}^{\mathfrak{a}}: \operatorname{Hom}_{\mathscr{C}}(\mathfrak{a}, \bar{H}) \to \operatorname{Hom}_{\mathscr{C}}(\mathfrak{a}, \bar{G})$ the morphism defined by $\bar{f}$. Passing to the inductive limit by $\mathfrak{a}$, the morphisms $f_X{}^{\mathfrak{a}}$ define a unique morphism $\bar{f}_X: H_F(X) \to G_F(X)$ such that $\bar{f}_X v_X{}^H = v_X{}^G f_X$. Therefore the morphisms $\{\bar{f}_X\}_{X \in \operatorname{Ob}\mathscr{C}}$ define a morphism $f_F: H_F \to G_F$ such that $f_F v^H = v^G f$.

LEMMA 10.3. *Let $H \in \operatorname{Ob}\mathscr{F}$; then for any object X of $\mathscr{C}$ we have:*

$$\varinjlim_{\mathfrak{a} \in F(X)} \operatorname{Hom}_{\mathscr{C}}(\mathfrak{a}, H) = 0$$

Proof. Indeed, if $f: \mathfrak{a} \to H$ is a morphism, then $\operatorname{im} f \in \operatorname{Ob}\mathscr{F}$ and thus $\ker f = \mathfrak{a}' \in F(X)$. Therefore the image of f in the inductive limit is null. ∎

THEOREM 10.4. *Let $\mathscr{C}$ be a small preadditive category and let $F \in L(\mathscr{C})$. With the above notations, the assignments $H \rightsquigarrow H_F$, $f \rightsquigarrow f_F$ define a functor $L': \operatorname{Mod}\mathscr{C} \to \operatorname{Mod}\mathscr{C}$ and the morphisms $\{v^H\}_{H \in \operatorname{ObMod}\mathscr{C}}$ define a functorial morphism $v: \operatorname{IdMod}\mathscr{C} \to L'$. Moreover there exists a functorial isomorphism $s: L' \to L$ such that $sv = u$.*

Proof. Using the above considerations it is sufficient to prove the last assertion. To do this we shall prove that for any $H \in \operatorname{ObMod}\mathscr{C}$, H_F is closed.

We begin by proving that for any $H \in \operatorname{ObMod}\mathscr{C}$, the morphism $t^H: \bar{H} \to H_F$ is a monomorphism. In fact, let X be an object of $\mathscr{C}$ and $x \in \bar{H}(X)$ such that $t_X{}^H(x) = 0$. By the definition of $t_X{}^H$, we see that $t_X{}^H(x) = u_x$ and so $\ker u_x \in F(X)$. Then since $\bar{H}$ is $\mathscr{F}$-torsionfree, we have necessarily $x = 0$.

Now we assume that $H_F \neq 0$. We consider the exact sequence:

$$0 \longrightarrow \ker v^H \longrightarrow H \xrightarrow{v^H} H_F \to \operatorname{coker} v^H \longrightarrow 0.$$

As mentioned above, we see that $\ker v^H = \ker p^H = H_{\mathscr{F}}$. In addition we claim that $\operatorname{coker} v^H \in \operatorname{Ob}\mathscr{F}$. For that, let $y = (\mathfrak{a}, f) \in H_F(X)$. Let us denote by $\mathfrak{a}'$ the left ideal on X generated by all the morphisms $g: X \to Y$, such that $H_F(g)(y) \in \operatorname{im} t_Y{}^H$. Then we have $\mathfrak{a} \subseteq \mathfrak{a}'$. Indeed, let $h: X \to Z$ be an element of $\mathfrak{a}$; then $H_F(h)(y) = ((\mathfrak{a}: h), f')$, where $f_K{}'(r) = f_K(rh)$, for any $(r, K) \in (\mathfrak{a}: h)$. If we take a closer look at the definition of $H_F(Z)$, we see that $((\mathfrak{a}: h), f') = (Z, u_{f_Z(h)})$, where, as always, $u_{f_Z(h)}: Z \to \bar{H}$ is the morphism associated with $f_Z(h) \in \bar{H}(Z)$. In conclusion, we obtain that for any $X \in \operatorname{Ob}\mathscr{C}$ and $x \in (\operatorname{coker} v^H)(X)$ we have $\operatorname{Ann}(x) \in F$, i.e. $\operatorname{coker} v^H \in \operatorname{Ob}\mathscr{F}$. Also it follows that the monomorphism t^H is essential and so H_F is $\mathscr{F}$-torsionfree.

Furthermore, let us consider the exact sequence:

$$0 \longrightarrow \bar{H} \xrightarrow{t^H} H_F \longrightarrow \operatorname{coker} v^H \longrightarrow 0$$

then for any $\mathfrak{a} \in F(X)$ we have the exact sequence:

$$0 \longrightarrow \operatorname{Hom}_{\mathscr{C}}(\mathfrak{a}, \bar{H}) \xrightarrow{(t^H)\mathfrak{a}} \operatorname{Hom}_{\mathscr{C}}(\mathfrak{a}, H_F) \longrightarrow \operatorname{Hom}_{\mathscr{C}}(\mathfrak{a}, \operatorname{coker} v^H)$$

where $(t^H)\,\mathfrak{a}$ is canonically induced by t^H. Passing to the inductive limit over $\mathfrak{a} \in F(X)$ and taking into account the above consideration, we have the isomorphism: $t_F{}^H : H_F \to (H_F)_F$.

On the other hand, we have the commutative diagram:

$$\begin{array}{ccc} \bar{H} & \xrightarrow{t^H} & H_F \\ {\scriptstyle t^H}\downarrow & & \downarrow{\scriptstyle t^{H_F}} \\ H_F & \xrightarrow{t^{H_F}} & (H_F)_F \end{array}$$

Then we obtain $(t^{H_F} - t_F{}^H)\,t^H = 0$. Therefore, since $\operatorname{coker} t^H \in \operatorname{Ob}\mathscr{F}$, we deduce that $t^{H_F} = t_F{}^H$ and so t^{H_F} is an isomorphism.

Now, the point is to prove that H_F is closed. We shall use Lemma 10.1. Let $\mathfrak{a} \in F(X)$ and let $f : \mathfrak{a} \to H_F$ be a morphism in $\operatorname{Mod}\mathscr{C}$. Then

$$(\mathfrak{a}, f) \in (H_F)_F\,(X)$$

and thus there exists an element $y \in H_F(X)$ such that $t^{H_F}(y) = (\mathfrak{a}, f)$, i.e. $(X, u_y) = (\mathfrak{a}, f)$. But then there exists an element $\mathfrak{a}' \in F(X)$ such that the restrictions of f and u_y to $\mathfrak{a}'$ are identical. Therefore, since X_F is $\mathscr{F}$-torsionfree we see that u_y is the unique extension of f to X.

To finish we consider the diagram:

$$\begin{array}{ccc} & & H_F \\ & \overset{v_H}{\nearrow} & \\ H & & \\ & \underset{u_H}{\searrow} & \\ & & L(H) = ST(H) \end{array}$$

in which the kernel and cokernel of the morphisms are in $\mathscr{F}$.

From Lemma 4.1 we deduce that there exists a unique morphism

$$s^H : H_F \to L(H)$$

which make the above diagram commutative. By a slight computation we prove that the morphisms s^H are isomorphisms and they define the promised morphism s. The proof is now complete. ∎

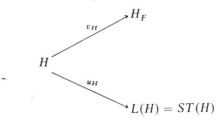

Frequently we shall identify the objects $L'(H) = H_F$ and $L(H) = ST(H)$, also the morphisms u and v. In this way, we have a practical procedure to define the localizing functor without having to use the quotient category $\mathrm{Mod}\mathscr{C}/\mathscr{F}$. This point is very useful in applications.

LEMMA 10.5. *For any two objects* H, R *of* $\mathrm{Mod}\mathscr{C}$, *the association* $f \rightsquigarrow fu_H$ *gives an isomorphism from the group* $\mathrm{Hom}_{\mathscr{C}}(H_F, R_F)$ *to the group* $\mathrm{Hom}_{\mathscr{C}}(H, R_F)$. *Particularly, if* $X \in \mathrm{Ob}\mathscr{C}$, *then the groups* $\mathrm{Hom}_{\mathscr{C}}(X_F, R_F)$ *and* $R_F(X)$ *are canonically isomorphic. (Frequently, for any* $X \in \mathrm{Ob}\mathscr{C}$, *we denote by* X *the object* $X_l = h^X$ *in* $\mathrm{Mod}\mathscr{C}$; *see Section* 3.4, *Chapter* 3). ∎

We denote by $\mathscr{C}_F'$ the full subcategory of $\mathrm{Mod}\,\mathscr{C}$ generated by the objects $\{X_F\}_{X \in \mathrm{Ob}\mathscr{C}}$. Obviously, the assignments $X \rightsquigarrow X_F$, $f \rightsquigarrow f_F$ define a contravariant additive functor $u_{\mathscr{C}}' : \mathscr{C} \to \mathscr{C}_F'$. Also we denote by $\mathscr{C}_F$ the dual of $\mathscr{C}_F'$ and by $u_{\mathscr{C}}$ the composition of the duality functor $\mathscr{C}_F' \to \mathscr{C}_F$ with $u_{\mathscr{C}}'$. Then we shall say that $\mathscr{C}_F$ is the *localization category* of $\mathscr{C}$ with respect to F, and $u_{\mathscr{C}}$ the *localization functor*. We call the reader's attention to the fact, that according to the conventions from Chapter 3, Section 3.4, an object X belonging to $\mathscr{C}$, considered in $\mathrm{Mod}\,\mathscr{C}$, is denoted also by X. But, if $f : X \to Y$ is a morphism in $\mathscr{C}$, then it induces canonically a morphism from Y to X in $\mathrm{Mod}\mathscr{C}$. That is why the above functor $u_{\mathscr{C}}'$ is necessarily contravariant.

Let H_F be a closed object in $\mathrm{Mod}\mathscr{C}$. We shall define a covariant additive functor $H_F' : \mathscr{C}_F \to \mathscr{Ab}$ as follows. For any $X_F{}^0$ of $\mathscr{C}_F$ we define $H_F'(X_F{}^0) = H_F(X)$. Furthermore, if $f^0 : X_F{}^0 \to Y_F{}^0$ is a morphism in $\mathscr{C}_F$, then $H_F(f^0) : H_F(X) \to H_F(Y)$ is defined in the following way. Firstly, we see that $f : Y_F \to X_F$ is a morphism in $\mathrm{Mod}\mathscr{C}$ and it defines the canonical morphism: $\bar{f} : \mathrm{Hom}_{\mathscr{C}}(X_F, H_F) \to \mathrm{Hom}_{\mathscr{C}}(Y_F, H_F)$. Then $\bar{f}$ defines a unique morphism of abelian groups $H_F'(f) : H_F(X) \to H_F(Y)$ such that the following diagram is commutative:

$$\begin{array}{ccc} \mathrm{Hom}_{\mathscr{C}}(X_F, H_F) & \xrightarrow{\bar{f}} & \mathrm{Hom}_{\mathscr{C}}(Y_F, H_F) \\ \big\updownarrow & & \big\updownarrow \\ H_F(X) & \xrightarrow{H'_F(f)} & H_F(Y) \end{array}$$

where the vertical isomorphisms are deduced by Lemma 10.5. It is easy to see that H_F' is as before a covariant additive functor.

Also let $h : H_F \to R_F$ be a morphism of $\mathscr{C}$-modules. Then a canonical morphism $h' : H_F' \to R_F'$ of $\mathscr{C}_F$-modules is defined such that for any

$$X^0{}_F{}' \in \mathrm{Ob}\mathscr{C}_F,$$

we have $h'_{X^0{}_F} = h_X$. It is easy to prove that h' is in fact a morphism of $\mathscr{C}_F$-modules.

Now, we have all the material to construct a covariant functor

$$S' : \mathrm{Mod}\mathscr{C}/\mathscr{F} \to \mathrm{Mod}\,\mathscr{C}_F.$$

Indeed, any object of $\mathrm{Mod}\mathscr{C}/\mathscr{F}$ is of the form $T(H)$ and thus, $S'(T(H)) = H_F{}'$. If $f : T(H) \to T(R)$ is a morphism in $\mathrm{Mod}\mathscr{C}/\mathscr{F}$, then $S(f) : H_F \to R_F$ is a morphism in $\mathrm{Mod}\mathscr{C}$, and thus $S'(f) = (S(f))'$.

Furthermore, the functor $u_{\mathscr{C}}$ defines a restriction functor

$$(u_{\mathscr{C}})_* : \mathrm{Mod}\,\mathscr{C}_F \to \mathrm{Mod}\,\mathscr{C}$$

and by the definition of the functor S', we see that $(u_{\mathscr{C}})_*\, S' = S$.

LEMMA 10.6. *Let H be a $\mathscr{C}$-module and R a $\mathscr{C}_F$-module. Then the natural morphism of abelian groups:*

$$r = (u_{\mathscr{C}})_* (R, H_F{}') : \mathrm{Hom}_{\mathscr{C}_F} (R, H_F{}') \to \mathrm{Hom}_{\mathscr{C}}((u_{\mathscr{C}})_* (R), (u_{\mathscr{C}})_* (H_F{}'))$$

is an isomorphism.

Proof. Clearly, the morphism considered is a monomorphism. Now let $f : (u_{\mathscr{C}})_* (R) \to (u_{\mathscr{C}})_* (H_F{}') = H_F$ be a morphism of $\mathscr{C}$-modules and

$$g : X_F{}^0 \to Y_F{}^0$$

a morphism in $\mathscr{C}_F$.

We must prove that $f_Y\, R(g) = H_F{}'(g)\, f_X$. To do this, let $x \in R(X)$ and let $f_1, f_2 : Y_F{}^0 \to H_F{}'$ be the morphisms $f_1 = f u_{R(g)(x)}$, $f_2 = u_{(H_F{}'(g) f_X)(x)}$ (as always, for any $y \in R(Y_F{}^0)$, $u_y : Y_F{}^0 \to R$ is the morphism associated with y). Then $(u_{\mathscr{C}})_* (f_1), (u_{\mathscr{C}})_* (f_2) : Y_F \to H_F$ are two morphisms which composed with the morphism $u_Y : Y \to Y_F$ become equal. But H_F being closed, we deduce that necessarily $(u_{\mathscr{C}})_* (f_1) = (u_{\mathscr{C}})_* (f_2)$ and so $f_1 = f_2$, $(u_{\mathscr{C}})_*$ being faithful. Finally, we see that $(f_Y\, R(g))\,(x) = (H_F{}'\,(g)\, f_X)\,(x)$, i.e. the morphism f can be considered as being of the form $(u_{\mathscr{C}})_*\,(f')$. ■

THEOREM 10.7. *Let $\mathscr{C}$ be a small preadditive category, F a left localizing system on $\mathscr{C}$, $\mathscr{C}_F$ the localizing category of $\mathscr{C}$ with respect to F and $u_{\mathscr{C}} : \mathscr{C} \to \mathscr{C}_F$ the localization functor. Thus one has the following diagram of categories and functors:*

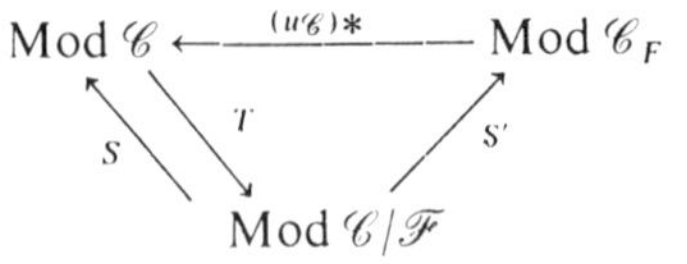

where T is the canonical functor, S a right adjoint of T, and S' being defined such that $(u_{\mathscr{C}})_*\, S' \simeq S$ (see above the definition of S'). Let us put $T' = T(u_{\mathscr{C}})_*$.

Then S' is full and faithful and T' is an exact left adjoint of S'. Moreover S' and T' define an equivalence between $\mathrm{Mod}\,\mathscr{C}_F/\ker T'$ *and* $\mathrm{Mod}\,\mathscr{C}/\mathscr{F}$. *Finally an object R of* $\mathrm{Mod}\,\mathscr{C}_F$ *belongs to* $\ker T'$ *if and only if* $(u_{\mathscr{C}})_*(R) \in \mathrm{Ob}\mathscr{F}$.

Proof. By Lemma 10.6 we check that S' is full and faithful. Furthermore, T' is exact as a composition of exact functors. Now let $T(H) \in \mathrm{Ob}(\mathrm{Mod}\,\mathscr{C}/\mathscr{F})$, and R a $\mathscr{C}_F$-module. From Lemma 10.6 we have that

$$(u_{\mathscr{C}})_*(R, S'T(H)) : \mathrm{Hom}_{\mathscr{C}_F}(R, S'T(H)) \to \mathrm{Hom}_{\mathscr{C}}((u_{\mathscr{C}})_*(R), (u_{\mathscr{C}})_* S'T(H))$$

is an isomorphism. Then by Lemma 4.1 the morphism:

$$T((u_{\mathscr{C}})_*(R), (u_{\mathscr{C}})_* S'T(H)) : \mathrm{Hom}_{\mathscr{C}}((u_{\mathscr{C}})_*(R), (u_{\mathscr{C}})_* S'T(H)) \to \mathrm{Hom}_{\mathrm{Mod}\mathscr{C}/\mathscr{F}}(T(u_{\mathscr{C}})_*(R), T(u_{\mathscr{C}})_* S'T(H))$$

is also a functorial isomorphism. But the last group is naturally isomorphic with the group $\mathrm{Hom}_{\mathrm{Mod}\mathscr{C}}\, T_{\mathscr{F}}(T'(R), T(H))$. By composition we obtain a functorial isomorphism which establishes an adjunction of T' with S'.

The other assertions are easy and left to the reader. ∎

Exercises

1. For any $H \in \mathrm{Ob}\mathrm{Mod}\mathscr{C}$ we denote by $M(H)$ the functor defined thus:

$$M(H)(X) = \varinjlim_{\mathfrak{a} \in F(X)} \mathrm{Hom}_{\mathscr{C}}(\mathfrak{a}, H),$$

for any $X \in \mathrm{Ob}\mathscr{C}$. Then M^2 is canonically isomorphic with the localization functor.

2. The isomorphisms indicated in Lemma 10.5, are functorial in each of their arguments.

3. Let P be a prelocalizing system of a small preadditive category $\mathscr{C}$ and $\bar{P}$ the smallest localizing system containing P. For an object H in $\mathrm{Mod}\mathscr{C}$, the following assertions are equivalent:

(a) H is $\bar{\mathscr{P}}$-closed;

(b) For any $X \in \mathrm{Ob}\mathscr{C}$, $\mathfrak{a} \in P(X)$, and any morphism $f : \mathfrak{a} \to H$ there exists a unique morphism $\bar{f} : X \to H$ such that the restriction of $\bar{f}$ to $\mathfrak{a}$ is exactly f.

References

Bucur–Deleanu [31]; Gabriel [59]; Goldman [76]; Gruson [81]; Hacque [83]; Hacque–Maury [85]; Lambek [114]; Maranda [121]; Morita [136], [137]; Popescu-Spircu [175], [176]; Roos [199]; Stenström [212], [213]; Tisseron [226], [227]; Walker–Walker [236].

4.11 Left exact functors. The embedding theorem

Let $\mathscr{C}$ be a small abelian category; we denote by M the set of all monomorphisms of $\mathscr{C}$. Obviously M is a left and right permutable multiplicative system. Also, we denote by L the left localizing system of $\mathscr{C}$ associated with M. By definition (see Section 4.9) a left ideal $\mathfrak{a}$ on X is an element of $L(X)$ if and only if for any morphism $f: X \to Y$ there exists a monomorphism $g: Y \to Z$ such that $(gf, Z) \in \mathfrak{a}$ and from Section 4.9 Exercise 4, if and only if $\mathfrak{a}$ contains an element (f, Y) with $f \in M$. For any object X of $\mathscr{C}$, we denote by $L'(X)$ the set of principal ideals of $L(X)$ generated by the elements (f, Y) with $f \in M$. Clearly, an object H in $\mathrm{Mod}\mathscr{C}$ is $\mathscr{L}$-closed if and only if for any $\mathfrak{a} \in L'(X)$ and any $h: \mathfrak{a} \to H$ there exists a unique morphism $\bar{h}: X \to H$ prolonging h. Let $\mathscr{L}$ be the localizing subcategory of $\mathrm{Mod}\mathscr{C}$ associated with L. An object H is in $\mathscr{L}$ if and only if for any $X \in \mathrm{Ob}\mathscr{C}$ and any $x \in H(X)$ we have $\ker u_x \in L(X)$, i.e. if and only if there exists a monomorphism $f: X \to Y$ in $\mathscr{C}$ such that $H(f)(x) = 0$.

LEMMA 11.1. *Any object X of $\mathscr{C}$, considered in* $\mathrm{Mod}\mathscr{C}$ *is $\mathscr{L}$-closed.*

Proof. Let $g: Y \to X$ be a morphism in $\mathrm{Mod}\mathscr{C}$ with $Y \in \mathrm{Ob}\mathscr{C}$, such that $\ker g \in L(Y)$. Then we have $g = u_y$, with $y \in X(Y) = \mathrm{Hom}_{\mathscr{C}}(X, Y)$ and there exists a monomorphism $f: Y \to Z$ such that $X(f)(y) = fy = 0$, hence $y = 0$ and so $g = 0$. Therefore X is $\mathscr{L}$-torsionfree. Now, let $f: Y \to Z$ be a monomorphism in $\mathscr{C}$ and $h: \langle f \rangle \to X$ a morphism in $\mathrm{Mod}\mathscr{C}$. Also let (p, R) be a cokernel of f in $\mathscr{C}$. Then $h_Z(f) \in X(Z) = \mathrm{Hom}_{\mathscr{C}}(X, Z)$ is a morphism such that $X(p)(h_Z(f)) = ph_Z(f) = h_R(pf) = 0$. Hence there exists a unique morphism $g: X \to Y$ such that $fg = h_Z(f)$. But g defines a unique morphism $\bar{g}: Y \to X$ in $\mathrm{Mod}\mathscr{C}$, prolonging h. Therefore X is $\mathscr{L}$-closed. ∎

LEMMA 11.2. *Let* $0 \to X' \xrightarrow{f} X \xrightarrow{p} X'' \to 0$ *be an exact sequence of $\mathscr{C}$. Then we have in* $\mathrm{Mod}\mathscr{C}$ *the corresponding exact sequence*:

$$0 \longrightarrow X'' \xrightarrow{\bar{p}} X \xrightarrow{\bar{f}} X'$$

in which $\mathrm{coker}\bar{f} \in \mathrm{Ob}\mathscr{L}$.

Proof. Obviously $\bar{p}$ and $\bar{f}$ are canonically induced, by p and f respectively. Let $Y \in \mathrm{Ob}\mathscr{C}$ and $y \in (\mathrm{coker}\bar{f})(Y)$; then an element

$$g \in X'(Y) = \mathrm{Hom}_{\mathscr{C}}(X', Y)$$

can be found such that $r_Y(g) = y$, where $r: X' \to \mathrm{coker}\bar{f}$ is the canonical morphism, and so $u_y = ru_g$. Consider in $\mathscr{C}$ the commutative diagram

$$\begin{array}{ccc} X' & \xrightarrow{f} & X \\ {\scriptstyle g}\downarrow & & \downarrow{\scriptstyle g'} \\ Y & \xrightarrow{f'} & X \coprod_{X'} Y \end{array}$$

in which f' is a monomorphism. But is clear that the left ideal on Y generated by f' is contained in $\ker(ru_g) = \ker u_y$, that is $\operatorname{coker} \bar{f} \in \mathrm{Ob}\mathscr{L}$. ∎

Now we are able to prove the following important result.

THEOREM 11.3. *Let $\mathscr{C}$ be a small abelian category, M the set of all monomorphisms in $\mathscr{C}$ and L the localizing system associated with M. For an object H in* Mod$\mathscr{C}$, *the following assertions are equivalent*:

(1) *H is L-closed*;

(2) *H is a left exact functor to $\mathscr{C}$ in $\mathscr{Ab}$.*

Proof. (1) ⇒ (2). Let H be a closed object in Mod$\mathscr{C}$. For any exact sequence:

$$0 \longrightarrow X' \xrightarrow{f} X \xrightarrow{p} X'' \longrightarrow 0$$

of $\mathscr{C}$ we have the following sequence of abelian groups:

$$H(X') \xrightarrow{H(f)} H(X) \xrightarrow{H(p)} H(X'')$$

for which we must check that it is left exact. Firstly, we claim that $H(f)$ is a monomorphism. Indeed, let $x' \in H(X')$ be such that $H(f)(x') = 0$. Then $\ker u_{x'} \in L(X')$ and so $x' = 0$, H being closed, i.e. $H(f)$ is a monomorphism.

Now let $x \in H(X)$ be such that $H(p)(x) = 0$; then $(u_x)_{X''}(p) = H(p)(x) = 0$. If $\bar{p}: X'' \to X$ is the morphism induced by p in Mod$\mathscr{C}$, then we see that $u_x \bar{p} = 0$ and from Lemma 11.2 there exists a unique morphism $t: X' \to H$ such that $t\bar{f} = u_x$. Finally, we see that $x = H(f)(x')$ for a suitable $x' \in H(X')$. The reader is invited to fill in the details.

(2) ⇒ (1). Let $H: \mathscr{C} \to \mathscr{Ab}$ be a covariant left exact functor, $X \in \mathrm{Ob}\mathscr{C}$ and $x \in H(X)$ such that $\ker u_x \in L(X)$. Then there exists a monomorphism $f: X \to Y$ in $\mathscr{C}$ such that $H(f)(x) = 0$. But $H(f)$ is also a monomorphism of abelian groups and thus $x = 0$. In conclusion H is $\mathscr{L}$-torsionfree.

Now, let $f: X \to Y$ be a monomorphism in $\mathscr{C}$, $\langle f \rangle$ the principal ideal generated by f and $g: \langle f \rangle \to H$ a morphism in Mod$\mathscr{C}$. Then f defines a morphism $\bar{f}: Y \to \langle f \rangle$ and thus $g\bar{f} = t: Y \to H$ is a morphism in Mod$\mathscr{C}$.

Let (p, Z) be a cokernel of f in $\mathscr{C}$; then, we have the commutative diagram of abelian groups:

$$\begin{array}{ccccccc} 0 \longrightarrow & h^Y(X) & \xrightarrow{h^Y(f)} & h^Y(Y) & \xrightarrow{h^Y(p)} & h^Y(Z) \\ & \downarrow{\scriptstyle t_X} & & \downarrow{\scriptstyle t_Y} & & \downarrow{\scriptstyle t_Z} \\ 0 \longrightarrow & H(X) & \xrightarrow{H(f)} & H(Y) & \xrightarrow{H(p)} & H(Z) \end{array}$$

So,

$$\begin{aligned}\big(H(p)\,t_Y)\,(1_Y)\big) &= \big(t_Z\,h^Y(p)\big)\,(1_Y) = \big(g_Z\,\bar{f}_Z\,h^Y(p)\big)\,(1_Y) \\ &= \big((g_Z\bar{f}_Z)\,(p_Y)\big) = g_Z(pf) = 0.\end{aligned}$$

Then there exists an element $x \in H(X)$ such that $H(f)\,(x) = t_Y(1_Y) = (g\bar{f})_Y\,(1_Y) = g_Y(f)$. We see that $u_x : X \to H$ is a morphism prolonging g. The uniqueness of u_x results from the fact that H is $\mathscr{L}$-torsionfree. ∎

The quotient category $\mathrm{Mod}\mathscr{C}/\mathscr{L}$ is denoted by $\mathrm{Lex}\mathscr{C}$ (left-exact functors). Also, we denote by $\bar{h} : \mathscr{C} \to \mathrm{Lex}\mathscr{C}$ the composition Th where

$$T : \mathrm{Mod}\mathscr{C} \to \mathrm{Lex}\mathscr{C}$$

is the canonical functor and $h : \mathscr{C} \to \mathrm{Mod}\mathscr{C}$ is obviously defined. From Lemma 11.1 and Lemma 11.2 we deduce the following result:

COROLLARY 11.4. *The functor* $\bar{h} : \mathscr{C} \to \mathrm{Lex}\mathscr{C}$ *is a contravariant exact full and faithful functor.* ∎

Later, we shall use the following result:

LEMMA 11.5. *Let* $\mathscr{D}$ *be an abelian category with a projective generator* U, *and let* A *be the ring of all endomorphisms of* U. *Then the functor*

$$S : \mathscr{D} \to \mathrm{Mod}A^0$$

defined canonically by $S(X) = \mathrm{Hom}_{\mathscr{D}}(U, X)$ *is exact and faithful. Moreover, if* X *is* U*-finitely generated, (i.e. it is a quotient object of* U^n *for a natural number* n*) then, for any object* Y *of* $\mathscr{D}$, *the homomorphism*

$$S(X, Y) : \mathrm{Hom}_{\mathscr{D}}(X, Y) \to \mathrm{Hom}_A\big(S(X), S(Y)\big)$$

induced by S *is an isomorphism.*

Proof. Since U is a projective generator, S is exact and faithful. Suppose that X is finitely generated. Then, we must show that every A^0-module morphism $g : S(X) \to S(Y)$ can be written as $S(g')$ for some $g' \in \mathrm{Hom}_{\mathscr{D}}(X, Y)$. In other words we must find g' such that $g(f) = g'f$ for all $f \in \mathrm{Hom}_{\mathscr{D}}(U, X)$. Firstly, we consider the case where $U = X$. Then, $g(f) = g(1_U f) = g(1_U) f$, since g is a morphism of A^0-modules. Therefore in this case, we can take $g' = g(1_U)$. Now, in general, if X is finitely generated, we can write an exact sequence:

$$0 \longrightarrow M \xrightarrow{j} U^n \xrightarrow{p} X \longrightarrow 0.$$

For $1 \leqslant i \leqslant n$ we denote by u_i the structural morphisms of the direct sum and by p_i the associated projections. Let us denote by g_i the composition $\mathrm{Hom}_{\mathscr{D}}(U, U) \xrightarrow{S(u_i)} \mathrm{Hom}_{\mathscr{D}}(U, U^n) \xrightarrow{S(p)} \mathrm{Hom}_{\mathscr{D}}(U, X) \xrightarrow{g} \mathrm{Hom}_{\mathscr{D}}(U, Y)$. Then g_i is a morphism of A^0-modules and so by the case already treated we have for each i a morphism $t_i : U \to Y$ such that $g_i(f) = t_i f$, for all

$$f \in \mathrm{Hom}_{\mathscr{D}}(U, U).$$

The morphisms t_i, $1 \leqslant i \leqslant n$ define a morphism $t = \sum_i t_i p_i : U^n \to Y$. If we denote the composition $gS(p)$ by $\bar{g}$, then for $f \in \mathrm{Hom}_{\mathscr{D}}(U, U^n)$ we have: $\bar{g}(f) = g(pf) = g(p(\sum_i u_i p_i) f) = \sum_i g(pu_i p_i f) = \sum_i g_i(p_i f) = \sum_i t_i p_i f = tf$.

We show that $tj = 0$. If $tj \neq 0$, then since U is a generator, we can find a morphism $s : U \to M$ such that $tjs \neq 0$. But we have

$$tjs = \bar{g}(js) = g(pjs) = g(0) = 0;$$

this is a contradiction. Therefore $tj = 0$, and so there is a morphism

$$g' : X \to Y \quad \text{such that} \quad t = g'p.$$

Let $f \in \mathrm{Hom}_{\mathscr{D}}(U, X)$. Since U is projective, we can write $f = ph$ for some $h : U \to U^n$. Then

$$g(f) = g(ph) = \bar{g}(h) = th = g'ph = g'f.$$

In other words, g' is the required morphism. ■

THEOREM 11.6 (Mitchell [132]). *Let $\mathscr{C}$ be a small abelian category. Then there exists a ring A and a full and faithful exact covariant functor $H: \mathscr{C} \to$* $\mathrm{Mod}A^0$.

Proof. Let $\mathscr{D} = (\mathrm{Lex}\,\mathscr{C})^0$ and let $D : \mathrm{Lex}\,\mathscr{C} \to \mathscr{D}$ be the duality functor. Obviously, $\mathscr{D}$ is an abelian category and D is exact. Furthermore, let V be an injective cogenerator of $\mathrm{Lex}\,\mathscr{C}$ and for any object X of $\mathscr{C}$ let Q_X be an injective object containing $\bar{h}(X)$. Then $V' = V \amalg (\prod_{X \in \mathrm{Ob}\,\mathscr{C}} Q_X)$ is also an injective cogenerator of $\mathrm{Lex}\mathscr{C}$ and so $D(V') = U$ is a projective generator of $\mathscr{D}$. Let $A = \mathrm{Hom}_{\mathscr{D}}(U, U)$ and let $S : \mathscr{D} \to \mathrm{Mod}\, A^0$ be the functor

$$S(R) = \mathrm{Hom}_{\mathscr{D}}(U, R).$$

Clearly, for any object X of $\mathscr{C}$, the object $D(\bar{h}(X))$ is U-finitely generated and so from Lemma 11.5, we deduce that the functor $H = SD\bar{h}$ is a covariant exact full and faithful functor. ■

PROPOSITION 11.7. *Let $\mathscr{D}_0$ be a small subcategory of an abelian category $\mathscr{D}$. Then, there exists a small full abelian subcategory $\mathscr{D}'$ of $\mathscr{D}$ such that $\mathscr{D}_0$ is a subcategory of $\mathscr{D}'$.*

Proof. We define inductively a sequence $\{\mathscr{D}_n\}_{n \geqslant 0}$ of subcategories of $\mathscr{D}$ as follows. The subcategory $\mathscr{D}_{n+1}$ is the full subcategory of $\mathscr{D}$ consisting of the objects in $\mathscr{D}_n$ together with single representatives for kernels and cokernels in $\mathscr{D}$ of every morphism in $\mathscr{D}_n$ and single representatives for the all finite products in $\mathscr{D}$ of objects in $\mathscr{D}_n$. If $\mathscr{D}_n$ is small, so is $\mathscr{D}_{n+1}$ for all $n \geqslant 0$. Consequently,

$$\mathscr{D}' = \bigcup_{n=0}^{\infty} \mathscr{D}_n$$

is small, and it is easy to show that $\mathscr{D}'$ is a full abelian subcategory of $\mathscr{D}$. ■

This result is very important, especially in homological algebra.

As a consequence of 11.7, we have

COROLLARY 11.8. *Let $\mathscr{D}$ be an arbitrary abelian category. If a theorem is of the form "P implies Q", where P is a categorical statement about a diagram in $\mathscr{D}$ with a finite number of edges and Q states that a finite number of additional morphisms exist between objects over designated vertices in the diagram, so as to make some categorical statements true for the extended diagram, and if the theorem is true when $\mathscr{D} =$ Mod$A$ for every ring A, then the theorem is true for any abelian category $\mathscr{D}$.* ■

A typical case of the above corollary is the following important result.

COROLLARY 11.9. (Snake Lemma). *Let $\mathscr{D}$ be an abelian category and consider the following diagram in $\mathscr{D}$:*

$$\begin{array}{ccccccc} & & X' & \longrightarrow & X & \longrightarrow & X'' & \longrightarrow & 0 \\ & & \downarrow{\scriptstyle f'} & & \downarrow{\scriptstyle f} & & \downarrow{\scriptstyle f''} & & \\ 0 & \longrightarrow & Y' & \longrightarrow & Y & \longrightarrow & Y'' & & \end{array}$$

where the rows are exact. Then, there exists a morphism θ : $\ker f'' \longrightarrow \operatorname{coker} f'$ *such that we have the exact sequence:*

$$\ker f' \longrightarrow \ker f \longrightarrow \ker f'' \xrightarrow{\theta} \operatorname{coker} f' \longrightarrow \operatorname{coker} f \longrightarrow \operatorname{coker} f''.$$

Proof. It is well known that the result is valid for the category Mod A ([33], Chapter 2). According to Proposition 11.7, there exists a small abelian full subcategory $\mathscr{C}$ of $\mathscr{D}$ which contains the diagram from the assumption. Let $H:\mathscr{C}\to \mathrm{Mod}\,A^0$ be an exact full and faithful covariant functor. To complete the proof, it is sufficient to notice that a sequence $X\to Y\to Z$ in $\mathscr{C}$ is exact if and only if the sequence $H(X)\to H(Y)\to H(Z)$ in Mod A^0 is exact. ∎

PROPOSITION 11.10. *Let $\mathscr{C}$ be a small abelian category. Any injective object in* Lex $\mathscr{C}$ *is an exact functor.*

Proof. Let F be an injective object in Lex $\mathscr{C}$. Also let

$$0\longrightarrow X'\longrightarrow X\longrightarrow X''\longrightarrow 0$$

be an exact sequence in $\mathscr{C}$. Then, in Lex $\mathscr{C}$ we have the exact sequence:

$$0\longrightarrow \bar{h}(X'')\longrightarrow \bar{h}(X)\longrightarrow \bar{h}(X')\longrightarrow 0$$

and it induces the following exact sequence of abelian groups:

$$0\to \mathrm{Hom}_{\mathrm{Lex}\mathscr{C}}(\bar{h}(X'),F)\to \mathrm{Hom}_{\mathrm{Lex}\mathscr{C}}(\bar{h}(X),F)\to \mathrm{Hom}_{\mathrm{Lex}\mathscr{C}}(\bar{h}(X''),F)\to 0.$$

Next, by adjunction we obtain the exact sequence of abelian groups

$$0\to \mathrm{Hom}_{\mathscr{C}}(X',S(F))\to \mathrm{Hom}_{\mathscr{C}}(X,S(F))\to \mathrm{Hom}_{\mathscr{C}}(X'',S(F))\to 0$$

where $S:\mathrm{Lex}\,\mathscr{C}\to \mathrm{Mod}\,\mathscr{C}$ is as always the canonical inclusion functor. To finish, it will be sufficient to use Chapter 3, Lemma 4.3. ∎

PROPOSITION 11.11. *Let $\mathscr{C}$ be a small abelian category and X any object of $\mathscr{C}$. Then, any subobject of $\bar{h}(X)$ in* Lex $\mathscr{C}$ *is the inductive limit of its set of subobjects of the form $\bar{h}(Y)$, with $Y\in \mathrm{Ob}\mathscr{C}$.*

Proof. Let $i:F\to \bar{h}(X)$ a monomorphism in Lex $\mathscr{C}$. Therefore since the set of objects $\{\bar{h}(Y)\}_{Y\in \mathrm{Ob}\mathscr{C}}$ is a set of generators of Lex $\mathscr{C}$, we deduce that there exists a non-zero morphism $t:\bar{h}(Y)\to F$. But then

$$it=\bar{h}(f),\quad \text{where}\quad f:X\to Y$$

is a morphism of $\mathscr{C}$ and so $\mathrm{im}\,t\simeq \mathrm{coker}\,\bar{h}(p)=\bar{h}(\mathrm{im}\,f)$, where $p:Y\to \mathrm{coker}\,f$ is the canonical epimorphism. Therefore F contains the non-zero subobjects of the form $\bar{h}(Z)$ with $Z\in\mathrm{Ob}\mathscr{C}$. Let F' be the sum of such subobjects of F. If $F'\neq F$, then there exists a non-zero morphism

$g : \bar{h}(Y) \to F$ which cannot be factored by the inclusion $F' \to F$. As above, we can assume that g is a monomorphism. The contradiction is obvious. ∎

Exercises

1. Let $\mathscr{C}$ be a small abelian category. An object F in Mod $\mathscr{C}$ is $\mathscr{L}$-torsion-free if and only if F is a monofunctor.

2. Let $0 \to F' \to F \to F'' \to 0$ be an exact sequence of covariant functors between two abelian categories:

(a) If F' is left exact and F a monofunctor, then F'' is also a monofunctor;

(b) If F'' is a monofunctor and F left exact, then F' is also left exact.

3. Let $\mathscr{C}$ be a small abelian category, F a $\mathscr{C}$-module and X an object of $\mathscr{C}$ such that there exists a monomorphism $i: X \to Q$ with Q injective. Then $F_{\mathscr{L}}(X)$ is the kernel of the morphism $F(i): F(X) \to F(Q)$.

4. Let us consider a covariant functor $F: \mathscr{D} \to \mathscr{D}'$ between two abelian categories. The right $0th$ *derived functor* of F is a pair $(u, R^0 F)$ where $R^0 F: \mathscr{D} \to \mathscr{D}'$ is a covariant left exact functor and $u: F \to R^0 F$ a functorial morphism such that any other functorial morphism from F to a left exact covariant functor $G: \mathscr{D} \to \mathscr{D}'$ factors uniquely through $R^0 F$. If $\mathscr{D}$ is small and $\mathscr{D}'$ is a Grothendieck category, then any covariant functor $F: \mathscr{C} \to \mathscr{Ab}$ has a right 0th derived functor. [Hint: Denote by $\mathscr{N}$ the full subcategory of $(\mathscr{D}, \mathscr{D}')$ consisting of all functors F such that for any $X \in \mathrm{Ob}\mathscr{D}$, $F(X)$ is the inductive limit of its subobjects R in $\mathscr{D}'$ for which a monomorphism $f: X \to Y$ exists in $\mathscr{D}$, with $F(f)(R) = 0$. Then $\mathscr{N}$ is a localizing subcategory and the quotient category $(\mathscr{D}, \mathscr{D}')/\mathscr{N}$ consists exactly of all the left exact functors of $\mathscr{D}$ into $\mathscr{D}'$.]

An alternative proof can be given in two steps. First we may assume that $\mathscr{D} = \mathrm{Mod} A^0$, where A is a ring and we make use of the idea of the proof of Theorem 11.3. Furthermore, we shall use Theorem 7.9 of Chapter 3.

5. If $\mathscr{D}$ is a category with sufficiently many injective objects, then any covariant functor $F: \mathscr{D} \to \mathscr{D}'$ has a right 0th derived functor. (Compare with Theorem 2.12, Chapter 3).

6. Let $\mathscr{C}$ be a small abelian category with sufficiently many injectives. Then any projective object in $\mathscr{C}$ becomes projective in Lex $\mathscr{C}$. Hence show that $\coprod_P \bar{h}(P)$ is a projective generator for Lex $\mathscr{C}$, where P runs through all projectives in $\mathscr{C}$.

7. Let $\mathscr{C}$ be a small abelian category with sufficiently many injectives. Then an object F in Mod $\mathscr{C}$ is in $\mathscr{L}$ if and only if $F(Q) = 0$ for any injective in $\mathscr{C}$.

8. Let $\mathscr{C}$ be a small abelian category. Given an epimorphism $F \to F''$ in Lex $\mathscr{C}$ and a morphism $X \to F''$, where $X \in \mathrm{Ob}\mathscr{C}$, show that these can be put into a commutative diagram

$$\begin{array}{ccc} Y & \xrightarrow{\bar{h}(f)} & X \\ \downarrow & & \downarrow \\ F & \longrightarrow & F'' \end{array}$$

where $f: X \to Y$ is a monomorphism in $\mathscr{C}$ and hence $\bar{h}(f)$ is an epimorphism in Lex $\mathscr{C}$. Then letting $F'' = X$ and taking $X \to F''$ as the identity on X, show that if

$$0 \to Z \to F \to X \to 0$$

is a short exact sequence in Lex $\mathscr{C}$, with Z, $X \in \mathrm{Ob}\mathscr{C}$ then $F \simeq Y' \in \mathrm{Ob}\mathscr{C}$. (Hint: Show that the canonical morphism $Y \amalg Z \to F$ is an epimorphism in Lex $\mathscr{C}$ and hence form a 3×3 exact commutative diagram in which all the functors are representable functors save possibly F).

9. Let $\mathscr{C}$ be an Ab 5 locally finitely presented category. Denote by $\mathscr{D}$ the full subcategory of $\mathscr{C}$ consisting of all finitely presented objects. $\mathscr{D}$ is a category with fibred sums and a morphism in $\mathscr{D}$ is an epimorphism if and only if it is an epimorphism in $\mathscr{C}$. Denote by H the set of all epimorphisms in $\mathscr{D}$ and by $\mathscr{F}$ the localizing subcategory of Mod $\mathscr{D}^0$ generated by all objects R such that for any $X \in \mathrm{Ob}\mathscr{D}$, $x \in R(X)$, there exists an epimorphism $f: Y \to X$ in $\mathscr{D}$ such that $R(f)(x) = 0$. Prove that an object K in Mod $\mathscr{D}^0$ is $\mathscr{F}$-closed if and only if for any exact sequence $X' \to X \to X'' \to 0$ in $\mathscr{D}$, we have the exact sequence of abelian groups:

$$0 \to K(X'') \to K(X) \to K(X'),$$

i.e. K is left exact. Now denote by $T: \mathscr{C} \to \mathrm{Mod}\, \mathscr{D}^0$ the functor $T(X) = h_X t$, where $t: \mathscr{D} \to \mathscr{C}$ is the inclusion functor. Prove that T defines an equivalence of categories between the category $\mathscr{C}$ and the full subcategory of Mod $\mathscr{D}^0$ consisting of all $\mathscr{F}$-closed objects.

References

Bucur–Deleanu [31]; Cartan–Eilenberg [33]; Freyd [57]; Gabriel [59]; Gruson [81]; Lambek [114]; Mitchell [132], [133]; Ouzilou [160]; Popescu [167]; Roos [198], [199].

4.12 The study of the localization ring of a ring

Let A be a ring, $F \in L(A)$ and $\mathscr{F}$ the localizing subcategory associated with F. As usual (see Section 4.10) we denote by $(A_l)_F$ the localization of the A-module A_l, and by A_F the ring $(\mathrm{End}_A((A_l)_F))^0$. Also we denote by

$$T_F : \mathrm{Mod}\, A \to \mathrm{Mod}\, A/\mathscr{F}$$

the canonical functor and by S_F a right adjoint of T_F. Frequently we shall write simply T and S if there is no danger of confusion. By $u : \mathrm{IdMod}A \to ST$ we denote an arrow of adjunction and by v a quasi-inverse of u. Sometimes we shall assume that $TS = \mathrm{Id}(\mathrm{Mod}A/\mathscr{F})$ and v is the identity of the latter functor. These notations are preserved in the rest of this Chapter.

Let X be an A-module, Q the injective envelope of $X/X_{\mathscr{F}}$ and $t : X \to Q$ the composition $X \to X/X_{\mathscr{F}} \to Q$. Then by Lemma 5.1. Q is F-closed and thus a unique morphism $s : X_F \to Q$ can be found such that $su_X = t$.

LEMMA 12.1. (a) *With the above notation s is a monomorphism. Moreover an element $x \in Q$ is in the image of s if and only if*

$$(t(X) : x) = \{a \in A | ax \in t(X)\} \in F.$$

(b) *If X' is a submodule of X, then X_F' is the submodule of X_F consisting of all elements x such that $\mathfrak{a}x \subseteq u_X(X')$ for an element $\mathfrak{a} \in F$.*

The proof is, in fact, a consequence of Theorem 4.5 and is left to the reader. ∎

Now, for any $f \in A_F$ we denote by $h(f)$ the element $f(\bar{1}) \in (A_l)_F$ where $\bar{1}$ is the image of 1 in $(A_l)_F$. We also denote by $u' : A \to A_F$ the ring morphism defined by the equality $u'(a) = (u_a)_F$, where $(u_a)_F$ is the localization of the morphism u_a of A-modules (that is, $(u_a)_F : (A_l)_F \to (A_l)_F$ is the unique morphism of A-modules such that $(u_a)_F u_{Al} = u_{Al} u_a$). Clearly, through u', A_F becomes an A-module.

PROPOSITION 12.2. *The above mapping $h : A_F \to (A_l)_F$ is an isomorphism of A-modules. Moreover, $hu' = u_A$.*

Proof. Obviously h is a morphism of abelian groups. Now if $h(f) = f(\bar{1}) = 0$, then for any $a \in A$ we have $f(a.\bar{1}) = f(u_A(a)) = af(\bar{1}) = 0$ and so $f = 0$, $(A_l)_F$ being F-closed. Therefore h is a monomorphism. Furthermore, if $x \in (A_l)_F$ denote by $u_x : A \to (A_l)_F$ the morphism associated with x. Then a unique morphism $(u_x)_F : (A_l)_F \to (A_l)_F$ may be found such that $(u_x)_F u_A = u_x$. But then $(u_x)_F(\bar{1}) = (u_x)_F(u_A(1)) = u_x(1) = x$; consequently, h is an isomorphism.

Finally, if $a \in A$, then $h(u'(a)) = h((u_a)_F) = (u_a)_F(\bar{1}) = u_A(a)$. If $f \in A_F$, then $h(a.f) = h(u'(a)f) = (u'(a)f)(\bar{1}) = ((u_a)_F f)(\bar{1}) = (u_a)_F(f(\bar{1})) = af(\bar{1}) = f(a.\bar{1}) = f(u_A(a))$ and also $ah(f) = af(\bar{1}) = f(a.\bar{1}) = f(u_A(a))$. ∎

The above lemma expresses the fact that $(A_l)_F$ has a ring structure such that the morphism u_A is, in fact, a morphism of rings. Later, we shall identify $(A_l)_F$ with A_F and u' with u_A. The ring A_F is called the *localization ring* of the ring A relative to F, and u_A, or simply u if there is no danger of confusion, is called the *localization morphism*.

Now let $\mathfrak{a}$ be a left ideal of A and $i : \mathfrak{a} \to A$ the canonical inclusion. As usual, we denote by $i_F : \mathfrak{a}_F \to A_F$ the localization of the morphism i. Then i_F is a monomorphism and so $\mathfrak{a}_F$ becomes an A-submodule of A_F. Moreover $\mathfrak{a}_F$ is, in fact, a left ideal of A_F (see Section 4.10). Denote by $\mathfrak{a}^*$ the F-saturation of $\mathfrak{a}$, that is the subset of elements $x \in A$ such that $\mathfrak{a}'x \subseteq \mathfrak{a}$ where $\mathfrak{a}' \in F$. The left ideal $\mathfrak{a}$ is called *F-saturated* if $\mathfrak{a}^* = \mathfrak{a}$. By $u(F)$ we denote the direct image of F by u (see Section 4.9).

PROPOSITION 12.3. *With the above notation we have:*

(a) *$i_F : \mathfrak{a}_F \to A_F$ is an isomorphism if and only if $\mathfrak{a} \in F$;*

(b) *$\mathfrak{a}_F$ is $u(F)$-saturated and $u^{-1}(\mathfrak{a}_F) = \mathfrak{a}^*$. Moreover, the assignment $\mathfrak{a} \rightsquigarrow \mathfrak{a}_F$ is a bijection from the set of F-saturated left ideals of A which are not in F to the set of all u(F)-saturated left ideals of A_F which are not in u(F).*

Proof. (a) If $\mathfrak{a} \in F$ then by the exact sequence:

$$0 \longrightarrow \mathfrak{a} \xrightarrow{i} A \longrightarrow A/\mathfrak{a} \longrightarrow 0$$

in which $A/\mathfrak{a} \in \text{Ob}\mathscr{F}$ we deduce that i_F is an isomorphism, the localization functor being left exact (see Section 4.10).

Conversely, if i_F is an isomorphism, then $i_F = ST(i)$ and so $T(i)$ is an isomorphism. Hence $T(A/\mathfrak{a}) = 0$, or equivalently $A/\mathfrak{a} \in \text{Ob}\mathscr{F}$.

(b) If $\mathfrak{a} \notin F$, then i_F is not an isomorphism and thus by Lemma 12.1, we deduce that $u^{-1}(\mathfrak{a}_F) \supseteq \mathfrak{a}$. Let $x \in u^{-1}(\mathfrak{a}_F)$; then $u(x) \in \mathfrak{a}_F$ and so an element $\mathfrak{a}' \in F$ can be found such that $\mathfrak{a}'u(x) \subseteq u(\mathfrak{a})$. Therefore, for any $y \in \mathfrak{a}'$, $u(yx) = u(z)$ with $z \in \mathfrak{a}$, so that $yx - z \in \mathfrak{r}_F$. Then there exists an element $\mathfrak{a}'' \in F$ such that $\mathfrak{a}''(yx - z) = 0$ or equivalently, $\mathfrak{a}''yx = \mathfrak{a}''z$. In other words $yx \in \mathfrak{a}^*$ for any $y \in \mathfrak{a}'$, i.e. $\mathfrak{a}'x \subseteq \mathfrak{a}^*$ and so $x \in \mathfrak{a}^{**} = \mathfrak{a}^*$. Conversely, if $x \in \mathfrak{a}^*$, then $\mathfrak{a}'x \subseteq \mathfrak{a}$ for a suitable $\mathfrak{a}' \in F$; hence $\mathfrak{a}'u(x) \subseteq u(\mathfrak{a})$, that is, $u(x) \in \mathfrak{a}_F$ and finally $x \in u^{-1}(\mathfrak{a}_F)$.

From the exact sequence:

$$0 \longrightarrow \mathfrak{a} \xrightarrow{k} \mathfrak{a}^* \longrightarrow \mathfrak{a}^*/\mathfrak{a} \longrightarrow 0$$

in which k is the canonical inclusion, we deduce that k_F is an isomorphism, by the definition of $\mathfrak{a}^*$. Therefore $\mathfrak{a}_F \equiv \mathfrak{a}_F^*$.

Now we shall show that $\mathfrak{a}_F$ is $u(F)$-saturated. To do this, let us consider the exact sequence:

$$0 \longrightarrow \mathfrak{a}_F \xrightarrow{i} \mathfrak{a}_F^* \longrightarrow \mathfrak{a}_F^*/\mathfrak{a}_F \longrightarrow 0$$

in which $\mathfrak{a}_F^*$ is the $u(F)$-saturation of $\mathfrak{a}_F$ and i the canonical inclusion. We see that i_F is an isomorphism and thus i is also an isomorphism since $\mathfrak{a}_F$ is $u(F)$-closed (see Section 4.10).

Finally, let $\mathfrak{b}$ be a left ideal in A_F which is not in $u(F)$ and $\mathfrak{b}^*$ the $u(F)$-saturation of $\mathfrak{b}$. Then, by Lemma 12.1 we deduce that $(u^{-1}(\mathfrak{b}^*))_F = \mathfrak{b}^*$. ∎

Let $F \subseteq F'$ be two elements of $L(A)$, $L = ST$, $L' = S'T'$ the corresponding localizing functors and u, respectively u', the arrow of adjunction of S with T and respectively of S' with T'. By Lemma 10.1 we deduce that any F'-closed module is also F-closed. Then there exists a unique morphism $s_X : X_F \to X_{F'}$ such that $s_X u_X = u_X'$. It is easy to see that the morphisms $\{s_X\}_{X \in \mathrm{Ob}\,\mathrm{Mod}\,A}$ define in fact, a functorial morphism $s : L \to L'$ such that $su = u'$. We want to determine $\ker s_X$ and $\mathrm{im}\, s_X$ in certain cases.

PROPOSITION 12.3. *Let $F \subseteq F'$ be two elements of $L(A)$, X an A-module and $s_X : X_F \to X_{F'}$ the morphism defined above. The following assertions are true*:

(a) $\ker s_X = (X_{\mathscr{F}'}/X_{\mathscr{F}})_F$;

(b) *If $X_{\mathscr{F}'} = X_{\mathscr{F}}$ or the functor $L = ST$ is exact, then* $\mathrm{im}\, s_X$ *consists of all elements $y \in X_{F'}$ for which an element $\mathfrak{a} \in F$ may be found such that $\mathfrak{a}y \subseteq u'(X)$.*

Proof. (a) From the exact sequence:

$$0 \longrightarrow X_{\mathscr{F}'}/X_{\mathscr{F}} \xrightarrow{i} X/X_{\mathscr{F}} \xrightarrow{p} X/X_{\mathscr{F}'} \longrightarrow 0 \tag{8}$$

we deduce, on passing to localization, the exact sequence:

$$0 \longrightarrow (X_{\mathscr{F}'}/X_{\mathscr{F}})_F \xrightarrow{i_F} (X/X_{\mathscr{F}})_F \xrightarrow{p_F} (X/X_{\mathscr{F}'})_F. \tag{9}$$

Let $h : (X/X_{\mathscr{F}'})_F \to X_{F'}$ be the unique morphism such that $hu_{X/X_{\mathscr{F}'}} = u_X'$. Then h is a monomorphism and $hp_F = s_X$. Hence $\ker s_X = \ker p_F$.

(b) First assume that $X_{\mathscr{F}'} = X_{\mathscr{F}}$; let Q be the injective envelope of $X/X_{\mathscr{F}}$ and $t : X \to Q$ the composition $X \to X/X_{\mathscr{F}} \to Q$. By Lemma 12.1 we deduce that X_F is canonically identified with the submodule of Q formed by elements x such that $(t(X) : x) \in F$. Analogously, $X_{F'}$ may be identified with the submodule of Q formed by the elements y with $(t(X) : y) \in F'$. Obvious $X_F \subseteq X_{F'}$ and an element $x \in X_{F'}$ belongs to X_F if and only if $(t(X) : x) \in F$.

Now we suppose that L is exact. Then from the exact sequence (8) we deduce the exact sequence (9) in which p_F is an epimorphism. Also the morphism h defined above is a monomorphism and $hp_F = s_X$. Again we deduce that $\operatorname{im} h = \operatorname{im} s_X$ consists of the elements $x \in X_{F'}$, such that an element $\mathfrak{a} \in F$ may be found for which $\mathfrak{a}x \subseteq u_X'(X)$. ∎

PROPOSITION 12.4. *Let $\{F_i\}_{i\in I}$ be a set of elements in $L(A)$, such that $F_i \subseteq F$ for a suitable $F \in L(A)$. Let $F' = \bigcap_{i\in I} F_i$. Assume that X is an A-module such that $X_{\mathscr{F}_i} = X_{\mathscr{F}}$ for any i. Then*

(a) $X_{\mathscr{F}'} = \bigcap_{i\in I} X_{\mathscr{F}_i}$;

(b) $X_{F_i} \subseteq X_F$ *for any i and* $X_{F'} = \bigcap_{i\in I} X_{F_i}$.

Proof. (a) If $x \in X_{\mathscr{F}}$, then $x \in X_{\mathscr{F}_i}$ and so an element $\mathfrak{a}_i \in F_i$ can be chosen with $\mathfrak{a}_i x = 0$. Then $\mathfrak{a} = \sum_{i\in I} \mathfrak{a}_i \in F'$ and obviously $\mathfrak{a}x = 0$, so that $x \in X_{\mathscr{F}'}$.

(b) By the above proposition we can assume that $X_{F'}$ and X_{F_i} are submodules of X_F. Clearly $X_{F'} \subseteq \bigcap_i X_{F_i}$. Furthermore, let $x \in \bigcap_i X_{F_i}$; then by the above proposition, an element $\mathfrak{a}_i \in F_i$ can be found such that $\mathfrak{a}_i x \subseteq u_X(X)$; let $\mathfrak{a} = \sum_i \mathfrak{a}_i$. Then $\mathfrak{a} \in F'$ and $\mathfrak{a}x \subseteq u_X(X)$. Hence $x \in X_{F'}$. ∎

NOTE. For any elements $F \subseteq F'$ in $L(A)$, it is easy to see that the morphism $s_A : A_F \to A_{F'}$ is a ring morphism.

We consider an ideal $\mathfrak{a}$ in A, and denote by $t : A \to A/\mathfrak{a}$ the canonical ring morphism. Then, for any $F \in L(A)$, the system $t(F) \in L(A/\mathfrak{a})$ consists precisely of the left ideals $\mathfrak{b}$ of $A/\mathfrak{a}$ for which $t^{-1}(\mathfrak{b}) \in F$. Also we observe that $t^{-1}(t(F)) = t^{-1}[t\{t^{-1}[t(F)]\}] \subseteq F$. In particular, let $\mathfrak{a} = \mathfrak{r}_F$. Then, the ideal $\mathfrak{r}_{t(F)}$ is null. Indeed, $t_*(\mathfrak{r}_{t(F)}) \in \operatorname{Ob}\mathscr{F}$ and so $t^{-1}(\mathfrak{r}_{t(F)})/\mathfrak{r}_F \in \operatorname{Ob}\mathscr{F}$. Hence in the exact sequence

$$0 \longrightarrow \mathfrak{r}_F \longrightarrow t^{-1}(\mathfrak{r}_{t(F)}) \longrightarrow t^{-1}(\mathfrak{r}_{t(F)})/\mathfrak{r}_F \longrightarrow 0$$

the ends are in $\mathscr{F}$, and thus the middle is also in $\mathscr{F}$, i.e. $\mathfrak{r}_F = t^{-1}(\mathfrak{r}_{t(F)})$ by the definition of $\mathfrak{r}_F$. Therefore $\mathfrak{r}_{t(F)} = 0$.

THEOREM 12.5. *Let $F \in L(A)$. Then the canonical morphism $t : A \to A/\mathfrak{r}_F$ defines an isomorphism $\bar{t}: A_F \to (A/\mathfrak{r}_F)_{t(F)}$.*

Proof. First we observe that t defines a natural isomorphism $t_F : A_F \to (A/\mathfrak{r}_F)_F$.

Denote by $t' : (A/\mathfrak{r}_F)_F \to (A/\mathfrak{r}_F)_{t(F)}$ the mapping defined as follows. If $\mathfrak{a} \in F$ and $f : \mathfrak{a} \to A/\mathfrak{r}_F$ is a morphism of A-modules, then $\mathfrak{b} = (\mathfrak{a} + \mathfrak{r}_F)/\mathfrak{r}_F$ is an element in $t(F)$ and by definition $t'(\mathfrak{a}, f) = (\mathfrak{b}, f')$ where $f' : \mathfrak{b} \to A/\mathfrak{r}_F$ is the morphism of $A/\mathfrak{r}_F$-modules canonically defined by f. It is easy to see that t' is a well defined morphism of A-modules. Moreover, t' is a monomorphism. In fact, let $(\mathfrak{a}, f) \in (A/\mathfrak{r}_F)_F$, such that $t'(\mathfrak{a}, f) = 0$. Then, an element $\mathfrak{b}' \in t(F)$ can be chosen with $\mathfrak{b}' \subseteq (\mathfrak{a} + \mathfrak{r}_F)/\mathfrak{r}_F$ and such that the restriction of f' to $\mathfrak{b}'$ is zero. Then, let $\mathfrak{a}' = t^{-1}(\mathfrak{b}')$; we see that $\mathfrak{a}' \subseteq \mathfrak{a} + \mathfrak{r}_F = t^{-1}(\mathfrak{b})$ and the restriction of f to $\mathfrak{a}'$ is also zero, that is $(\mathfrak{a}, f) = 0$.

Finally, let $(\mathfrak{b}, g) \in (A/\mathfrak{r}_F)_{t(F)}$; then $t^{-1}(\mathfrak{b}) \in F$ and the composed morphism $gt : t^{-1}(\mathfrak{b}) \to A/\mathfrak{r}_F$ is such that $t'(t^{-1}(\mathfrak{b}), gt) = (\mathfrak{b}, g)$. To complete the proof, we shall take $\bar{t} = t' t_F$. The reader is invited to show that $\bar{t}$ is a morphism of rings. ∎

Let $x \in A_F$; then by the Lemma 12.1 the ideal $(u(A) : x)$ belongs to F. Moreover we have:

LEMMA 12.6. (a) $x = ((u(A) : x), m_x)$, *where* $m_x : (u(A) : x) \to u(A) = A/\mathfrak{r}_F$ *is the mapping* $m_x(a) = u(a)\,x$.

(b) *For any* $a \in A$ *and* $x \in A_F$ *we have*:

$$(u(A) : u(a)\,x) = ((u(A) : x) : a),$$
$$(u(A) : xu(a)) \supseteq (u(A) : x).$$

The proof is easy and is left to the reader. ∎

COROLLARY 12.7. *If* $a \in Z(A)$, *the centre of* A, *then* $u(a) \in Z(A_F)$ *the centre of* A_F.

Proof. Let $x \in A_F$. Then by Lemma 12.6 we deduce that:

$$(u(A) : x) \subseteq ((u(A) : x) : a) = (u(A) : u(a)\,x).$$

On the other hand we have:

$$(u(A) : x) \subseteq (u(A) : xu(a)).$$

Then, for any $a' \in (u(A) : x)$, we obtain that $m_{u(a)x}(a') = u(a')\,u(a)\,x = u(a'\,a)\,x = u(a)\,u(a')\,x$; also $m_{xu(a)}(a') = u(a')\,xu(a)$. But $u(a')\,x = u(a'')$ and thus $u(a')\,xu(a) = u(a'')\,u(a) = u(a)\,u(a'') = u(a)\,u(a')\,x$. Again, by Lemma 12.6 we deduce that $xu(a) = u(a)\,x$. ∎

COROLLARY 12.8. *If* A *is a commutative ring, then* A_F *is also a commutative ring.*

Proof. Let $x \in A_F$; denote by $f : A_F \to A_F$ the mapping $f(x') = xx' - x'\,x$. It is easy to see that f is a morphism of A-modules and by the above corollary $u(A) \subseteq \ker f$. Then $f = 0$, since A_F is F-closed. ∎

As a consequence of the above corollary and of Proposition 12.3 we deduce the following result.

PROPOSITION 12.9. *Let A be a commutative ring and $F \in L(A)$. Then the assignment $\mathfrak{a} \rightsquigarrow \mathfrak{a}_F$ defines a bijection between the prime ideals of A which are not in F and the prime ideals of A_F which are not in $u(F)$.*

Proof. First, we see that a prime ideal in A is F-saturated. Therefore, in order to complete the proof, it will be sufficient to show that $\mathfrak{a}_F$ is prime, whereas $\mathfrak{a}$ is prime and not in F. In fact, if $x, x' \in A$ are such that $xx' \in \mathfrak{a}_F$, then by Lemma 12.1 we deduce the inclusions:

$$\big(u(A):x\big)\big(u(A):x'\big) \subseteq \big(u(A):xx'\big) \subseteq \mathfrak{a}.$$

But $\mathfrak{a}$ being prime, we deduce that, for example, $\big(u(A):x\big) \subseteq \mathfrak{a}$, that is, again by Lemma 12.1 we deduce that $x \in \mathfrak{a}_F$. ∎

Exercises

1. Let $F \in L(A)$ be such that $\mathfrak{r}_F = 0$. For any A-submodule Y of A_F we put $Y^* = \{x \in A_F | xY \subseteq u(A)\}$.

(i) If $A_F/Y \in \operatorname{Ob} u(\mathscr{F})$, show that $Y^* \simeq \operatorname{Hom}_A(Y, u(A))$.

(ii) Call Y *F-invertible* if there exist $y_1, \ldots y_n \in Y$ and $x_1 \ldots, x_n \in Y^*$ such that $\sum_i x_i y_i = 1$. Show that the following properties of Y are equivalent:

(a) Y is F-invertible;

(b) $Y^* Y = \{x \in A_F \mid Yx \subseteq Y\}$;

(c) Y is a finitely generated projective A-module and $A_F/Y \in \operatorname{Ob} u(\mathscr{F})$.

(Hint: use Chapter 3, Section 3.9).

2. Let P be a prelocalizing system on A and $\bar{P}$ the closure of P. Prove that for any A-module X one has:

$$X/X_{\bar{P}} = \varinjlim_{\mathfrak{a} \in P} \operatorname{Hom}_A(\mathfrak{a}, X).$$

Moreover, if Q is the injective envelope of X, then

$$Q/X_E = \varinjlim_{\mathfrak{a} \in E} \operatorname{Hom}_A(\mathfrak{a}, X),$$

E being the Goldie prelocalizing system.

3. For any $\mathfrak{a} \in F$, the A-module $A_F \otimes_A (A/\mathfrak{a})$ is an object in $\mathscr{F}$.

4. Let $v: A \to B$ be a ring morphism such that A is commutative and $v(A) \subseteq Z(B)$, the centre of B. If $F \in L(A)$, then a B-module Y is $v(F)$-closed if and only if $v_*(Y)$ is F-closed. Deduce that for any A-algebra B, the module B_F is canonically an A_F-algebra.

5. Let A be a ring, $F \in L(A)$ and A_F the localization ring. Assume that the natural addition of A_F together with a multiplication "o" gives a ring structure on A_F such that the morphism $u: A \to A_F$ is also a ring morphism. Then the multiplication "o" coincides with the natural multiplication of A_F. That is, in some sense, the ring structure of A_F is unique.

6. Consider the additive group $\mathbf{Q}/\mathbf{Z}$, where $\mathbf{Q}$ is the additive group of rational numbers. Then it is impossible to introduce a multiplication on the additive group $\mathbf{Q}/\mathbf{Z}$, such that this set becomes a ring.

7. The hypotheses are as in Exercise 5. Prove that $(A_F)_{u(F)} \equiv A_F$.

8. Let $\mathfrak{a} \in F$; then $A_F\, u(\mathfrak{a}) \subseteq \mathfrak{a}_F$. In particular, if $A_F\, u(\mathfrak{a}) = A_F$, then $\mathfrak{a} \in F$.

9. Let $F \in L(A)$ and let X be such that X_F is an injective A_F-module. Then X_F is the injective envelope of $X/X_{\mathscr{F}}$ in Mod A.

10. Let F, $F' \in L(A)$ and let $u: A \to A_F$, $u': A \to A_{F'}$ be the canonical morphisms. Then the rings $(A_F)_{u(F')}$ and $(A_{F'})_{u'(F)}$ are canonically isomorphic with $A_{F''}$, where F'' is the smallest localizing system containing F and F'.

11. Let $F \in L(A)$ and let $A \xrightarrow{v} A' \xrightarrow{s} A_F$ be ring morphisms such that $sv = u$ and $\ker v = \mathfrak{r}_F$. Then, there exists a canonical isomorphism $\bar{s}: A'_{v(F)} \to A_F$ such that $\bar{s}u' = s$, where $u': A' \to A'_{v(F)}$ is defined canonically.

References

Bourbaki [26]; Bucur–Deleanu [31]; Gabriel [59]; Goldman [76]; Hacque–Maury [85]; Lambek [113], [114]; Maranda [121]; Morita [136], [137]; Năstăsescu–Popescu [144]; Ouzilou [160]; Popescu [167]; Popescu–Spircu [176]; Roos [199]; Stenström [213].

4.13 The complete ring of quotients

Let A be a ring, $F \in (L(A))$, A_F the localization ring,

$$T: \text{Mod } A \to \text{Mod } A/\mathscr{F}$$

the canonical functor and S a right adjoint of T. Also, by

$$T': \text{Mod } A_F \to \text{Mod } A/\mathscr{F}$$

denote the functor Tu_* ($u : A \to A_F$ is the morphism of localization, i.e. the canonical morphism which derives from an arrow of adjunction

$$u : \mathrm{Id}\,\mathrm{Mod}\, A \to ST)$$

and S' a right adjoint of T'. By Section 4.10, it follows that S' and T' define an equivalence between $\mathrm{Mod}\, A/\mathcal{F}$ and $\mathrm{Mod}\, A_F/u(\mathcal{F})$. Let I be an injective cogenerator of $\mathrm{Mod}\, A/\mathcal{F}$ and $V = S(I)$. By the result of Section 4.6 we can show that $F = F(V)$, i.e. a left ideal $\mathfrak{a}$ of A belongs to F if and only if $\mathrm{Hom}_A(A/\mathfrak{a}, V) = 0$.

Lemma 13.1. *Let X be an A-module and Q an injective envelope of $X/X_{\mathcal{F}}$. Then X_F is a submodule of Q and we have the inclusion $X/X_{\mathcal{F}} \subseteq X_F \subseteq Q$. Moreover, for an element $x \in Q$ the following assertions are equivalent:*

(1) $x \in X_F$;

(2) *For any morphism $f : Q \to V$ such that $X/X_{\mathcal{F}} \subseteq \ker f$ it follows that $x \in \ker f$.*

Proof. The implication (1) ⇒ (2) follows by Proposition 4.3 and Lemma 5.1.

(2) ⇒ (1). Let x be an element as in the hypothesis; we want to show that the left ideal $\mathfrak{a} = (X/X_{\mathcal{F}} : x)$ is in F, i.e. to show that $\mathrm{Hom}_A(A/\mathfrak{a}, V) = 0$. If $g : A/\mathfrak{a} \to V$ is given, we may extend it to give a commutative diagram

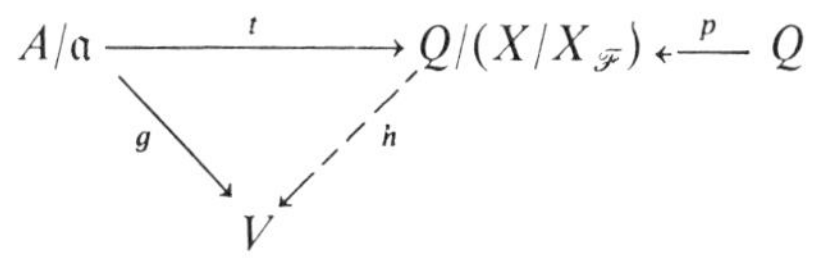

where $t(\bar{a}) = ap(x)$, ($a \in A$ and $\bar{a}$ is the image of a in $A/\mathfrak{a}$) and p is canonical. Since $hp(X/X_{\mathcal{F}}) = 0$, we have $hp(x) = 0$ which implies $g = 0$. ■

Corollary 13.2. *With the notation of the above lemma, we have:*

$$X_F = \textstyle\bigcap_f \ker f,$$

where f runs through all morphisms $f : Q \to V$, with $X/X_F \subseteq \ker f$. ■

Now we shall give an alternative description of the ring A_F.

Lemma 13.3. *Let B be the bicentralizer of V. For any $a \in A$ denote by*

$$m_a : V \to V$$

the morphism of abelian groups defined by $m_a(x) = ax$. It is clear that the mapping $a \leadsto m_a$ is, in fact, a ring morphism $v : A \to B$. Then $\ker v = \mathfrak{r}_F$.

Proof. If $a \in \mathfrak{r}_F$, then for any $x \in V$ the element $v(a)(x) = m_a(x) = ax$ has the annihilator in F, hence $ax = 0$, V being F-closed. Therefore $\mathfrak{r}_F \subseteq \ker v$.

Now let $a \in \ker v$. Then $v(a)(x) = ax = 0$ for any $x \in V$. We shall prove that $\mathrm{Ann}(a) \in F$ or equivalently $\mathrm{Hom}_A(A/\mathrm{Ann}(a), V) = 0$. Indeed, let $f: Aa \to V$ be a morphism of A-modules. By the injectivity of V, the morphism f can be extended to a morphism $\bar{f}: A \to V$. Then for any $a' \in A$ we have by hypothesis:

$$f(a'a) = \bar{f}(a'a) = a'a\bar{f}(1) = 0.$$

Hence $f = 0$ and so $a \in \mathfrak{r}_F$. ∎

Let us put $V' = S'(I)$; then $u_*(V') = V$, and $\mathrm{End}_{A_F}(V') \simeq \mathrm{End}_A(V)$. A slight investigation shows that the structures of the $\mathrm{End}_{A_F}(V')$-module V' and $\mathrm{End}_A(V)$-module V are identical, and so the ring B becomes the bicentralizer of the A_F-module V'. Furthermore the multiplication on the left with an element of A_F defines a ring morphisms $\bar{v}: A_F \to B$ such that $\bar{v}u = v$. Then we check that $\bar{v}$ is, in fact, a monomorphism. Also we observe that V' is an object of definition of $u(F)$.

THEOREM 13.4. *Assume that V contains $A/\mathfrak{r}_F$ as a submodule. Then with the above notations the ring morphism $\bar{v}: A_F \to B$ is an isomorphism.*

Proof. Let Q be an injective envelope of $A/\mathfrak{r}_F$. We can always assume that there exists a monomorphism $i: Q \to V$ (if necessary we replace V by $V \amalg Q$). Let p be a retraction of i, and $f: V \to V$ a $\mathrm{End}_A(V)$-homomorphism. Also let $a = pf(\bar{1})$, where $\bar{1} \in A/\mathfrak{r}_F \subseteq Q$ is the image of 1. We want to show that $a \in A_F$ by using Lemma 13.1. Let $g: Q \to V$ be any morphism of A-modules such that $A/\mathfrak{r}_F \subseteq \ker g$ or equivalently $g(\bar{1}) = 0$. Extend g to $\bar{g}: V \to V$ by defining $\bar{f}(L) = 0$, where $V = Q \amalg L$. Then $g(pf(\bar{1})) = \bar{g}(ipf(\bar{1})) = \bar{g}f(\bar{1}) = f(g(\bar{1})) = 0$ since f is $\mathrm{End}_A(V)$-linear. Hence

$$pf(\bar{1}) = a \in A_F.$$

Now the point is to show that $\bar{v}(a) = f$. For that, let $x \in V$ be any element and $u_x: A/\mathfrak{r}_F \to V$ the morphism associated with x, i.e. $u_x(\bar{a}) = \bar{a}x$. Let h' be an extension of u_x to Q and $h: V \to V$, a morphism such that $hi = h'$ and $L \subseteq \ker h$. Then $hip = h$ and $h(\bar{1}) = x$. Furthermore, we have: $f(x) = f(\bar{1}.x) = f(u_x(\bar{1})) = f(h(\bar{1})) = hf(\bar{1})$ (f being a morphism of $\mathrm{End}_A(V)$-modules). Then: $hf(\bar{1}) = hipf(\bar{1}) = hi(a) = h'(a.\bar{1}) = ah'(\bar{1}) = ax$. ∎

COROLLARY 13.5. *Let $F \in \mathrm{L}(A)$ and $\bar{F} = F_{\mathfrak{r}_F}$ the saturation of F. Then the ring $A_{\bar{F}}$ is canonically isomorphic with the ring $\mathrm{Cent}_A(Q)$, where Q is an injective envelope of $A/\mathfrak{r}_F$. Moreover $A/\mathfrak{r}_F \subseteq A_F \subseteq A_{\bar{F}}$ and an element $x \in A_{\bar{F}}$ belongs to A_F if and only if $(A/\mathfrak{r}_F : x) \in F$.* ∎

Now, let us consider the system D of all dense left ideals of A. We know that $D = F(Q)$ (see Section 4.9), where Q is injective envelope of A. The ring A_D is frequently denoted by $Q_l(A)$ or simply $Q(A)$ if there is no confusion and is called the *left complete ring of quotients of A* or simply, *the complete ring of quotients of A*, if there is no danger of confusion.

We have some remarks to make about the ring $Q(A)$. For that purpose we shall identify A with a subring of $Q(A)$ and $Q(A)$ with an A-(or $Q(A)$-) submodule of Q. Clearly, an element $x \in Q$ belongs to $Q(A)$ if and only if $(\mathfrak{r}A : x)$ is a dense left ideal of A.

PROPOSITION 13.6. *Q is the injective envelope of $Q(A)$ in the category* Mod $Q(A)$. *Moreover* $Q(Q(A)) = Q(A)$.

Proof. First we observe that the functor $S' : \text{Mod } A/\mathscr{D} \to \text{Mod } Q(A)$ preserves the injective envelopes (Chapter 3, Theorem 2.8 and Lemma 10.11) having an exact left adjoint.

Further we see that $\text{Cent}_A(Q) \simeq \text{Cent}_{Q(A)}(Q)$; now, we can use Theorem 13.4. ■

PROPOSITION 13.7. *For any ring the following assertions are equivalent*:

(1) *The canonical morphism of* $H = \text{End}_A(Q)$*-modules* $H \xrightarrow{p} Q$ *with* $p(h) = h(1)$ *is an ismorphism*;

(2) *The morphism* $s : Q(A) = \text{Cent}_A(Q) \to Q$ *with* $s(q) = q(1)$ *is an isomorphism*;

(3) *The rings $Q(A)$ and H are canonically isomorphic*;

(4) *$Q(A)$ is an injective A-module*;

(5) *$Q(A)$ is an injective $Q(A)$-module.*

Proof Assume (1), then the mapping p has the kernel zero that is, for any $h \in H$, $h(1) \neq 0$, whenever $h \neq 0$ Then by Lemma 13.1, (2) follows.

Conversely, if s is surjective and $h \in H$ is such that $ph = h(1) = 0$ then for any $x \in Q$, $x = s(q)$ and so $h(x) = h(s(q)) = h(q(1)) = q(h(1)) = 0$. Consequently $h = 0$, hence (1) $\Leftrightarrow$ (2).

Assume (1) and (2). Tracing the given isomorphisms $H \xrightarrow{p} Q \xleftarrow{s} Q(A)$ we find that $h \in H$ corresponds to $q \in Q$ if and only if $h(1) = q(1)$. Suppose also $h'(1) = q'(1)$. Then $(hh')(1) = h(h'(1)) = h(q'(1)) = q'(h(1)) = q'(q(1)) = (q'q)(1)$. Thus (1) and (2) $\Rightarrow$ (3).

Assume (3), then the relation $h(1) = q(1)$ is an isomorphism between H and Q. Now for any $x \in Q$ there exists $h \in H$ such that $h(1) = x$ (see Exercise 2). Hence by assumption there exists $q \in Q(A)$ such that $x = q(1)$, that is (2). Thus (3) $\Rightarrow$ (2).

Assume (4); then $Q(A)$ is injective as an A-submodule. But Q is an essential extension of $Q(A)$, hence $Q(A) = Q$, that is (2). Since clearly (2) $\Rightarrow$ (4) we have (2) $\Leftrightarrow$ (4).

We have established the equivalence of (1), (3) and (5). ∎

NOTE. By abuse of notation we have identified the ring $Q(A)$ with the A-module $(A_l)_D$ (see Section 4.12), which is a submodule of Q. We shall always mean that $Q(A)$ and $(A_l)_D$ are identified by the help of the morphism

$$s: Q(A) \to Q, \quad s(q) = q(1).$$

LEMMA 13.8. *Let A be a ring and X an A-module. Then X is injective if and only if any morphism from a large left ideal $\mathfrak{a}$ of A can be extended to A.*

Proof. Let $\mathfrak{a}'$ be any left ideal of A. Then another left ideal $\mathfrak{a}''$ may be found such that $\mathfrak{a}' + \mathfrak{a}''$ is large and $\mathfrak{a} \cap \mathfrak{a}' = 0$. Then any morphism $f: \mathfrak{a}' \to X$ can be extended to $\mathfrak{a}' + \mathfrak{a}''$ (setting $f'(x' + x'') = f(x')$, for any $x' \in \mathfrak{a}'$ and $x'' \in \mathfrak{a}''$), and so to A, by assumptions. The converse assertion is obvious. ∎

PROPOSITION 13.9. *Let E be the prelocalizing system of left large ideals of A and $\bar{E} = E^2$ the closure of E. Then*

(1) *Any $\bar{E}$-closed module is injective;*

(2) *The category* Mod $A/\bar{\mathscr{E}}$ *is a spectral category and the functor*

$$S: \text{Mod } A/\bar{\mathscr{E}} \to \text{Mod } A$$

is exact;

(3) *The ring $G(A) = A_{\bar{E}}$ is a regular ring.*

Proof. The assertions (1) and (2) are direct consequences of Lemma 13.8. Finally the assertion (3) follows by (2) and Lemmas 13.1 and 3.8. ∎

The ring $G(A)$ is called the *left regular ring of quotients* of A or simply the *regular ring of quotients of A*.

We know that $D \subseteq \bar{E}$ and thus a unique ring morphism $t: Q(A) \to G(A)$ can be found such that the following diagram is commutative:

$$\begin{array}{ccc} & & Q(A) \\ & \nearrow & \downarrow t \\ A & \longrightarrow & G(A) \end{array}$$

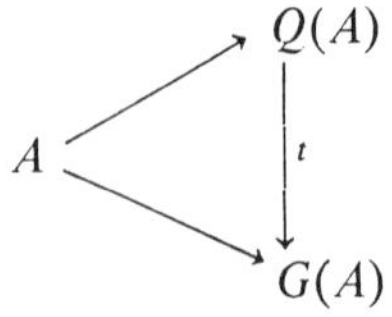

the other morphisms being canonical.

Now we shall investigate an important case, when t is an isomorphism.

THEOREM 13.10. *Let A be a ring. The following assertions are equivalent:*

(1) *The canonical ring morphism $t: Q(A) \to G(A)$ is an isomorphism;*

(2) $J(A) = 0$;

(3) $Q(A)$ *is a regular ring;*

(4) $J(Q) = 0$.

Proof. (2) $\Rightarrow$ (1). By Proposition 9.7 it follows that $E = D$ and so t is an isomorphism.

(2) $\Rightarrow$ (3). Again by Proposition 9.7 one has that $Q(A) = G(A)$. Now we utilize Proposition 13.9.

(2) $\Rightarrow$ (4). Then $J(Q) \cap A \subseteq J(A) = 0$, in other words $J(Q) = 0$ (A being essential in Q). Conversely, if $J(Q) = 0$, then $J(A) = J(Q) = 0$. Hence (4) $\Rightarrow$ (2).

(3) $\Rightarrow$ (2). Let $\mathfrak{a}$ be a large left ideal of A, hence a large A-submodule of Q and so of $Q(A)$ and assume that $\mathfrak{a}x = 0$ for $0 \neq x \in A$. If $Q(A)$ is regular, there exists $q \in Q(A)$ such that $xqx = x$. Therefore $xq \neq 0$ and so $Axq \cap \mathfrak{a} \neq 0$. Hence we can find $s \in A$ such that $0 \neq sxq$ and $sxqx = sx = 0$, a contradiction. ∎

If A is a left Ore domain, then the set of all non-zero left ideals is an element of $L(A)$ (moreover any non-zero left ideal is essential) which coincides with D. Later we shall see that $Q(A)$ is in fact, the field of fractions of A.

Exercises

1. Let $F \in L(A)$ and let $\bar{F} = F_{\mathfrak{r}_F}$ the saturation of F. Then the ring A_F is the complete ring of quotients of $A/\mathfrak{r}_F$.

2. Let $F \in L(A)$, Q the injective envelope of $A/\mathfrak{r}_F$, $H = \mathrm{End}_A(Q)$ and $C = \mathrm{Cent}_A(Q)$. Then the mapping $h \rightsquigarrow h(\bar{1})$ of H into Q is an epimorphism of H-modules. Also the mapping $g \rightsquigarrow g(\bar{1})$ of C into Q is a monomorphism of C-modules.

3. Let $J^2(A)$ be the bisingular ideal of A, that is $J^2(A) = \mathfrak{r}_E$. Then $G(A)$ is the total ring of quotients of $A/J^2(A)$.

4. The assertions of Proposition 13.7 are also equivalent to the fact that $\mathfrak{R}(H) = 0$.

5. A ring A is called *left self-injective* if it is injective as an A-module. For any ring A, the ring $G(A)$ is left self-injective. Any regular ring can be embedded in a left self-injective regular ring.

6. Let $(\mathrm{Mod}\, A)_E$ be the spectral category of $\mathrm{Mod}\, A$ (see Section 4.2) and $P : \mathrm{Mod}\, A \to (\mathrm{Mod}\, A)_E$ the canonical functor. Then $(\mathrm{Mod}\, A)_E$ is equivalent with the product of categories $(\mathrm{Mod}\, A/\overline{\mathscr{E}}) \times P(\overline{\mathscr{E}})$, where $P(\overline{\mathscr{E}})$ is the image of $\overline{\mathscr{E}}$ by P.

7. Let I be an injective object of a Grothendieck category $\mathscr{C}$, and $\mathfrak{a}$ the ideal of all $h \in \mathrm{End}_{\mathscr{C}}(I)$ such that $\ker h$ is essential. Then:

(a) $\mathfrak{a}$ is the Jacobson radical of $\mathrm{End}_{\mathscr{C}}(I)$;

(b) $\mathrm{End}_{\mathscr{C}}(I)/\mathfrak{a}$ is regular;

(c) Idempotent module $\mathfrak{a}$ can be *lifted* in $\mathrm{End}_{\mathscr{C}}(I)$.

(The last assertion means that if e is an idempotent in $\mathrm{End}_{\mathscr{C}}(I)/\mathfrak{a}$, then there exists an idempotent in $\mathrm{End}_{\mathscr{C}}(I)$ mapping canonically onto e.)

8. Let $\{A_i\}_{i \in I}$ be a set of rings, $A = \prod_i A_i$ the direct product and

$$p_i : A \to A_i$$

the associated projections. A left ideal $\mathfrak{a}$ of A is dense if and only if $p_i(\mathfrak{a})$ is dense for any i. Deduce that $Q(A) = \prod_i Q(A_i)$.

9. Let F be a localizing system on A. Assume that $\mathfrak{r}_F = 0$ and A_F is regular and flat as A^0-module. Then the following assertions are equivalent:

(i) A is left semihereditary;

(ii) Any $\mathscr{F}$-torsionfree A-module is flat.

10. If $A = R_n$ the ring of all $n \times n$ matrices over A, show that $Q(A) = Q(R)_n$

11. A mapping $d : A \to A$ is called a *derivation* of A if

$$d(x + y) = d(x) + d(y), \qquad d(xy) = d(x)y + xd(y)$$

for all $x, y \in A$. Show that every derivation d on A can be extended to a unique derivation d' of $Q(A)$.

12. For any ring A, the following statements are equivalent.

(a) $J(A) = 0$ and every finitely generated non-singular A-module is projective;

(b) A is left semi-hereditary, $Q(A)$ is flat and $J(Q(A) \otimes_A Q(A)) = 0$.

13. Let A be a ring and Q the injective envelope of A. Then $Q(A)$ is left self-injective if and only if $\mathrm{Hom}_A(Q/A, Q) = 0$.

14. Let X be a non-singular injective A-module. Then the ring $\mathrm{End}_A(X)$ is regular.

15. Show that if E is an injective module and M is non-singular, then every exact sequence $0 \to K \to E \to M \to 0$ splits. Using this, show that if E and E' are injective submodules of a non-singular module, then $E + E'$ is also injective.

16. If A is left semi-hereditary, then $J(A) = 0$.

References

Desq [41]; Gabriel [59]; Gabriel–Oberst [61]; Hacque–Maury [85]; Harada [87]; Hudry [94]; Johnson [99]; Kato [104]; Lambek [112], [113], [114]; Leclerc [118]; Popescu [167]; Popescu–Spircu [176]; Roos [193], [194], [195], [199]; Stenström [213]; Tachikawa [219]; Teply [223]; Tisseron [227]; Utumi [232].

4.14 Some remarks on Grothendieck categories

The aim of this paragraph is to give an alternative proof for Theorem 7.9 of Chapter 3 and also to make some useful remarks on localization in categories of modules.

PROPOSITION 14.1. *Let A be a ring and $\mathscr{M}$ a class of A-modules. There exists a localizing system $F \in L(A)$ such that all modules in $\mathscr{M}$ are F-closed and for any $G \in L(A)$ such that each $M \in \mathscr{M}$ is G-closed one has $G \subseteq F$. Moreover, an A-module X is in $\mathscr{F}$ if and only if for any $x \in X$ the inclusion* $\mathrm{Ann}(x) \to A$ *induces the isomorphism* $\mathrm{Hom}_A(A, M) \simeq \mathrm{Hom}_A(\mathrm{Ann}(x), M)$ *for every $M \in \mathscr{M}$.*

Proof. If $G \in L(A)$, then an A-module M is G-closed, if and only if M and $E(M)/M$ are $\mathscr{G}$-torsionfree (see Theorem 4.5 or Lemmas 5.1 and 12.1). It follows that the localizing system F consisting of all left ideals $\mathfrak{a}$ such that $\mathrm{Hom}_A(A/\mathfrak{a}, E(M) \amalg E(E(M)/M)) = 0$ for every $M \in \mathscr{M}$ is the strongest localizing system for which all modules in $\mathscr{M}$ are closed (see Section 4.6).

It remains to determine the localizing subcategory $\mathscr{F}$, associated with F. If $X \in \mathrm{Ob}\,\mathscr{F}$, then $\mathrm{Ann}(x) \in F$ for each $x \in X$ and hence the inclusion $\mathrm{Ann}(x) \to A$ induces the isomorphism $\mathrm{Hom}_A(A, M) \simeq \mathrm{Hom}_A(\mathrm{Ann}(x), M)$ for all $M \in \mathscr{M}$. Conversely, if a module X satisfies this latter condition, then we shall show that $X \in \mathrm{Ob}\,\mathscr{F}$. We begin by observing that $\mathrm{Hom}_A(Ax, M) = 0$ for any $x \in M$. Furthermore, if $f : Ax \to E(M)$ is a non-zero morphism, then an element $a \in A$ may be chosen such that $0 \neq f(ax) \in M$. Then the mapping $a'ax \to a'f(ax)$ is a non-zero morphism on Aax into M, a contradiction. To complete the proof, it will be sufficient to prove that

$$\mathrm{Hom}_A(Ax, E(M)/M) = 0$$

for all $M \in \mathcal{M}$. For that purpose we consider the following commutative diagram, canonically constructed:

$$\begin{array}{ccccccccc}
 & & 0 & & 0 & & 0 & & \\
 & & \downarrow & & \downarrow & & \downarrow & & \\
0 & \to & \mathrm{Hom}_A(Ax, M) & \to & \mathrm{Hom}_A(A, M) & \to & \mathrm{Hom}_A(\mathrm{Ann}(x), M) & \to & 0 \\
 & & \downarrow & & \downarrow & & \downarrow & & \\
0 & \to & \mathrm{Hom}_A(Ax, E(M)) & \to & \mathrm{Hom}_A(A, E(M)) & \to & \mathrm{Hom}_A(\mathrm{Ann}(x), E(M)) & \to & 0 \\
 & & \downarrow{\scriptstyle h} & & \downarrow & & \downarrow & & \\
0 & \to & \mathrm{Hom}_A(Ax, E(M)/M) & \to & \mathrm{Hom}_A(A, E(M)/M) & \to & \mathrm{Hom}_A(\mathrm{Ann}(x), E(M)/M) & & \\
 & & & & \downarrow & & & & \\
 & & & & 0 & & & &
\end{array}$$

It is clear that $h = 0$ and $\mathrm{coker}\, h = \mathrm{Hom}_A(Ax, E(M)/M)$. By the snake lemma the following sequence is exact:

$$\mathrm{Hom}_A(A, M) \to \mathrm{Hom}_A(\mathrm{Ann}(x), M) \xrightarrow{d} \mathrm{Hom}_A(Ax, E(M)/M) \to 0$$

where d is canonically constructed. Then obviously, the last term of the sequence is zero, as required. ∎

Let $\mathcal{M}$ be a class of A-modules. Then the system F defined as in the previous proposition is called the *strongest localizing system* on A for which all modules in $\mathcal{M}$ are F-closed.

Now we are able to give a new proof for a substantial part of Theorem 7.9 of Chapter 3.

THEOREM 14.2. (Gabriel–Popescu [60]). *Let $\mathscr{C}$ be an* Ab 5-*category, U an object of $\mathscr{C}$, $A = \mathrm{End}_{\mathscr{C}}(U)$, $S : \mathscr{C} \to \mathrm{Mod}\, A^0$ the functor $S(X) = \mathrm{Hom}_{\mathscr{C}}(U, X)$ and T a left adjoint of S. The following assertions are equivalent*:

(1) *U is a generator of $\mathscr{C}$*;

(2) *S is full and faithful and T exact.*

Proof. (1) ⇒ (2). As in the proof of Theorem 7.9 of Chapter 3 we show that S is full and faithful. Therefore, to complete the proof, it remains to prove that T is exact.

Let F be the strongest localizing system of right ideals on A for which all modules $S(X)$, $X \in \mathrm{Ob}\mathscr{C}$ are F-closed. Then we may construct a diagram of functors:

$$\mathscr{C} \underset{T}{\overset{S}{\rightleftarrows}} \mathrm{Mod}\, A^0 \underset{S'}{\overset{T'}{\rightleftarrows}} \mathrm{Mod}\, A^0/\mathscr{F}$$

where T' and S' are canonically defined. It is sufficient now to show that $T'S$ is an equivalence and thus T will be exact, S' being full and faithful. Therefore the single point is to show that every F-closed object is isomorphic to a module of the form $S(X)$, $X \in \mathrm{Ob}\mathscr{C}$.

Thus, let M be an F-closed module. Choose an exact sequence:

$$A^{(I)} \xrightarrow{f} A^{(J)} \longrightarrow M \longrightarrow 0 \tag{10}$$

in $\mathrm{Mod}\, A^0$. Since the functor T has a right adjoint, it preserves inductive limits and therefore carries (10) into an exact sequence

$$U^{(I)} \xrightarrow{T(f)} U^{(J)} \longrightarrow T(M) \longrightarrow 0 \tag{11}$$

in $\mathscr{C}$.

LEMMA 14.3. *The functor $T'S$ is exact and preserves direct sums.*

We conclude the proof of the theorem before we prove the lemma. By applying the lemma to (11) and noting that $A = S(U)$ is F-closed, we obtain the upper exact row of the following diagram in $\mathrm{Mod}\, A^0/\mathscr{F}$:

$$\begin{array}{ccccccc}
T'(A)^{(I)} & \xrightarrow{T'ST(f)} & T'(A)^{(J)} & \longrightarrow & T'ST(M) & \longrightarrow & 0 \\
{\scriptstyle m}\uparrow & & {\scriptstyle n}\uparrow & & & & \\
T'(A)^{(I)} & \xrightarrow{T'(f)} & T'(A)^{(J)} & \longrightarrow & T'(M) & \longrightarrow & 0
\end{array}$$

The lower row is obtained by applying T' to (10) and is also exact. The diagram commutes, because $A = S(U)$ implies that f has the form $f = S'(f')$ and one has $T'STS'(f') = T'(f)$; also, if $u : \mathrm{IdMod}\, A^0 \to ST$ is an arrow of adjunction, then $m = T'(\mu_{A^{(I)}})$, $n = T'(\mu_{A^{(J)}})$, that is the isomorphisms. We conclude from the diagram that $T'(M) \simeq T'ST(M)$ and so $M \simeq ST(M)$, M being F-closed. ■

Proof of Lemma 14.3. We already know the functor $T'S$ to be left exact, so to prove exactness it will suffice to show that it preserves epimorphisms. This means that if $f : M' \to M''$ is an epimorphism in $\mathscr{C}$, then we must show that coker $S(f)$ is an object of $\mathscr{F}$. By Proposition 13.1 this is equivalent to showing that for each $x \in S(M'')$ we have

$$\mathrm{Hom}_A(A, S(M)) \simeq \mathrm{Hom}_A((\mathrm{im}\, S(f) : x), S(M))$$

for all $M \in \mathrm{Ob}\mathscr{C}$. Define a morphism $h: U \to M''$ such that $S(h): A \to S(M'')$ maps 1 onto x. From the diagram in $\mathscr{C}$, in which the right-hand square is cartesian:

$$\begin{array}{ccccccccc} 0 & \longrightarrow & K & \xrightarrow{k} & P & \xrightarrow{g} & U & \longrightarrow & 0 \\ & & \| & & \downarrow & & \downarrow{\scriptstyle h} & & \\ 0 & \longrightarrow & K & \longrightarrow & M' & \xrightarrow{f} & M'' & \longrightarrow & 0 \end{array}$$

We get the following diagram in Mod A^0 in which the right-hand square is also cartesian:

$$\begin{array}{ccccccc} 0 & \longrightarrow & S(K) & \xrightarrow{S(k)} & S(P) & \xrightarrow{S(g)} & A \\ & & \downarrow & & \downarrow & & \downarrow{\scriptstyle S(h)} \\ 0 & \longrightarrow & S(K) & \longrightarrow & S(M') & \xrightarrow{S(f)} & S(M'') \end{array}$$

and $(\mathrm{im}\, S(f) : x) = \mathrm{im}\, S(g)$. Note that $T(\mathrm{coker}\, S(g)) = 0$, that is,

$$T(A) = \mathrm{coker}\, TS(k).$$

It follows that if we have a morphism $p: \mathrm{im}\, S(g) \to S(M)$ for some $M \in \mathrm{Ob}\mathscr{C}$, then it factors uniquely over A, S being full and faithful and this is precisely what we wanted to show.

It remains to show that $T'S$ preserves direct sums. Actually we prove a little more, namely that $T'S$ commutes with the sums of a direct set of subobjects of any object M in $\mathscr{C}$, that is, if $\{M_i\}_i$ is a direct set of subobjects of M, then the cokernel of the canonical morphism:

$$f: \textstyle\sum_i S(M_i) \to S(\sum_i M_i)$$

is an object in $\mathscr{F}$. By Proposition 14.1 this means that for each $x \in S(\sum_i M_i)$, must show that $\mathrm{Hom}_A(A, S(M')) \simeq \mathrm{Hom}_A((\mathrm{im}\, f : x), S(M'))$ for all

$$M' \in \mathrm{Ob}\mathscr{C}.$$

Again, define a morphism $h: U \to \sum_i M_i$ such that $S(h): A \to S(\sum_i M_i)$ maps 1 onto x.

From the cartesian diagram:

$$\begin{array}{ccc} N_i & \longrightarrow & A \\ \downarrow & & \downarrow{\scriptstyle h} \\ M_i & \xrightarrow{u_i} & \sum_i M_i \end{array}$$

in which u_i is the structural morphism, we obtain a commutative diagram in Mod A^0:

$$\begin{array}{ccc} \sum_i S(N_i) & \xrightarrow{g} & A \\ \downarrow & & \downarrow S(h) \\ \sum_i S(M_i) & \xrightarrow{f} & S(\sum_i M_i) \end{array}$$

which is a cartesian square, because cartesian squares are preserved both by S and when taking inductive limits in Mod A^0, (see Chapter 2, Theorem 8.6). Therefore we have: $(\operatorname{im} f : x) = \operatorname{im} g \simeq \sum_i S(N_i)$. Now $\operatorname{Hom}_A(\sum_i S(N_i),$

$$S(M')) \simeq \varinjlim_i \operatorname{Hom}_A(S(N_i), S(M')) \simeq \varinjlim_i \operatorname{Hom}_{\mathscr{C}}(N_i, M') \simeq \operatorname{Hom}_{\mathscr{C}}(\textstyle\sum_i N_i, M')$$

$$= \operatorname{Hom}_{\mathscr{C}}(U, M') \simeq \operatorname{Hom}_A(A, S(M')),$$

where we have utilized the fact that exactness of the inductive direct limit implies that $\sum_i N_i = U$. Now the proof is complete. ∎

Let A be a ring. The strongest localizing system F on A, such that A is F-closed is called the *canonical localizing system* on A. Proposition 13.1 gives a complete description of the localizing subcategory $\mathscr{F}$.

It is clear that the canonical system is smaller than the system of dense ideals. It is an important problem to determine the ring A for which the two systems coincide. (An example of such a ring is a self-injective ring).

Now we shall see that the canonical system is connected with the study of certain generators in Grothendieck categories.

A generator U of a Grothendieck category $\mathscr{C}$ is called a *proper generator*, if U has the property that a monomorphism $U' \xrightarrow{i} U$ induces an isomorphism $\operatorname{Hom}_{\mathscr{C}}(U, U) \simeq \operatorname{Hom}_{\mathscr{C}}(U', U)$ if and only if i is an isomorphism.

LEMMA 14.4. *Any Grothendieck category $\mathscr{C}$ has a proper generator.*

Proof. Let U be a generator of $\mathscr{C}$, V an injective envelope of the direct sum $\coprod U/U'$ where U' runs through all subobjects of U and $\bar{U} = U \amalg V$. Then $\bar{U}$ is a proper generator of $\mathscr{C}$. In fact, if $U_1 \xrightarrow{i} \bar{U}$ is a monomorphism such that $h_{\bar{U}}(i)$ is a monomorphism, then in particular $\operatorname{Hom}_{\mathscr{C}}(\bar{U}/U_1, \bar{U}) = 0$. But this is impossible if i is not an isomorphism, $\bar{U}$ being a cogenerator of $\mathscr{C}$. ∎

A ring can be the endomorphism ring of generators U and U' of Grothendieck categories $\mathscr{C}$ and $\mathscr{C}'$ without $\mathscr{C}$ and $\mathscr{C}'$ being equivalent. For example, if K is a division ring and U is an infinite dimensional vector space in Mod K, then its endomorphism ring $A = \operatorname{End}_K(U)$ is also the endomorphism ring of the generator $A \in \operatorname{Ob}\operatorname{Mod} A$. However by Chapter 3, Corollary 7.7, we deduce that Mod A and Mod K are not equivalent. The point is that A is not a proper generator in Mod A. The main objective here is to show that if U and U' are proper generators of Grothendieck categories $\mathscr{C}$ and $\mathscr{C}'$, and if U and U' have isomorphic endomorphism rings, then $\mathscr{C}$ and $\mathscr{C}'$ are equivalent.

LEMMA 14.5. *Let A be a ring and let G be the set of right ideals $\mathfrak{a}$ of A such that the inclusion $\mathfrak{a} \subseteq A$ induces an isomorphism* $\mathrm{Hom}_A(A, A) \simeq \mathrm{Hom}(\mathfrak{a}, A)$. *If G is a localizing system, then A is a proper generator of the quotient category* $\mathscr{C} = \mathrm{Mod}\, A^0/\mathscr{G}$.

Proof. Let $T: \mathrm{Mod}\, A^0 \to \mathscr{C}$ be the canonical functor and $\bar{f}: M \to T(A)$ an inclusion in $\mathscr{C}$, which induces an isomorphism

$$\mathrm{Hom}_{\mathscr{C}}(T(A), T(A)) \simeq \mathrm{Hom}_{\mathscr{C}}(M, T(A)).$$

But it is clear that A is G-closed and by Lemma 4.1 $\bar{f} = T(f)$, where $f: M' \to A$ is a morphism in Mod A^0 (in fact we assume that $T(M') = M$). Let

$$\mathfrak{a} = f(M').$$

Then $M \simeq T(A)$ in $\mathscr{C}$ if and only if $\mathfrak{a} \in G$, for since $\bar{f}$ is a monomorphism $\ker f \in \mathrm{Ob}\mathscr{G}$, and $M \simeq T(\mathfrak{a})$. Consider the exact commutative diagram of abelian groups, canonically constructed:

$$\begin{array}{ccccccc}
0 \longrightarrow \mathrm{Hom}_A(A/\mathfrak{a}, A) & \longrightarrow & \mathrm{Hom}_A(A, A) & \xrightarrow{\quad m \quad} & \mathrm{Hom}_A(\mathfrak{a}, A) \\
& & \Big\downarrow{\scriptstyle h} & & \Big\downarrow{\scriptstyle n} \\
& & \mathrm{Hom}_{\mathscr{C}}(T(A), T(A)) & \xrightarrow{\ d\ } & \mathrm{Hom}_{\mathscr{C}}(T(\mathfrak{a}), T(A))
\end{array}$$

in which the vertical morphisms are induced by T. Clearly, h and n are isomorphisms, A being G-closed (see Lemma 4.1) and d is an isomorphism by hypothesis. Consequently, m is also an isomorphism, that is $\mathfrak{a} \in G$. ■

NOTE 1. The above localizing system G is, in fact, the canonical localizing system. We have moreover a converse result.

PROPOSITION 14.6. *Let $\mathscr{C}$ be a Grothendieck category, U a generator of $\mathscr{C}$ and* $A = \mathrm{End}_{\mathscr{C}}(U)$. *Denote by K the localizing system in A^0 associated with the canonical functor* $T: \mathrm{Mod}\, A^0 \to \mathscr{C}$ *(see Theorem* 14.2). *Then U is a proper generator if and only if $K = G$, the canonical localizing system.*

Proof. If $K = G$, then the conclusion follows by the above Lemma and Theorem 14.2.

Conversely, assume that U is a proper generator of $\mathscr{C}$. If $\mathfrak{a}$ is a right ideal of A, the sequences

$$0 \to T(\mathfrak{a}) \to T(A) \to T(A/\mathfrak{a}) \to 0$$

and

$$0 \to \mathfrak{a}(U) \to U \to U/\mathfrak{a}(U) \to 0$$

are equivalent, by Exercise 1. (We have used the notations of Exercise 3, Chapter 3, Section 3.7). Thus the adjoint maps lead to a commutative diagram:

$$\begin{array}{ccccccc}
0 \longrightarrow & \mathrm{Hom}_{\mathscr{C}}(U/\mathfrak{a}(U), U) & \longrightarrow & \mathrm{Hom}_{\mathscr{C}}(U, U) & \xrightarrow{t} & \mathrm{Hom}_{\mathscr{C}}(\mathfrak{a}(U), U) \\
& \downarrow & & \downarrow & & \downarrow \\
0 \longrightarrow & \mathrm{Hom}_A(A/\mathfrak{a}, A) & \longrightarrow & \mathrm{Hom}_A(A, A) & \longrightarrow & \mathrm{Hom}_A(\mathfrak{a}, A)
\end{array}$$

with the vertical arrows isomorphisms and the rows exact. Hence $\mathfrak{a} \in G$ if and only if t is an isomorphism and this map is an isomorphism if and only if $\mathfrak{a}(U) = U$, that is, $\mathfrak{a} \in K$, since U is proper. ■

From the two previous results one has immediately

COROLLARY 14.7. *A ring A is the endomorphism ring of a proper generator of a Grothendieck category if and only if the set of right ideals $\mathfrak{a}$ of A for which the map* $\mathrm{Hom}_A(A, A) \to \mathrm{Hom}_A(\mathfrak{a}, A)$ *induced by the inclusion $\mathfrak{a} \to A$ is an isomorphism, is a localizing system.* ■

NOTE 2. Let A be a ring. Denote by G' the set of right ideals $\mathfrak{a}$ of A such that one has the isomorphism $\mathrm{Hom}_A(A, A) \simeq \mathrm{Hom}_A(\mathfrak{a}, A)$. Then G' is a localizing system (namely identical with the canonical localizing system) if and only if for any $\mathfrak{a} \in G'$ and $x \in A$ we have that $(\mathfrak{a} : x) \in G'$. As a direct consequence we have the following result.

THEOREM 14.8. *Any commutative ring is the endomorphism ring of a proper generator of a suitable Grothendieck category.*

Proof. Let $\mathfrak{a} \in G'$, $x \in A$. Since $\mathfrak{a} \subseteq (\mathfrak{a} : x)$, $\mathrm{Hom}_A(A/(\mathfrak{a} : x), A) = 0$. For $y \in (\mathfrak{a} : x)/\mathfrak{a}$ we deduce that $(Ay' + \mathfrak{a})/\mathfrak{a} \simeq A/(\mathfrak{a} : y')$ (where y' is an element in $(\mathfrak{a} : x)$ whose image in $(\mathfrak{a} : x)/\mathfrak{a}$ is y), so $\mathrm{Hom}_A((Ay' + \mathfrak{a})/\mathfrak{a}, A) = 0$. It follows that $\mathrm{Hom}_A((\mathfrak{a} : x)/\mathfrak{a}, A) = 0$.

Now let $f : (\mathfrak{a} : x) \to A$ be a morphism of A-modules and f' the restriction to $\mathfrak{a}$. Then a unique morphism $\bar{f} : A \to A$ may be found which extends f'. If $\bar{f}'$ is the restriction of $\bar{f}$ to $(\mathfrak{a} : x)$ then $\mathfrak{a} \subseteq \ker(f - \bar{f}')$ and so necessarily $f = \bar{f}'$ by the above results. ■

THEOREM 14.9. *Let $\mathscr{C}$ and $\mathscr{C}'$ be Grothendieck categories, and let U and U' be proper generators of $\mathscr{C}$ and $\mathscr{C}'$, respectively. If the rings* $\mathrm{Hom}_{\mathscr{C}}(U, U)$ *and* $\mathrm{Hom}_{\mathscr{C}'}(U', U')$ *are isomorphic, then there exists a categorical equivalençe $F : \mathscr{C} \to \mathscr{C}'$ such that $F(U) = U'$.*

The proof follows by the Theorem 14.2 and Proposition 14.6. ■

Exercises

1. Let $\mathscr{C}$ be a Grothendieck category, U a generator of $\mathscr{C}$. Using the notations of the Theorem 14.2 and Exercise 3, Section 3.7, Chapter 3, we have that a right ideal $\mathfrak{a}$ of A is in F if and only if $\mathfrak{a}(U) = U$. Also any $\mathfrak{a} \in F$ is essential in A and for any $x \in A$ such that $x\mathfrak{a} = 0$ we deduce that $x = 0$.

2. If A is a *principal ideal domain* (that is an integral domain in which any ideal is principal), then A is a proper generator of Mod A.

3. Let k be a field and $A = k[X, Y]$. Prove that A is not a proper generator of Mod A.

4. Let A be a *right hereditary ring* (that is a ring in which any right ideal is projective). Then A is the endomorphism ring of a proper generator of a Grothendieck category if and only if $\mathfrak{a} \in G'$ and $\mathfrak{a}' \supseteq \mathfrak{a}$, imply $\mathfrak{a}' \in G'$. (Use the notations of Theorem 14.8).

5. Let $\mathscr{Ab}_p$ be the full subcategory of $\mathscr{Ab}$ consisting of all abelian groups X such that Ann(x) is a power of p for any $x \in X$, p being a prime number (an object in $\mathscr{Ab}_p$ is called a p-group). Let G be an object of $\mathscr{Ab}_p$ with maximum divisible subgroup D (see Section 4.8). If G/D is unbounded, then G is flat as a module over its endomorphism ring. (Hint: $\mathscr{Ab}_p$ is a Grothendieck category, and G/D is a generator of it. Then we may utilize Theorem 14.2.)

6. Let U be a generator in the category $\mathscr{Ab}_p$. If U is a direct sum of cyclic groups, then U is a proper generator.

7. Let $\mathscr{C}$ be a spectral Grothendieck category and U a generator of $\mathscr{C}$. Then $\mathscr{C}$ is equivalent with $\mathrm{Mod}(\mathrm{End}_{\mathscr{C}}(U))/\bar{\mathscr{E}}$, $\bar{E}$ being the Goldie system.

8. Let $\mathscr{M}$ be a class of objects of a Grothendieck category $\mathscr{C}$. Prove that there exists a localizing subcategory $\mathscr{F}$ of $\mathscr{C}$ such that any object in $\mathscr{M}$ is $\mathscr{F}$-closed and $\mathscr{F}$ is strongest with this property. (Hint: Use Theorem 14.2 and Proposition 14.1).

References

Bucur–Deleanu [31]; Cartan–Eilenberg [33]; Freyd [57]; Gabriel [59]; Gabriel–Popescu [60]; Hacque–Maury [85]; Harada [86]; Lambek [113], [114]; Popescu [167]; Roos [199]; Roux [200]; Stenström [213]; Walker–Walker [235], [236].

4.15 Finiteness conditions on localizing systems

In this paragraph we will consider two kinds of finiteness conditions on the localizing system $F \in L(A)$.

THEOREM 15.1. *Let A be a ring and $F \in L(A)$. The following assertions are equivalent:*

(1) *If $\mathfrak{a}_1 \subseteq \mathfrak{a}_2 \subseteq \ldots \subseteq \mathfrak{a}_n \subseteq \ldots$ is a countable ascending chain of left ideals of A such that*

$$\bigcup_n \mathfrak{a}_n \in F,$$

then $\mathfrak{a}_n \in F$ for some n.

(2) *Every direct sum of F-closed modules is F-closed;*

(2′) *Every direct sum of countably many F-closed modules is F-closed;*

(3) *The functor $S: \text{Mod } A/\mathscr{F} \to \text{Mod } A$ commutes with direct sums;*

(3′) *The localizing functor commutes with direct sums.*

Proof. (1) ⇒ (2). Let $\{X_i\}_{i\in I}$ be a family of F-closed modules. Then $\coprod_{i\in I} X_i$ is F-torsion-free and we must show that it is also F-closed. Let $f: \mathfrak{a} \to \coprod_{i\in I} X_i$ be any morphism with $\mathfrak{a} \in F$. Considering $\coprod_{i\in I} X_i$ as a submodule of the F-closed module $\prod_{i\in I} X_i$, there exists $x = (x_i) \in \prod_{i\in I} X_i$ such that $f(a) = ax$ for all $a \in \mathfrak{a}$. We only have to show that $x \in \coprod X_i$. If this were not true, there would exist an infinite set $i_1, \ldots i_n, \ldots$ of indices i for which $x_{i_n} \neq 0$. Put $\mathfrak{a}_n = \{a \in \mathfrak{a} \mid ax_{i_m} = 0 \text{ for } m \geqslant n\}$. Since $f(\mathfrak{a}) \subseteq \coprod_{i\in I} X_i$, we have $\mathfrak{a} = \bigcup_n \mathfrak{a}_n$. But (1) then implies that $\mathfrak{a} = \mathfrak{a}_n$ for some n. Now X_{i_n} has no non-zero sub-objects in $\mathscr{F}$ so $\mathfrak{a}x_{i_n} = 0$ implies $x_{i_n} = 0$, which is a contradiction.

(2) ⇒ (3). Let $\{X_i\}_{i\in I}$ be a family of objects in $\text{Mod } A/\mathscr{F}$. The direct sum of the objects X_i is $\coprod X_i = T(\coprod_{i\in I} S(X_i))$. So if $\coprod_{i\in I} S(X_i)$ is F-closed, then $S(\coprod_{i\in I} X_i) = \coprod_{i\in I} S(X_i)$.

The implications (3) ⇒ (3′) ⇒ (2) ⇒ (2′) are rather obvious.

(2′) ⇒ (1). Let $\mathfrak{a}_1 \subseteq \mathfrak{a}_2 \subseteq \ldots$ be an ascending chain with

$$\mathfrak{a} = \bigcup_n \mathfrak{a}_n \in F.$$

There is a well-defined canonical morphism $\mathfrak{a} \xrightarrow{g} \coprod_n A/\mathfrak{a}_n$. Since $\mathfrak{a} \in F$, one obtains a commutative diagram

$$\begin{array}{ccc} \mathfrak{a} & \longrightarrow & A \\ {\scriptstyle g}\downarrow & & \downarrow{\scriptstyle f} \\ \coprod_n A/\mathfrak{a}_n & \longrightarrow & \coprod_n (A/\mathfrak{a}_n)_F \end{array}$$

where $f(a) = ax$ for some $x = (x_n) \in \coprod_n (A/\mathfrak{a}_n)_F$. There exists m such that $x_n = 0$, for $n \geqslant m$. The image of $\mathfrak{a}$ in $A/\mathfrak{a}_m$ then lies in the kernel of the canonical morphism $A/\mathfrak{a}_m \to (A/\mathfrak{a}_m)_F$, so $\mathfrak{a}/\mathfrak{a}_m$ is an object of $\mathscr{F}$. The exact sequence

$$0 \to \mathfrak{a}/\mathfrak{a}_m \to A/\mathfrak{a}_m \to A/\mathfrak{a} \to 0$$

where also $A/\mathfrak{a} \in \operatorname{Ob}\mathscr{F}$ shows that $A/\mathfrak{a}_m \in \operatorname{Ob}\mathscr{F}$ and hence $\mathfrak{a}_m \in F$. ∎

We strengthen the finiteness condition somewhat by considering the finitely generated systems.

THEOREM 15.2. *The following assertions are equivalent for a system* $F \in L(A)$.

(1) *F contains a coinitial family of finitely generated left ideals, i.e. F is finitely generated*;

(2) *If* $\{X_i\}_i$ *is a complete direct set of F-closed submodules of X, then X also is F-closed*;

(3) *If* $\{X_i\}_i$ *is a direct complete set of subobjects of an object X in* Mod $A/\mathscr{F}$, *then* $\sum_i S(X_i) = S(X)$, *that is the direct set* $\{S(X_i)\}_i$ *is also complete*;

(4) *If* $(X_i, f_{ij})_{i \in I}$ *is a direct system in* Mod A, *then*

$$(\varinjlim_i X_i)_{\mathscr{F}} = \varinjlim_i (X_i)_{\mathscr{F}}$$

(*i.e. the canonical functor* $H : \text{Mod}\, A \to \mathscr{F}$ *commutes with direct limits*).

Proof. (1) ⇒ (2). Let $\{X_i\}_i$ be a direct complete set of F-closed subobjects of X. Consider any morphism $f : \mathfrak{a} \to X$ where $\mathfrak{a} \in F$. $\mathfrak{a}$ contains a finitely generated left ideal $\mathfrak{a}' \in F$. f maps $\mathfrak{a}'$ into some X_i, so by hypothesis, there exists $x \in X_i$ such that $f(a') = a'x$ for any $a' \in \mathfrak{a}'$. Since X is $\mathscr{F}$-torsionfree one then has $f(a) = ax$ also for all $a \in \mathfrak{a}'$. The fact (2) ⇒ (3) is proved similarly to the preceding theorem.

(2) $\Rightarrow$ (1). Let $\{\mathfrak{a}_i\}_i$ the direct complete system of finitely generated left ideals contained in an ideal $\mathfrak{a} \in F$. Then $ST(A) = ST(\mathfrak{a}) = ST(\sum_i \mathfrak{a}_i) = \sum_i ST(\mathfrak{a}_i)$, and so the canonical morphism $u : A \to A_F$ factors as

$$\begin{array}{ccc} A & \xrightarrow{\;u\;} & A_F = ST(A) \\ {\scriptstyle f}\downarrow & & \downarrow{\scriptstyle s} \\ ST(\mathfrak{a}_j) & \xrightarrow{\;t\;} & \sum_i ST(\mathfrak{a}_i) \end{array}$$

where t and s are canonical. Let $g : ST(\mathfrak{a}_j) \to ST(A) = A_F$ be the canonical morphism. We want to show that g is an isomorphism, because this will imply $\mathfrak{a}_j \in F$. Since $sg = t$, g is obviously a monomorphism. We have $sgf = tf = su$, so $gf = u$. It follows that $\operatorname{im} g$ is an F-closed submodule of A_F containing $\operatorname{im} u$, and we conclude that g is an epimorphism.

(1) $\Rightarrow$ (4). Let $\{X_i, f_{ij}\}$ be a direct system of modules. The inclusions $(X_i)_{\mathscr{F}} \to X_i$ induce in the direct limit an inclusion

$$\varinjlim_i (X_i)_{\mathscr{F}} \to \varinjlim_i X_i.$$

The subcategory $\mathscr{F}$ is closed under inductive limits, since it is closed under direct sums and quotients,

$$\varinjlim_i (X_i)_{\mathscr{F}}$$

is therefore a submodule of

$$(\varinjlim_i X_i)_{\mathscr{F}}.$$

To show that we actually have equality, suppose $x \in (\varinjlim X_i)_{\mathscr{F}}$. Then $\mathfrak{a}x = 0$, for some finitely generated $\mathfrak{a} \in F$. Since $\mathfrak{a}$ is finitely generated, it is clear that we may represent x by some $x_i \in X_i$, such that still $\mathfrak{a}x_i = 0$. Then $x_i \in (X_i)_{\mathscr{F}}$ and

$$x \in \varinjlim_i (X_i)_{\mathscr{F}}.$$

(4) $\Rightarrow$ (1). Write $\mathfrak{a} \in F$ as a sum of a direct system $\{\mathfrak{a}_i\}_i$ of finitely generated left ideals. Then $A/\mathfrak{a} = \varinjlim A/\mathfrak{a}_i \in \mathrm{Ob}\mathscr{F}$ so $A/\mathfrak{a} = (A/\mathfrak{a})_{\mathscr{F}} = \varinjlim \big(A(\mathfrak{a}_i)_{\mathscr{F}}\big)$. In particular, the element $\bar{1} \in A/\mathfrak{a}$ comes from some $(A/\mathfrak{a}_i)_{\mathscr{F}}$ which means that there exist $a \in A$ and $\mathfrak{a}' \in F$ such that $\mathfrak{a}'a \subseteq \mathfrak{a}_i$ and $1 - a \in \mathfrak{a}$. We may choose j such that $1 - a \in \mathfrak{a}_j$, and then $\mathfrak{a}' \subseteq \mathfrak{a}_j$. Hence $\mathfrak{a}_j \in F$. ∎

It is clear that any finitely generated system satisfies Theorem 15.1. The converse holds e.g. when all left ideals in $\mathfrak{a}$ are countably generated.

COROLLARY 15.3. *The following assertions are equivalent for the Goldie system* $\bar{E}$.

(1) *Every direct sum of non-singular injective modules is injective*;

(2) $\bar{E}$ *is finitely generated.*

Proof. The implication (2) ⇒ (1) results by Theorem 15.1 and Proposition 13.9. The implication (1) ⇒ (2) results by Lemma 13.8 and Theorem 15.2(c). ■

Exercises

1. Show that the conditions of Corollary 15.3 are equivalent to:

Every non-singular module contains a unique maximal injective submodule.

2. Let A be a commutative ring and $F \in L(A)$ a system as in Theorem 15.1. Then $F = \bigcap_{\mathfrak{p}} F_{\mathfrak{p}}$, where $\mathfrak{p}$ runs over the set H of all prime ideals $\mathfrak{p}$ of A which are not in F. Moreover, a subset H' of H can be chosen such that $F = \bigcap_{\mathfrak{p} \in H'} F_{\mathfrak{p}}$ and this intersection is reduced.

3. Show that the following properties of a system $F \in L(A)$ are equivalent:

(a) If $\mathfrak{a}_1 \subseteq \mathfrak{a}_2 \subseteq \ldots \subseteq \mathfrak{a}_n \subseteq \ldots$ is a countable ascending chain of left ideals such that

$$\bigcup_n \mathfrak{a}_n \in F,$$

then some $\mathfrak{a}_n \in F$;

(b) If $\mathfrak{a}_1 \subseteq \ldots \subseteq \mathfrak{a}_n \subseteq \ldots$ is a countable ascending chain of left F-saturated ideals, then

$$\bigcup_n \mathfrak{a}_n$$

is also F-saturated.

References

Gabriel [59]; Goldman [76]; Roos [199]; Stenström [213]; Teply [222], [223], [224]; Walker–Walker [236].

4.16 Flat epimorphisms of rings

The (left) flat epimorphisms of rings are special cases of the localization ring of a ring and are used to calculate the quotient category Mod $A/\mathscr{A}$ for a suitable localizing subcategory $\mathscr{A}$. First we make some useful observations.

Let $u : A \to B$ be a ring morphism. We recall that u is an *epimorphism of rings* (that is, an epimorphism in the category of rings) if for any ring C and any two ring morphisms $t, s : B \to C$, the equality $tu = su$ implies $t = s$. More generally, we say that $b \in B$ is *dominated* by u if $su = tu$ always implies $t(b) = s(b)$. The set of elements of B dominated by u is a subring of B, called the *dominion* of u. Then u is an epimorphism if and only if its dominion is equal to B.

LEMMA 16.1. *Let $u : A \to B$ be a ring morphism and M a (B, B^0)-bimodule. If $x \in M$ has the property that $u(a)x = xu(a)$ for all $a \in A$, then $bx = xb$ for all b in the dominion of u.*

Proof. We make $B \amalg M$ into a ring by defining

$$(b, y) + (b', y') = (b + b', y + y')$$
$$(b, y)(b', y') = (bb', by' + yb').$$

The ring axioms are easily verified; in particular $B \amalg M$ has a unit-element, namely $(1, 0)$. Let us define two maps $s, t : B \to B \amalg M$, as $s(b) = (b, 0)$ and $t(b) = (b, bx - xb)$. Both s and t are ring morphisms and $su = tu$. So if b is dominated by u, then $s(b) = t(b)$, and thus $bx = xb$. ∎

PROPOSITION 16.2. *A ring morphism $u : A \to B$ dominates $b \in B$ if and only if $b \otimes 1 = 1 \otimes b$ in $B \otimes_A B$.*

Proof. Suppose $b \otimes 1 = 1 \otimes b$ and let s, $t : B \to C$ be morphisms of rings such that $su = tu$. Define a morphism of (A, A^0)-bimodules $h : B \otimes_A B \to C$ setting $h(b \otimes b') = s(b)t(b')$ and extending by linearity. Then the equality $b \otimes 1 = 1 \otimes b$ implies that $s(b) = h(b \otimes 1) = h(1 \otimes b) = t(b)$.

Assume on the other hand that u dominates b. The assertion now follows by applying the Lemma 16.1 to $x = 1 \otimes 1 \in B \otimes_A B$. ∎

We will now elaborate a consequence of Proposition 16.2.

PROPOSITION 16.3. *The following properties of a ring morphism $u : A \to B$ are equivalent:*

(a) *u is an epimorphism of rings;*

(b) *The canonical map $m : B \otimes_A B \to B$, with $m(b \otimes b') = bb'$ is bijective;*

(c) *The functor $u_* :$ Mod $B \to$ Mod A is full.*

Proof. (a) ⇒ (c). Suppose that M and N are B-modules and

$$f : u_*(M) \to u_*(N)$$

is a morphism of A-modules. For each $x \in M$, consider the map

$$g : B \otimes_A B \to N$$

given by $g(b \otimes b') = bf(b'x)$. Note that this is a well defined map. Since $1 \otimes b = b \otimes 1$, by Proposition 16.2, we have $f(bx) = bf(x)$ and therefore f is a morphism of B-modules.

(c) ⇒ (b). If $v : u^* u_* \to \operatorname{Id} \operatorname{Mod} B$ is a morphism of adjunction one sees, that $v_B = m$. Then we can apply the Theorem 5.2 of Chapter 1.

The implication (b) ⇒ (a) follows from Proposition 16.2. ■

For the proof of the next theorem we need the following result.

LEMMA 16.4. *Let A be a ring, X an A-module and Y an A^0-module. The following assertions are equivalent*:

(a) $(yA) \otimes_A X = 0$ *for every* $y \in Y$.

(b) *For every* $x \in X$, *and* $y_1, \dots y_n \in Y$, *there exist* $x_1, \dots x_m \in X$ *and*

$$a_1, \dots, a_m \in A$$

such that

$$x = \sum_{i=1}^{m} a_i x_i \quad \text{and} \quad y_j a_i = 0$$

for all i and j;

(c) *For every* $x \in X$ *and* $y \in Y$, *there exist* $x_1, \dots, x_n \in X$, *and* $a_1, \dots, a_m \in A$ *such that*

$$x = \sum_{i=1}^{m} a_i x_i \quad \text{and} \quad y a_j = 0$$

for all $1 \leq j \leq m$.

Proof. The equivalence (a) ⇔ (c) results by Lemma 8.2 of Chapter 3. Furthermore (a) implies that $Y' \otimes_A X = 0$ for every submodule Y' of Y, and one easily proves that this implies $Z \otimes_A X = 0$ for every submodule Z of a direct sum Y^n of copies of Y. (b) is then obtained from (a) by considering $(y_1, \dots, y_n)$ as an element of Y^n. The remaining assertion (b) ⇒ (c) is trivial. ■

A ring morphism $u : A \to B$ is called *left* (*respectively, right*) *flat epimorphism* if u is an epimorphism and B is a right (respectively, left) flat A-module.

As a consequence of Theorem 2.8 of Chapter 3 we have the following result:

COROLLARY 16.5. *Let $u: A \to B$ be a ring morphism. Then u is a left flat epimorphism if and only if u is an epimorphism and for any injective B-module Q, it follows that $u^*(Q)$ is also injective in* Mod A. ■

THEOREM 16.6. *Let $u: A \to B$ be a ring morphism. The following assertions are equivalent:*

(a) *u is a left flat epimorphism;*

(b) *The family F_u of left ideals $\mathfrak{a}$ of A such that $Bu(\mathfrak{a}) = B$ is a localizing system on A, and there exists a ring isomorphism $t: B \to A_{F_u}$ such that $tu = u_A$ ($u_A: A \to A_{F_u}$ is canonical);*

(c) *The following two conditions are satisfied:*

(i) *for every $b \in B$ there exist $s_1, \ldots, s_n \in A$ and $b_1 \ldots, b_n \in B$ such that $u(s_i)\, b \in u(A)$ for any $1 \leqslant i \leqslant n$ and*

$$\sum_{i=1}^{n} b_i\, u(s_i) = 1;$$

(ii) *if $u(a) = 0$, then there exist $s_1, \ldots, s_n \in A$, and $b_1, \ldots, b_n \in B$ such that $s_i\, a = 0$ for any $1 \leqslant i \leqslant n$ and $\Sigma b_i\, u(s_i) = 1$.*

Proof. (a) ⇒ (b). The restriction functor u_*: Mod B → Mod A makes Mod B equivalent to a full subcategory of Mod A, by Proposition 16.3. The functor u_* has a left adjoint $u^* = B \otimes_A ?$ which is exact, since B is a flat A^0-module. Then by Theorem 4.9, Mod B is equivalent with the category Mod $A/\ker u^*$. Now we calculate $\ker u^*$. An A-module X is in $\ker u^*$ if and only if

$$u^*(X) = B \otimes_A X = 0.$$

Let F_u be the localizing system associated with $\ker u^*$. A left ideal $\mathfrak{a}$ is a member of F_u if and only if $B \otimes_A (A/\mathfrak{a}) = 0$, that is, if and only if the morphism $B \otimes_A \mathfrak{a} \to B \otimes_A A = B$ defined by the natural inclusion $\mathfrak{a} \to A$ is an isomorphism. In conclusion $\mathfrak{a} \in F_u$ if and only if $Bu(\mathfrak{a}) = B$.

The ring A_{F_u} is now identified with the ring $(\mathrm{Hom}_B(B, B))^0$, that is, with the ring B itself.

We leave it to the reader to fill in the details.

(b) ⇒ (c). If $b \in B = A_{F_u}$, there exists $\mathfrak{a} \in F_u$ such that $\mathfrak{a}b \subseteq u(A)$ (see Lemma 12.1). But $Bu(A) = B$, and so the elements

$$s_1, \ldots, s_n \in \mathfrak{a} \quad \text{and} \quad b_1, \ldots, b_n \in B,$$

can be found such that $\sum_i b_i u(s_i) = 1$; obviously $u(s_i) b \in u(A)$ for $i = 1, \ldots, n$, and (i) is verified.

If $u(a) = 0$, then $a \in \ker u_A$ so there exists $\mathfrak{a} \in F_u$ such that $\mathfrak{a}a = 0$. As before we choose $s_1, \ldots, s_n \in \mathfrak{a}$ such that $\sum_i b_i u(s_i) = 1$, for suitable $b_1, \ldots, b_n$ in B. Clearly, thus (ii) is also verified.

(c) $\Rightarrow$ (a). From Exercise 2 we have that $B \otimes_A X = 0$ for any A-submodule of $B/u(A)$ and thus u is an epimorphism.

It remains to see that B is flat as an A^0-module. For this, we use Theorem 8.3 of Chapter 3. Suppose we have $s_1, \ldots, s_n \in A$ and $b_1, \ldots, b_n \in B$ such that

$$\sum_{i=1}^{n} b_i u(s_i) = 0.$$

Applying Lemma 16.4 to $1 \in B$ and $\bar{b}_1, \ldots, \bar{b}_n$, the images of $b_1 \ldots, b_n$ in $B/u(A)$, we obtain $a_1, \ldots, a_m \in A$ and $b_1', \ldots, b_m' \in B$ such that

$$1 = \sum_{j=1}^{m} b_j' u(a_j),$$

$u(a_j) b_i = u(c_{ji})$ for some $c_{ji} \in A$.

Then

$$\sum_{i=1}^{n} u(a_j) b_i u(s_i) = \sum_{i=1}^{n} u(c_{ji} s_i) = 0$$

and so

$$\sum_{i=1}^{n}$$

$c_{ji}s_i \in \ker u$ for any $1 \leqslant j \leqslant m$. Now we use the condition (ii) and for each j, we obtain elements $t_{j1}, \ldots, t_{jr} \in A$ and $b''_{j1}, \ldots, b''_{jr} \in B$ such that

$$\sum_{i=1}^{n} t_{jk} c_{ji} s_i = 0 \quad \text{for any} \quad j \quad \text{and any} \quad k,$$

$$\sum_{k=1}^{r} b''_{jk} u(t_{jk}) = 1 \quad \text{for any} \quad j.$$

Then we have:

$$b_i = \sum_{j=1}^{m} b_j' u(a_j) b_i = \sum_{j=1}^{m} b_j' u(c_{ji}) = \sum_{j,k} b_j' b''_{jk} u(t_{jk} c_{ji})$$

and

$$\sum_{i=1}^{n} t_{jk} c_{ji} s_i = 0,$$

for any j and k, i.e. the given relation comes from a relation in A, as was to be shown. ■

NOTES. 1. Note that condition (c) of the theorem implies in particular that each $b \in B$ may be written as

$$b = \sum_{i=1}^{n} b_i\, u(s_i)\, b = \sum_{i=1}^{n} b_i\, u(a_i),$$

with

$$\sum_{i=1}^{n} b_i\, u(s_i) = 1$$

and $u(s_i)\, b = u(a_i)$.

2. As was noted in the proof of (c) ⇒ (a), the condition (i) of (c) may be restated as:

(i′) For every family $b_1, \ldots, b_n \in B$, there exist $s_1, \ldots, s_m \in A$ and

$$b_1', \ldots, b_m' \in B$$

such that $u(s_i)\, b_j \in u(A)$ for any $1 \leqslant i \leqslant n$, $1 \leqslant j \leqslant m$ and

$$\sum_{i=1}^{n} b_i'\, u(s_i) = 1.$$

3. The condition (ii) of (c) may be formulated as:

(ii′) For any $a \in \ker u$, $B \otimes_A (Aa) = 0$.

Now let $F \in L(A)$ and let $u : A \to A_F$ be the canonical ring morphism. Then by Proposition 9.8 we deduce that $F \supseteq u^{-1}(u(F))$. A very interesting case is when we have $F = u^{-1}(u(F))$.

THEOREM 16.7. (Walker–Walker [236]). *Let A be a ring $F \in L(A)$. With the usual notations, the following assertions are equivalent*:

(1) *For any* $\mathfrak{a} \in F$ *we have* $A_F\, u(\mathfrak{a}) = A_F$;

(2) $u(F) = \{A_F\}$ *or equivalently* $u(\mathscr{F})$ *is null*;

(3) *F is a finitely generated system and the localization functor* (*associated with F*) *is exact*;

(4) *For any A-module X, the composition of morphism*

$$\theta_X : A_F \otimes_A X \xrightarrow{A_F \otimes u_X} A_F \otimes_A X_F \xrightarrow{m_{X_F}} X_F$$

(*where* $m = m_{X_F}$ *is defined by* $m(b \otimes x) = bx$) *is an isomorphism of A-modules*;

(5) *The functor* $T' : \mathrm{Mod}\, A_F \to \mathrm{Mod}\, A/\mathscr{F}$, $T' = Tu_*$ *is an equivalence of categories.*

Proof. (1) ⇒ (2). Let F' be the set of left ideals $\mathfrak{a}$ of A such that $A_F\, u(\mathfrak{a}) = A_F$. It is clear that $\mathfrak{a} \in F'$ if and only if $A_F \otimes_A (A/\mathfrak{a}) = 0$. From Exercise 8 of

Section 4.12 we deduce that $F' \subseteq F$ and by hypothesis we see that $F' = F$. According to Theorem 16.6 we deduce that the morphism u is a left flat epimorphism of rings and u_* is full and u^* exact. Then an A_F-module X is in $u(F)$ if and only if $u_*(X) \in \mathrm{Ob}\,\mathscr{F}$, that is necessarily $X = 0$.

The implications (1) ⇒ (3), (1) ⇒ (5) and (5) ⇒ (2) are immediate consequences of the Theorem 16.6. Also the implication (5) ⇒ (4) is true because the functor $u_* u^*$ is canonically isomorphic with the localization functor.

(4) ⇒ (1). Let $\mathfrak{a} \in F$; from the exact sequence

$$0 \longrightarrow \mathfrak{a} \xrightarrow{\;i\;} A \longrightarrow A/\mathfrak{a} \longrightarrow 0$$

we may deduce the canonically constructed commutative diagram:

$$\begin{array}{ccccccc}
A_F \otimes_A \mathfrak{a} & \xrightarrow{A_F \otimes i} & A_F \otimes_A A & \longrightarrow & A_F \otimes_A (A/\mathfrak{a}) & \longrightarrow & 0 \\
\downarrow{\scriptstyle \theta_{\mathfrak{a}}} & & \downarrow{\scriptstyle \theta_A} & & \downarrow{\scriptstyle \theta_{A/\mathfrak{a}}} & & \\
\mathfrak{a}_F & \xrightarrow{\quad i_F \quad} & A_F & \longrightarrow & 0 & &
\end{array}$$

in which necessarily $A_F \otimes i$ is an isomorphism. Then necessarily

$$A_F \otimes_A (A/\mathfrak{a}) = 0.$$

(3) ⇒ (5). By Theorem 15.2 the functor S commutes with direct sums. But the functor S is also exact; then we deduce that the localization functor commutes with inductive limits. Also, the canonical functor

$$S : \mathrm{Mod}\, A/\mathscr{F} \longrightarrow \mathrm{Mod}\, A$$

is exact (see Section 4.4) and commutes with inductive limits.

Now, by adjunction, we deduce that the functor $\mathrm{Hom}_{\mathrm{Mod}A/\mathscr{F}}(T(A), ?)$ is canonically isomorphic with the functor $\mathrm{Hom}_A(A, S(?))$ and S being exact, we check that $T(A)$ is a projective generator of $\mathrm{Mod}\, A/\mathscr{F}$. Moreover, $T(A)$ is also small, by the above considerations. In conclusion we deduce by Chapter 3, Section 3.7, Exercise 4, that the functor $S' : \mathrm{Mod}\, A/\mathscr{F} \to \mathrm{Mod}\, A_F$ is an equivalence of categories.

The remaining implication (2) ⇒ (5) results by Theorem 4.9, observing that $\ker T' = 0$. ∎

A localizing system F on A is called *perfect* if it has the properties listed in Theorem 16.7.

COROLLARY 16.8. *Let F be a perfect system on A. Then A_F is flat as A^0-module and $u : A \to A_F$ is a ring epimorphism in other words, u is a left flat epimorphism of rings.* ∎

We shall now give a class of rings in which certain systems are perfect. It would be interesting to know whether for an arbitrary ring there exist non-trivial perfect systems, or, at least, to obtain sufficient conditions to this effect.

PROPOSITION 16.9. *If A is a left hereditary ring then the functor*

$$S : \mathrm{Mod}\, A/\mathscr{F} \to \mathrm{Mod}\, A$$

is exact for all localizing systems F on A (i.e. any system on A is exact).

Proof. Let

$$0 \longrightarrow X \xrightarrow{f} Y \xrightarrow{g} Z \longrightarrow 0$$

be an exact sequence of $\mathrm{Mod}\, A/\mathscr{F}$. Thus we have in $\mathrm{Mod}\, A$ the following commutative diagram canonically constructed:

$$\begin{array}{ccccccccc} 0 & \longrightarrow & S(X) & \xrightarrow{S(f)} & S(Y) & \xrightarrow{p} & \mathrm{Coker}\, S(f) & \longrightarrow & 0 \\ & & & & {\scriptstyle S(g)}\searrow & & \swarrow{\scriptstyle u} & & \\ & & & & & S(Z) & & & \end{array}$$

in which u is a monomorphism, so that coker $S(f)$ is F-torsionfree. We shall prove that coker $S(f)$ is F-closed or equivalently u is an isomorphism (because coker $u \in \mathrm{Ob}\, \mathscr{F}$). Indeed, let $\mathfrak{a} \in F$ and $t : \mathfrak{a} \to$ coker $S(f)$ be a morphism of A-modules. By hypothesis $\mathfrak{a}$ is projective, so that there exists a morphism $s : \mathfrak{a} \to S(Y)$ with $ps = t$. Since $S(Y)$ is F-closed, we may extend s to

$$s' : A \to S(Y).$$

The composition ps' then extends t uniquely, and so coker $S(f)$ is F-closed. ∎

COROLLARY 16.10. *If A is left semi-hereditary and F finitely generated then F is perfect.*

Proof. As in the proof of above proposition we can check that F is exact. Now we can utilize Theorem 16.7. ∎

NOTES 4. The conclusion of Corollary 16.10 remains valid for a system F in a ring having a coinitial set of projective left ideals.

5. If A is left hereditary noetherian ring then every system is perfect.

LEMMA 16.11. *Let $\mathfrak{a}$, $\mathfrak{a}'$ be two left ideals of a ring A such that $\mathfrak{a} + \mathfrak{a}' = A$ and $\mathfrak{a} \cap \mathfrak{a}' = 0$. If A is reduced (that is any nilpotent element of A is zero), then $\mathfrak{a}\,\mathfrak{a}' = \mathfrak{a}'\,\mathfrak{a} = 0$. Moreover $\mathfrak{a}$ and $\mathfrak{a}'$ are two-sided ideals.*

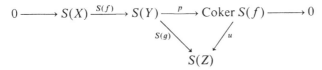

Proof. Let $e \in \mathfrak{a}$, $e' \in \mathfrak{a}'$ be such that $e + e' = 1$. Then $ee' = e'e = 0$ because $ee' \in \mathfrak{a} \cap \mathfrak{a}'$. If $x \in \mathfrak{a}$, then $x = ye$, and $xe' = yee' = 0$. Now

$$e'x = e'ye = x - ex \in \mathfrak{a},$$

that is $e'xe' = 0$ and thus $(e'x)^2 = 0$. But then $e'x = 0$, A being reduced. Analogously, we deduce that for any $x' \in \mathfrak{a}'$ we have that $x'e = ex' = 0$. In conclusion for any $x \in \mathfrak{a}$, $x' \in \mathfrak{a}'$ we have $xx' = x'x = 0$. Particularly, we check that $\mathfrak{a}$ and $\mathfrak{a}'$ are two-sided ideals. ■

THEOREM 16.12. *If A is a reduced ring, then any simple projective A-module is also injective.*

Proof. Let S be a simple projective A-module. Then $A = S + \mathfrak{m}$ and $S \cap \mathfrak{m} = 0$, $\mathfrak{m}$ being a maximal left ideal of A. By Lemma 16.11 we see that $\mathfrak{m}$ is a two-sided ideal, $A/\mathfrak{m}$ is a ring and the canonical morphism $u : A \to A/\mathfrak{m}$ is a left flat epimorphism of rings. By the Theorem 16.6, $A/\mathfrak{m} = A_{F_u}$. Now we shall calculate the system F_u. A left ideal $\mathfrak{a}$ is an element of F_u if and only if $(A/\mathfrak{m})\,u(\mathfrak{a}) = A/\mathfrak{m}$, that is if and only if $\mathfrak{a} \not\subseteq \mathfrak{m}$, $A/\mathfrak{m}$ being a division ring. But it is clear that any left essential ideal of A is an element of F_u. To complete the proof, we apply the definition of an F_u-closed module and Lemma 13.8. ■

COROLLARY 16.13. *Let A be a commutative ring. Then any projective simple module is also injective. In particular, if A is a regular commutative ring, then any minimal ideal is injective.*

Proof. We see that Lemma 16.11 is valid for any commutative ring. Now, we can apply Theorem 16.12. If A is regular, any minimal ideal is finitely generated, and so, projective (see Chapter 3, Section 3.8). ■

NOTES 6. The last part of Corollary 16.13 is valid for any reduced regular ring.

COROLLARY 16.14. *Let A be a reduced ring. The following assertions are equivalent:*

(a) *Any left minimal ideal of A is projective;*

(b) *Any left minimal ideal of A is injective.* ■

NOTES 7. A localizing system $F \in L(A)$ is called *exact* if the localization functor associated with F is exact. Proposition 16.9 and Exercise 9 give examples of exact systems. Then condition (3) of Theorem 16.7 can be formulated as: (3′), F is finitely generated and exact.

Now we give an important example of a flat epimorphism of rings.

THEOREM 16.15. *Let $u: A \to B$ be a flat epimorphism of commutative rings and X an A-finitely generated A-module. Then the canonical morphism*:

$$v: \mathrm{End}_A(X) \to \mathrm{End}_B(B \otimes_A X)$$

with $v(f) = B \otimes f$ is a left and right flat epimorphism of rings.

Proof. For any $a \in A$ we denote by $m_a: X \to X$ the morphism $m_a(x) = ax$; then $v(m_a) = m_{u(a)}: B \otimes_A X \to B \otimes_A X$ is defined as

$$m_{u(a)}(b \otimes x) = u(a)\, b \otimes x = b \otimes ax.$$

Now let $f: B \otimes_A X \to B \otimes_A X$ be a morphism of B-modules, and $x_1, \ldots, x_n$ the generators of the A-module X; then $1 \otimes x_1, \ldots 1 \otimes x_n$ are generators of B-module $B \otimes_A X$ and so

$$f(1 \otimes x_i) = \sum_{j=1}^{n} b_{ij}(1 \otimes x_j).$$

Furthermore, let $b_1, \ldots, b_m \in B$ and $s_1, \ldots, s_m \in A$ such that

$$\sum_{k=1}^{m} u(s_k)\, b_k = 1$$

and $u(s_k)\, b_{ij} = u(a_{ijk})$ [see Theorem 16.6, (c)]. We see that

$$\sum_{k=1}^{m} m_{u(s_k)} m_{b_k} = 1,$$

the identity morphism of $B \otimes_A X$; moreover, $m_{u(s_k)} f \in \mathrm{im}\, v$. Indeed, for any $1 \leqslant i \leqslant n$, we have

$$(m_{u(s_k)} f)(1 \otimes x_i) =$$

$$u(s_k)\left(\sum_{j=1}^{n} b_{ij} \otimes x_j\right) = \sum_{j=1}^{n} u(a_{ijk}) \otimes x_j = \sum_{j=1}^{n} 1 \otimes a_{ijk} x_j \in t(X),$$

where $t: X \to B \otimes_A X$ is defined as $t(x) = 1 \otimes x$. In conclusion

$$m_{u(s_k)} f = v(m_{a_{ijk}}).$$

Now let $f \in \ker v$; then $B \otimes f = 0$, so for any $1 \leqslant i \leqslant n$, $1 \otimes f(x_i) = 0$, in other words $f(x_i) \in X_{\mathscr{F}}$ where F is the localizing system consisting of all ideals $\mathfrak{a}$ of A such that $Bu(\mathfrak{a}) = B$. Let $\mathfrak{a} \in F$ such that $\mathfrak{a} f(x_i) = 0$ for any i. Also let $s_1, \ldots, s_m \in \mathfrak{a}$, $b_1, \ldots, b_m \in B$ with

$$\sum_{j=1}^{m} b_j\, u(s_j) = 1.$$

Then

$$\sum_{j=1}^{m} m_{b_j} m_{u(s_j)} = \sum_{j=1}^{m} b_j\, v(m_{s_j}) = 1$$

and $m_{s_j} f = 0$ for any $1 \leqslant i \leqslant n$. The proof is now complete by Theorem 16.6. ■

Later, in Section 4.18 we shall see that the result is not valid in the non-commutative case.

NOTE 8. Theorem 16.12 is also a consequence of Theorem 16.6 and Theorem 2.8 of Chapter 3.

Exercises

1. Let $u: A \to B$ be an epimorphism of rings. If M is a B-module such that $u_*(M)$ is injective, then M is also injective.

2. Let $u: A \to B$ be a ring morphism. For any $b \in B$ denote by $\bar{b}$ the image of b in the A-module $B/u(A)$. The following assertions are equivalent:

(a) For any $b \in B$, $B \otimes_A (A\bar{b}) = 0$;

(b) For $b_1, \ldots, b_n \in B$, there exist $b_1', \ldots, b_m' \in B$, $a_1, \ldots, a_m \in A$ such that

(i) $\sum_{i=1}^{m} b_i'\, u(a_i) = 1$;

(ii) $u(a_i)\, b_j \in u(A)$, $\quad (1 \leqslant i \leqslant m,\ 1 \leqslant j \leqslant n)$.

Such a morphism is called *preflat*. Prove that any preflat morphism is an epimorphism.

3. The conditions of the Theorem 16.7 are also equivalent with the following conditions:

(i) The functor $S: \text{Mod } A/\mathscr{F} \to \text{Mod } A$ has a right adjoint;

(ii) The functor T is isomorphic with $T'\, u_*$;

(iii) For any $X \in \text{Ob Mod } A$ we have $X_{\mathscr{F}} = \ker t_X$, where $t_X : X \to A_F \otimes_A X$ is canonically defined: $t_X(x) = 1 \otimes x$;

(iv) For any A_F-module Y, the A-module $u^*(Y)$ is $\mathscr{F}$-torsionfree;

(v) An A-module X is in $\mathscr{F}$ if and only if $A_F \otimes_A X = 0$.

4. Let $u: A \to B$ be a morphism of commutative rings. Denote by F_u the set of ideals $\mathfrak{a}$ of A such that $Bu(\mathfrak{a}) = B$. Also denote by B_u the subset of B consisting of all elements b such that $B(u(A) : b) = B$, that is such that

$(u(A):b) \in F_u$. Prove that B_u is a subring of B containing $u(A)$, namely the localization of $u(A)$ relatively to F_u. Moreover u is preflat if and only if $B_u = B$, that is $u(A)_{F} \equiv B$. Particularly, if u is a monomorphism, then u is a flat epimorphism if and only if $B = B_u$.

5. If in the commutative diagram of rings

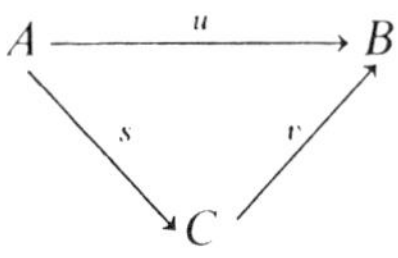

B is flat as A-module and s an epimorphism of rings, then B is also flat as C-module.

6. If $a: A \to B$ is an epimorphism of rings and A semisimple, then u is a surjection, that is, B is a quotient ring of A. Conversely, any quotient ring of a semisimple ring is flat as left (or right)-module on A. Moreover if A is a division ring, and $u: A \to B$ a ring epimorphism then u is an isomorphism.

7. Let A be a commutative ring and $\mathfrak{a}$ an ideal of A such that any finitely generated ideal contained in $\mathfrak{a}$ is generated by an idempotent. Then the canonical morphism $u: A \to A/\mathfrak{a}$ is a flat epimorphism of rings.

8. Prove that any commutative regular Noetherian ring is semisimple. (Hint: use Exercise 7, Exercise 10 of Section 3.11 and Theorem 8.6, both Chapter 3).

9. The system $F \in L(A)$ is perfect if and only if the following condition is fulfilled:

(P) For any $\mathfrak{a} \in F$, we can find an element $\mathfrak{a}' \in F$, elements $x_1, \ldots, x_n \in \mathfrak{a}$ and morphisms $f_1, \ldots, f_n \in \mathrm{Hom}_A(\mathfrak{a}', A/\mathfrak{r}_F)$ such that for any $x' \in \mathfrak{a}'$ we have:

$$\sum_{i=1}^{n} f_i(x')\, u'(x_i) = u'(x'),$$

$u': A \to A/\mathfrak{r}_F$ being canonical.

10. Let $u: A \to B$ be a flat epimorphism of commutative rings. Then the canonical morphism $u_n: A[X_1, \ldots, X_n] \to B[X_1, \ldots, X_n]$ induced by u is also a flat epimorphism of rings. Is the result valid for an infinite set of indeterminates?

11. If $u: A \to B$ is a left flat epimorphism of rings, then $\mathfrak{b} = Bu(u^{-1}(\mathfrak{b}))$ for any left ideal $\mathfrak{b}$ of B.

12. Let A be a ring, $\mathscr{C}$ an Ab 5-category, $T : \text{Mod } A^0 \to \mathscr{C}$ a covariant functor and S a full and faithful right adjoint of T. Prove that a ring B and a ring morphism $u : A \to B$ can be found such that:

(i) There exists a localizing subcategory $\mathscr{F}$ in $\text{Mod } B^0$ such that

$$\mathscr{C} = \text{Mod } B^0/\mathscr{F};$$

(ii) $u_* S' = S$, and $T' u^* = T$, where $T' : \text{Mod } B^0 \to \mathscr{C}$ is the canonical functor and S' a right adjoint of T'.

Give sufficient conditions such that u is an epimorphism of rings.

13. Let $u : A \to B$ be a ring morphism of commutative rings. Then u is an epimorphism if and only if for any $b \in B$ there exist elements $\{a_{ij}\}$, $1 \leqslant i \leqslant n$, $1 \leqslant j \leqslant m$ of A and elements $b_1, \ldots, b_m$ and $\beta_1 = b, \beta_2, \ldots, \beta_m$ of B such that:

(a) $\sum_{j=1}^{m} u(a_{ij})\, b_j = \begin{cases} 1 & \text{if } i = 1 \\ 0 & \text{if } i \neq 1 \end{cases}$ for any i,

(b) $\sum_{i=1}^{n} u(a_{ij})\, \beta_i \in u(A)$, for any j.

References

Bănică–Popescu [13]; Bkouche [18]; Elizarov [50]; Gabriel [59]; Goldman [76]; Hacque [82], [84]; Hacque–Maury [85]; Hudry [93]; Knight [107]; Lambek [114]; Lazard [116], [117]; Marot [122], [123]; Mazet [128]; Morita [138]; Năstăsescu–Popescu [144]; Olivier [156]; Popescu–Spircu [175], [176]; Popescu–Spulber [177]; Ribenboim [184]; Roby [188], [189]; Storrer [215], [216]; Walker–Walker [236].

4.17 Left quasi-orders of a ring

Let $u : A \to B$ be a left flat epimorphism of rings. We shall say that u defines A as a *left quasi-order* of B if $\ker u = 0$, that is if u is a left flat bimorphism of rings. Sometimes we shall say that a subring A of a ring B is a left quasi-order of B if the inclusion $A \to B$ is a left flat bimorphism of rings. If $u : A \to B$ defines A as a left quasi-order of B, then we shall say that B is a *left flat epimorphic extension* of A.

First we shall give the following result.

THEOREM 17.1. *For any ring A, there exists a universal left flat bimorphic extension i.e. a ring $M_l(A)$ (we shall write frequently $M(A)$) and a left flat* *epimorphism of rings* $\phi : A \to M(A)$ *such that*:

(i) ϕ *defines* A *as a left quasi-order of* $M(A)$;

(ii) *for every left flat bimorphism of rings* $v : A \to B$, *there exists a unique ring morphism* $\bar{v} : B \to M(A)$ *such that* $\bar{v}v = \phi$. *Moreover,* $\bar{v}$ *is also a monomorphism of rings.*

Proof. Every left flat epimorphism is obtained as the canonical homomorphism $u : A \to A_F$ for a perfect localizing system (Theorem 16.6). u is injective if and only if $\mathfrak{r}_F = 0$, i.e. if and only if $F \subseteq D$, the system of dense left ideals (see Section 4.9). Thus if $M(A)$ exists, it should be a subring of $Q(A)$ (see Section 4.13). This leads us to consider the family P of all subrings B of $Q(A)$ such that $A \subseteq B$ and the inclusion is a left flat epimorphism. Note that an inclusion of subrings in P corresponds to the inclusion of the corresponding perfect systems (see Section 4.12).

LEMMA 17.2. *The family* P *is directed under inclusion.*

Proof. We will show that if B and C are members of P, then the smallest subring H of $Q(A)$ containing both B and C is also a member of P. Every element of H is a sum of elements of the form

$$h = b_1 c_1 b_2 c_2 \ldots b_n c_n \quad \text{with} \quad b_i \in B,\ c_i \in C \tag{12}$$

and we may assume that each h appearing in a given sum has the same length n. We will verify the condition (i) of Theorem 16.6, (c), i.e. given $h_1, \ldots, h_r$ of length n, there exist $h_1' \ldots, h_m' \in H$ and $s_1, \ldots s_m \in A$ such that $s_k h_i \in A$ for all i, k and $\sum_k h_k' s_k = 1$.

(We identify A with a subring of $Q(A)$, in a canonical way). We do this by induction on the length n. For $n = 0$, i.e. $h = 1$, the condition is clearly satisfied. Suppose $n > 1$ and the condition has been verified for $n - 1$. To simplify the notation somewhat, we only consider the case $r = 1$; it is easy to see that our argument extends immediately to the case of a family

$$h_1, \ldots, h_r \in H.$$

So suppose we are given h of the form (12). By the induction hypothesis there exist $h_1', \ldots, h_m' \in H$ and $t_1, \ldots, t_m \in A$ such that $x_i = t_i b_1 c_1 \ldots b_{n-1} c_{n-1} \in A$ for all i, and

$$\sum_{i=1}^{m} h_i' t_i = 1.$$

By Note 2 of Section 4.16, applied to B, there exist $b_1', \ldots, b_q' \in B$ and $r_1, \ldots, r_q \in A$ such that:

$$x_{ki} = r_k x_i b_n \in A \quad \text{for all} \quad i, k, \quad \text{and} \quad \sum_{k=1}^{p} b_k' r_k = 1.$$

Similarly, there exist $c_1', \ldots, c_p' \in C$ and $s_1, \ldots, s_p$ in A such that

$$s_j x_{ki} c_n \in A, \quad \text{for all} \quad j, k, i \quad \text{and} \quad \sum_{j=1}^{p} c_j' s_j = 1.$$

Thus we have elements $h_i' b_k' c_j' \in H$ and $r_k s_j t_i \in A$ such that

$$s_j r_k t_i h \in A \quad \text{and} \quad \sum_{i,j,k} h_i' b_k' c_j' s_j r_k t_i = 1.$$

This completes proof of the lemma and we may continue the proof of the theorem. Define $M(A)$ as the union of all rings in P. It is obvious that $\phi : A \to M(A)$ is an epimorphism, and $M(A)$ is flat as A^0-module, since it is a direct limit of flat A^0-modules (see Chapter 3, Section 3.8).

Suppose $v : A \to B$ is any other injective left flat epimorphism. There is a corresponding perfect system F_v and as we have noticed above we have $F_v \subseteq D$, and hence a commutative diagram:

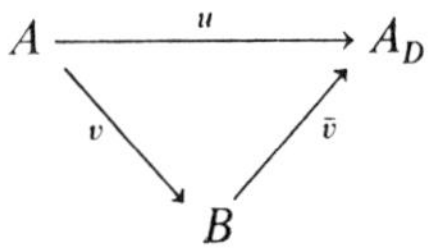

But also $\bar{v} : B \to Q(A)$ must be injective, because $\bar{v}(b) = 0$ would mean that there exists a morphism $f : \mathfrak{a} \to A$, with $\mathfrak{a} \in F_v$ such that $f(\mathfrak{a}') = 0$ for some $\mathfrak{a}' \subseteq \mathfrak{a}$ and $\mathfrak{a}' \in D$; for each $a \in \mathfrak{a}$ we have $(\mathfrak{a}' : a) \in D$, and thus it follows that $f = 0$, since A is D-torsion free. We conclude that the image of $\bar{v}$ lies within $M(A)$. Since $\bar{v}$ obviously is unique (v being an epimorphism) we have proved the theorem.

Alternative proof. For any ordinal α we shall define a system $D_\alpha \subseteq D$ as follows:

$D_0 = D$.

If $\alpha = \beta + 1$, then D_α consists of all left ideals $\mathfrak{a}$ of A such that

$$A_{D_\beta}(\mathfrak{a} : x) = A_{D_\beta}$$

for any $x \in A$. (Since $D_\beta \subseteq D$, then A is identified as a subring of A_{D_β}, and A_{D_β} as a subring of $A_D = Q(A)$).

If α is a limit ordinal then $D_\alpha = \bigcap_{\beta < \alpha} D_\beta$.

Now let α_0 be an ordinal number such that $D_{\alpha_0} = D_{\alpha'}$, for any ordinal $\alpha' \geqslant \alpha_0$, and write $D' = D_{\alpha_0}$. Then $A_{D'} = M(A)$. Indeed, if $\mathfrak{a} \in D'$, then $A_{D'}(\mathfrak{a} : x) = A_{D'}$, according to the definition of D'. Thus by Theorem 16.6 we find that the canonical inclusion $A \subseteq A_{D'}$ is a left flat epimorphism of rings.

Finally, let $v: A \to B$ be a left flat bimorphism of rings thus $B \simeq A_F$ where F is a suitable system on A such that $F \subseteq D$. (Theorem 16.6) so that B can be identified with a subring of $Q(A)$. Now if $\mathfrak{a} \in F$ then $A_F\,\mathfrak{a} = A_F$, so that $Q(A)\,\mathfrak{a} = Q(A)$, in other words $F \subseteq D_1$, and necessarily $B \simeq A_F \subseteq A_{D_1}$. Furthermore, it is obvious that $F \subseteq D_\alpha$ for any ordinal number α, so as before, $F \subseteq D'$, and necessarily $B \simeq A_F \subseteq A_{D'}$.

Some simple details are left to the reader. ∎

The ring $M(A)$ is called the *maximal left flat epimorphic extension* of A.

COROLLARY 17.3. *A left ideal $\mathfrak{b}$ of $M(A)$ is finitely generated if and only if $\mathfrak{b} \cap A$ is finitely generated. In particular, if A is left noetherian or left artinian, then so is $M(A)$. Moreover if A is commutative, then $M(A)$ is also commutative.*

The proof is a direct consequence of Exercise 11 of Section 4.16 and Lemma 17.8. ∎

We note that the result is valid for any left flat epimorphism $u: A \to B$.

For any ring A we have the natural inclusions $A \subseteq M(A) \subseteq Q(A)$. We shall investigate some cases when $M(A) = Q(A)$ or $A = M(A)$.

PROPOSITION 17.4. *If A is regular then $A = M(A)$.*

Proof. By Exercise 2 of Section 4.16 we check that $M(A) \otimes_A (M(A)/A) = 0$. But since A is regular, we deduce that $M(A)/A$ is flat as an A-module, and thus the inclusion $A \subseteq M(A)$ gives as the inclusion

$$M(A)/A = A \otimes_A (M(A)/A) \to M(A) \otimes_A (M(A)/A) = 0,$$

that is $M(A)/A = 0$. ∎

LEMMA 17.5. *Let A be a left non-singular ring. Then for any non-singular A-module X we have: that $J(Q(A) \otimes_A X)$ is equal to the kernel of the composite morphism:*

$$h: Q(A) \otimes_A X \to Q(A) \otimes_A X_E \to X_E$$

where, as always, X_E is the localization of X relative to E, the system of essential left ideals.

Proof. Assume that X is identified to a submodule of X_E; then

$$h(\textstyle\sum_i q_i \otimes x_i) = \sum_i q_i x_i.$$

Let $y = \sum_i q_i \otimes x_i$ such that $h(y) = 0$ and $\mathfrak{a}$ an essential left ideal such that $\mathfrak{a} q_i \subseteq A$ for every i. Then $\mathfrak{a} y = \sum_i \mathfrak{a} q_i \otimes x_i = 1 \otimes \sum_i \mathfrak{a} q_i x_i = 0$, that is $y \in J(Q(A) \otimes_A X)$. Conversely if $\mathrm{Ann}(y)$ is essential, then for any $a \in \mathrm{Ann}(y)$

we have $0 = ay = \sum_i aq_i \otimes x_i = \sum_i aq_i x_i$. But then $h(y) = 0$, because X_E is E-closed. ■

COROLLARY 17.6. *Let A be a left non-singular ring. Then the natural inclusion $A \subseteq Q(A)$ is an epimorphism of rings if and only if $J(Q(A) \otimes_A Q(A)) = 0$.*

The proof follows by Lemma 17.5 and the Proposition 16.3. ■

THEOREM 17.7. (Cateforis [35]). *Let A be a left non-singular ring. The following statements are equivalent:*

(1) $M(A) = Q(A)$;

(2) *For any finitely generated non-singular module X we have that*

$$J(Q(A) \otimes_A X) = 0;$$

(3) *For any $q \in Q(A)$, there is a finitely generated essential left ideal $\mathfrak{a}$ such that $\mathfrak{a}q \subseteq A$.*

Proof. (1) ⇒ (2). Then $Q(A)$ is flat as an A^0-module, so that for any non-singular A-module X, the inclusion $X \subseteq X_E$ gives rise to the exact sequence:

$$0 \to Q(A) \otimes_A X \to Q(A) \otimes_A X_E \simeq X_E.$$

Now we can use Lemma 17.5.

(2) ⇒ (1). Let $\mathfrak{b}$ be a finitely generated left ideal of A; then $\mathfrak{b}$ is non-singular, so that in the commutative diagram:

$$\begin{array}{ccccc} & & Q(A) \otimes_A \mathfrak{b} & \xrightarrow{i} & Q(A) \otimes_A A \\ & & \downarrow h & & \downarrow \wr \\ 0 & \longrightarrow & \mathfrak{b}_E & \longrightarrow & Q(A) \end{array}$$

h is a monomorphism. Then i is necessarily a monomorphism. Hence, by Theorem 8.3 of Chapter 3 we check that $Q(A)$ is flat as A^0-module. Thus to complete the proof that (2) implies (1) it remains to prove that

$$J(Q(A) \otimes_A Q(A)) = 0$$

(see Corollary 17.6). In fact if $y = \sum q_i \otimes p_i$ is such that $\mathrm{Ann}(y) = 0$ then a finitely generated A-submodule X of $Q(A)$ can be calculated such that in the commutative diagram:

$$\begin{array}{ccccc} & & Q(A) \otimes_A X & \xrightarrow{t} & Q(A) \otimes_A Q(A) \\ & & \downarrow h' & & \downarrow h \\ 0 & \longrightarrow & X_E & \longrightarrow & Q(A) \end{array}$$

h' is a monomorphism by Lemma 17.4 and $y \in \operatorname{im} t$. Then necessarily $y = 0$, so that h is an isomorphism.

(3) ⇒ (1). We shall prove that the condition (c) of Theorem 16.6 is satisfied. In fact, let $q \in Q(A)$ and $\mathfrak{a}$ a finitely generated essential left ideal such that $\mathfrak{a}q \subseteq A$. Then $\mathfrak{a} \subseteq Q(A)\,\mathfrak{a} \subseteq Q(A)$. By Theorem 13.10 $Q(A)$ is a regular ring and thus $Q(A)\,\mathfrak{a}$ is a direct summand (Chapter 3, Theorem 8.6). Therefore necessarily $Q(A)\,\mathfrak{a} = Q(A)$, because $\mathfrak{a}$ is essential in $Q(A)$ as A-module.

The implication (1) ⇒ (3) is obvious. ∎

Now we shall give some useful definitions. If X is an A-module, a *left X-annihilator* is a left ideal of the form

$$l_X(Y) = \{a \in A \mid aY = 0\}$$

where Y is a subset of X. A ring A has the *ascending chain condition on left X-annihilators*, if any increasing sequence $l_X(Y_1) \subseteq l_X(Y_2) \subseteq \dots$ of left X-annihilators becomes stationary (i.e. $l_X(Y_n) = l_X(Y_m)$ if $n \leqslant m$, for a suitable n). We shall write: A has the *ACC* on left X-annihilators. For any set Y of A we shall put $l_A(Y) = l(Y)$ and say simply *left annihilator*. Analogously, we shall define the right X-annihilators, $r_X(Y)$ and right annihilators $r(Y)$.

An A-module X is called *finite dimensional* or of *finite Goldie dimension* if it does not contain any infinite family of non-zero submodules X_i such that their sum $\sum_i X_i$ is direct. A ring A is called *left finite dimensional* if A is finite dimensional as A-module. Similarly one has the notion of right finite dimension.

LEMMA 17.8. *Let B be a ring and A a left quasi-order of B. Then*:

(1) *A is an essential A-submodule of B*;

(2) *For any left ideal $\mathfrak{b}$ of B one has $\mathfrak{b} = B(\mathfrak{b} \cap A)$*;

(3) *If A is left finite dimensional, then B is also left finite dimensional*;

(4) *Conversely, if B is left finite dimensional, then A has the same property.*

(5) *If B is left noetherian then A has ACC on left X-annihilators for every B-module X.*

Proof. The assertions (1) and (2) are easy to prove.

(3) Let $I = \{\mathfrak{b}_i\}$ be a set of non-zero left ideals of B such that the sum $\sum_i \mathfrak{b}_i$ is direct. Then the sum of left ideals $\sum_i (A \cap \mathfrak{b}_i)$ is also direct in A, and so the set $\{\mathfrak{b}_i \cap A\}$ contains only a finite number of elements. By (1) we check that I is necessarily a finite set.

(4) Let $I = \{\mathfrak{a}_i\}$ be a set of left ideals of A such that the sum $\sum_i \mathfrak{a}_i$ is direct. Then the sum $\sum_i B\mathfrak{a}_i \simeq B \otimes_A(\sum_i \mathfrak{a}_i)$ is also direct in B. In conclusion, the set I is finite.

(5) Let $l_X(X_1) \subseteq l_X(X_2) \subseteq \ldots l_X(X_n) \subseteq \ldots$ be an increasing sequence of left X-annihilators of A, X being an A-module. Then:

$$Bl_X(X_1) \subseteq Bl_X(X_2) \subseteq \ldots Bl_X(X_n) \subseteq \ldots$$

is an increasing sequence of left ideals of B. By hypothesis

$$Bl_X(X_n) = Bl_X(X_{n+1}) = \ldots$$

for a suitable index n. Let $a \in l_X(X_{n+1})$; then $a \in Bl_X(X_n)$, hence $a = \sum_i b_i a_i$ with $b_i \in B$ and $a_i \in l_X(X_n)$, for any i. Also for any $x \in X_n$ one has:

$$ax = (\textstyle\sum_i b_i a_i)\, x = \sum_i b_i(a_i x) = 0$$

or equivalently $a \in l_X(X_n)$. ∎

Now we shall characterize the rings A which can be the left quasi-orders in a semisimple ring. Obviously, for these rings one has $M(A) = Q(A)$.

LEMMA 17.9. *Any A-module X of finite dimension contains a finitely generated large submodule.*

Proof. We shall define for any natural number n, a finitely generated submodule X_n, such that $X_n \cap (\sum_{i<n} X_i) = 0$. Indeed, X_1 is defined as being Ax_1 for a suitable non-zero element $x_1 \in X$. Assuming X_{n-1} to be defined, we shall define $X_n = 0$ if $\sum_i X_i$ is essential; if the latter submodule is not essential, then it has a complement X' and so we may take $X_n = Ax_n$ for a non-zero element X_n of X'. Let us put $X'' = \sum_n X_n$. By hypothesis $X_n = 0$ for all but a finite number of indices, and so X'' is finitely generated and essential. ∎

THEOREM 17.10. *Let A be a ring. The following assertions are equivalent*:

(1) $J(A) = 0$ *and A is left finite dimensional*;

(2) *A is left quasi-order of a semisimple ring.*

Proof. (1) ⇒ (2). Clearly, $E = D$ and by Proposition 13.9, E is an exact system. Moreover, by Lemma 17.9, E is in fact a perfect system, so that the inclusion $A \subseteq Q(A)$ is a left flat epimorphism of rings. Again, by Proposition 13.9 we check that Mod $Q(A)$ is in fact a spectral category (being equivalent with Mod $A/\mathscr{E}$), so that $Q(A)$ is semisimple.

(2) $\Rightarrow$ (1). Now assume that A is a left quasi-order in a semisimple ring B. It is trivial that any semisimple ring has a finite dimension; then by Lemma 17.8, A also is left finite dimensional.

Furthermore, if $\mathfrak{a}$ is a left essential ideal of A, then $B\mathfrak{a}$ is a left essential ideal of B, hence $B\mathfrak{a} = B$, by hypothesis. Finally if $x \in A$ is such that $\mathfrak{a}x = 0$ for an essential left ideal $\mathfrak{a}$, then $Bx = B\mathfrak{a}x = 0$ so that $x = 0$. In other words, $J(A) = 0$. It is also clear that B is the injective envelope of A, so that $M(A) = Q(A) = B$. ∎

COROLLARY 17.11. *Let A be a left quasi-order of a semisimple ring. Then A has ACC on left X-annihilators for any $Q(A)$-module X.*

The proof is checked by Theorem 17.10 and Lemma 17.8. ∎

A ring A is called a *Quasi-Frobenius* ring (written *QF*-ring) if A has *ACC* on left annihilators and A is left self-injective.

THEOREM 17.12. *Let A be a left quasi-order on a ring B. The following assertions are equivalent:*

(1) *B is a QF-ring;*

(2) *A has the following conditions:*

(a) *A has ACC on left Q-annihilators, Q being the injective envelope of A;*

(b) $B \otimes_A (Q/A) = 0$.

Proof. (1) $\Rightarrow$ (2). We may assume that $B = Q$; the proof is obtained by Lemma 17.8 and Exercise 2 of Section 4.16.

(2) $\Rightarrow$ (1). By Theorem 16.6 we deduce that $B = A_F$ where F is a suitable perfect localizing system on A. Then by (b) we check that $Q/A \in \mathrm{Ob}\,\mathscr{F}$ so that $B = Q$, i.e. B is left self injective ring.

Now let X be a subset of B. We put $\bar{l}(X) = \{b \in B \mid bX = 0\}$, hence

$$l(X) = \bar{l}(B) \cap A.$$

By hypothesis, if $\bar{l}(X_1) \subseteq \bar{l}(X_2) \subseteq \ldots$ is an increasing sequence of left annihilators in B, then $l(X_n) = l(X_{n+1}) = \ldots$ for a suitable n, so that $\bar{l}(X_n) = \bar{l}(X_{n+1}) = \ldots$ according to Lemma 17.8. In conclusion B is a *QF*-ring. ∎

We finish this paragraph by proving the following interesting result:

PROPOSITION 17.13. *The following assertions are equivalent for a ring A.*

(1) $Q(A) = M(A)$;

(2) *$Q(A)$ has no proper dense left ideals (i.e. $D = \{Q(A)\}$).*

Proof. (1) ⇒ (2). If $Q(A) = M(A)$, then the system D of all dense left ideals is perfect. If $\mathfrak{b}$ is a dense left ideal of $Q(A)$, then $\mathfrak{b} \cap A \in D$. Therefore $Q(A) = Q(A)(\mathfrak{b} \cap A) \subseteq \mathfrak{b}$ and so $\mathfrak{b} = Q(A)$.

(2) ⇒ (1). If $\mathfrak{a} \in D$, and $Q(A)\mathfrak{a} \neq Q(A)$, then $r(Q(A)\mathfrak{a}) \neq 0$, by Exercise 12. Thus there is $0 \neq q \in Q(A)$ such that $\mathfrak{a}q \neq 0$. But this is impossible, because $Q(A)$ is D-closed. ∎

This result permits us to make certain assertions about the rings $Q(A)$ whenever $M(A) = Q(A)$.

Exercises

1. Let A, B, C be three rings such that $A \subseteq B \subseteq C$ (the inclusions are morphisms of rings). Assume that $A \subseteq C$ is a left epimorphism of rings.

(i) If B is a left flat A-module, and C is also A^0-flat, then $A \subseteq B$ is a left flat epimorphism.

(ii) If B is A and A^0-flat, then $A \subseteq B$ is a left and right flat epimorphism.

2. Let A be a ring. If $A \subseteq B \subseteq M(A)$ are such that the condition (i) or (ii) of the previous exercise is fulfilled, then $A \subseteq B$ is a left flat epimorphism of rings.

3. Let $u: A \to B$ be a ring morphism. If X is a flat A^0-module, then $X \otimes_A B$ is also a flat B^0-module. Moreover, if u is a left flat epimorphism and Y a flat B^0-module, then $u_*(Y)$ is also a flat A^0-module. Also, in this case if Y is a B^0-module, such that $u_*(Y)$ is flat as A^0-module, then Y is a flat B^0-module.

4. For any ring A, $M(M(A)) = M(A)$.

5. A ring A is called *left coherent* if every direct product of flat A^0-modules is flat (see [37]). If A is left coherent, then so is $M(A)$.

6. If A is a left semi-hereditary ring, then A is non-singular and $Q(M)$ is flat as A^0-module.

7. The following assertions are equivalent for a ring A:

(a) Any finitely generated non-singular A-module is projective;

(b) $J(A) = 0$ and every finitely generated A-module is projective;

(c) A is left semi-hereditary and $Q(A) = M(A)$.

8. Let A be a ring. Then

(a) If $X \subset Y$ are subsets of A, then $l(X) \supseteq l(Y)$ and $r(X) \supseteq r(Y)$.

(b) For any subset X we have

$$X \subseteq l(r(X)), \quad X \subseteq r(l(X)), \quad l(X) = l(r(l(X)))$$

and

$$r(X) = r(l(r(X))).$$

(c) If $\mathfrak{a}$ is a left (respectively, right) ideal, then $l(\mathfrak{a})$ (respectively, $r(\mathfrak{a})$) is a two-sided ideal.

9. Let A be a left quasi-order of a semisimple ring. Then $J(X) = J^2(X)$ for any A-module X.

10. A ring A is left quasi-order in a semisimple ring if and only if $J(A) = 0$ and $J(X) = \ker(X \to Q(A) \otimes_A X)$ for any A-module X.

11. If A is a left quasi-order of a semisimple ring, then $J(X)$ is a direct summand of X for any injective A-module X.

12. The following assertions are equivalent for a ring A:

(a) The injective envelope Q of A is a cogenerator of Mod A.

(b) Every simple A-module is isomorphic to a minimal left ideal of A.

(c) $\mathrm{Hom}_A(X, A) \neq 0$ for every cyclic non-zero A-module X.

(d) $r(\mathfrak{a}) \neq 0$ for every left ideal $\mathfrak{a} \neq A$.

(e) A has no proper dense left ideals (i.e. $D = \{A\}$).

A ring A which has the above conditions (a)–(e) is called a *left S-ring*.

13. Show that if A is regular and $Q(A)$ is projective as a A-module, then $A = Q(A)$. (Hint: an embedding $Q \subseteq X$ with X free, and $Q = E(A)$ induces $A \subseteq X'$, X' finitely generated and free; then use the fact that a finitely presented flat module is projective).

14. Let A be a left finite dimensional and non-singular ring. Then $Q(A)$ is also the maximal right ring of quotients of A (i.e. $Q_l(A) = Q_r(A)$) if and only if $Q(A)$ is flat as a A-module.

15. Let F be a system such that $F \subseteq D$. If A is left finite dimensional, then so is also A_F.

16. For a ring A the following assertions are equivalent:

(a) Any injective module is projective;

(b) Any projective module is injective;

(c) A is a QF-ring.

(Hint: See also Chapter 6, Section 6.3, Exercise 9).

17. The ring A is called *semiperfect* if $A/\mathfrak{R}(A)$ is semisimple, where $\mathfrak{R}(A)$ is the Jacobson radical of A and idempotents may be lifted mod $\mathfrak{R}(A)$. The endomorphism ring of an injective module E is semiperfect if and only if E is of finite dimension. Show that a commutative ring is semiperfect if and only if it is a direct product of finitely many local rings. Show that a semiperfect ring is a direct sum of indecomposable left ideals.

18. For a regular ring A the following assertions are equivalent:

(a) A is left self-injective;

(b) Every finitely generated non-singular A-module is projective;

(c) $Q(A)$ is projective as A-module.

19. Assume that A is a left semi-hereditary ring. The following statements are equivalent:

(a) $Q(A)$ is semisimple;

(b) $J(Q(A) \otimes_A Q(A)) = 0$.

References

Cateforis [34], [35]; Cateforis–Sandomierski [36]; Chase [37]; Desq [41]; Findlay [54]; Gabriel [59]; Goldman [76]; Hacque–Maury [85]; Knight [107]; Lazard [116], [117]; Morita [137]; Năstăsescu–Popescu [144]; Popescu–Spircu [175], [176]; Popescu–Spulber [177]; Roos [199]; Sandomierski [203], Stenström [213]; Van der Water [237].

4.18 Rings of fractions

Let A be a ring. A multiplicative system S in A is called *saturated* if for any $x, y \in A$ such that $xy \in S$ it follows that $y \in S$.

By a 1-*system* in A we mean a system F containing a coinitial family of principal left ideals. A 1-system F is determined by the set

$$S(F) = \{s \in A \mid As \in F\}.$$

PROPOSITION 18.1. *The map $F \rightsquigarrow S(F)$ defines a bijection between* 1-*systems on A and subsets S of multiplicative left permutable saturated systems on A.*

Proof. Let F be a 1-system. If $s, s' \in S(F)$, then $(Ass' : as') \supseteq (As : a)$ for any $a \in A$, hence $Ass' \in F$, i.e. $S(F)$ is multiplicative. Now let $s \in S(F)$ and $a \in A$. Then $(As : a) \supseteq As'$ for a suitable $s' \in S(F)$; therefore $s'a = a's$, and so $S(F)$ is left permutable. Finally, if $xy \in S(F)$ then $Ay \supseteq Axy$ so, $y \in S(F)$, that is $S(F)$ is saturated.

Conversely, if S is left permutable and saturated, then the system F_S (see Section 4.9) is a 1-system and $S(F_S) = S$. Finally it is obvious that $F_{S(F)} = F$. ∎

For a 1-system F one can describe, for any module X, the module X_F in a rather explicit way.

PROPOSITION 18.2. *Let F be a 1-system on A, X an A-module and $S = S(F)$. Denote $\bar{X}$ the subset of the direct product of sets $S \sqcap X$ defined as follows. A couple (s, x) lies in $\bar{X}$ if for any $a \in A$ such that $as = 0$ there exists $t \in S$ so that $tax = 0$. Then $X_F = \bar{X}/\sim$ where "$\sim$" is the equivalence relation given by $(s, x) \sim (s', x')$ if and only if there exist $a, a' \in A$ such that $as = a's' \in S$ and $ax = a'x'$.*

Proof. Recall that we have

$$X_F = \varinjlim_{s \in S} \operatorname{Hom}_A(As, X/X_{\mathscr{F}})$$

An A-morphism $f: As \to X/X_{\mathscr{F}}$ is determined by an element $x \in X$ such that $as = a's$ in A implies $ax - a'x \in X_{\mathscr{F}}$, i.e. $s'(ax - a'x) = 0$ for some $s' \in S$. (We observe that an element $y \in X$ is in $X_{\mathscr{F}}$ if and only if $sy = 0$ for an element $s \in S$). In the inductive limit f gives the same element of X_F as $f': As' \to X/X_{\mathscr{F}}$ determined by an element $x' \in X$, if and only if f and f' coincide on some $As'' \subseteq As \cap As'$ with $s'' \in S$, i.e. if and only if there exist $a, a' \in A$ such that $s'' = as = a's' \in S$, and $as - a's' \in X_{\mathscr{F}}$. This clearly corresponds to the relation $\sim$. ∎

It is clear that under the isomorphism described in the previous proposition, the module operations in X_F take the form:

$$(s, x) + (s', x') = (s'', ax + a'x') \quad \text{where} \quad as = a's' = s'' \in S$$

$(s, a)(s', x) = (s''s', a'x)$ for some $a' \in A$, $s'' \in S$ such that $s''a = s'a'$.

PROPOSITION 18.3. *Let F be a 1-system on A and $S = S(F)$. The following assertions are equivalent:*

(1) *F is a perfect system;*

(2) *S satisfies the condition:*

(*R*) *For every $s \in S$ there exists $a \in A$ such that $as \in S$ and such that $a'as = 0$ implies $s'a'a = 0$ for some $s' \in S$.*

Proof. (1) ⇒ (2). By Theorem 16.6, (c) for each $s \in S$ one has $A_F u(As) = A_F$ or equivalently, there exists $q \in A_F$ such that $qu(s) = 1$. Suppose that this is so. Then again by Theorem 16.6 we can find $s' \in S$ and $q' \in A_F$ such that

$q'u(s') = 1$ and $u(s')q = u(a')$. Then $u(a's) = u(s')$ so that $a's - s' \in \ker u$. By Theorem 16.6, (c) there exists an element $s'' \in S$ such that $s''a's = s''s \in S$; hence we may take $a = s''a'$. Indeed if $a'as = 0$, then $u(a'a)u(s) = 0$. Using the flatness of A_F we can find elements $q_1, \ldots, q_n$ of A_F and $a_1, \ldots, a_n$ of A such that $\sum_i q_i u(a_i) = u(s)$ and $a_i a' a = 0$ for every i. Then $1 = qu(s) = \sum_i qq_i u(a_i)$, i.e. the left ideal $\langle a_1, \ldots, a_n \rangle \in F$, by Exercise 8 of Section 4.12. Then $As' \subseteq \langle a_1, \ldots, a_n \rangle$ with $s' \in S$ and it is clear that $s'a'a = 0$. Hence the perfectness of F implies the condition (R).

(2) ⇒ (1). Let us assume that the condition (R) is verified. Firstly it is clear that F is a finitely generated system, so by Theorem 16.7 it will be sufficient to prove that the localization functor is exact, or equivalently, if

$$0 \longrightarrow X' \longrightarrow X \xrightarrow{p} X'' \longrightarrow 0$$

is an exact sequence of A-modules such that X' and X are F-closed then X'' is also F-closed. Indeed, let $f: As \to X''$ be a morphism and $a \in A$ as in the condition (R). Denote by $f': Aas \to X$ the map: $f'(a'as) = a'ax$, where $x \in X$ is such that $p(x) = f(s)$. If $a'as = 0$, then $s'a'a = 0$, so that $s'a'ax = 0$ and thus $a'ax = 0$, X being F-closed. In other words, the map f' considered is in fact a morphism of modules. Furthermore, there exists a unique morphism $\bar{f}': A \to X$ which prolongs f' and thus $p\bar{f}'$ is the unique morphism which prolongs f, because X'' is F-torsionfree.

An alternative proof of the last part is as follows. If for each $s \in S$ there exists $a \in A$, as in the condition (R), then (as, a) represents an element in A_F and one deduces $(1, s)(as, a) = 1$. Now we can use Theorem 16.6. ∎

COROLLARY 18.4. *Let F be a* 1*-system on A such that $S(F)$ is left simplifiable. Then F is perfect.*

Proof. It is easy to check that any left simplifiable system verifies the condition R. Thus the proof follows by the above proposition. ∎

NOTE. The condition (R) for a multiplicative system S is a weakened form of the better known condition: if $as = 0$ with $s \in S$, then $s'a = 0$ for some $s' \in S$, which expresses the fact that S is left simplifiable.

Now we shall establish the connection with the additive categories of fractions. For that, we give a more special definition of rings of fractions. Let S be a multiplicatively closed subset of A. A *classical left ring of fractions* of A with respect to S is a ring $A[S^{-1}]$ and a ring morphism $\phi: A \to A[S^{-1}]$ which satisfy:

($CF1$) $\phi(s)$ is invertible for every $s \in S$;

($CF2$) Every element in $A[S^{-1}]$ has the form $\phi(s)^{-1}\phi(a)$ with $s \in S$;

($CF3$) $\phi(a) = 0$ if and only if $sa = 0$ for some $s \in S$.

LEMMA 18.5. *If* $A[S^{-1}]$ *exists, it is a ring of additive fractions of* A *relatively to* S, *i.e.* $A[S^{-1}] = A_S$.

Proof. Let $\psi : A \to B$ be a ring morphism such that $\psi(s)$ is invertible in B for every $s \in S$. We define a morphism $\bar{\psi} : A[S^{-1}] \to B$ as $\bar{\psi}(\phi(s)^{-1}\phi(a)) = \psi(s)^{-1}\psi(a)$. We have then to verify that this is well defined. So suppose $\phi(s)^{-1}\phi(a) = \phi(s')^{-1}\phi(a')$. Then $\phi(a) = \phi(s)\phi(s')^{-1}\phi(a') = \phi(s'')^{-1}\phi(b)\phi(a')$ for some $b \in A$, $s'' \in S$ by $(CF2)$. Thus $\phi(s'')\phi(a) = \phi(b)\phi(a')$ and from $(CF3)$ this implies that $ts''a = tba'$ for some $t \in S$. Then $\psi(s'')\psi(a) = \psi(b)\psi(a)$ since $\psi(t)$ is invertible, and we may reverse the argument to obtain

$$\psi(s)^{-1}\psi(a) = \psi(s')^{-1}\psi(a').$$

We leave it to the reader to verify that $\bar{\psi}$ is a ring morphism. It is clear that $\bar{\psi}\phi = \psi$ and that $\bar{\psi}$ is unique. ∎

THEOREM 18.6. *Let* S *be a multiplicatively closed system in* A. *The following conditions are equivalent*:

(1) *The classical ring of left fractions* $A[S^{-1}]$ *exists*;

(2) S *is a left calculable system*;

(3) *If* $F = F_S$, *then the image of any element* $s \in S$ *by the canonical morphism* $u : A \to A_F$ *is an invertible element.*

Under these conditions A_F *is canonically isomorphic with* A_S, *and* F *is a perfect system.*

Proof. (1) ⇒ (2). Let $a \in A$, $s \in S$ and $\phi : A \to A[S^{-1}]$ the canonical ring morphism. From $(CF2)$ we have $\phi(a)\phi(s)^{-1} = \phi(s')^{-1}\phi(a')$, so that $s'a - a's \in \ker\phi$. Then by $(CF3)$ there exists an $s'' \in S$ such that $s''s'a = s''a's$. Hence S is left permutable. Now, if $s \in S$ and a is an element, such that $as = 0$, then $\phi(a)\phi(s) = 0$, i.e. $\phi(a) = 0$, $\phi(s)$ being invertible. By $(CF3)$ one has that $s''a = 0$ for some $s'' \in S$. In conclusion, S is left calculable.

(2) ⇒ (3). By Corollary 18.4, we deduce that F is a perfect system and so, for any $s \in S$, the element $u(s)$ is left invertible. Moreover, the element $u(s)$ is also right invertible.

In fact if $qu(s) = 1$ for an element $q \in A_F$, then $\big(u(s)q - 1\big)\,u(s) = 0$. Then by Theorem 8.3, Chapter 3, there exist elements $q_1, \ldots, q_n \in A_F$ and

$$a_1, \ldots, a_n \in A$$

such that:

$$\sum_{i=1}^{n} q_i\,u(a_i) = u(s)q - 1$$

and $a_i s = 0$ for every i.

Let $s' \in S$ be an element such that $s'a_i = 0$ for any i (see Exercise 5). But then $a_i \in \ker u = \mathfrak{r}_F$ for any i and so $u(s)q = 1$.

(3) ⇒ (1). We shall take $A[S^{-1}] = A_F$ and $\phi = u$. Then the condition ($CF1$) is obvious. Now let $a \in A$ be such that $u(a) = 0$. Then, an element $\mathfrak{a} \in F$ can be chosen such that $\mathfrak{a}a = 0$. But $\mathfrak{a} \cap S \neq \varnothing$ and so $sa = 0$ for some $s \in S$. Hence the condition ($CF3$) is satisfied.

Moreover, if $q \in A_F$, then $(t(A):q) \in F$ and thus there exists an element $s \in S$ with $u(s)q = u(a)$. Then $q = u(s)^{-1}u(a)$, hence the condition ($CF2$) is satisfied.

Finally the proof follows by Proposition 18.3 and Lemma 18.5. ∎

NOTES 2. If S is a left permutable system on A, verifying the condition (R) and $F = F_S$, then for any A-module X we have $X_F = A_S \otimes_A X$; moreover, this module is denoted by X_S and is called sometimes the *module of fractions* of X with respect to S.

3. If S is a multiplicative system contained in the centre $Z(A)$ of a ring A, then obviously S is left calculable so that the ring A_S always exists. Moreover, if A is a commutative ring and S a multiplicative system on A, then the ring A_S can always be calculated.

Now we shall give a counterexample promised at the end of Section 4.16. Let K be a field and A the ring of all 2×2 matrices of the form

$$\begin{pmatrix} f(X) & 0 \\ g(X) & f(0) \end{pmatrix} \text{ where } f(X),\ g(X) \in K[X].$$

Let $\mathfrak{a}$ be the left ideal of A consisting of all elements of the form

$$\begin{pmatrix} 0 & 0 \\ k & 0 \end{pmatrix}, \qquad k \in K$$

and $M = A/\mathfrak{a}$. Then $B = \mathrm{End}_A(M)$ is canonically isomorphic with the ring $\bar{A}/\mathfrak{a}$, where $\bar{A}$ is the subring of A consisting of all elements $a \in A$ such that $\mathfrak{a}a \subseteq \mathfrak{a}$. But an element

$$a = \begin{pmatrix} f(X) & 0 \\ g(X) & f(0) \end{pmatrix}$$

belongs to $\bar{A}$ if and only if $f(X) = k \in K$.

Furthermore, let S be the multiplicative system on A consisting of all elements of the form:

$$\begin{pmatrix} f(X) & 0 \\ 0 & f(0) \end{pmatrix} \qquad 0 \neq f(X) \in K[X].$$

Let

$$x = \begin{pmatrix} f(X) & 0 \\ g(X) & f(0) \end{pmatrix} \quad \text{and } s = \begin{pmatrix} h(X) & 0 \\ 0 & h(0) \end{pmatrix} \in S$$

then $x' s = s' x$, where

$$x' = \begin{pmatrix} f(X)\,X & 0 \\ 0 & 0 \end{pmatrix}; \quad s' = \begin{pmatrix} h(X)\,X & 0 \\ 0 & 0 \end{pmatrix}$$

Hence S is a left permutable system. Moreover, if $x \in A$ and $s \in S$ are such that $xs = 0$, then $x = 0$ so that as before S is left calculable. Let us put $F = F_S$. By a slight calculation we can prove that $\mathfrak{r}_F$ consists of all elements of the form:

$$\begin{pmatrix} 0 & 0 \\ g(X) & 0 \end{pmatrix}; \; g(X) \in K[X]$$

and $A_S \simeq M_S \simeq K(X)$ and the localizing ring of B always contains zero divisors.

Exercises

1. Show that if A has no non-zero nilpotent elements, then any set of elements of A is left simplifiable.

2. Let A be a ring and S a left calculable system on A. An A-module X is called *S-divisible* if for any $x \in X$ and any $s \in S$ there exists a unique element $y \in X$ such that $sy = x$. The module X is F_S-closed, if and only if it is S-divisible.

3. Let A be a regular ring and let $S = \{a \in A \mid ab = 0 \text{ implies } b = 0\}$. Show that S is a saturated left permutable system and satisfies the condition (R). Moreover, S is left permutable if and only if all elements in S are invertible.

4. Let S be a multiplicative system in a commutative regular ring A. Then $A_S = A/\mathfrak{r}_{F_S}$.

5. Let S be a left calculable system of a ring A and $q_1, \ldots, q_n$ elements of A_S. Then an element $s \in S$ can be found such that $sq_i \in u(A)$ for any i (as always $u: A \longrightarrow A_S$ is the canonical morphism).

6. An ideal $\mathfrak{a}$ in a commutative ring A is called *invertible* in a ring B containing A as subring if there is an A-submodule $\mathfrak{b}$ of B such that $\mathfrak{a}\mathfrak{b} = A$. Show that the following statements are equivalent:

(a) $\mathfrak{a}$ is invertible in some ring B (containing A);

(b) $\mathfrak{a}$ is invertible in $Q(A)$;

(c) $\mathfrak{a}$ is dense, finitely generated and projective.

7. Let A be a commutative ring. Show that the product of any two invertible ideals is also invertible. Denote by H the set of all ideals of A containing an invertible ideal. Then H is a perfect localizing system.

8. Let A be a ring and S a multiplicative system on A. Then there exist a ring A^S and a ring morphism $u : A \to A^S$ such that:

(a) u is a left flat epimorphism of rings;

(b) If $b \in A^S$, then an element $s \in S$ can be found such that s is left regular, and $u(s)\, b \in u(A)$;

(c) If $v : A \to B$ is a left flat epimorphism of rings such that the condition (b) is verified for any element $b \in B$, then a unique ring morphism

$$\bar{v} : A^S \to B$$

can be found such that $\bar{v}u = v$.

The ring A^S is called ring of *pseudo-fractions* of A relative to S. Give a description of A^S using Theorem 16.6.

This result is due to Ribenboim [184].

References

Akiba [1], [2], [3]; Almkvist [7]; Bkouche [17], [18]; Bourbaki [26]; Elizarov [49]; Gabriel [59]; Gabriel–Zisman [62]; Goldie [72], [73], [75]; Grothendieck–Dieudonné [80]; Lambek [113]; Morita [138]; Ore [158]; Popescu [169]; Ribenboim [184]; Roos [195], [199]; Samuel–Zariski [202]; Stenström [213]; Storrer [216].

4.19 Left orders

Recall that an element $r \in A$ is regular if for any $x \in A$ the relation "$xr = 0$ or $rx = 0$" implies that $x = 0$. It is clear that the set of all regular elements of A is a multiplicative system and is left or right simplifiable. By a *left calculable regular system* of elements of A we mean a multiplicative left permutable system consisting of regular elements.

A ring A is called a *left order* in a ring B containing A, if there exists a set S of elements of A such that:

(α) Any element of S is invertible in B.

(β) Any element of B is of the form $s^{-1}a$ with $s \in S$ and $a \in A$.

COROLLARY 19.1. *Let A be a left order of B. Write $S = \{s \in A \mid s$ is invertible in $B\}$. Then S is a left calculable regular system and $B = A_S$. Moreover A is left quasi-order of B.*

The proof is left to the reader. ∎

NOTE 1. By the above corollary it follows that any left order is a left quasi-order. The converse assertion is not true, i.e. there exist left quasi-orders which are not left orders. In fact, let A be the matrix ring

$$x = \begin{pmatrix} a & q \\ 0 & b \end{pmatrix} \quad \text{with} \quad a, b \in \mathbf{Z} \quad \text{and} \quad q \in \mathbf{Q}.$$

Thus A is a subring of $\mathbf{Q}_2$ and A is not a left order in $\mathbf{Q}_2$. Indeed, an element x of A is invertible in $\mathbf{Q}_2$ if and only if $ab \neq 0$. Assume, if possible, that A is a left order in $\mathbf{Q}_2$, and let

$$y = \begin{pmatrix} a_{11} & a_{12} \\ a_{21} & a_{22} \end{pmatrix}$$

be an element of $\mathbf{Q}_2$ such that $a_{21} \neq 0$. Thus there exists an element $x \in A$, invertible in $\mathbf{Q}_2$, so that $xy \in A$. But then necessarily $ba_{21} = 0$, a contradiction.

On the other hand A is a left quasi-order of $\mathbf{Q}_2$. Indeed, let y be as above an element of $\mathbf{Q}_2$, a, a multiple of the denominator of a_{11} and q a multiple of the denominator of a_{21}. Thus:

$$\begin{pmatrix} 1/a & 0 \\ 0 & 0 \end{pmatrix} \begin{pmatrix} a & 0 \\ 0 & 0 \end{pmatrix} + \begin{pmatrix} 0 & 0 \\ 1/q & 0 \end{pmatrix} \begin{pmatrix} 0 & q \\ 0 & 0 \end{pmatrix} = \begin{pmatrix} 1 & 0 \\ 0 & 1 \end{pmatrix}$$

and

$$\begin{pmatrix} a & 0 \\ 0 & 0 \end{pmatrix} y \in A, \quad \begin{pmatrix} 0 & q \\ 0 & 0 \end{pmatrix} y \in A.$$

Now we utilize Theorem 16.6, (c).

For left orders there is an analogous result to Theorem 17.1.

THEOREM 19.2. *For every ring A there exists a ring $O(A)$ containing A as subring and such that:*

(i) *A is a left order of $O(A)$;*

(ii) *For every ring B such that A is a left order of B, there exists a unique ring morphism $t: B \to O(A)$ such that the following diagram is commutative:*

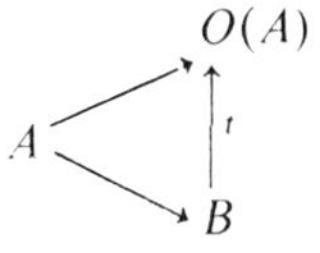

Proof. It is clear that if the ring $O(A)$ exists then $O(A) = A_F$ for some perfect system $F \subseteq D$, so that $O(A) \subseteq Q(A)$. Let C be the set of all subrings B of $Q(A)$ containing A and such that A is a left order of B. It is clear that $B = A_S$, where S is a left calculable regular system. We claim that the set C is in fact a direct set. In fact, if S and S' are two left calculable regular systems on A denote by SS' the set of all elements $s_1 s_1' s_2 s_2' \ldots s_n s_n'$ with $s_i \in S$ and $s_i' \in S'$, where n is a natural number. We see that SS' is a multiplicative regular system and $SS' = S'S$. We shall prove that SS' is in fact left calculable.

Denote by B the smallest subring of $Q(A)$ containing at the same time A_S and $A_{S'}$. An element of B has the form:

$$b = \sum_{i=1}^{m} s_{1i}^{-1} a_{1i} s_{1i}'^{-1} a_{1i}' \ldots s_{ni}^{-1} a_{ni} s_{ni}'^{-1} a_{ni}'$$

where $s_{1i}, \ldots, s_{ni} \in S$, $s_{1i}', \ldots, s_{ni}' \in S'$ and all the other elements are in A. In order to simplify assume first that $n = m = 1$.

Then we have:

$$b = s_1^{-1} a_1 s_1'^{-1} a_1' = s_1^{-1} (a_1 s_1'^{-1}) a_1'.$$

But $a_1 s_1'^{-1} = s_2'^{-1} a_2$, where $s_2' \in S'$. Then

$$b = s_1^{-1} s_2'^{-1} a_2 a_1' = (s_2' s_1)^{-1} a_3, \text{ with } s_2' s_1 \in SS'.$$

Furthermore, if $m = 1$ and $n \geqslant 1$, then by induction over n we deduce that any element of B of the form

$$b = s_1^{-1} a_1 s_1'^{-1} a_1' \ldots s_n^{-1} a_n s_n'^{-1} a_n'$$

can be expressed as

$$b = s^{-1} a \quad \text{with} \quad s \in SS'.$$

Hence an element in B has the form

$$b = \textstyle\sum_i s_i^{-1} a_i \quad \text{with} \quad s_i \in SS'.$$

Let us assume that $i = 2$. Then

$$b = s_1^{-1} a_1 + s_2^{-1} a_2$$

with $s_1, s_2 \in SS'$. Then we have:

$$s_1 b = a_1 + s_1 s_2^{-1} a_2.$$

For simplicity we suppose that $s_1 = s_{11} s_{11}'$ and $s_2 = s_{21} s_{21}'$ with $s_{11}, s_{21} \in S$, $s_{11}', s_{21}' \in S'$. Then $s_1 b = a_1 + s_{11} s_{11}' s_{21}'^{-1} s_{21}^{-1} a_2$.

As the above, $s_{11}' s_{21}'^{-1} = \bar{s}'^{-1} a_3$, with $\bar{s}' \in S'$ and $a_3 s_{21}^{-1} = \bar{s}^{-1} a_4$ with $\bar{s} \in S$. Therefore, we have:

$$s_1 b = a_1 + s_{11} \bar{s}'^{-1} \bar{s}^{-1} a_4 a_2.$$

Again $s_{11} \bar{s}'^{-1} = s'^{-1} a_5$, with $s' \in S'$ and $a_5 \bar{s}^{-1} = s^{-1} a_6$ with $s \in S$, hence $s_1 b = a_1 + s'^{-1} s^{-1} a_7 = a_1 + s_3^{-1} a_7$ where $s_3 = ss' \in SS'$, $a_7 = a_6 a_4 a_2$.

Furthermore, we can write

$$s_3 s_1 b = s_3 a_1 + a_7 = a_8 \in A,$$

i.e.

$$b = s_4^{-1} a_8, \quad \text{where} \quad s_4 = s_3 s_1 \in SS'.$$

Using induction on n, we deduce that any element in B may be written as:

$$b = s^{-1} a \quad \text{where} \quad s \in SS',$$

and the conditions $(CF1)$—$(CF3)$ of Section 4.18 are obviously satisfied. In conclusion $B = A_{SS'}$, so that SS' is left calculable by Theorem 18.6, as claimed.

Now the rest of the proof becomes clear: $O(A) = \bigcup_{B\in C} B_S$ or equivalently, the set of left calculable regular systems on A has a greatest element S, and $O(A) = A_S$. The other assertions are obvious. ∎

The ring $O(A)$ (or $O_l(A)$ if there is a risk of confusion) is called the *total classical left ring of fractions* of A. It is clear that for any ring A we have the inclusions of rings:

$$A \subseteq O(A) \subseteq M(A) \subseteq Q(A).$$

We shall denote by T the multiplicative system of all elements $s \in A$ which become invertible in $O(A)$.

NOTE 2. In frequent usage, by the total classical left ring of fractions one means a ring Q_{cl} which contains A as a left order and such that any regular element of A is invertible in Q_{cl}. It is clear that Q_{cl} exists if and only if T is precisely the set of all regular elements of A, or equivalently, if the set of regular elements is left calculable. In this case $Q_{cl} = O(A)$.

Now we shall make some observations on left orders and we shall "calculate" $O(A)$ for some rings.

The next result is due to Jans [97], and Mewborn and Winton [129].

THEOREM 19.3. *Let A be a ring and Q the injective envelope of A. The following assertions are equivalent:*

(1) *A is a left order in a QF-ring;*

(2) *A satisfies the following conditions:*

(a) *A has finite left dimension;*

(b) *A has ACC on left Q-annihilators;*

(c) *For any element $q \in Q$ there exists a regular element $a \in A$ such that $aq \in A$.*

Moreover, then $O(A) = M(A) = Q(A)$.

Proof. According to Theorem 17.12 it will be sufficient to prove that (2) ⇒ (1). First we observe that $\mathrm{Hom}_A(Q/A, Q) = 0$. Indeed, if $f: Q/A \to Q$ is a non-zero morphism then an element $q \in Q$ can be chosen such that $0 \neq f(\bar{q}) \in A$, $\bar{q}$ being the image of q in Q/A; let $a \in A$ be a regular element such that $aq \in A$ Then $af(\bar{q}) = f(a\bar{q}) = 0$, that is $f(\bar{q}) = 0$, a contradiction. Then by Exercise 13 of Section 4.13 we check that $Q(A)$ is self injective. Furthermore, we see that $Q(A)$ is also finite dimensional as an A-module. Thus if s is a regular element of A, then s becomes also left regular in $Q(A)$ so that the A-submodule $Q(A)s^n$ of $Q(A)$ is isomorphic with $Q(A)$ for any natural number n. Hence $Q(A)s^n$ is a direct summand of $Q(A)s^{n-1}$ for $n > 1$. But $Q(A)$ being finite dimensional we saw that $Q(A)s^n = Q(A)s^{n+1}$ for some natural number n. Thus $s^n = qs^{n+1}$, that is $1 = qs$. By Theorem 16.6 we may conclude that A is in fact a left quasi-order of $Q(A)$ and thus by an argument already used in the proof of Theorem 18.6, implication (2) ⇒ (3) we deduce that any regular element of A is invertible in $Q(A)$. Therefore A is a left order of $Q(A)$ and by Theorem 17.12, $Q(A)$ is a QF-ring. The proof is now complete. ∎

LEMMA 19.4. *If A satisfies ACC on left annihilators, then the singular ideal of A is nilpotent.*

Proof. We will show that the ascending chain $l(J(A)) \subseteq l((J(A)^2) \subseteq \ldots$ would be strictly ascending if $J(A)$ were not nilpotent. If $(J(A))^n \neq 0$, choose an element $a \in J(A)$ with $a(J(A))^{n-1} \neq 0$ and the largest possible left annihilator. For each $b \in J(A)$ we have $l(b) \cap Aa \neq 0$ since $l(b)$ is essential in A. So there exists $c \in A$ such that $ca \neq 0$, but $cab = 0$, which means that $l(ab)$ is strictly larger than $l(a)$, and by the choice of a, we must therefore have $ab(J(A))^{n-1} = 0$. Since $b \in J(A)$ is arbitrary, we get $a(J(A))^n = 0$ and hence $l((J(A))^{n-1})$ is strictly contained in $l((J(A))^n)$. ∎

We may now characterize those rings for which $O(A)$ is semisimple, or equivalently, which are left orders in a semisimple ring.

THEOREM 19.5 (Goldie [73]). *The following properties of a ring A are equivalent:*

(1) *$O(A)$ is a semisimple ring;*

(2) *A is left finite-dimensional and non-singular, and has no nilpotent non-zero ideals;*

(3) *A is left finite dimensional, satisfies ACC on left annihilators, and has no nilpotent non-zero ideals;*

(4) *A is left finite dimensional and every essential left ideal contains a regular element.*

Proof. (1) ⇒ (4). Since necessarily $O(A) = Q(A)$ is semisimple, we know from Theorem 17.10 that for every essential left ideal $\mathfrak{a}$ of A we have $O(A)\mathfrak{a} = O(A)$. Write $1 = \sum q_i a_i$ with $a_1, \ldots, a_n \in \mathfrak{a}$. By Exercise 5 of Section 4.18, there exists a regular element $s \in A$ such that $sq_i \in A$ for every i. Then

$$s = \sum sq_i a_i \in \mathfrak{a},$$

so that any essential left ideal of A contains a regular element. By Lemma 17.8, we know that A is left finite dimensional.

(4) ⇒ (1). It is clear that any essential left ideal of A is dense so that $Q(A)$ is a left self injective ring. We claim that any regular element $s \in A$ becomes left invertible in $Q(A)$. In fact, $Q(A)$ is also left finite-dimensional and s is left regular in $Q(A)$. Using an argument already used in the proof of Theorem 19.3 we check that $qs = 1$ for a suitable $q \in Q(A)$, as claimed. As so far A is a left quasi-order in $Q(A)$ and again as in the proof of Theorem 18.6, implication (2) ⇒ (3), we find that s is in fact invertible in $Q(A)$. Now it is clear that $O(A) = Q(A)$. Finally by Theorem 17.10 we deduce that $Q(A)$ is in fact semisimple.

(1) ⇒ (3). By Lemma 17.8 it follows that we have to show only that every nilpotent ideal $\mathfrak{a}$ is zero. In fact $r(\mathfrak{a})$ is an essential left ideal, for if $x \neq 0$ is an element of A, let n be the smallest integer such that $\mathfrak{a}^n x = 0$; then there exists $b \in \mathfrak{a}^{n-1}$ such that $bx \neq 0$ and $bx \in r(\mathfrak{a})$. But since (1) ⇒ (4), every essential left ideal contains a regular element and therefore $\mathfrak{a}$ must be zero.

(3) ⇒ (2). $J(A)$ is a nilpotent ideal by Lemma 19.4. By hypothesis it must be zero, so A is non-singular.

(2) ⇒ (3). Indeed, by Theorem 17.10 $Q(A)$ is semisimple and from Lemma 17.8 this implies ACC on left annihilators.

(3) ⇒ (4). We first prove:

LEMMA 19.6. *If A satisfies ACC on left annihilators and has no nilpotent non-zero ideal, then every left or right nilideal is zero.*

Proof. Since Aa is a nilideal if and only if aA is a nilideal it is sufficient to consider a nilideal aA. Assume $aA \neq 0$. Among the non-zero elements of aA, we choose one b as with maximal left annihilator. For each $c \in A$, let $(bc)^k = 0$ and $(bc)^{k-1} \neq 0$. Since $l(b) \subseteq l((bc)^{k-1})$, we must have equality by maximality. Hence $bc \in l(b)$ and $bAb = 0$, which gives $(bA)^2 = 0$. Since every nilpotent ideal is zero, we get $b = 0$. This is a contradiction and so $aA = 0$.

We return to proof of the implication (3) ⇒ (4). Let $\mathfrak{a}$ be an essential left ideal. Since A has ACC on left annihilators and $\mathfrak{a}$ is not a nilideal (by Lemma 19.4), there exists $a_1 \neq 0$ in $\mathfrak{a}$ such that $l(a_1) = l(a_1{}^2)$. If

$$l(a_1) \cap \mathfrak{a} \neq 0,$$

we continue and choose $a_2 \in \mathfrak{a} \cap l(a_1)$ such that $a_2 \neq 0$ and $l(a_2) = l(a_2{}^2)$. If then $\mathfrak{a} \cap l(a_2) \cap l(a_1) \neq 0$, we continue and get $a_3 \in \mathfrak{a} \cap l(a_2) \cap l(a_1)$, and so on. At each step we obtain a direct sum $Aa_1 \amalg \dots \amalg Aa_k$. This is proved by induction: let us suppose that $Aa_1 \amalg \dots \amalg Aa_{k-1}$ is a direct sum and $ba_k = b_1 a_1 + \dots + b_{k-1} a_{k-1}$; since for each $i < k$ we have $a_k a_i = 0$, we get $b_i \in l(a_i^2) = l(a_i)$ and hence

$$\sum_{i=1}^{k-1} b_i a_i = 0.$$

But A is left finite dimensional, so the process must stop at some stage, where we have $\mathfrak{a} \cap l(a_1) \cap \dots \cap l(a_k) = 0$. Then $l(a_1) \cap \dots \cap l(a_k) = 0$ so if $s = a_1 + \dots + a_k \in \mathfrak{a}$, then $l(s) = 0$. Furthermore, by Theorem 17.10 $Q(A)$ is semisimple and s becomes again left regular in $Q(A)$. Then s is invertible by Exercise 1. The proof is now complete. ■

COROLLARY 19.7. *Let A be a commutative ring. The following assertions are equivalent:*

(1) *A is an order of a semisimple ring;*

(2) *A is finite-dimensional and reduced (i.e. has no non-zero nilpotent elements).*

Proof. It will be sufficient to prove that (2) ⇒ (1). Indeed, by Exercise 11 of Section 4.9, A is non-singular and thus $Q(A)$ is semisimple by the Theorem 17.10. Finally, by Lemma 17.8 we see that conditions of Theorem 19.5 are fulfilled. ■

NOTES 3. Corollary 19.7 proves that for a commutative reduced ring of finite dimension one obtains: $O(A) = M(A) = Q(A)$. Particularly, this is the case for a commutative noetherian and reduced ring.

4. For the rings considered in Theorem 19.3 and 19.5 we find that $Q_{cl}(A)$ exists and moreover $Q_{cl}(A) = O(A)$.

COROLLARY 19.8. *Let A be a left Ore domain. Then $O(A) = Q(A)$. Moreover $Q(A)$ is a division ring, namely the division ring of additive left fractions of A.*

Proof. It is clear that A is left one-dimensional and any essential left ideal of A contains a regular element. Then by Theorem 19.5, $Q(A)$ is a semisimple left one-dimensional, that is a division ring. ■

COROLLARY 19.9. *Let A be a left Ore domain and $F \in L(A)$. Then A_F is the subring of $Q(A)$ consisting of all elements b such that $(A : b) \in F$.* ■

Now we shall give a general form of a theorem of Endo (see [51], Theorem 1, see also [206]).

THEOREM 19.10. *Let A be a ring and S be a left permutable regular system on A. Then a finitely generated A-module M is projective if and only if M is flat and the module $A_S \otimes_A M$ is projective as an A_S-module.*

Proof. If M is projective, then M is flat (see Chapter 3, Section 3.8) and $A_S \otimes_A M$ is also projective by the dual of the Theorem 2.8 of Chapter 3. Conversely, let M be a flat A-module generated by elements $m_1, \ldots, m_n$. The sequence $0 \to M \to A_S \otimes_A M$ induced by the inclusion $A \subseteq A_S$ is exact, hence we may identify elements m and $1 \otimes m$. Let us assume, moreover, that $A_S \otimes_A M$ is projective as A_S-module. We shall show that the condition (b) of Lemma 9.1, Chapter 3 is satisfied by the A-module M. By this Lemma applied to $A_S \otimes_A M$ there exist elements $q_{ij} \in A_S$, $i, j = 1, \ldots, n$ such that

$$m_i = \textstyle\sum_j q_{ij} m_j \qquad i = 1, \ldots, n$$

and if

$$\sum r_{ik} m_i = 0 \quad \text{for} \quad k \in K, \qquad r_{ik} \in A \subseteq A_S$$

then

$$\text{(a)} \quad \textstyle\sum_i r_{ik} q_{ij} = 0 \quad \text{for all} \quad j, k.$$

By Exercise 5 of Section 4.18 an element $s \in S$ can be found such that $sq_{ij} \in A$ for all i, j. Then:

$$\textstyle\sum_j s(\delta_{ij} - q_{ij}) m_j = 0 \quad \text{for} \quad j = 1, \ldots, n, \tag{13}$$

and $s(\delta_{ij} - q_{ij}) \in A$ (here δ_{ij} is the well-known Kronecker symbol). If the A-module M is flat, then by Exercise 1 of Section 3.8, Chapter 3 applied to (13) there exist $a_{lj} \in A$, $l, j = 1, \ldots, n$ such that

$$m_j = \sum_{l=1}^{n} a_{lj} m_l, \qquad \sum_{j=1}^{n} s(\delta_{ij} - q_{ij}) a_{lj} = 0 \quad \text{for all} \quad i = 1, \ldots, n.$$

Since s is regular we have

$$a_{li} = \sum_{j=1}^{n} q_{ij} (a)_{lj}.$$

Suppose now that $\sum_i r_{ik} m_i = 0$ for $k \in K$, $r_{ik} \in A$. Then (a) implies:

$$\sum r_{ik} a_{li} = \sum_j \sum_i r_{ik} a_{lj} q_{ij} = 0$$

for all $k \in K$, $l = 1, \ldots, n$. Then M is projective as an A-module by Lemma 9.1, Chapter 3. ∎

Exercises

1. Let A be a semisimple ring. The following assertions are equivalent for an element $a \in A$.

(i) a is left regular;

(ii) a is right regular;

(iii) a is left invertible;

(iv) a is right invertible;

(v) a is invertible.

Moreover the equivalence (i) $\Leftrightarrow$ (iii) is valid if A is only assumed to be left self-injective and left finite-dimensional.

2. Let S be a left calculable regular system on A and B a ring such that $A \subseteq B \subseteq A_S$. If B is flat as A^0-module then the inclusion $A \subseteq B$ is a right flat epimorphism of rings.

3. Prove that for any ring A one has $O(O(A)) = O(A)$.

4. A domain A is a left Ore domain if and only if the intersection of any two non-zero left ideals is also non-zero. Moreover, a subring A of a division ring K, which is essential as an A-module is a left Ore domain.

5. If A is a left Ore domain and B a subring of $O(A)$ containing A, then B is also a left Ore domain and $O(B) = O(A)$.

6. Let A be a left Ore domain and X a *torsion free* A-module (i.e. for any $x \in X$, $x \neq 0$, $\mathrm{Ann}(x) = 0$). Then X is injective if and only if it is divisible.

7. Let A be a domain. The following assertions are equivalent:

(a) A is a left Ore domain;

(b) The injective envelope of A is a torsion-free A-module. (Hint: If Q is the injective envelope of A, then any non-zero left ideal of A is an element of $F(Q) = D$).

8. A semi-hereditary integral domain is called also a *Prüfer ring*. Show that the following statements are equivalent:

(a) A is a Prüfer ring.

(b) Any subring B of $O(A)$ containing A is flat as A-module (or equivalently, by Exercise 2), the inclusion $A \subseteq B$ is a flat epimorphism of rings).

9. The following assertions are equivalent for a ring A:

(a) $O(A)$ is a left self injective semi-perfect ring.

(b) A is left finite dimensional and $(A:q)$ contains a regular element for each $q \in E(A)$.

10. Suppose that A satisfies:

(a) A is left finite-dimensional;

(b) A has ACC on left annihilators;

(c) $(A:q)$ contains a regular element for each $x \in E(A)$. Then $O(A)$ is a semi-primary left self-injective ring.

11. Let A be the ring of matrices:

$$\begin{pmatrix} a & q \\ 0 & b \end{pmatrix} \text{ with } a, b \in \mathbf{Z} \quad \text{and} \quad q \in \mathbf{Q}.$$

Show that the ring of upper triangular matrices over $\mathbf{Q}$ is equal to $O(A)$, while $\mathbf{Q}_2$ is equal to $Q(A)$. (Note that A has nilpotent non-zero ideals).

References

Bourbaki [27]; Cohn [39]; Endo [51]; Gabriel [59]; Goldie [72], [73]; Goldman [76]; Jacobson [96]; Jans [97]; Lazard [117]; Mewborn–Winton [129]; Ore [158]; Popescu–Spulber [177]; Simson [206]; Stenström [213]; van der Water [237].

4.20 The left spectrum of a ring

By a *local Grothendieck category* we shall mean a Grothendieck category $\mathscr{C}$ having a simple object S whose injective envelope $E(S)$ is a cogenerator of $\mathscr{C}$. A ring A is called *left local* if the category Mod A is a local Grothendieck category. Analogously, we have the concept of right local ring.

PROPOSITION 20.1. (1) *In a local Grothendieck category $\mathscr{C}$ any two simple objects are isomorphic.*

(2) *A ring A is left local if and only if for any two maximal left ideals $\mathfrak{m}_1$ and $\mathfrak{m}_0$ there exist two elements $x_1 \notin \mathfrak{m}_1$, $x_0 \notin \mathfrak{m}_0$ such that* $(\mathfrak{m}_1 : x_1) = (\mathfrak{m}_0 : x_0)$.

Proof (1). Let S be a simple object of $\mathscr{C}$ such that $E(S)$ is a cogenerator of $\mathscr{C}$; if S' is another simple object of $\mathscr{C}$, then $\operatorname{Hom}_{\mathscr{C}}(S', E(S)) \neq 0$ and thus S' can be regarded as a subobject of $E(S)$. Then $S \cap S' \neq 0$, so that $S \simeq S'$.

(2) Let $\mathfrak{m}$ be a left maximal ideal of A such that $E(A/\mathfrak{m})$ is an injective cogenerator of Mod A. By (1) for any left maximal ideal $\mathfrak{m}_1$ of A we have

that $A/\mathfrak{m}_1 \simeq A/\mathfrak{m}$ so that $\mathfrak{m}_1 = (\mathfrak{m} : x_1)$ for an element $x_1 \notin \mathfrak{m}$. Now, if $\mathfrak{m}_0$ is another left maximal ideal of A, then $\mathfrak{m} = (\mathfrak{m}_0 : x)$ with $x \notin \mathfrak{m}_0$ so that $\mathfrak{m}_1 = (\mathfrak{m} : x_1) = ((\mathfrak{m}_0 : x) : x_1) = (\mathfrak{m}_0 : x_1 x)$ and $x_1 x \notin \mathfrak{m}_0$.

Conversely, we deduce that for any two left maximal ideals $\mathfrak{m}_1$, $\mathfrak{m}_0$ of A we have that $A/\mathfrak{m}_1 \simeq A/\mathfrak{m}_0$, so by Exercise 6 of Section 3.10, Chapter 3, we check that A is left local. ■

COROLLARY 20.2. (1) *Any local ring is left and right local.*

(2) *A commutative ring A is left or right local if and only if it is local.*

Proof. Point (1) is a direct consequence of Proposition 20.1.

(2) Let A be a left local commutative ring and $\mathfrak{m}$ a maximal ideal of A. Then $\mathfrak{m} = (\mathfrak{m} : x)$ for any $x \notin \mathfrak{m}$; so that any two maximal ideals of A are identical by the Proposition 20.1. ■

NOTE 1. There exist of course, left local rings which are not local. Such a ring is, for example, a full matrix ring over a field (see Chapter 3, Section 3.11).

Another example can be obtained as follows. Let A be a commutative local ring. Then by Theorem 11.4 of Chapter 3, Mod A and Mod A_n are equivalent for any natural number n. Thus we find that A_n is a left local ring for each n. Moreover if $n > 1$, then A_n is a true left local ring, i.e. it is not a local ring. Indeed, let $\mathfrak{a}$ be the left ideal of A_n consisting of all one-column matrices of the form

$$\begin{pmatrix} a_{11} & 0\ldots0 \\ a_{21} & 0\ldots0 \\ \ldots & \ldots \\ a_{n1} & 0\ldots0 \end{pmatrix}$$

It is easy to see that $\mathfrak{a}A_n = A_n$, so that $\mathfrak{a}$ is not contained in any two-sided ideal of A_n.

A non-zero object X of an abelian category $\mathscr{C}$ is called *coirreducible* if it is an essential extension of any of its non-zero subobjects. Obviously, X is coirreducible if and only if for any two non-zero subobjects X', X'' of X it follows that $X' \cap X'' \neq 0$. An *injective indecomposable* object is by definition an injective coirreducible object.

LEMMA 20.3. *For an injective object Q of an abelian category $\mathscr{C}$ the following assertions are equivalent:*

(1) *Q is an injective indecomposable object;*

(2) *Any direct summand of Q is zero or Q;*

(3) *The ring* $\mathrm{End}_{\mathscr{C}}(Q)$ *is local.*

Proof. The equivalence (1) $\Leftrightarrow$ (2) is obvious.

(1) $\Rightarrow$ (3). Let f, g two non-units of $\mathrm{End}_{\mathscr{C}}(Q)$; then $\ker f$ and $\ker g$ are non-zero, so that $0 \neq \ker f \cap \ker g \subseteq \ker\ (f + g)$, that is $f + g$ is also non-invertible. Therefore $\mathrm{End}_{\mathscr{C}}(Q)$ is local by the Proposition 9.5 of Chapter 3.

(3) $\Rightarrow$ (2). Assume if possible that Q has proper direct summands. Then by Theorem 1.1, Chapter 2, $1_Q = u + v$ where u and v are suitable non-units of $\mathrm{End}_{\mathscr{C}}(Q)$; a contradiction. ∎

NOTE 2. It is clear that any essential extension of a coirreducible object is also coirreducible. Particularly, the injective envelope of a coirreducible object is an injective indecomposable object.

LEMMA 20.4. *Let $\mathscr{C}$ be an abelian category, S a simple object of $\mathscr{C}$ and Q the injective envelope of S. Then Q is an injective indecomposable object and the division ring* $\mathrm{End}_{\mathscr{C}}(S)$ *is canonically isomorphic with* $\mathrm{End}_{\mathscr{C}}(Q)/\mathfrak{m}$, $\mathfrak{m}$ *being the maximal ideal of* $\mathrm{End}_{\mathscr{C}}(Q)$.

Proof. Let $f \in \mathrm{End}_{\mathscr{C}}(Q)$; we assume that $f(S) \neq 0$. Then, $f(S)$ is also simple and $S \cap f(S) \neq 0$, i.e. $S = f(S)$. Hence f defines an endomorphism $\alpha(f)$ of S. It is clear that the mapping $f \rightsquigarrow \alpha(f)$ defines a ring morphism $\alpha: \mathrm{End}_{\mathscr{C}}(Q) \to \mathrm{End}_{\mathscr{C}}(S)$. We see that α is onto, Q being injective. Finally, $\alpha(f) = 0$ if and only if $\ker f \neq 0$, i.e. $f \in \mathfrak{m}$ so that $\ker \alpha = \mathfrak{m}$. ∎

By a *left prime* of a ring A, we shall mean an element $F \in L(A)$ such that the category $\mathrm{Mod}\, A/\mathscr{F}$ is a local Grothendieck category. Sometimes we shall say that F is a "left prime system" or simply "prime" if there is no danger of confusion.

PROPOSITION 20.5. *Let F be a left prime of A, U a simple object of* $\mathrm{Mod}\, A/\mathscr{F}$, *$E = E(U)$ and $Q = S(E)$ ($S: \mathrm{Mod}\, A/\mathscr{F} \to \mathrm{Mod}\, A$ is the canonical functor). Then $F = F(Q)$, that is Q is an injective object defining F.*

Proof. It is clear that an A-module X is in $\mathscr{F}(Q)$ if and only if

$$\mathrm{Hom}_A(X, Q) = 0.$$

Then $\mathrm{Hom}_A(X, S(E)) \simeq \mathrm{Hom}_{\mathrm{Mod}\, A/\mathscr{F}}(T(X), E) = 0$ so that $T(X) = 0$, E being a cogenerator of $\mathrm{Mod}\, A/\mathscr{F}$. Hence $\mathscr{F}(Q) \subseteq \mathscr{F}$. Conversely, if $T(X) = 0$, then $\mathrm{Hom}_A(X, S(E)) \simeq \mathrm{Hom}_{\mathrm{Mod}\, A/\mathscr{F}}(T(X), E) = 0$, i.e. $\mathscr{F} \subseteq \mathscr{F}(Q)$. ∎

THEOREM 20.6. *Let $F \in L(A)$, U a simple object of* $\mathrm{Mod}\, A/\mathscr{F}$, *$E$ the injective envelope of U and $Q = S(E)$. Then $F(Q)$ is a left prime of A.*

Proof. Let $\mathscr{A}$ be the localizing subcategory of Mod $A/\mathscr{F}$ consisting of all objects X such that $\mathrm{Hom}_{\mathrm{Mod}\,A/\mathscr{F}}(X, E) = 0$. Then $\mathscr{F}(Q)$ consists exactly of all A-modules M such that $T(M) \in \mathrm{Ob}\,\mathscr{A}$. Indeed, if $\mathrm{Hom}_A(M, Q) = 0$, then $\mathrm{Hom}_{\mathrm{Mod}\,A/\mathscr{F}}(T(M), E) = 0$ so that $T(M) \in \mathrm{Ob}\,\mathscr{A}$. Conversely, if

$$T(M) \in \mathrm{Ob}\,\mathscr{A}$$

then obviously $M \in \mathrm{Ob}\ \mathscr{F}\ (Q)$. Furthermore, it is easy to check that $\mathrm{Mod}A/\mathscr{F}(Q)$ is canonically equivalent with

$$(\mathrm{Mod}\ A/\mathscr{F})/\mathscr{A} = \mathscr{C}.$$

Now the point is to check that Mod $A/\mathscr{F}(Q)$ is local. To that end, let

$$T' : \mathrm{Mod}\ A/\mathscr{F} \to \mathscr{C}$$

be the canonical functor, $U' = T'(U)$ and $E' = E(U')$. It is clear that U' is simple and $E' = T'(E)$ (see Lemma 5.1). Furthermore, if X' is a non-zero object of $\mathscr{C}$, then $X' = T'(X)$ and $\mathrm{Hom}_{\mathrm{Mod}\,A/\mathscr{F}}(X, E) \neq 0$, hence

$$\mathrm{Hom}_{\mathscr{C}}(T(X), E') \simeq \mathrm{Hom}_{\mathscr{C}}(X', E') \neq 0,$$

E being $\mathscr{A}$-closed. Therefore E' is a cogenerator of $\mathscr{C}$, i.e. $\mathscr{C}$ is local. ∎

Let A be a ring and $\mathfrak{m}$ a left maximal ideal of A; then $F\mathfrak{m}$ is a left prime of A, according to the previous theorem. For any ring A, we shall denote by Spel(A) the set of all left primes of A. From the above considerations we deduce the following result:

COROLLARY 20.7. *For any ring A the set* Spel(A) *is non empty.* ∎

NOTE 3. Let $\mathscr{C}$ be a Grothendieck category, denote by $L(\mathscr{C})$ the set of all localizing subcategories of $\mathscr{C}$ (see Theorem 14.2 and Exercise 6 of Section 4.3). An element $\mathscr{A} \in L(\mathscr{C})$ is called a *prime localizing subcategory* or a *prime of* $\mathscr{C}$ if the category $\mathscr{C}/\mathscr{A}$ is a local category. If $T : \mathrm{Mod}\ A^0 \to \mathscr{C}$ is an exact functor having a full and faithful right adjoint S, then $\mathscr{A}$ is a prime of $\mathscr{C}$ if and only if the right localizing system F on A consisting of all right ideals $\mathfrak{a}$ such that $T(A/\mathfrak{a}) \in \mathrm{Ob}\,\mathscr{A}$ is a right prime of A. From these considerations we derive that the study of primes of a Grothendieck category can be reduced to the study of certain right primes of a suitable ring. Moreover, the problem of existence of primes for a Grothendieck category is still open.

From the Proposition 20.5 we derive that any left prime of a ring A is of the form $F(Q)$, Q being a suitable injective indecomposable A-module; the next result proves the uniqueness of Q.

THEOREM 20.8. *Let F be a left prime for a ring A. If Q and Q' are two injective indecomposable A-modules, such that $F = F(Q) = F(Q')$, then Q is isomorphic with Q'.*

Proof. It is clear that Q and Q' are $\mathscr{F}$-closed. Let U be a simple object of $\operatorname{Mod} A/\mathscr{F}$; then $\operatorname{Hom}_{\operatorname{Mod} A/\mathscr{F}}(U, T(Q)) \neq 0$, so that $T(Q)$ is an injective envelope of U (see Exercise 1). Similarly $T(Q')$ is an injective envelope of U, hence $T(Q) \simeq T(Q')$ and finally $Q \simeq Q'$. ∎

Now, let Q be an injective indecomposable object of $\operatorname{Mod} A$; then for any non-zero subobject X of Q we have $F(X) = F(Q)$. Particularly if

$$0 \neq x \in Q,$$

then $F_{\operatorname{Ann}(x)} = F(Ax) = F(Q)$, i.e. $F(Q)$ has a left ideal of definition. Moreover any left prime of A has a left ideal of definition. Now, we shall determine all left ideals $\mathfrak{a}$ such that $F_{\mathfrak{a}}$ is a left prime of A.

THEOREM 20.9. *Let $F \in L(A)$. The following assertions are equivalent:*

(1) *F is a left prime of A;*

(2) *$F = F_{\mathfrak{a}}$ where $\mathfrak{a}$ is a left ideal of A such that for any left ideal $\mathfrak{b}$, with $\mathfrak{a} \subset \mathfrak{b}$ (and $\mathfrak{a} \neq \mathfrak{b}$), it follows that $\mathfrak{b} \in F_{\mathfrak{a}}$.*

Proof. (1) ⇒ (2). Let U be a simple object of $\operatorname{Mod} A/\mathscr{F}$, $0 \neq x \in S(U)$ and $\mathfrak{a} = \operatorname{Ann}(x)$. By Proposition 20.5 and the above considerations, we have: $F = F(S(U)) = F_{\mathfrak{a}}$. Moreover, if X is a non-zero subobject of $S(U)$, then $S(U)/X \in \operatorname{Ob} \mathscr{F}$. Indeed, from the exact sequence:

$$0 \longrightarrow X \xrightarrow{i} S(U) \longrightarrow S(U)/X \longrightarrow 0$$

in which i is a canonical injection, we have that $T(i)$ is an isomorphism because $U = TS(U)$ is simple. Therefore $T(S(U)/X) = 0$. Now, let $\mathfrak{b}$ be a left ideal of A such that $\mathfrak{a} \subset \mathfrak{b}$ ($\mathfrak{a} \neq \mathfrak{b}$). Then $0 \neq \mathfrak{b}/\mathfrak{a} = X \subseteq A/\mathfrak{a} \subseteq S(U)$. As above $S(U)/X \in \operatorname{Ob} \mathscr{F}$ so that from the exact sequence

$$0 \longrightarrow \mathfrak{b}/\mathfrak{a} \longrightarrow A/\mathfrak{a} \longrightarrow A/\mathfrak{b} \longrightarrow 0$$

we deduce that $A/\mathfrak{b}$ is isomorphic with a subobject of $S(U)/X$, hence $\mathfrak{b} \in F$.

(2) ⇒ (1). First we shall prove that $\mathfrak{a}$ is irreducible (see Exercise 2). Let $\mathfrak{b}_1, \mathfrak{b}_2$ be two left ideals of A such that $\mathfrak{a} = \mathfrak{b}_1 \cap \mathfrak{b}_2$. If $\mathfrak{b}_1 \neq \mathfrak{a} \neq \mathfrak{b}_2$, then $\mathfrak{b}_1, \mathfrak{b}_2 \in F_{\mathfrak{a}}$ and so $\mathfrak{a} = \mathfrak{b}_1 \cap \mathfrak{b}_2 \in F_{\mathfrak{a}}$, a contradiction. Hence $\mathfrak{a}$ is irreducible, i.e. $A/\mathfrak{a}$ is coirreducible. Now let $Q = E(A/\mathfrak{a})$; then Q is an injective indecomposable object. Also, let $U = T(A/\mathfrak{a}) \subseteq T(Q)$. It is clear that $U \neq 0$;

moreover U is simple. In fact, if X is a non-zero subobject of U, then we have in Mod A the following diagram:

$$\begin{array}{c} A/\mathfrak{a} \subset S(U) \simeq ST(A/\mathfrak{a}) \\ \cup \\ S(X) \end{array}$$

in which we derive that $S(X) \cap A/\mathfrak{a} = Y \neq 0$ (see Lemma 4.6). Therefore $Y = \mathfrak{b}/\mathfrak{a}$ where $\mathfrak{b}$ is a left ideal containing $\mathfrak{a}$ and different from $\mathfrak{a}$. By hypothesis, $\mathfrak{b} \in F_{\mathfrak{a}}$. But then, if $i: \mathfrak{b}/\mathfrak{a} \to A/\mathfrak{a}$ is the inclusion we derive that $T(i)$ is an isomorphism. Thus we find that $X = U$, i.e. U is a simple object.

Finally we see that $T(Q)$ is an injective cogenerator of Mod $A/\mathcal{F}$ and an injective envelope of U. ∎

NOTE 4. Condition (2) of Theorem 20.9 is verified for a left ideal $\mathfrak{a}$ if and only if for any $x, y, z \in A$ such that $x, z \notin \mathfrak{a}$ there exists $u \in A$ with $ux \notin \mathfrak{a}$ and $uy \in \mathfrak{a} + Az$. A left ideal $\mathfrak{a}$ as in Theorem 20.9 is called a *left super-prime* ideal of A and $F_{\mathfrak{a}}$ the left prime associated with it.

LEMMA 20.10. *Let A be a commutative ring. An ideal $\mathfrak{a}$ of A is super-prime if and only if it is prime.*

Proof. Let $\mathfrak{a}$ be super-prime and $x, y \in A$ such that $xy \in \mathfrak{a}$. Assume that $x\mathfrak{a}, y \notin \mathfrak{a}$. Then $a + Ax$ and $\mathfrak{a} + Ay$ are elements of $F_{\mathfrak{a}}$. Then their product is also an element of $F_{\mathfrak{a}}$ (see Exercise 3, Section 4.9), a contradiction. Hence $x \in \mathfrak{a}$ or $y \in \mathfrak{a}$. Conversely, let us assume that $\mathfrak{a}$ is prime and $\mathfrak{b} \supset \mathfrak{a} (\mathfrak{b} \neq \mathfrak{a})$. Then $\mathfrak{b}$ contains an element x not in $\mathfrak{a}$ or equivalently $\mathfrak{b} \in F_{\mathfrak{a}}$ (see Proposition 9.11). Hence $\mathfrak{a}$ is super-prime. ∎

COROLLARY 20.11. *Let A be a commutative ring. Then an element $F \in L(A)$ is prime if and only if $F = F_{\mathfrak{a}}$ where $\mathfrak{a}$ is a prime ideal.*

Proof. From the above lemma and Theorem 20.9, $F_{\mathfrak{a}}$ is prime whenever $\mathfrak{a}$ is a prime ideal. Conversely if F is prime, then $F = F_{\mathfrak{a}}$ where $\mathfrak{a}$ is a super-prime ideal, i.e. a prime ideal. ∎

NOTE 5. It is easy to check that the map $\mathfrak{a} \rightsquigarrow F_{\mathfrak{a}}$ is a bijection between the set of prime ideals of a commutative ring and the set of prime systems. Moreover, for a commutative ring A we shall write Spec(A) = Spel(A) = Sper(A) {= the set of right primes of A}.

Let F be a left prime of A; we denote by U_F a simple object of Mod $A/\mathcal{F}$, $K_F = S(U_F)$, E_F an injective envelope of U_F and $Q_F = S(E_F)$. It is clear that these objects are defined up an isomorphism. If there is no danger of confusion, we shall write simply U, E, K, Q. Also we shall denote by k_F or simply k the division ring $\mathrm{End}_{\mathrm{Mod}\,A/\mathscr{F}}(U_F)$; k_F is called the *residue division ring* associated with F. Always U_F is considered as being a subobject of E_F and also K_F a subobject of Q_F.

THEOREM 20.12. *Let* $F \in L(A)$. *The following assertions are equivalent*:

(1) $F \in \mathrm{Spel}(A)$;

(2) *F has a module of definition X such that for any non-zero submodule X' of X, any non-zero morphism $f : X' \to X$ is a monomorphism.*

Proof. (1) ⇒ (2). We shall take $X = K$; let X' be a non-zero submodule of X and $f : X' \to K$ a non-null morphism. If $\ker f \neq 0$ from the sequence:

$$0 \longrightarrow \ker f \xrightarrow{\ i\ } X' \longrightarrow \mathrm{im}\, f \longrightarrow 0$$

we deduce that $\mathrm{im}\, f \in \mathrm{Ob}\,\mathscr{F}$; $T(i)$ being an isomorphism, because $T(X') = T(K) = U$. Hence K contains non-zero subobjects of $\mathscr{F}$, a contradiction. Consequently, it follows necessarily that $\ker f = 0$.

(2) ⇒ (1). First we check that X is coirreducible. Indeed, if X_1, X_2 are non-zero subobjects of X such that $X_1 \cap X_2 = 0$, then the sum $X_1 + X_2$ is direct and thus a suitable non-null morphism $f : X_1 + X_2 \to X$ can be chosen with $\ker f \neq 0$, a contradiction.

Now let Q be injective envelope of X. We shall prove that $F = F(Q)$ is a left prime. For this, it will be sufficient to show that $T(X)$ is a simple object of $\mathrm{Mod}\, A/\mathscr{F}$, because $T(Q)$, the injective envelope of $T(X)$ is a cogenerator. Furthermore, to prove that $T(X)$ is simple, it will be enough to check that $X/X' \in \mathrm{Ob}\,\mathscr{F}$ for any non-zero subobject X' of X, i.e. that $\mathrm{Hom}_A(X/X', Q) = 0$. By *reductio ad absurdum*, let us assume that there exists a non-null morphism $f : X/X' \to Q$. Let $p : X \to X/X'$ be the canonical epimorphism. Then $\ker(fp) \supseteq X'$ and $\mathrm{im}(fp) \cap X = Z \neq 0$. Then $Z = (fp)(X_1)$, where X_1 is a suitable subobject of X containing X'. The restriction of f to X_1/X' defines a non-zero morphism $t : X_1/X' \to Z$. Also the restriction of p to X_1 defines a morphism $h : X_1 \to X_1/X'$. Finally, th is a non-zero morphism of X_1 into X which is not a monomorphism. The contradiction is obvious. ∎

COROLLARY 20.13. (Goldman [76]). *Let* $F \in L(A)$. *The following assertions are equivalent*:

(1) *F is a left prime of A*;

(2) *$F = F(X)$ where X is such that for any non-zero submodule X' of X one has $\mathrm{Hom}_A(X/X', Q) = 0$, Q being the injective envelope of X.*

Proof. From the Theorem 20.12 it will be sufficient to prove that (2) ⇒ (1). Indeed, let X be as in (2), X' a non-zero submodule of X and $f: X' \to X$ a non-null morphism. If f is not a monomorphism, then from the exact sequence:

$$0 \longrightarrow \ker f \longrightarrow X' \longrightarrow \operatorname{im} f \longrightarrow 0$$

we find that $\operatorname{im} f \in \operatorname{Ob} \mathscr{F}$, a contradiction. ∎

Let $F \in \operatorname{Spel}(A)$, and let $X' \subseteq X \subseteq K_F$ be such that $X' \neq 0$. Then from the exact sequence:

$$0 \longrightarrow X' \xrightarrow{\ i\ } X \xrightarrow{\ p\ } X/X' \longrightarrow 0$$

we deduce the exact sequence:

$$0 \longrightarrow \operatorname{Hom}_A(X/X', X) \xrightarrow{p^*} \operatorname{Hom}_A(X, X) \xrightarrow{i^*} \operatorname{Hom}_A(X', X)$$

where i^* and p^* are canonically induced by i and p. By the above corollary we deduce that $p^* = 0$ so that i^* is a monomorphism. Moreover, if X' runs through the set of all non-zero submodules of X, then we have the canonical direct system of abelian groups

$$\{\operatorname{Hom}_A(X', X)\}_{X'}. \tag{14}$$

THEOREM 20.14. *Let F be a left prime of A. Then*: (1) *If* $\mathfrak{m}$ *is the maximal ideal of the local ring* $\operatorname{End}_A(Q_F)$, *then one has the canonical isomorphisms*:

$$k_F \simeq \operatorname{End}_{\operatorname{Mod} A/\mathscr{F}}(U_F) \simeq \operatorname{End}_A(K_F) \simeq \operatorname{End}_A(Q_F)/\mathfrak{m}.$$

(2) *For any non-zero submodule X of* $K_F = K$, *the system* (14) *has in a canonical way* k_F *as inductive limit.*

Proof. (1) The first assertion is a direct consequence of Lemma 20.4 and the fact that the functor S is full and faithful. However, for greater clarity, we shall give an alternative proof. Moreover we shall prove that the ring $\operatorname{End}_A(Q)/\mathfrak{m}$ is isomorphic with $\operatorname{End}_A(K)$. Let $f \in \operatorname{End}_A(Q)$ such that $f \in \mathfrak{m}$ that is $\ker f \neq 0$. Then we prove that $K \subseteq \ker f$. Indeed, let $X = \ker f \cap K$; then $X \neq 0$. If $X \neq K$, we see that $T(K/X) = 0$, according to Corollary 20.13. But this is a contradiction, Q being F-closed. Hence, necessarily $K \subseteq \ker f$.

Furthermore, let f be an automorphism of Q; then $f(K) = K$. Indeed, from the exact sequence of $\operatorname{Mod} A/\mathscr{F}$:

$$0 \longrightarrow U \longrightarrow E \longrightarrow E/U \longrightarrow 0$$

we deduce in $\operatorname{Mod} A$ the exact sequence:

$$0 \longrightarrow K \longrightarrow Q \longrightarrow S(E/U).$$

Thus we check that Q/K is $\mathcal{F}$-torsionfree (Q/K being a subobject of $S(E/U)$). Hence, if $K \not\subseteq f(K)$ we see that $K/(K \cap f(K))$ is an object of $\mathcal{F}$, a contradiction. Therefore, necessarily $K \subseteq f(K)$. Analogously, $K \subseteq f^{-1}(K)$ so that $f(K) \subseteq K$ and finally $K = f(K)$. Thus any automorphism f of Q defines an automorphism $\bar{f}$ of K. Moreover, we can define a ring morphism $\alpha : \mathrm{End}_A(Q) \to \mathrm{End}_A(K)$ such that $\alpha(f) = 0$ if $f \in \mathfrak{m}$ and $\alpha(f) = \bar{f}$, if f is an automorphism.

(2) The second assertion is a direct consequence of the construction of the category Mod $A/\mathcal{F}$ (see Section 4.3). Again, for greater clarity we shall give another independent proof. First we see that for any non-zero submodule X of K, we have $ST(X) = K$ and the canonical inclusion $X \to K$ can be considered as the adjunction morphism (see Section 4.4). Then by adjunction, we deduce that for any morphism $f : X' \to X$ there exists a unique morphism $\bar{f} : K \to K$ such that the following diagram is commutative:

$$\begin{array}{ccc} X' & \xrightarrow{f} & X \\ \downarrow & & \downarrow \\ K & \xrightarrow{\bar{f}} & K \end{array}$$

In this way a canonical injective map $u_{X'} : \mathrm{Hom}_A(X', X) \to k_F$ can be calculated for any non-zero submodule X' of X. Furthermore, if $h : K \to K$ is a non-zero morphism, then $h(X) \cap X = X' \neq 0$, so that we have the commutative diagram:

$$\begin{array}{ccc} X' & \xrightarrow{h'} & X \\ \downarrow & & \downarrow \\ K & \xrightarrow{h} & K \end{array}$$

where $h'(x) = h(x)$, for any $x \in X'$. We leave it to the reader to verify as before that k_F is the inductive limit of the considered system (14) and $u_{X'}$ the structural morphism. ∎

If A is a commutative ring, then an ideal $\mathfrak{a}$ of A is prime if and only if $A/\mathfrak{a}$ is an integral domain, i.e. if and only if any non-zero morphism

$$f : X \to A/\mathfrak{a},$$

where X is a non-zero submodule of $A/\mathfrak{a}$ is a monomorphism. The same result is also valid in the non-commutative case, namely:

COROLLARY 20.15. *Let $F \in L(A)$. The following.assertions are equivalent:*

(1) *F is a left prime of A;*

(2) *There exists a left ideal $\mathfrak{a}$ of A such that $F = F_{\mathfrak{a}}$ and any non-zero morphism $f: X \to A/\mathfrak{a}$ where X is a submodule of $A/\mathfrak{a}$, is a monomorphism.* ∎

THEOREM 20.16. *The following assertions are equivalent for a ring A;*

(1) *The zero ideal of A is left super-prime:*

(2) *A is a left Ore domain.*

Proof. (1) ⇒ (2). In fact A is a coirreducible A-module so that $Ax \cap Ay \neq 0$ for any two non-zero elements x, y of A.

(2) ⇒ (1). Then the ring $Q(A)$ is a division ring and injective as A-module. To finish it will be enough to prove that $\mathrm{Hom}_A(A/\mathfrak{a}, Q(A)) = 0$ for any non-zero left ideal of A. But this fact is obvious. ∎

NOTE 6. Let $F \in \mathrm{Spel}(A)$; then in [171, (Theorem 2.12)] it is claimed that for any non-zero submodule X of X_F, the ring $\mathrm{End}_A(X)$ is right Ore domain, but the proof given at that place is incomplete. Up to the present we do not know if this result is true, but in our opinion, this is likely. (It is easy to see that the result is valid in the commutative case). Moreover there exists a more general conjecture. Let X be an A-module and Q the injective envelope of X. Denote by $\mathfrak{a}(X)$ the ideal of $\mathrm{End}_A(X)$ consisting of all elements f such that $\ker f$ is essential. Also, we denote $H(X) = \mathrm{End}_A(X)/\mathfrak{a}(X)$. By a slight computation we can prove that $H(X)$ is canonically a subring of $H(Q)$ (the ring $H(Q)$ is regular according to Exercise 7 of Section 4.13). Our conjecture is that $H(Q)$ is the right total ring of quotients of $H(X)$. (It will be enough to prove that $H(X)$ is essential as a $H(X)^0$-submodule of $H(Q)$). Of course, this last conjecture contains as a particular case the above conjecture.

In what follows, we shall consider a sort of special left prime of a ring A.

LEMMA 20.17. *Let $\mathfrak{a}$ be a left super-prime of A and $F = F_{\mathfrak{a}}$. The following assertions are equivalent:*

(1) *F is a perfect system;*

(2) *A_F is a left local ring and $\mathfrak{a}_F$ a maximal left ideal.*

Proof. The implication (1) ⇒ (2) is obvious, because $\mathrm{Mod}\, A/\mathscr{F}$ is a local Grothendieck category and is equivalent with $\mathrm{Mod}\, A_F$ by the Theorem 16.7.

(2) ⇒ (1). Let $u: A \to A_F$ be the canonical morphism; then by Exercise 5 we check that $u(F) = F_{\mathfrak{a}_F} = \{A_F\}$ by hypothesis. Again, by Theorem 16.7, we deduce that F is perfect. ∎

A left prime F of A is called a *perfect left prime* if it is perfect as a left localizing system.

THEOREM 20.18. *Let A be a ring such that $J(A) = 0$. The following assertions are equivalent*:

(1) *The system E of all essential left ideals (ie. the Goldie system) is a perfect left prime*;

(2) *The ring $Q(A)$ is a simple ring.*

Proof. (1) $\Rightarrow$ (2). It is clear that $Q(A)$ is the injective envelope of A (Proposition 13.7) and $E = F(Q(A))$. By hypothesis, for any left ideal $\mathfrak{a}$ of $Q(A)$ we deduce that $\mathrm{Hom}_A(Q(A)/\mathfrak{a}, Q(A)) \neq 0$, so that $Q(A)$ contains simple left ideals. We claim that in fact $Q(A)$ is semisimple, that is $Q(A) = \mathrm{So}(Q(A))$. Indeed, if $\mathrm{So}(Q(A)) \neq Q(A)$, let $\mathfrak{m}$ be a maximal left ideal of $Q(A)$ such that $\mathrm{So}(Q(A)) \subseteq \mathfrak{m}$. Then $Q(A)/\mathfrak{m}$ can be viewed as a left ideal of $Q(A)$, so that a suitable non-zero element $x \in \mathrm{So}(Q(A))$ can be calculated such that $\mathfrak{m}x = 0$.

Because $J(Q(A)) = 0$, $\mathfrak{m}$ cannot be essential, so that $\mathfrak{m}$ is a direct summand. But this contradicts the definition of $\mathfrak{m}$. Finally $Q(A) = \mathrm{So}(Q(A))$ as claimed. Now we use Lemma 20.17 and Exercise 4.

The implication (2) $\Rightarrow$ (1) results easily and we leave it to the reader. ∎

It is clear that if A is a commutative ring, then any prime of A is perfect prime. In fact, if $\mathfrak{p}$ is a prime ideal of A, then $F_{\mathfrak{p}} = F_{A-\mathfrak{p}}$. In this case, the ring $A_{F_{\mathfrak{p}}}$ is denoted by $A_{\mathfrak{p}}$ and is called the *local ring associated with* $\mathfrak{p}$.

Exercises

1. Let $T: \mathscr{C} \to \mathscr{C}'$ be an exact functor and S a full and faithful right adjoint of T, $\mathscr{C}$ and $\mathscr{C}'$ being abelian categories. If X is a $\ker T$-torsionfree coirreducible object, then $T(X)$ is also coirreducible. In addition, if X is injective (i.e. is an injective indecomposable object) then $T(X)$ is also injective indecomposable. Furthermore, if X' is coirreducible in $\mathscr{C}'$, then $S(X')$ is also coirreducible in $\mathscr{C}$. Moreover, S preserves injective indecomposable objects.

2. A subobject X' of an object X in an abelian category is called *irreducible* if for any two subobjects X'', X''' of X, with $X'' \cap X''' = X'$ we deduce necessarily that $X' = X''$ or $X' = X'''$. Then an object X is coirreducible if and only if the null subobject of X is irreducible. Moreover, the object X/X' is coirreducible if and only if X' is irreducible.

3. Let X, Y be two A-modules as in Theorem 20.12. Then their injective envelopes are isomorphic.

4. Let F be a left prime on A, $u: A \to A_F$ the canonical ring morphism. Then $u(F)$ is also a left prime on A. Moreover, if $\mathfrak{a}$ is a left super-prime ideal of A, then $\mathfrak{a}_F$ is again a left super-prime ideal of A_F and $F_{\mathfrak{a}_F} = u(F)$.

5. The system D of left dense ideals of A is a left perfect prime and $Q(A)$ has no nilpotent ideals if and only if $Q(A)$ is a simple ring.

6. A morphism $u: A \to B$ of commutative rings is a preflat (respectively, flat) epimorphism if and only if for any prime ideal $\mathfrak{p} \in \operatorname{Spec}(A)$ such that $\mathfrak{p} \notin F_u$, it follows that the canonical morphism

$$u_{\mathfrak{p}} : A \otimes_A A_{\mathfrak{p}} \to B \otimes_A A_{\mathfrak{p}}$$

is a surjection (respectively, an isomorphism). Here F_u is the system of all ideals $\mathfrak{a}$ such that $Bu(\mathfrak{a}) = B$.

The same result is valid for the non-commutative case. (Of course, with the obvious changes).

7. If $u: A \to B$ is a surjective ring morphism, then the assignment

$$F \rightsquigarrow u^{-1}(F)$$

defines an injective mapping of $\operatorname{Spel}(B)$ into $\operatorname{Spel}(A)$. The same result is valid if u is a left flat epimorphism of rings.

8. Let A be a commutative ring. For any ideal $\mathfrak{a}$ of A, we denote by $D(\mathfrak{a})$ the set of all prime ideals $\mathfrak{p}$ which do not contain $\mathfrak{a}$. If $\mathfrak{a}$ runs through all ideals of A, the subsets $D(\mathfrak{a})$ define the collection of open sets in a topology on $\operatorname{Spec}(A)$, called the *Zariski topology*. For any ideal $\mathfrak{a}$ let us write

$$G_{(\mathfrak{a})} = \bigcap_{\mathfrak{p} \in D(\mathfrak{a})} F_{\mathfrak{p}}.$$

If X is an A-module, then the assignment

$$X \rightsquigarrow X_{G(\mathfrak{a})}$$

defines a presheaf of abelian groups on the topological space $\operatorname{Spec}(A)$. Denote by $\tilde{X}$ the associated sheaf. Particularly $\tilde{A}$ becomes a sheaf of rings and any $\tilde{X}$ is an $\tilde{A}$-module (see [80] Chapter 1 for details). Denote by Mod $\tilde{A}$ the category of all $(\operatorname{Spec}(A), \tilde{A})$-modules. Then the assignment $X \rightsquigarrow \tilde{X}$ is an exact full and faithful functor, having a right adjoint. If A is an integral domain, then for any ideal $\mathfrak{a}$ we have $\tilde{A}(D(\mathfrak{a})) = A_{G(\mathfrak{a})} = \bigcap_{\mathfrak{p} \in D(\mathfrak{a})} A_{\mathfrak{p}}$.

9. A two-sided ideal $\mathfrak{a}$ of A is (left)-super-prime if and only if it is completely prime and $A/\mathfrak{a}$ is a left Ore domain. In this case $F_{\mathfrak{a}}$ is called a *special left prime*.

References

Bkouche [17], [18]; Bourbaki [26]; Goldman [76]; Grothendieck–Dieudonné [80]; Hacque–Maury [85]; Koch [108]; Marot [122], [123]; Michler [130]; Năstăsescu–Popescu [144]; Popescu [171], [172], [173]; Stenström [213].

4.21 Bilocalizing subcategories

Let $\mathscr{C}$ be an abelian category, $\mathscr{A}$ a localizing subcategory of $\mathscr{C}$ and $T : \mathscr{C} \to \mathscr{C}/\mathscr{A}$ the canonical functor. We shall say that $\mathscr{A}$ is a *bilocalizing subcategory* if the functor T has a left adjoint H.

THEOREM 21.1. *Let A be a ring and $F \in L(A)$. The following assertions are equivalent:*

(1) *The subcategory $\mathscr{F}$ is bilocalizing;*

(2) *Any intersection of elements of F is also an element of F;*

(3) *Any direct product of objects of $\mathscr{F}$ is also an object of $\mathscr{F}$ (i.e. $\mathscr{F}$ is closed under direct products, which are calculated in* Mod A);

(4) *F is the set of all left ideals which contains an idempotent two-sided ideal $\mathfrak{b}$ (that is $\mathfrak{b}^2 = \mathfrak{b}$);*

(5) *The functor T commutes with direct products;*

(6) *Any A-module X has a largest quotient module which is an object of $\mathscr{F}$.*

Proof. (2) ⇒ (3). Let $\{X_i\}_i$ be a set of objects of $\mathscr{F}$ and $x \in \Pi X_i$; then $x = (x_i)_i$ and $\mathrm{Ann}(x) = \bigcap_i \mathrm{Ann}(x_i) \in F$ by hypothesis. Then $\prod_i X_i \in \mathrm{Ob}\,\mathscr{F}$.

(3) ⇒ (2). Let $\{\mathfrak{a}_i\}_i$ be a set of elements of F.

From the exact sequence:

$$0 \to \bigcap_i \mathfrak{a}_i \to A \to \prod_i (A/\mathfrak{a}_i)$$

we deduce that $\bigcap_i \mathfrak{a}_i \in F$.

(2) ⇒ (4). Let $\mathfrak{b} = \bigcap_{\mathfrak{a}\in F} \mathfrak{a}$; then $\mathfrak{b} \in F$ and so $\mathfrak{b} \subseteq (\mathfrak{b} : x)$ for any $x \in A$ (see the definition of a localizing system), so that $\mathfrak{b}$ is a two-sided ideal. Finally $\mathfrak{b}^2 \supseteq \mathfrak{b}$ since $\mathfrak{b}^2 \in F$ according to Exercise 3 of Section 4.9.

(4) ⇒ (2). Let $\mathfrak{b}$ be an idempotent two-sided ideal of A, and F the set of all left ideals $\mathfrak{a}$ which contain $\mathfrak{b}$. An easy calculation proves that F is in fact a localizing system which satisfies (2).

The implications (1) ⇒ (5) and (5) ⇒ (3) are obvious.

(1) ⇒ (6). Let H be a left adjoint of T and $w: HT \to \mathrm{Id}\,\mathrm{Mod}\,A$ an arrow of adjunction. By Exercise 1, H is full and faithful. Then a quasi-inverse of w is a functorial isomorphism so that $T(w_X)$ is an isomorphism for any module X. Hence, $\ker w_X$ and $\operatorname{coker} w_X$ are objects of $\mathscr{F}$, by the exactness of T.

Now, let $p: X \to X'$ be an epimorphism in $\mathrm{Mod}\,A$ such that $X' \in \mathrm{Ob}\,\mathscr{F}$; then from the following, canonically constructed diagram

$$\begin{array}{ccccc} HT(X) & \xrightarrow{w_X} & X & \xrightarrow{q} & \operatorname{coker} w_X \\ \downarrow & & \downarrow{\scriptstyle p} & \swarrow{\scriptstyle t} & \\ XT(X') & \xrightarrow{w_{X'}} & X' & & \end{array}$$

we deduce that $w_{X'} = 0$, so that $pw_X = 0$. Hence a unique morphism

$$t : \operatorname{coker} w_X \to X'$$

can be constructed such that $tq = p$. Therefore $\operatorname{coker} w_X$ is the greatest quotient object of X in $\mathscr{F}$.

(6) ⇒ (4). Let $\mathfrak{b}$ be a left ideal of A such that $A/\mathfrak{b}$ is the greatest quotient object of A in $\mathscr{F}$. Then for any $\mathfrak{a} \in F$ there exists a unique morphism

$$r : A/\mathfrak{b} \to A/\mathfrak{a}$$

rendering commutative the following diagram:

$$\begin{array}{ccc} A & \longrightarrow & A/\mathfrak{b} \\ & \searrow \quad \swarrow{\scriptstyle r} & \\ & A/\mathfrak{a} & \end{array}$$

We deduce that $\mathfrak{b} \subseteq \mathfrak{a}$ so that F has a smallest element, which, as above is an idempotent ideal.

(4) ⇒ (1). Let $L' : \mathrm{Mod}\,A \to \mathrm{Mod}\,A$ be the functor

$$L'(X) = \varinjlim_{\mathfrak{a} \in F} \mathrm{Hom}_A(\mathfrak{a}, X)$$

(see Section 4.10). Then $L = L'\,L'$, where as always, $L = ST$ is the localization functor. But by hypothesis we have:

$$L'(X) = \mathrm{Hom}_A(\mathfrak{b}, X)$$

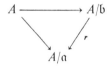

$\mathfrak{b}$ being the smallest element of F. Since $\mathfrak{b}$ is a two-sided ideal, it is in fact an (A, A^0) bimodule. Then by Corollary 7.5 of Chapter 3, L' has a left adjoint denoted by H'. Hence the functor $K = H'\,H'$ is a left adjoint of the functor $L' L' = L$. Then for any object $T(X)$ of Mod $A/\mathscr{F}$ and any A-module Y we have the functorial isomorphism

$$\mathrm{Hom}_A(Y, ST(X)) \simeq \mathrm{Hom}_A(K(Y), X)$$

Now let $H = KS$. Then for any object $T(X)$ of Mod $A/\mathscr{F}$ and any A-module Y, we have the functorial isomorphisms:

$$\mathrm{Hom}_A(K\,ST(X), Y) \simeq \mathrm{Hom}_A(ST(X), L(Y)) \simeq \mathrm{Hom}_{\mathrm{Mod}\,A/\mathscr{F}}(T(X), T(Y))$$

(where the first is induced by the adjunction of K and L and the latter results by Lemma 4.1). The proof is now complete, because any object of Mod $A/\mathscr{F}$ is of the form $T(X)$, where X is an object of Mod A (see Sections 4.1 and 4.3). ∎

A system $F \in L(A)$ is called *bilocalizing* if the subcategory $\mathscr{F}$ is bilocalizing By the above Theorem we deduce that there exists a bijection between the set of all left (or right) bilocalizing systems on A and the set of all idempotent two-sided ideals of A.

PROPOSITION 21.2. *Let F be a bilocalizing system,* $\mathfrak{b}$, *the smallest idempotent two-sided ideal contained in F and $t: A \to A/\mathfrak{b}$ the canonical epimorphism. Then the functor $t_*: \mathrm{Mod}(A/\mathfrak{b}) \to \mathrm{Mod}\,A$ establishes an equivalence between* $\mathrm{Mod}(A/\mathfrak{b})$ *and the subcategory* $\mathscr{F}$.

Proof. Let X be an $A/\mathfrak{b}$-module; then for any $x \in t_*(X)$, we see that

$$\mathrm{Ann}(x) \supseteq \mathfrak{b} \quad \text{so that} \quad t_*(X) \in \mathrm{Ob}\,\mathscr{F}.$$

Furthermore, if Y is an object of $\mathscr{F}$, then it is easy to see that the canonical morphism: $Y \to (A/\mathfrak{b}) \otimes_A Y$ is an isomorphism, hence Y can be regarded as an $A/\mathfrak{b}$-module. ∎

Now we shall give some results about Grothendieck categories, which will be useful in what follows.

Let $\mathscr{C}$ be a Grothendieck category, U a generator of $\mathscr{C}$, $A = \mathrm{End}_{\mathscr{C}}(U)$ and $S: \mathscr{C} \to \mathrm{Mod}\,A^0$ the functor defined by $S(X) = \mathrm{Hom}_{\mathscr{C}}(U, X)$. Further, let T be a left adjoint of S, $u: \mathrm{Id\,Mod}\,A^0 \to ST$ an arrow of adjunction; as always, it will be assumed that $\mathrm{Id}\mathscr{C} = TS$ and that a quasi-inverse of u is the identity of $\mathrm{Id}\mathscr{C}$ (see Theorem 7.9 of Chapter 3 or Theorem 14.2). Finally let $\mathscr{F} = \ker T$.

LEMMA 21.3. *Let $X \in \mathrm{Ob}\,\mathscr{C}$, and M a submodule of $S(X) = \mathrm{Hom}_{\mathscr{C}}(U, X)$. We have $X = \sum_{f \in M} \mathrm{im}\, f$ if and only if $S(X)/M \in \mathrm{Ob}\,\mathscr{F}$.* ∎

NOTE 1. Generally, let G be a subset of $S(X)$ and M the submodule generated by G. Then $\sum_{f \in G} \mathrm{im} f = X$ if and only if $S(X)/M \in \mathrm{Ob}\,\mathscr{F}$.

LEMMA 21.4. *For any $X \in \operatorname{Ob} \mathscr{C}$, there exists an object $X' \in \operatorname{Ob} \mathscr{C}$ with the property that for each epimorphism $f : Y \to X$, there exists a morphism $g : X' \to Y$ such that fg is an epimorphism.*

Proof. Put $X' = U^{(S(X))}$. Now if $f : Y \to X$ is an epimorphism, and $M = \operatorname{im} S(f)$, then $S(X)/M \in \operatorname{Ob} \mathscr{F}$. Furthermore, for any element $m \in M$ we choose an element $m' \in S(Y)$ such that $S(f)(m') = m$; denote by G the set of all elements m' where m runs over M. Denote by $p : U^{(G)} \to Y$ the unique morphism such that $pu_{m'} = m'$ where u_m are the structural inclusions, and let $t : U^{(M)} \to U^{(G)}$, be the unique morphism with the conditions: $tu_m = u_{m'}$, where $S(f)(m') = m$, and $u_m, u_{m'}$ are the structural morphisms. Finally we consider the diagram:

$$X' = U^{(S(X))} \xrightarrow{q} U^{(M)} \xrightarrow{t} U^{(G)} \xrightarrow{p} Y$$

where q is the natural projection, and write $g = ptq$. We claim that fg is an epimorphism. To prove this it will be sufficient to check that

$$\operatorname{im} S(fg) \supseteq M.$$

In fact let $m \in M$ and let $m' \in G$ such that $S(f)(m') = fm' = m$; also let $u_m' : U \to U^{(S(X))}$ be the structural morphisms. Thus $S(fg)(u_m') = fgu_m' = fgu_m' = fptq\, u_m' = fptu_m = fpu_{m'} = fm' = m$. Hence $\operatorname{im} S(fg) \supseteq M$, so that $\operatorname{coker} S(fg) \in \operatorname{Ob} \mathscr{F}$, or equivalently fg is an epimorphism. ■

NOTE 2. The above inclusion $M \subseteq \operatorname{im} S(fg)$ is in fact an equality.

LEMMA 21.5. *Let X be an object of $\mathscr{C}$. Denote by $R(X)$ the A^0-submodule of $S(X)$ consisting of all elements $f : U \to X$ such that for any epimorphism $g : Y \to X$ there exists $h : U \to Y$ so that $gh = f$ or equivalently $S(g)(h) = f$.*

If $\mathscr{C}$ is an Ab4* *category, then for any $X \in \operatorname{Ob} \mathscr{C}$ one has*:

$$X = \textstyle\sum_{f \in R(X)} \operatorname{im} f.$$

Proof. By Lemma 21.3, it will be sufficient to check that $S(X)/R(X) \in \operatorname{Ob} \mathscr{F}$.

In fact, let X' be as in the previous Lemma and E be the set of all epimorphisms from X' to X. According to Lemma 21.4 one has:

$$R(X) = \textstyle\bigcap_{p \in E} \operatorname{im} S(p).$$

By the Ab4* condition we check that the product morphism

$$s = \textstyle\prod_{p \in E} p : X'^{E} \to X^{E}$$

is also an epimorphism. Since S commutes with direct products, we also check that $S(s) = \prod_{p\in E} S(p)$, so that $\operatorname{im} S(s) = \prod_{p\in E} \operatorname{im} S(p)$. Now if $\Delta : S(X) \to S(X)^E$ is the diagonal morphism, then we have, by Lemma 21.4:

$$\Delta(R(X)) = \Delta(S(X)) \cap \operatorname{im} S(s)$$

and finally:

$$S(X)/R(X) \simeq \Delta(S(X))/\Delta(R(X)) \subseteq \operatorname{coker} S(s).$$

Hence $S(X)/R(X) \in \operatorname{Ob} \mathscr{F}$, because s is an epimorphism. ∎

Now we are able to prove the following result due to Roos:

THEOREM 21.6. (Roos, [190]). *Let $\mathscr{C}$ be a Grothendieck category, U a generator of $\mathscr{C}$, $A = \operatorname{End}_{\mathscr{C}}(U)$, $S = \operatorname{Hom}_{\mathscr{C}}(U, ?) : \mathscr{C} \to \operatorname{Mod} A^0$, T a left adjoint of S and $\mathscr{F} = \ker T$. The following assertions are equivalent*:

(1) *$\mathscr{F}$ is a bilocalizing subcategory of* Mod A^0;

(2) *$\mathscr{C}$ is an* Ab6 *and* Ab4* *category.*

Proof. The implication (1) ⇒ (2) is a direct consequence of Theorem 21.1, since Mod A^0 is an Ab6 and Ab4* category.

(2) ⇒ (1). Let X be an object of $\mathscr{C}$ and let B be the set of all subobjects of X relatively of finite type. Then B is a direct and complete set of subobjects. Indeed, let J be the set of all direct and complete set of subobjects of X. For every $j \in J$ one has a direct and complete set $\{X_{i(j)}\}_{i\in I_j}$ of subobjects, and so by the Ab6 condition one has:

$$X = \bigcap_{j\in J} \left(\textstyle\sum_{i\in I_j} X_{i(j)}\right) = \textstyle\sum_{(i(j))\in\prod_{j\in J} I_j} \left(\bigcap_{(i(j))} X_{i(j)}\right).$$

It is clear that for any element $(i(j)) \in \prod_{j\in J} I_j$ the subobject $\bigcap_{(i(j))} X_{i(j)}$ is relatively of finite type, and the last equality proves that $\sum_{X'\in B} X' = X$. Finally, B is direct, since the sum of a finite set of objects which are relatively of finite type is also a subobject, relatively of finite type.

Now we shall define:

$$E = \varinjlim_{X'\in B} R(X')$$

where $R(X')$ is defined as in the previous lemma. It is obvious that E is canonically a submodule of $S(X)$. We shall prove that E is the smallest submodule P of $S(X)$ such that $S(X)/P \in \operatorname{Ob} \mathscr{F}$.

First we shall check that $S(X)/E \in \operatorname{Ob} \mathscr{F}$. Indeed from the canonically constructed commutative diagram

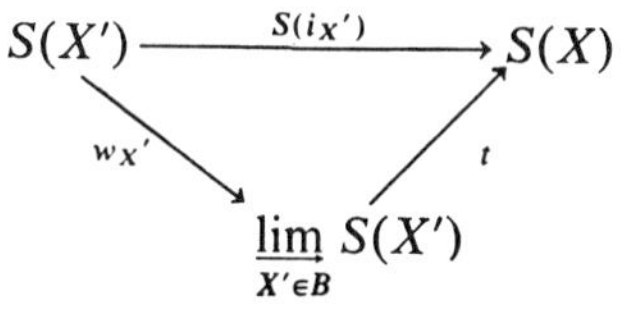

where $i_{X'}\ X' \to X$ is the canonical inclusion and $w_{X'}$ the structural morphisms. We obtain that t is a monomorphism and $T(t)$ is an isomorphism, so that $\operatorname{coker} t \in \operatorname{Ob} \mathscr{F}$. According to Lemma 21.5, we have also $S(X')/R(X') \in \operatorname{Ob} \mathscr{F}$ for each $X' \in B$, so that from the exact sequence:

$$0 \to \varinjlim_{X' \in B} S(X') \Big/ \varinjlim_{X' \in B} R(X') \to S(X) \Big/ \varinjlim_{X' \in B} R(X') \to \operatorname{coker} t \to 0$$

it follows that $S(X)/E \in \operatorname{Ob} \mathscr{F}$.

Furthermore, let M be a submodule of $S(X)$ so that $S(X)/M \in \operatorname{Ob} \mathscr{F}$. Thus by Lemma 21.3 we have the canonical epimorphism of $\mathscr{C}$:

$$p : U^{(M)} \to X.$$

Now, let $u \in E$; then there exists $X' \in B$, such that $u \in R(X')$. We have also $X = \sup_F p(U^F)$, where F runs over all finite sets of M, and thus $X' \subseteq p(U^F)$ for a suitable F. Let V be a subobject of U^F such that $p(V) = X'$. By the definition of $R(X')$, there exists a morphism $f : U \to V$ such that $p' if = u$, where $i : V \to U^F$ is the canonical inclusion and p' is the restriction of p to U^F. Finally we can find a morphism $v : U \to U^{(M)}$ such that $pv = u$. Since $M = \operatorname{im} S(p)$ we have $u \in M$, so that $E \subseteq M$, as claimed.

Now we are able to prove that Condition (6) of Theorem 21.1 is fulfilled. In fact let X be an A^0-module, $u_X : X \to ST(X)$ the canonical morphism, E the smallest element in the set of all submodules Y of $ST(X)$ such that $S\,T(X)/Y \in \operatorname{Ob} \mathscr{F}$ and $E' = u_X{}^{-1}(E)$. Now if X' is a submodule of X such that $X/X' \in \operatorname{Ob} \mathscr{F}$, then from the exact sequence

$$0 \to u_X(X)/u_X(X') \to ST(X)/u_X(X') \to ST(X)/u_X(X) \to 0$$

we see that the middle object is in $\mathscr{F}$, so that $u_X(X') \supseteq E$, and thus $X' \supseteq E'$. Finally, X/E' is the greatest quotient object of X in $\mathscr{F}$ and so $\mathscr{F}$ is bilocalizing by Condition (6) of Theorem 21.1. ∎

Now let $\mathscr{C}$ be a Grothendieck category. By Theorem 14.2 we can define the "exact sequence" of categories and functors:

$$\mathscr{F}_0 \underset{K_0}{\overset{I_0}{\rightleftarrows}} \operatorname{Mod} A_0^0 \underset{S_0}{\overset{T_0}{\rightleftarrows}} \mathscr{C}$$

whose meaning is that S_0 is a full and faithful functor, T_0 a left exact adjoint of S_0, I_0 define an equivalence between $\ker T_0$ and $\mathscr{F}_0$, and K_0 is a right adjoint of I_0 (obviously I_0 is full and faithful). Furthermore $\mathscr{F}_0$ is again a Grothendieck category, so that also by Theorem 14.2 we can write the following exact sequence of categories and functors in the same manner as above

$$\mathscr{F}_1 \underset{K_1}{\overset{I_1}{\rightleftarrows}} \operatorname{Mod} A_1^0 \underset{S'_1}{\overset{T'_1}{\rightleftarrows}} \mathscr{F}_0.$$

Thus $T_1 = I_0 T_1'$ is a left adjoint of $S_1' K_0 = S_1$, hence we have the exact sequence of categories and functors:

$$\mathscr{F}_1 \underset{K_1}{\overset{I_1}{\rightleftarrows}} \operatorname{Mod} A_1^0 \underset{S_1}{\overset{T_1}{\rightleftarrows}} \operatorname{Mod} A_0^0 \underset{S_0}{\overset{T_0}{\rightleftarrows}} \mathscr{C}.$$

By iteration we can write the following "exact sequence" of categories and functors:

$$\cdots \underset{S_{n+1}}{\overset{T_{n+1}}{\rightleftarrows}} \operatorname{Mod} A_n^0 \underset{S_n}{\overset{T_n}{\rightleftarrows}} \cdots \underset{S_1}{\overset{T_1}{\rightleftarrows}} \operatorname{Mod} A_0^0 \underset{S_0}{\overset{T_0}{\rightleftarrows}} \mathscr{C} \qquad (15)$$

Such a diagram gives a *modular resolution* of $\mathscr{C}$. Obviously, a Grothendieck category has in general many modular resolutions. We shall say that the modular resolution (15) of $\mathscr{C}$ has length n if T_{n+1} defines an equivalence between $\ker T_n$ and $\operatorname{Mod} A^0{}_{n+1}$.

The least such integer n is called the *modular dimension* of $\mathscr{C}$ and denoted by $\mu(\mathscr{C})$. If no such integer exists, the modular dimension is defined to be ∞. In this way the category of modules Mod A where A is a ring has the modular dimension just zero. By Theorem 21.6 and Proposition 21.2 if $\mathscr{C}$ is an Ab6 and Ab4* Grothendieck category then $\mu(\mathscr{C}) \leqslant 1$. It would be interesting to characterize the Grothendieck categories for which the modular dimension is $\leqslant n$ for $n = 2, 3, \ldots$ (for $n = 2$ see Exercise 4).

NOTE 3. Let $\mathscr{C}$ be a bilocalizing subcategory of Mod A, where A is a suitable ring. Then $\mathscr{C}$ is an Ab4* but not an Ab5* category. In this way we check the independence of Ab4 and Ab5.

Exercises

1. Let $\mathscr{C}$ be a locally small abelian category and $\mathscr{A}$ a dense subcategory of $\mathscr{C}$. We shall say that $\mathscr{A}$ is a *colocalizing subcategory* if the canonical functor $T : \mathscr{C} \to \mathscr{C}/\mathscr{A}$ has a left adjoint H. The reader is invited to transpose for the colocalizing subcategory the result of Section 4, using the dual of $\mathscr{C}$. In particular H is full and faithful.

2. A full subcategory of Mod A is bilocalizing if and only if it is dense and closed under direct products. In particular, if P is a projective A-module

then the full subcategory consisting of all modules X such that

$$\mathrm{Hom}_A(P, X) = 0$$

is bilocalizing.

3. Let $\mathfrak{b}$ be an idempotent two-sided ideal of a ring A and F the bilocalizing system defined of $\mathfrak{b}$ (i.e. a left ideal $\mathfrak{a}$ belongs to F if and only if $\mathfrak{b} \subseteq \mathfrak{a}$). Let H be a left adjoint of the canonical functor $T : \mathrm{Mod}\, A \to \mathrm{Mod}\, A/\mathscr{F}$. An A-module X is called *coclosed* if the canonical morphism: $v_X : HT(X) \to X$ is an isomorphism. Prove that an object X is coclosed if and only if it is closed. Moreover, for any A-module X one has: $HT(X) \simeq (\mathfrak{b} \otimes_A \mathfrak{b}) \otimes_A X$; this provides a natural description of the arrow of adjunction

$$v : HT \to \mathrm{Id}\, \mathrm{Mod}\, A.$$

Finally, prove that the localization functor associated with F is isomorphic to the functor $X \rightsquigarrow \mathrm{Hom}_A(\mathfrak{b} \otimes_A \mathfrak{b}, X)$.

4. The following assertions are equivalent for a Grothendieck category $\mathscr{C}$:

(a) $\mu(\mathscr{C}) \leqslant 2$;

(b) There exists a ring A, a right localizing system F on A such that for any $\mathfrak{a} \in F$ there exists $\mathfrak{a}^* \in F$ such that $\mathfrak{a}^* \subseteq \mathfrak{a}$ and for every $\mathfrak{b} \in F$, there exists $x \in \mathfrak{a}$ such that $\mathfrak{a}^*(1 + x) \subseteq \mathfrak{b}$ and $\mathscr{C}$ is equivalent to $\mathrm{Mod}\, A^0/\mathscr{F}$.

5. Let A be a ring such that for any non-zero two-sided ideal $\mathfrak{a}$ of A one has $\mathfrak{a}^2 \neq \mathfrak{a}$. Then any non-zero projective A-module is a generator of $\mathrm{Mod}\, A$.

6. Let A be a ring and U a generator of $\mathrm{Mod}\, A^0$. Let $B = \mathrm{End}_A(U)$, $S : \mathrm{Mod}\, A^0 \to \mathrm{Mod}\, B^0$ the functor: $S(X) = \mathrm{Hom}_A(U, X)$ and T a left adjoint of S (see Corollary 7.5 of Chapter 3). Thus S is full and faithful and T is exact. Moreover $\ker T$ is a bilocalizing subcategory, so that the right localizing system F on B associated with $\ker T$ has a smallest ideal $t(U)$. Prove that $t(U)$ is the image of the morphism:

$$U \otimes_A \mathrm{Hom}_B(U, B) \xrightarrow{p} B$$

defined by $p(u \otimes f) = f(u)$. Prove also that $t(U)$ is a faithful B-module, finitely generated as a two-sided ideal and essential as a right ideal.

7. Let K be a division ring, U a K-module and $\mathfrak{a}$, the two-sided ideal of $A = \mathrm{End}_K(U)$ consisting of all elements $f \in A$ such that $\mathrm{im}\, f$ is finitely generated as a K-module. Then $\mathfrak{a}^2 = \mathfrak{a}$. Moreover let F be the right localizing system on A generated by all right ideals $\mathfrak{b}$ which contain $\mathfrak{a}$. Prove that the categories $\mathrm{Mod}\, K$ and $\mathrm{Mod}\, A^0/\mathscr{F}$ are canonically equivalent.

8. A ring A is called *left perfect* if $R(A)$ is right T-nilpotent and $A/R(A)$ is semisimple. If A is left perfect then any localizing subcategory of **Mod** A is bilocalizing.

References

Alin [4]; Alin–Dickson [6]; Bass [16]; Chase [37]; Dlab [45], [46]; Goblot [67], [68], [69], [71], [private letter to the author]; Popescu [167]; Roos [190], [199].

5. The Krull-Remak-Schmidt Theorem and Decomposition Theories

A classical result due to Steinitz asserts that any vector space over a division ring K has a basis, i.e. it is a direct sum of copies of K.

This result, which makes the study of vector spaces very easy, has continually attracted the attention of algebraists, who looked for generalizations in ever larger classes of modules. Thus, at present, one of the important methods of algebra consists of the study of an object belonging to a category by "decomposing" it into objects which may be investigated more easily.

A basic tool of study was the K.R.S.-decompositions (see Section 5.1). An important contribution to the study of the K.R.S.-decompositions was accomplished by Azumaya [11] (see Theorem 1.3), Matlis [125] and Gabriel [59] (see Theorem 8.11). Matlis stated Theorem 8.11 for injective objects over left noetherian rings, whereas Gabriel transferred this result to l.n.-categories.

In his thesis, Gabriel [59] introduced a new type of K.R.S.-decomposition, which we call K.R.S.G.-decomposition (see Section 5.3). The fact that to any Grothendieck category a spectral category may be associated (see Chapter 4, Section 4.3) allows one to remark that a close connection exists between the K.R.S.-decompositions and the K.R.S.G.-decompositions (see Corollary 3.2). Section 3 is devoted to the study of the l.c.-categories. These are a large class of categories in which any object admits a K.R.S G.-decomposition. The l.c.-categories have been introduced and studied by Năstăsescu and Popescu [141], [142] and independently by Fort [55], [56].

A particular class of l.c.-categories are the s.n.-categories, a study of which is made in Section 5.5. These categories have been introduced by Gabriel under the name of categories with definite Krull dimension. Theorem 5.9 gives a characterization of the categories with the help of the notion of prime localizing category. In this way, we exhibit a class of rings, the s.n.-rings. The study of these rings shows that between these rings and the noetherian rings there is a certain connection.

In Section 5.6 we study a particular class of s.n.-categories, namely the s.a.-categories. Their study was commenced by Năstăsescu and Popescu in [143], who introduced the semi-artinian rings. An important result is Theorem 6.9 which has a corollary exhibiting the structure of the commutative perfect rings.

Another class of s.n.-categories are the l.n.-categories introduced by Gabriel in [59]. An important characterization of these has been formulated by Nouazé [153] (see Theorem 8.7). Theorem 8.9, due to Gabriel [59], gives an elegant characterization of the l.n.-categories with the help of left exact functors.

In Section 5.4 we study the decomposition theory. The first decomposition has been produced by Euclid, who showed that any natural number is a unique product of prime numbers. An important generalization of the Euclidean decomposition is the so-called Lasker–Noether decomposition [202]. An extension of the Lasker–Noether decomposition is the tertiary decomposition introduced by Lesieur and Croisot [40]. Gabriel [59] remarked that the theory of the tertiary decomposition is closely connected with the theory of the K.R.S.-decompositions. In [186], [187] Riley gave an axiomatization of the decomposition theory and showed that the Lasker–Noether and the tertiary decompositions may be obtained as particular cases. By using the axiomatization given by Riley and the characterization given by Gabriel for the l.n.-categories, Bucur [29] proved a result which characterizes all the decomposition theories of Riley on an l.n.-category.

Popescu [170] showed that the decomposition theories defined by Riley are, in some way, a specific feature of the l.n.-categories. Therefore, he proposed a new axiomatization of the decomposition theories (see Section 5.4). Theorem 4.12 gives an exhaustive description for the decomposition theories on an l.c.-category. This result shows that between the K.R.S.G.-decompositions and the decomposition theories there exists a close connection.

In Section 10, we introduce the notion of $\mathscr{P}$-coprimary object of a Grothendieck category, where $\mathscr{P}$ is a prime localizing subcategory. Theorem 10.1 gives a characterization of the $\mathscr{P}$-coprimary objects. In this way, the theory of primary decomposition is introduced. Indeed, later we show that the theory of the Lasker–Noether decomposition for the noetherian commutative rings is a particular case of this primary decomposition (Theorem 10.8).

This primary decomposition imposes the introduction of a new type of primary ideal; the connection between these primary ideals and the classical primary ideals for the commutative semi-noetherian rings is described in Theorems 10.6 and 10.7.

Finally, Section 11 is devoted to the theory of the tertiary decomposition. We study the conditions under which the tertiary decomposition theory coincides with the primary decomposition theory. One shows that these results are closely connected with the stability with respect to the injective envelope and the Artin–Rees lemma (Theorem 11.14).

5.1 The classical Krull–Remak–Schmidt theorem

An object X of a preadditive category $\mathscr{C}$ is called *indecomposable* if from an isomorphism of the form $X \simeq X_1 \amalg X_2$ we deduce that $X_1 = 0$ or $X_2 = 0$.

It is clear that a coirreducible object in an abelian category is indecomposable. Moreover, an injective object X in an abelian category is indecomposable if and only if it is coirreducible, that is, an injective indecomposable.

We shall say that an object X in an abelian category $\mathscr{C}$ has a *Krull–Remak–Schmidt decomposition* (we shall write simply a *K.R.S.–decomposition*) if it is isomorphic with a direct sum $\coprod_i X_i$ where all X_i are indecomposable. Furthermore, we shall say that an object X has a unique K.R.S.-decomposition if it has at least one K.R.S.-decomposition and if when

$$X \simeq \coprod_{i\in I} X_i \simeq \coprod_{j\in J} X_j$$

are two such decompositions, then one can find a bijection $f: I \to J$ such that $X_i \simeq X_{f(i)}$ for any $i \in I$.

NOTE. In fact, the names of Krull, Remak and Schmidt are attached to the uniqueness of a K.R.S.-decomposition for the objects of a category which is not necessarily abelian. Theoretically, the notion of K.R.S.-decomposition may be introduced for any category, but substantial results have been obtained only for abelian categories.

In the sequel, we intend to study two problems:

(1) To give sufficiently wide conditions in order that K.R.S.-decompositions in abelian categories should exist.

(2) To study the uniqueness of K.R.S.-decompositions in some Abelian categories.

First, we shall investigate the uniqueness of the K.R.S.-decomposition. The proof of the uniqueness theorem will be preceded by a few lemmas.

LEMMA 1.1. *Let $\mathscr{C}$ be an abelian category, $X = \coprod_{i\in I} X_i$ and $f, g \in \mathrm{End}_{\mathscr{C}}(X)$ such that $f + g = 1_X$. Suppose that for each $i \in I$ the ring $\mathrm{End}_{\mathscr{C}}(X_i)$ is a local ring. Under these conditions for any finite system $i_1, i_2, \ldots, i_s$ of elements of I, there exist subobjects $P_1, P_2, \ldots, P_s$ with the following properties*:

(1) *For each $k = 1, 2, \ldots, s$ either f or g induces an isomorphism of P_k onto X_{i_k}.*

(2) $X = (P_1 \amalg P_2 \amalg \ldots \amalg P_s) \amalg (\coprod_{i\in I'} X_i)$ *where we denote by I' the difference set I—$\{i_1, \ldots i_s\}$.*

Proof. Denote by $u_i : X_i \to X$ the structural morphisms and by p_i the associated projections. Clearly:

$$p_i = p_i(f + g) = p_i f + p_i g$$
$$1_{X_i} = p_i u_i = p_i f u_i + p_i g u_i$$

for any $i \in I$. This implies (since the ring of endomorphisms of X_i is local) that either $p_i f u_i$ or $p_i g u_i$ is an isomorphism. Let us consider the index i_1 and assume that $p_{i_1} f u_{i_1} : X_{i_1} \to X_{i_1}$ is an isomorphism. Denote by P_1 the subobject $f u_{i_1}(X_{i_1})$. It follows evidently that P_1 is isomorphic with X_{i_1} and the restriction of p_{i_1} to P_1 is an isomorphism of P_1 with X_{i_1}. We may apply Exercise 1. It follows that $X = P_1 \amalg (\coprod_{i \in I - \{i_1\}} X_i)$. We now repeat the argument for the index i_2. We thus obtain after a finite number of steps the objects $P_1, \ldots, P_s$ with the required properties. ■

LEMMA 1.2. *Let $\mathscr{C}$ be an* Ab 5-*category, $X = \coprod_{i \in I} X_i$ a K.R.S.-decomposition of X such that for each index $i \in I$ the ring* $\mathrm{End}_{\mathscr{C}}(X_i)$ *is a local ring. Under these conditions, if $f : X \to X$ is such that $f^2 = f$, then the following assertions are true*:

(1) *There exists at least one $i \in I$ such that f induces an isomorphism of X_i onto $f(X_i)$ and $f(X_i)$ is a direct summand of X;*

(2) *Any indecomposable direct summand of X is isomorphic with an X_i, for a suitable i.*

Proof. If we set $f' = 1 - f$ then we obviously have the following decomposition:

$$X = f(X) \amalg f'(X)$$

(see Chapter 2, Corollory 7.5). Then, according to condition Ab 5, we have:

$$f(X) = \Sigma_J \left(f(X) \cap \coprod_{j \in J} X_j\right)$$

where J runs over all finite subsets of I. But $f(X) \neq 0$, so there exists a finite subset $i_1, \ldots, i_s$ of I such that

$$f(X) \cap (X_{i_1} \amalg \ldots \amalg X_{i_s}) \neq 0.$$

We now apply Lemma 1.1 with f and $g = f'$ and with the set of indices $\{i_1, \ldots, i_s\}$. We get the decomposition:

$$X = (P_1 \amalg \ldots \amalg P_s) \amalg (\coprod_{i \in I'} X_i)$$

where $I' = I - \{i_1, \ldots, i_s\}$. As on the other hand

$$f(X) \cap (X_{i_1} \amalg \ldots \amalg X_{i_s})$$

belongs to the kernel of f', it follows that f' induces an isomorphism between $X_{i_1} \amalg \dots \amalg X_{i_s}$ and $P_1 \amalg \dots \amalg P_s$. Hence there exists at least one index i_k such that f induces an isomorphism between X_{i_k} and P_k whence the assertion (a).

To prove the assertion (b), let P be an indecomposable direct summand of X, i.e. $X = P \amalg Q$. There exists an idempotent endomorphism p of X such that $p(X) = P$. From the assertion (a) it follows therefore that there exists an index i such that p induces an isomorphism of X_i onto $p(X_i) \subseteq P$. Since $p(X_i)$ is a direct summand of X and P is an indecomposable direct summand, we infer that $p(X_i) = P$. Thus the assertion (b) is proved. ∎

We are now in a position to state and prove the part concerning the uniqueness of the K.R.S.-decomposition for suitable categories.

THEOREM 1.3. (Azumaya, [11]). *Assume that*:

$$X = \coprod_{i \in I} X_i = \coprod_{j \in J} Y_j$$

are two K.R.S.-decompositions of an object X in an Ab 5*-category $\mathscr{C}$ such that the rings* $\mathrm{End}_{\mathscr{C}}(X_i)$ *and* $\mathrm{End}_{\mathscr{C}}(Y_j)$ *are local for any $i \in I$ and any $j \in J$. Under these conditions there exists a bijection $\alpha : I \to J$, such that X_i is isomorphic with $Y_{\alpha(i)}$ for any i, in other words, the object X has a unique K.R.S.-decomposition.*

Proof. By applying the Lemma 1.2 it follows immediately that any object X_i is isomorphic with an object Y_j and conversely. On the set of indices I we now introduce the following equivalence relation. We say that i_1 is equivalent to i_2 if X_{i_1} is isomorphic with X_{i_2}. An equivalence relation is similarly introduced in the set J. Let K, respectively L, be the quotient set obtained from the set I, respectively the set J, by this equivalence relation. From the above remark, it follows that K and L are equivalent. We shall therefore identify L with K.

Let now $k \in K$. Denote by $I(k)$ the set of all the elements of the set I which belong to the class k, and by $J(k)$ all the elements of the set J which belong to the class k. It is sufficient to prove that the sets $I(k)$ and $J(k)$ have the same cardinal for any k. Let c_1, respectively c_2 be the cardinal of the set $I(k)$, respectively of the set $J(k)$. Since the sets I and J play a symmetric role, it is sufficient to prove that $c_1 \geqslant c_2$. We distinguish two cases:

(1) c_1 is finite. Let $j_1 \in J(k)$ and $p_{j_1} : X \to X$ be an idempotent endomorphism such that $p_{j_1}(X) = Y_{j_1}$. By Lemma 1.2 there exists $i_1 \in I(k)$ such that p_{j_1} induces an isomorphism of X_{i_1} with Y_{j_1} and by Lemma 1.1 we have the following decomposition:

$$X = X_{i_1} \amalg (\textstyle\coprod_{j \neq j_1} Y_j). \tag{1}$$

If j_1 is the only element of the set $J(k)$, then the relation $c_1 \geqslant c_2$ is evidently proved. If $J(k)$ contains also the element j_2, then by considering the projection p_{j_2} in the decomposition (1) it follows that there exists an $i_2 \in I(k)$ such that hp_{j_2} induces an isomorphism of X_{i_2} with Y_{j_2}, $h: Y_{j_2} \to X$ being the structural inclusion. All now reduces to proving that $i_1 \neq i_2$. But

$$p_{j_2}(X_{i_1}) = 0,$$

which shows that $i_1 \neq i_2$. But continuing in this manner, we clearly achieve, after a finite number of steps, the proof of the relation $c_1 \geqslant c_2$.

(2) c_1 is infinite. Let $j \in J$ and $p_j : X \to Y_j$ the canonical projection. For any $i \in I$ we have the relation:

$$X_i = \sum_F (X_i \cap (\coprod_{j \in F} Y_j)),$$

where F runs through the finite subsets of J. Thus $X_i \cap \ker p_j$ is null only for a finite number of j's.

For a given i there exist only a finite number of j's such that p_j induces an isomorphism of X_i with Y_j. Since c_1 is infinite, it follows that $c_1 \geqslant c_2$. ■

The theorem given above is concerned with uniqueness of decompositions into indecomposable objects for objects in some abelian categories. We propose now to prove a Krull–Remak–Schmidt type theorem which involves the existence and uniqueness of K.R.S.-decompositions for the objects of certain categories. The conditions we are going to impose on the category $\mathscr{C}$ are similar to the classical conditions of Noetherian and Artinian type in the theory of modules.

A *double chain* of an abelian category $\mathscr{C}$ is a sequence of systems

$$(X_n, i_n, p_n)_{n \geqslant 0}$$

with the following properties:

(i) X_n are objects of $\mathscr{C}$;

(ii) $i_n : X_n \to X_{n-1}$ is a monomorphism for any $n \geqslant 1$;

(iii) $p_n : X_{n-1} \to X_n$ is an epimorphism for any $n \geqslant 1$.

A double chain $(X_n, i_n, p_n)_{n \geqslant 0}$ is said to be *stationary* if there exists an integer n_0 such that for any $n \geqslant n_0$, i_n and p_n are isomorphisms.

THEOREM 1.4. *If in the abelian category $\mathscr{C}$ any double chain is stationary, then any object X of $\mathscr{C}$ admits a unique K.R.S.-decomposition*

$$X = \coprod_{i=1}^{n} X_i.$$

Proof. Existence. Assume that we have an unlimited sequence of decompositions:

$$X = X_1 \amalg Y_1; \quad X_1 = X_2 \amalg Y_2, \ldots, X_n = X_{n+1} \amalg Y_{n+1}, \ldots$$

and hence the canonical morphisms:

$$i_n : X_n \to X_{n-1}$$
$$p_n : X_{n-1} \to X_n.$$

Thus we obtain the double chain (X_n, i_n, p_n). But from the hypothesis it follows that this double chain is stationary. Therefore X_n is indecomposable for a suitable n, so that any object contains an indecomposable direct summand. Let us write $X = X_1 \amalg Y_1$ where X_1 is indecomposable. If $Y_1 \neq 0$, then $Y_1 = X_2 \amalg Y_2$ where X_2 is again indecomposable. In this way, for any n, we can write:

$$X = X_1 \amalg X_2 \amalg \ldots \amalg X_n \amalg Y_n$$

where $X_1, \ldots, X_n$ are indecomposable. As above, we shall deduce that Y_n is indecomposable for a suitable n, so that any object of $\mathscr{C}$ has a K.R.S.-decomposition.

For the part of uniqueness we shall use the following lemma and Theorem 1.3.

LEMMA 1.5. *Let X be an indecomposable object of an Abelian category $\mathscr{C}$, such that any double chain $(X_n, i_n, p_n)_{n \geq 0}$ with $X_0 = X$ is stationary. Then the ring* $\mathrm{End}_{\mathscr{C}}(X)$ *is local.*

Proof. Let $f, g \in \mathrm{End}_{\mathscr{C}}(X)$, such that $f + g = 1_X$. Then we must prove that f or g are isomorphisms. Put $X_n = \mathrm{im}\, f^n$, and let $j_n \colon X_n \to X$ be the canonical inclusion. Let $p_n \colon X_n \to X_{n+1}$ be the unique epimorphism such $j_{n+1} p_n = f j_n$; also let i_n: $X_{n+1} \to X_n$ be the canonical inclusion. Then we have the double sequence $(X_n, i_n, p_n)_{n \geq 0}$ with $X_0 = X$. By hypothesis, an integer n_0 can be found such that p_n and i_n are isomorphisms for any $n \geqslant n_0$. Let $q_n = p_{n-1} p_{n-2} \cdots p_1 p_0 j_n$. If $n \geqslant n_0$, q_n is an isomorphism, so that X_n is a direct summand of X. But X is indecomposable, so that $X = X_n$, in which case f is an isomorphism, or $X_n = 0$ and thus $f^n = 0$. But then $g = 1_X - f$ is an invertible element having $1_X + f + \ldots + f^{n-1}$ as inverse. ■

COROLLARY 1.6. *Let X be an object as in Lemma 1.5. Then any element of* $\mathrm{End}_{\mathscr{C}}(X)$ *is either invertible, or nilpotent.* ■

Exercises

1. Let us consider in an abelian category an object $X = X_1 \amalg X_2$ and denote by u_1, u_2 the structural morphisms and p_1, p_2 the associated pro-

jections. If $h: P \to X$ is a monomorphism such that $p_1 h: P \to X_1$ is an isomorphism, then $X = h(P) \amalg X_2$. The structural morphisms are then $u_1 p_1 h$ and u_2, and the associated projections are $(p_1 h)^{-1} p_1$ and p_2.

2. Any local ring A is indecomposable as an A-module. In this way, we have an example of indecomposable module which is not coirreducible. Moreover any two bases of a free module over a local ring have the same cardinal. (See Chapter 3, Section 3.9).

3. Let $\mathscr{C}$ be an Ab 6 locally small category and X an object of $\mathscr{C}$ such that the sum of any direct set of direct summands of X is also a direct summand of X. Then X is a direct sum of indecomposable objects.

4. Let $\mathscr{C}$ be an Ab 6-category in which any direct inductive limit of projective objects is also a projective object. Then, any projective object of $\mathscr{C}$ is a direct sum of indecomposable projective objects.

5. Let us assume that an object X of a Grothendieck category has a K.R.S.-decomposition. Then, any decomposition of X as a direct sum of subobjects can be refined to a direct decomposition into indecomposable objects. Particularly, any direct summand of X has a K.R.S.-decomposition.

References

Atiyah [9]; Azumaya [11]; Bourbaki [21], [23]; Bucur–Deleanu [31]; Dickson [42], [43]; Gabriel [59]; Pareigis [162].

5.2 The structure of spectral Grothendieck categories

Let $\mathscr{C}$ be a spectral Grothendieck category. We shall denote by $\mathrm{Dis}(\mathscr{C})$ the smallest localizing subcategory of $\mathscr{C}$ generated by all the simple objects of $\mathscr{C}$.

LEMMA 2.1. *For the object X of $\mathscr{C}$, the following assertions are equivalent:*

(1) *Any non-zero quotient object of X contains a simple object;*

(2) *X is semisimple (that is a direct sum of simple objects);*

(3) *X is an object of* $\mathrm{Dis}(\mathscr{C})$.

Proof. The equivalence (1) $\Leftrightarrow$ (3) is obvious. For the other implications, we can utilize the same proof as in Lemma 11.1 of Chapter 3. However, we shall give an independent proof. Of course, it will be sufficient to prove that (1) $\Rightarrow$ (2). Let us denote by M the set of all direct sums of simple objects, which are included in X. If $A = \coprod_{i \in I} X_i$, $B = \coprod_{j \in J} X_j$ are two elements of

M, then we put $A \leqslant B$ if and only if there exists an injection $f: I \to J$ such that $X_i = X_{f(i)}$. M is an inductive set. In fact, if $\{A_\alpha = \coprod_{i \in I_\alpha} X_{i_\alpha}\}$ is a totally ordered subset of M, then for any $A_\alpha \leqslant A_{\alpha'}$ there exists an injection $f_{\alpha\alpha'}: I_\alpha \to I_{\alpha'}$ such that $X_{i_\alpha} = X_{f_{\alpha\alpha'}(i_\alpha)}$ for any $i_\alpha \in I_\alpha$. But then we have the direct set $\{I_\alpha, f_{\alpha\alpha'}\}_\alpha$ of sets and mappings to which we shall associate the inductive limit $I = \varinjlim I_\alpha$. Let $u_\alpha: I_\alpha \to I$ be the structural morphisms. For any $i \in I$ let us put $X_i = X_{i_\alpha}$ for any $i_\alpha \in I_\alpha$ such that $u_\alpha(i_\alpha) = i$. The Ab 5-condition proves that the sum $A = \coprod_{i \in I} X_i$ is direct and by its construction we see that $A_\alpha \leqslant A$ for any α. Now the proof will proceed in an obvious manner. ∎

For any object X of $\mathscr{C}$ we shall denote by So(X) (the *socle* of X) the greatest subobject of X in Dis($\mathscr{C}$). It is clear that an object X will be Dis($\mathscr{C}$)-closed if and only if So(X) = 0 (see Lemma 5.1 of Chapter 4). A Dis($\mathscr{C}$)-closed object of $\mathscr{C}$ is called a *continuous object*. We shall denote by Cont($\mathscr{C}$) the full subcategory of $\mathscr{C}$ generated by the continuous objects. A slight computation shows that Cont($\mathscr{C}$) is in fact a localizing subcategory and any object X of $\mathscr{C}$ is a direct sum $X = \mathrm{So}(X) \amalg \mathrm{Co}(X)$, where Co($X$) is the continuous part of X, defined in obvious fashion. Finally, we can define a covariant functor $R: \mathrm{Dis}(\mathscr{C}) \sqcap \mathrm{Cont}(\mathscr{C}) \to \mathscr{C}$, such that $R((X, Y)) = X \amalg Y$. Obviously, R is an equivalence of categories. By a *discrete* (*respectively, continuous*) spectral category, we mean **a** spectral category $\mathscr{C}$ in which any object is discrete (respectively, continuous). Hence we have proved the following result:

THEOREM 2.2. *Any spectral Grothendieck category is a unique direct product of a discrete and a continuous spectral category.* ∎

Now we shall give a complete description of the discrete spectral Grothendieck categories.

THEOREM 2.3. *Let $\mathscr{C}$ be a spectral Grothendieck category. The following assertions are equivalent*:

(1) *$\mathscr{C}$ is a discrete category* (*i.e.* Cont($\mathscr{C}$) = 0).

(2) *There exists a unique set $\{K_i\}_{i \in I}$ of division rings such that $\mathscr{C}$ is equivalent to the category* $\prod_{i \in I} \mathrm{Mod}\, K_i^0$.

Proof. The implication (2) ⇒ (1) is clear, the category Mod K^0 being discrete for any division ring K.

(1) ⇒ (2). Let I be the set of all the types of simple objects of $\mathscr{C}$. For any $i \in I$ we shall fix a simple object U_i of the type "i" and denote by K_i the division ring $\mathrm{End}_{\mathscr{C}}(U_i)$.

By Theorem 1.3, any object X of $\mathscr{C}$ can be written in a unique way as a direct sum $\coprod_{j \in J} X_j$ of simple objects. Denote by X_i the sum of all simple subobjects of X of type "i". It is clear that $X = \coprod_i X_i$ and the sum is direct. Furthermore, we denote by $\mathscr{C}_i$ the full subcategory of $\mathscr{C}$ generated by all objects X, which are "i"—isotypical, that is any simple subobject of X is of the type "i" and by $H : \mathscr{C} \to \prod_{i \in I} \mathscr{C}_i$ the functor $H(X) = (X_i)_i$, where, as above, X_i is the isotypical component of X of type "i". It is clear that the functor H is, in fact, an equivalence of categories. Now the proof is reduced to proving that the categories $\mathscr{C}_i$ and Mod $K_i{}^0$ are equivalent. In fact, the functor $S_i : \mathscr{C}_i \to \text{Mod } K_i{}^0$ defined by $S_i(X) = \text{Hom}_{\mathscr{C}_i}(U_i, X)$ has a left adjoint and commutes with direct sums, U_i being small. These arguments are enough to show that S_i is an equivalence of categories. The uniqueness of the set I and the division rings K_i are obvious. ∎

NOTES 1. To describe a discrete spectral Grothendieck category, it is sufficient to give a set $\{K_i\}_i$ of division rings, which generally are not distinct. For example, if K is a division ring, then the spectral category

$$\mathscr{C} = \text{Mod } K^0 \amalg \text{Mod } K^0$$

has two types of distinct simple objects U_1, U_2 although $\text{End}_{\mathscr{C}}(U_1)$ and $\text{End}_{\mathscr{C}}(U_2)$ are in fact isomorphic.

2. To describe an object X in a spectral Grothendieck category

$$\mathscr{C} = \textstyle\prod_{i \in I} \text{Mod } K_i{}^0$$

it will be sufficient to give for any i, a set J_i. In fact $X = \coprod_{i \in I} X_i$ where X_i are isotypic components and $X_i = \coprod_{j(i) \in J_i} X_{j(i)}$, $X_{j(i)} \simeq U_i$ for any $j(i)$.

Exercises

1. Prove that any spectral Grothendieck category is discrete if and only if it satisfies the condition Ab 6. (Hint: Use Exercise 3 of Section 5.1).

2. Let X be an object in a spectral Grothendieck category $\mathscr{C}$. Then $\text{Co}(X) = \bigcap Y$ where Y runs over the set of all maximal subobjects of X. The ring $\text{End}_{\mathscr{C}}(X)$ is the direct product

$$\text{End}_{\mathscr{C}}(X) = \text{End}_{\mathscr{C}}(\text{So}(X)) \amalg \text{End}_{\mathscr{C}}(\text{Co}(X)).$$

3. Let $\mathscr{C}$ be a Grothendieck category. The canonical functor $P : \mathscr{C} \to \mathscr{C}_E$ is exact unless it is an equivalence of categories. (Here $\mathscr{C}_E$ is the spectral category associated with $\mathscr{C}$ (See Section 2 of Chapter 4)).

4. Let G be a generator of a spectral category $\mathscr{C}$ and $A = \mathrm{End}_{\mathscr{C}}(G)$. Then we have the canonical functor $S : \mathscr{C} \to \mathrm{Mod}\, A^0$ (see Chapter 3, Section 3.7) and T a left exact adjoint of S. Then S and T define an equivalence between $\mathscr{C}$ and $\mathrm{Mod}\, A^0/\mathscr{E}$, where E is the system of essential right ideals of A.

5. A ring A is said to be *strongly regular* if for any $a \in A$ there is an $x \in A$ such that $a = a^2x$. A commutative regular ring as well as an arbitrary product of division rings are strongly regular rings. Moreover, A is strongly regular if, and only if, every principal right (respectively, left) ideal of A is generated by an idempotent that belongs to the centre of A.

6. We shall say that an object X in a Grothendieck category $\mathscr{C}$ is *distributive* if the lattice of subobjects of X is a distributive lattice. We shall say that $\mathscr{C}$ is *locally distributive* if $\mathscr{C}$ has a set of generators consisting of distributive objects. For an object X in a spectral category $\mathscr{C}$, the following assertions are equivalent:

(a) X is distributive in $\mathscr{C}$;

(b) For every $X_3 \subseteq X$ and every decomposition $X = X_1 \amalg X_2$, the natural morphism:

$$(X_3 \cap X_1) \amalg (X_3 \cap X_2) \to X_3$$

is an isomorphism;

(c) For any decomposition $X = X_1 \amalg X_2$ we have $\mathrm{Hom}_{\mathscr{C}}(X_1, X_2) = 0$.

(d) If $X_1 \subseteq X$, $X_2 \subseteq X$ are such that $X_1 \cap X_2 = 0$, then

$$\mathrm{Hom}_{\mathscr{C}}(X_1, X_2) = 0.$$

(e) The endomorphism ring of X is a strongly regular ring.

Moreover the direct sum of any two distributive objects X_1, X_2 is distributive if and only if $\mathrm{Hom}_{\mathscr{C}}(X_1, X_2) = 0$.

7. Let $\mathscr{C}$ be a spectral category. Then $\mathscr{C}$ admits a distributive object U, that is maximum in the sense that every distributive object in $\mathscr{C}$ is isomorphic to a subobject of U. The object U is unique up to isomorphisms. Let $\mathscr{C}_a$ denote the localizing subcategory of $\mathscr{C}$ formed by the objects X such that $\mathrm{Hom}_{\mathscr{C}}(X, U) = 0$ and let $\mathscr{C}_d$ be the smallest localizing subcategory of $\mathscr{C}$ that contains U. Then $\mathscr{C}_d$ is a locally distributive spectral category, and $\mathscr{C}_a$ is a spectral category that does not contain any non-zero distributive object and the natural functor:

$$\mathscr{C} \to \mathscr{C}_d \amalg \mathscr{C}_a$$

is an equivalence of categories. These results are due to Roos [194].

NOTE. If $\mathscr{C}'$ is a localizing subcategory of a spectral category and $\mathscr{C}''$ the full subcategory of all $\mathscr{C}'$-torsion-free objects, then the natural functor:

$$\mathscr{C} \to \mathscr{C}' \Pi \mathscr{C}''$$

is an equivalence of categories.

(Hint: First prove that the inductive limit of a direct system of distributive objects in a spectral category is again a distributive object. Furthermore, if G is a generator of a spectral category $\mathscr{C}$ then every distributive object of $\mathscr{C}$ is isomorphic to a subobject of G.)

8. A strongly regular ring A is right self-injective if and only if it is left self-injective.

9. Let $\mathscr{C}$ be a locally distributive spectral category, the modular dimension $\mu(\mathscr{C})$ of which satisfies $\mu(\mathscr{C}) \leqslant 2$. Then $\mathscr{C}$ is discrete.

References

Bourbaki [23]; Gabriel–Oberst [61]; Johnson [99]; Roos [192], [193], [194], [195], [196], [199]; Tisseron [227].

5.3 Locally coirreducible categories

Gabriel [59] has given a much wider notion of K.R.S.-decomposition, but this has an essential connection with K.R.S.-decompositions, as they have been defined above. We shall say that an object X of a Grothendieck category $\mathscr{C}$ admits of a *K.R.S.G.-decomposition* (that is a generalized K.R.S.-decomposition or Krull–Remak–Schmidt–Gabriel-decomposition !) if X is the essential extension of a direct sum of coirreducible objects. Thus, by K.R.S.G.-decomposition we shall understand a direct sum $\coprod_i X_i$ of co-irreducible objects, which is essential in X.

Now we shall characterize the Grothendieck categories in which any object has a K.R.S.G.-decomposition. First, we shall give some preliminary, but interesting results. As always, if $\mathscr{C}$ is an abelian category, we denote by $\mathscr{C}_E$ the associated spectral category and by $P: \mathscr{C} \to \mathscr{C}_E$ the canonical functor (see Chapter 4, Section 4.2).

LEMMA 3.1. *An object X in a Grothendieck category $\mathscr{C}$ is coirreducible if and only if the object $P(X)$ is simple in $\mathscr{C}_E$.*

Proof. Let us assume that X is coirreducible and let $t: P(Y) \to P(X)$ be a monomorphism of $\mathscr{C}_E$. In order to check that $P(X)$ is simple it will be sufficient to prove that any such monomorphism is also an epimorphism. By the construction of the category $\mathscr{C}_E$ (see Chapter 4, Section 4.2),

$$t = P(f)\,P(s)^{-1},$$

where $f: Z \to X$, $s: Z \to Y$ are morphisms of $\mathscr{C}$ and s is an essential monomorphism. By hypothesis, we can assume that f is also a monomorphism and thus Y and Z are coirreducible objects. If t is not an epimorphism, then $P(f)$ also is not an epimorphism so that $\operatorname{im} P(f)$ is a direct summand of $P(X)$. Hence, there exists a non-zero subobject Z' of X such that

$$P(Z') \cap \operatorname{im} P(f) = 0.$$

Let us put $U = Z' \cap \operatorname{im} f$; then $U \neq 0$, because X is coirreducible, so that $P(U) \subseteq P(Z') \cap \operatorname{im} P(f) \neq 0$, a contradiction. Hence t is necessarily an isomorphism, or equivalently, $P(X)$ is simple.

Conversely, let us assume that $P(X)$ is simple, and let X_1, X_2 be two non-zero subobjects of X such that $X_1 \cap X_2 = 0$. Then $P(X_1) \cap P(X_2) = 0$, the functor P being left exact and $P(X_1)$, $P(X_2)$ are non-zero subobjects of $P(X)$, a contradiction. ∎

Corollary 3.2. *An object X of a Grothendieck category $\mathscr{C}$ has a K.R.S.G.-decomposition if and only if $P(X)$ is a semisimple object of $\mathscr{C}_E$.*

Proof. First, let us assume that X has a K.R.S.G.-decomposition, that is X is the essential extension of the direct sum $\coprod_i X_i$ of its coirreducible subobjects. Then $P(X) \simeq P(\coprod_i X_i) = \coprod_i P(X_i)$, according to Theorem 2.6, Chapter 4. Now we can apply Lemma 3.1.

Conversely, let us assume now that $P(X)$ is semisimple. Then any non-zero subobject X' of X contains a coirreducible subobject Y'. Indeed, $P(X')$ is a subobject of $P(X)$ and so it is also semisimple. Let $t: P(Z) \to P(X')$ a non-zero morphism, such that $P(Z)$ is simple. Then $t = P(f)\,P(s)^{-1}$ and we have in $\mathscr{C}$ the diagram:

$$Z \xleftarrow{\ s\ } Z' \xrightarrow{\ f\ } X'$$

f being a monomorphism and s an essential extension. Then by Lemma 3.1, Z is coirreducible and $Y' = f(Z')$ is a coirreducible subobject of X'.

Now for any ordinal α we define a subobject X_α of X which is a direct sum of coirreducible subobjects, in the following manner:

X_1 is a coirreducible subobject of X.

If $\alpha = \beta + 1$, that is α has a predecessor, then $X_\alpha = X_\beta$ if X_β is essential in X; otherwise, $X_\alpha = X_\beta \amalg Y_\alpha$ where Y_α is a suitable coirreducible subobject of X such that $Y_\alpha \cap X_\beta = 0$.

If α is a limit ordinal, then $X_\alpha = \sum_{\beta<\alpha} X_\beta$.

Now the point is to prove that X_α is a direct sum of coirreducible subobjects for any ordinal α. For that, let α be the first limit ordinal. It is clear then that X_β is a direct sum of coirreducible objects, for any $\beta < \alpha$. Also for any $\beta < \alpha$, $\beta > 1$, one has $\beta = \gamma + 1$ and so $X_\beta = X_\gamma \amalg Y_\beta$, Y_β being coirreducible. Then we shall prove that $X_\alpha = \amalg_{\beta<\alpha} Y_\beta$. First, it is clear that $X_\alpha = \sum_{\beta<\alpha} Y_\beta$ and according to Proposition 8.12 of Chapter 2, it will be sufficient to check that $(\sum_{\beta' \neq \beta} Y_{\beta'}) \cap Y = 0$ for any $\beta < \alpha$. Indeed, if this last intersection is not zero, then by the Ab 5-condition, we can choose the ordinals $\beta_1, \ldots, \beta_n < \alpha$, such that

$$\left(\sum_{i=1}^{n} Y_{\beta_i}\right) \cap Y_\beta \neq 0.$$

Then let $\beta' = \sup(\beta, \beta_1, \ldots, \beta_n)$. By the above definition, $Y_\beta, Y_{\beta_1}, \ldots, Y_{\beta_n}$ are the components of the direct sum $X_{\beta'}$, so that we have a contradiction. Hence X_α is a direct sum of coirreducible subobjects $\{Y_\beta\}_{\beta<\alpha}$. Furthermore, we can use transfinite induction on the set of limit ordinals to prove that any X_α is a direct sum of coirreducible subobjects.

Finally, there exists an ordinal α such that $X_\alpha = X_\beta$ for any $\beta \geqslant \alpha$, so that X_α is an essential subobject of X, by hypothesis. ∎

THEOREM 3.3. *Let $\mathscr{C}$ be a Grothendieck category. The following assertions are equivalent:*

(1) *Any non-zero object of $\mathscr{C}$ contains a coirreducible subobject;*

(2) *Any non-zero injective of $\mathscr{C}$ contains an indecomposable injective object;*

(3) *Any object of $\mathscr{C}$ has a K.R.S.G.-decomposition;*

(4) *Any non-zero injective of $\mathscr{C}$ is the essential extension of a direct sum of indecomposable injectives;*

(5) *The spectral category $\mathscr{C}_E$ is discrete.*

Proof. (1) ⇒ (2). If Q is a non-zero injective of $\mathscr{C}$, then it contains a coirreducible subobject X and so the injective envelope of X, which is obviously indecomposable.

The implication (2) ⇒ (1) is obvious, because any subobject of an indecomposable injective is coirreducible.

(1) ⇒ (3) is similar to the proof of the last part of Corollary 3.2.

(3) ⇒ (4) Let Q be an injective of $\mathscr{C}$; then Q is an essential extension of a direct sum $\coprod_i X_i$ of coirreducible subobjects. Let Q_i be the injective envelope of X_i, considered as a subobject of Q. It is clear that the sum $\sum_i Q_i$ is essential

in Q, so that in order to finish, it will be sufficient to prove that this sum is direct. For that, we claim that for any index i_0, we have

$$Q_{i_0} \cap (\sum_{i \neq i_0} Q_i) = 0.$$

If this is not the case, then, by the Ab 5-condition we can choose the indices $i_1, \ldots, i_n \in I - \{i_0\}$ such that

$$Q_{i_0} \cap \left(\sum_{j=1}^{n} Q_{i_j} \right) \neq 0.$$

But this is in contradiction with Exercise 2.

(1) ⇒ (5). It will be sufficient to prove that any non-zero subobject of $\mathscr{C}_E$ contains a simple object (see Lemma 2.1). But this follows by Lemma 3.1.

The other implications are obvious. ■

A Grothendieck category as in Theorem 3.3 is called a *locally coirreducible category* or, abbreviated, a *l.c.-category*.

In connection with the uniqueness of the K.R.S.G.-decomposition we have the following result:

LEMMA 3.4. *Let X be an object of a Grothendieck category $\mathscr{C}$ and let*

$$\coprod_{i \in I} X_i, \quad \coprod_{j \in J} Y_j$$

be two K.R.S.G.-decompositions of X. Then there exists abije ction $f : I \to J$ such that the objects X_i and $Y_{f(i)}$ have isomorphic injective envelopes.

Proof. It is clear that the objects $P(\coprod_i X_i)$ and $P(\coprod_j Y_j)$ are isomorphic. Then by the Theorem 1.3, a bijection $f : I \to J$ can be chosen such that $P(X_i)$ and $P(Y_{f(i)})$ are isomorphic. Then by Chapter 4, Exercise 3 of Section 4.2 we deduce that X_i and $Y_{f(i)}$ have isomorphic injective envelopes. ■

COROLLARY 3.5. *Any object of an l.c.-category has a unique K.R.S.G.-decomposition.* ■

In the sequel we shall give some results on l.c.-categories.

PROPOSITION 3.6. *Let $\mathscr{C}$ be a Grothendieck category and $\mathscr{F}$ a localizing subcategory of $\mathscr{C}$. Then $\mathscr{C}$ is an l.c.-category if and only if $\mathscr{F}$ and $\mathscr{C}/\mathscr{F}$ are l.c.-categories.*

Proof. It is clear that if $\mathscr{C}$ is an l.c.-category then so is $\mathscr{F}$ by point (1) of Theorem 3.3. Furthermore, let $T : \mathscr{C} \to \mathscr{C}/\mathscr{F}$ be the canonical functor, S a right adjoint of T, $X \in \mathrm{Ob}(\mathscr{C}/\mathscr{F})$ and Y a coirreducible subobject of $S(X)$. Then $T(Y)$ is a coirreducible subobject of $TS(X) \simeq X$. Indeed, if U, V are two non-zero subobjects of $T(Y)$ such that $U \cap V = 0$, then

$$S(U) \cap S(V) = 0,$$

so that $S(U) \cap S(V) \cap Y = \big(S(U) \cap Y\big) \cap \big(S(V) \cap Y\big) = 0$, a contradiction (see Chapter 4, Lemma 4.6). Hence by Theorem 3.3, we deduce that $\mathscr{C}/\mathscr{F}$ is an l.c.-category.

Conversely, assume that $\mathscr{F}$ and $\mathscr{C}/\mathscr{F}$ are l.c.-categories, and let X be an object of $\mathscr{C}$. Assume that $X_{\mathscr{F}} = 0$ (otherwise X has coirreducible subobjects); then X is an essential subobject of $ST(X)$. If Y is a coirreducible subobject of $T(U)$, then $S(Y)$ is clearly coirreducible in $\mathscr{C}$ and $S(Y) \cap X$ is coirreducible subobject of X. Again, we use Theorem 3.3. ■

Let $\mathscr{C}$ be any Grothendieck category. We denote by $\mathrm{Sp}(\mathscr{C})$ and call the *spectrum of* $\mathscr{C}$, the set of all types of injective indecomposable objects. ($\mathrm{Sp}(\mathscr{C})$ is a set, $\mathscr{C}$ having a generator U and any indecomposable injective object being the injective envelope of a suitable quotient object of U).

PROPOSITION 3.7. *Let $\mathscr{F}$ be a localizing subcategory of an l.c.-category $\mathscr{C}$. An object X is in $\mathscr{F}$ if and only if* $\mathrm{Hom}_{\mathscr{C}}(X, Q) = 0$ *for any indecomposable injective $\mathscr{F}$-closed object Q.*

Proof. It is clear that if $X \in \mathrm{Ob}\,\mathscr{F}$ then $\mathrm{Hom}_{\mathscr{C}}(X, Y) = 0$ for any $\mathscr{F}$-closed object X. Conversely, let X be as in the statement and assume that $T(X) \neq 0$. Then, by Proposition 3.6, the injective envelope of $T(X)$ contains an indecomposable injective object Q, and $T(X) \cap Q \neq 0$. Let, as always,

$$u_X : X \to ST(X)$$

be an adjunction arrow, and $X' = u_X{}^{-1}(ST(X) \cap S(Q))$. Then $X' \neq 0$ (u_X being essential) so that a suitable non-zero morphism $f : X \to S(Q)$ can be found. The contradiction is obvious. ■

If $\mathscr{F}$ is a localizing subcategory in a Grothendieck category $\mathscr{C}$, then the mapping $Q \rightsquigarrow S(Q)$ defines an injection of $\mathrm{Sp}(\mathscr{C}/\mathscr{F})$ into $\mathrm{Sp}(\mathscr{C})$. We shall identify $\mathrm{Sp}(\mathscr{C}/\mathscr{F})$ with a subset of $\mathrm{Sp}(\mathscr{C})$, by means of the above map. Moreover if Q is an indecomposable injective in $\mathscr{C}$, then two cases are possible. Firstly $Q_{\mathscr{F}} \neq 0$, so that the injective envelope of this object in the subcategory $\mathscr{F}$ is an indecomposable injective, or $Q_{\mathscr{F}} = 0$ so that $Q \simeq ST(Q)$. Also, if Q' is an indecomposable injective of $\mathscr{F}$, then its injective envelope $\bar{Q}'$ in $\mathscr{C}$ is again an indecomposable injective. Hence the mapping $Q' \rightsquigarrow \bar{Q}'$ is an injection of $\mathrm{Sp}(\mathscr{F})$ into $\mathrm{Sp}(\mathscr{C})$. Again we shall identify $\mathrm{Sp}(\mathscr{F})$ with the image of this map. Also so far we see that $\mathrm{Sp}(\mathscr{C}) = \mathrm{Sp}(\mathscr{C}/\mathscr{F}) \cup \mathrm{Sp}(\mathscr{F})$ and $\mathrm{Sp}(\mathscr{F}) \cap \mathrm{Sp}(\mathscr{C}/\mathscr{F}) = \varnothing$. Generally speaking, the set $\mathrm{Sp}(\mathscr{F})$ does not determine the subcategory $\mathscr{F}$. But for the l.c.-categories we have the following result.

COROLLARY 3.8. *Let $\mathscr{C}$ be an l.c.-category and $\mathscr{F}$ and $\mathscr{F}'$ two localizing subcategories of $\mathscr{C}$. The following assertions are equivalent:*

(1) $\mathscr{F} \equiv \mathscr{F}'$;

(2) $\mathrm{Sp}(\mathscr{F}) \equiv \mathrm{Sp}(\mathscr{F}')$;

(3) $\mathrm{Sp}(\mathscr{C}/\mathscr{F}) \equiv \mathrm{Sp}(\mathscr{C}/\mathscr{F}')$.

Proof. The implications (1) ⇒ (2) ⇒ (3) ⇒ (2) are obvious. The implication (3) ⇒ (1) can be checked by the Proposition 3.7. ∎

COROLLARY 3.9. *For any subset R of the set* $\mathrm{Sp}(\mathscr{C})$, *where $\mathscr{C}$ is an l.c.-category, denote by $\mathscr{F}(R)$ the localizing subcategory of $\mathscr{C}$ generated by all objects X such that* $\mathrm{Hom}_{\mathscr{C}}(X, Q) = 0$ *for all $Q \in R$. Let $\bar{R} = \mathrm{Sp}(\mathscr{C}/\mathscr{F}(R))$. Then the mapping $R \rightsquigarrow \bar{R}$ is a closure operator on the set* $\mathrm{Sp}(\mathscr{C})$.

Proof. It is clear that $R \subseteq \bar{R}$, therefore any $Q \in R$ is $\mathscr{F}(R)$-closed. Also by Proposition 3.7, we check that $\bar{R} = \bar{\bar{R}}$. To finish the proof, it will be sufficient to prove that $\overline{R \cup R'} = \bar{R} \cup \bar{R}'$ for any two subsets of $\mathrm{Sp}(\mathscr{C})$. In fact, let $Q \in \overline{R \cup R'}$ be such that $Q \notin \bar{R} \cup \bar{R}'$; then Q contains a non-null subobject X of $\mathscr{F}(R)$ and a non-null subobject X' of $\mathscr{F}(R')$. Furthermore, $X \cap X' \neq 0$, that is Q has non-zero subobjects in $\mathscr{F}(R \cup R')$, a contradiction.

Now let $Q \in \bar{R} \cup \bar{R}'$; then $Q \in \bar{R}$ for example, so that Q is $\mathscr{F}(R)$-closed. Therefore, since $\mathscr{F}(R \cup R') \subseteq \mathscr{F}(R)$, we check that Q is necessarily $\mathscr{F}(R \cup R')$-closed, hence $Q \in \overline{R \cup R'}$. ∎

THEOREM 3.10. (Exchange theorem). *Let $\mathscr{C}$ be an l.c.-category $\{Q_i\}_{i \in I}$ a set of indecomposable injectives of $\mathscr{C}$, and Q the injective envelope of $\coprod_i Q_i$. If Q' is a direct summand of Q, then a subset I' of I can be determined such that any injective envelope of $\coprod_{i' \in I'}^{Q_i}$ contained in Q, is a supplement of Q'.*

Proof. Let I' be a maximal element of the set of all subsets I'' of I, subject to the condition that $Q' \cap (\coprod_{i'' \in I''} Q_{i''}) = 0$. Then Q is an essential extension of the direct sum $Q' \amalg (\prod_{i' \in I'} Q_{i'})$. Indeed, let Q_1 be a maximal essential extension of the direct sum $Q' + \coprod_{i' \in I'} Q_{i'}$ in Q (that is an injective envelope contained in Q). By the way in which I' has been chosen, we deduce that

$$Q_i \cap (Q' + (\coprod_{i' \in I'} Q_{i'})) \neq 0$$

so that $Q_i \cap Q_1 \neq 0$ for any $i \in I - I'$. We see that the injective envelope of $Q_i \cap Q_1$ is in fact Q_i. On the other hand, the injective envelope of $Q_i \cap Q_1$

is also included in Q_1, so that $Q_i \subseteq Q_1$, for any i. Finally we see that

$$\coprod_{i\in I} Q_i \subseteq Q_1,$$

and thus $Q_1 = Q$, since it is injective. ∎

A ring A is called left *l.c.-ring* if the category Mod A is an l.c.-category.

PROPOSITION 3.11. *Let A be a ring. The following assertions are equivalent:*

(1) *A is a left l.c.-ring;*

(2) *Any left ideal of A is either irreducible or the intersection of two left ideals of which one is irreducible;*

(3) *For any left ideal $\mathfrak{a}$ of A, there exists $x \in A$ such that $(\mathfrak{a} : x)$ is irreducible.*

Proof. (1) ⇒ (2). Let $\mathfrak{a}$ be a left ideal of A. Two cases are possible: $\mathfrak{a}$ is irreducible or $\mathfrak{a}$ is not irreducible. In the last case, the module $A/\mathfrak{a}$ contains a non-zero coirreducible submodule X, which is not essential. Let Y be a complement of X, $\mathfrak{b} = p^{-1}(X)$, $\mathfrak{c} = p^{-1}(Y)$, where $p : A \to A/\mathfrak{a}$ is canonical. Then $\mathfrak{b}$ and $\mathfrak{c}$ are left ideals of A containing $\mathfrak{a}$ and $\mathfrak{a} = \mathfrak{b} \cap \mathfrak{c}$. It is clear that $A/\mathfrak{c}$ is isomorphic with $(A/\mathfrak{a})/Y$. By the construction, this last module contains X as an essential submodule, i.e. is coirreducible. Then $\mathfrak{c}$ is necessarily irreducible.

(2) ⇒ (1). Let X be an A-module; we can assume that $X = A/\mathfrak{a}$. Then $\mathfrak{a}$ is irreducible or $\mathfrak{a} = \mathfrak{b} \cap \mathfrak{c}$, with $\mathfrak{c}$ irreducible. In the last case

$$A/\mathfrak{a} = A/\mathfrak{b} \cap \mathfrak{c} \subseteq A/\mathfrak{b} \amalg A/\mathfrak{c}$$

and obviously $A/\mathfrak{a} \cap A/\mathfrak{c} \neq 0$. In any case, X contains a non-zero coirreducible submodule.

(1) ⇒ (3). Let X be a coirreducible submodule of $A/\mathfrak{a}$ and $x \in A$ such that $0 \neq \bar{x} \in X$. Then $A\bar{x}$ is coirreducible so that $\operatorname{Ann}(\bar{x}) = (\mathfrak{a} : x)$ is irreducible.
The other implications are obvious. ∎

Now we shall give an example of a ring B such that the category Mod B is not an l.c.-category.

THEOREM 3.12. (Fort [56]). *Let $A = \prod_{i\in I} A_i$ be a direct product of an infinite set of rings, and $\mathfrak{a}$ the ideal of A consisting of all elements $x = (x_i)_{i\in I}$ with $x_i = 0$ for all i but a finite number of indices and let $B = A/\mathfrak{a}$. Then the category* Mod B *is not an l.c.-category.*

Proof. We shall prove that the ring B does not contain a coirreducible left ideal. In fact, let $\mathfrak{b}$ be a left ideal of A, which strongly contains $\mathfrak{a}$. In order to prove the theorem, it will be sufficient to establish that there exist two left ideals $\mathfrak{b}_1, \mathfrak{b}_2$ of A, such that $\mathfrak{a} \subset \mathfrak{b}_1 \subseteq \mathfrak{b}$; $\mathfrak{a} \subset \mathfrak{b}_2 \subseteq \mathfrak{b}$ and $\mathfrak{a} = \mathfrak{b}_1 \cap \mathfrak{b}_2$ (the inclusion of $\mathfrak{a}$ in $\mathfrak{b}_1$ and $\mathfrak{b}_2$ is strict). Therefore, since $\mathfrak{a} \neq \mathfrak{b}$, then we can find an element $b = (b_i)_{i \in I}$ of $\mathfrak{b}$ such that the set $J = \{i | i \in I, b_i \neq 0\}$ is infinite. Let H, K be two infinite subsets of J such that $H \cap K = \varnothing$ and $H \cup K = J$.

Let us define the elements $h = (h_i)_{i \in I}$, and $k = (k_i)_{i \in I}$ as follows:

$$h_i = 1 \quad \text{if} \quad i \in H \quad \text{and} \quad h_i = 0 \quad \text{if} \quad i \notin H.$$
$$k_i = 1 \quad \text{if} \quad i \in K \quad \text{and} \quad k_i = 0 \quad \text{if} \quad i \notin K.$$

It is clear that the elements $r = hb = bh$ and $s = kb = bk$ are not in $\mathfrak{a}$. Let us put:

$$\mathfrak{a}' = (\mathfrak{a} + Ar) \cap (\mathfrak{a} + As).$$

Furthermore, let $x = (x_i)_{i \in I}$ be an element of $\mathfrak{a}'$; then:

$$x = a + ur = a' + vs$$

where a, a' are in $\mathfrak{a}$ and $u, v \in A$. If $y = (y_i)_{i \in I}$ is an element of A, we denote by $\operatorname{Supp}(y)$ the subset of I consisting of all indices i, such that $y_i \neq 0$. Then the elements of $\mathfrak{a}$ are just all the elements y for which $\operatorname{Supp}(y)$ is finite. By hypothesis, it is clear that the sets $\operatorname{Supp}(a) \cap K$, $\operatorname{Supp}(a') \cap H$ and $\operatorname{Supp}(a) \cap (I - J) = \operatorname{Supp}(a') \cap (I - J)$ are finite sets. But then we check that the set $\operatorname{Supp}(x)$ is also finite. Indeed:

$$i \in \operatorname{Supp}(x) \cap H \quad \text{implies} \quad s_i = k_i b_i = 0, \quad a_i' \neq 0, \quad i \in \operatorname{Supp}(a') \cap H,$$
$$i \in \operatorname{Supp}(x) \cap K \quad \text{implies} \quad r_i = h_i b_i = 0, \quad a_i \neq 0, \quad i \in \operatorname{Supp}(a) \cap K,$$
$$i \in \operatorname{Supp}(x) \cap (I - J) \quad \text{implies} \quad s_i = r_i = 0, \quad a_i = a_i' \neq 0 \quad i \in \operatorname{Supp}(a) \cap (I - J).$$

Finally, we deduce that $\mathfrak{a} = (\mathfrak{a} + Ar) \cap (\mathfrak{a} + As)$ so that it will be sufficient to consider $\mathfrak{b}_1 = \mathfrak{a} + Ar$ and $\mathfrak{b}_2 = \mathfrak{a} + As$. ∎

NOTE 1. Now we are able to prove the independence of the conditions Ab 5 and Ab 6. Indeed, according to Theorem 3.12 let A be a ring which is not an l.c.-ring. Then the category $(\operatorname{Mod} A)_E$ the spectral category of Mod A is not an Ab 6 category (see Theorem 3.3 and Exercise 1 of Section 5.2).

Exercises

1. Let X and Y be two coirreducible objects in an abelian category with injective envelopes. Then X and Y have isomorphic injective envelopes if and only if a non-zero subobject X' of X is isomorphic with a subobject of Y.

2. Let $\mathscr{C}$ be an abelian category with injective envelopes. Then any complements of a subobject X' of an object X have isomorphic injective envelopes. Also, let $X_1, \ldots, X_n$ be subobjects of an injective object Q such that the sum $\sum_i X_i$ is direct. If Q_i is the injective envelope of X_i, considered as a subobject of Q, then the sum $\sum_i Q_i$ is also direct. If Q is an essential extension of $\sum_i X_i$ then $Q = \coprod_i Q_i$.

3. Let $\mathscr{C}$ be a Grothendieck category and $\{X_i\}_i$ a set of subobjects of an object X of $\mathscr{C}$ such that the sum $\sum_i X_i$ is direct. Assume that for any i there exists a subobject Y_i which is an essential extension of X_i. Then the sum $\sum_i Y_i$ is also direct. Now, if Y is a subobject of X such that

$$Y \cap (\textstyle\sum_i X_i) \neq 0,$$

then an X_{i_0} can be found such that a non-zero subobject of X_{i_0} is isomorphic with a subobject of Y.

4. An object X of a Grothendieck category $\mathscr{C}$ is called *quasi-continuous*, if it does not contain non-zero coirreducible subobjects. Then:

(a) Any subobject or any essential extension of a quasi-continuous object is also quasi-continuous;

(b) An object X is quasi-continuous if and only if the object $P(X)$ is continuous in $\mathscr{C}_E$.

5. An object X of a Grothendieck category $\mathscr{C}$ is called *l.c.-object* if it is an essential extension of a direct sum of coirreducible objects. Then:

(a) Any non-zero subobject and an essential extension of a an l.c.-object is also an l.c.-object.

(b) X is an l.c.-object if and only if $P(X)$ is a semisimple object.

(c) X is an l.c.-object if and only if any non-zero subobject of X contains a coirreducible subobject.

(d) Any object X of $\mathscr{C}$ is the essential extension of a direct sum $X_e \amalg X_c$, where X_e is a l.c.-object and X_c a quasi-continuous object.

(e) Any injective object Q of $\mathscr{C}$ is a direct sum $Q = Q_e \amalg Q_c$ of an injective l.c.-object and a quasi-continuous injective object. Moreover, if

$$Q = Q_e' \amalg Q_c'$$

(where Q_e' is again a l.c.-object and Q_c' a quasi-continuous object) then an automorphism u of Q can be defined such that $u(Q_e) = Q_e'$ and $u(Q_c) = Q_c'$).

NOTE 2. The assertions (d) and (e) are some generalizations of the K.R.S.G.-decomposition for the objects of any Grothendieck category.

6. An integral domain A is called a *valuation ring* if the set of its ideals is totally ordered, that is for any two ideals $\mathfrak{a}$, $\mathfrak{b}$ of A we have $\mathfrak{a} \subseteq \mathfrak{b}$ or $\mathfrak{b} \subseteq \mathfrak{a}$. Any valuation ring is an l.c.-ring. Prove that for any cardinal α there exists a valuation ring A such that Sp(Mod A) has cardinality α.

7. If $\mathfrak{a}$ is a two-sided ideal of a left l.c.-ring A, then $A/\mathfrak{a}$ is also a l.c.-ring.

8. Let A be a ring such that A is a l.c.-object of Mod A. Then for any two elements $x, y \in A$ the relation $Ax \cap Ay = 0$ implies $xy = yx = 0$.

9. The direct product of a finite number of left l.c.-rings is also a left l.c.-ring. If K is a division ring and X a K-module, then $\mathrm{End}_K(X)$ is a l.c.-ring if and only if the dimension of X (as a left vector space) is finite.

10. An object X in a Grothendieck category $\mathscr{C}$ is l.c.-object if and only if there exist a set $\{X_i\}_i$ of its subobjects, so that

$$0 = \bigcap_i X_i$$

and the intersection is reduced.

References

Fort [55], [56]; Gabriel [59]; Gabriel–Oberst [61]; Hudry [95]; Năstăsescu–Popescu [141], [142], [143]; Popescu [168]; Roos [192], [199].

5.4 Decomposition theory

Let $\mathscr{P}$ be a prelocalizing subcategory of a Grothendieck category $\mathscr{C}$ and M a set. By a *predecomposition theory* on $\mathscr{P}$, we understand a function Γ which associates to each object X of $\mathscr{P}$ a subset $\Gamma(M)$ of M such that the following conditions are satisfied:

(P1) $\Gamma(X) = \varnothing$ if and only if $X = 0$;

(P2) For any exact sequence

$$0 \to X' \to X \to X'' \to 0$$

of $\mathscr{C}$, such that $X \in \mathrm{Ob}\,\mathscr{P}$, we have

$$\Gamma(X') \subseteq \Gamma(X) \subseteq \Gamma(X') \cup \Gamma(X'');$$

(P3) If $X \in \mathrm{Ob}\,\mathscr{P}$ and $\{X_i\}_i$ is a direct complete set of subobjects of X, (that is $\sum_i X_i = X$) then

$$\Gamma(X) = \bigcup_i \Gamma(X_i);$$

(P4) If X' is an essential subobject of an object $X \in \mathrm{Ob}\,\mathscr{P}$, then

$$\Gamma(X') = \Gamma(X).$$

In the following we shall denote simply by $\Gamma : \mathscr{P} \to M$ a predecomposition theory on $\mathscr{P}$ with values in M. For any object X of $\mathscr{P}$, the elements of the set $\Gamma(X)$ are called Γ-*associates* with X, or simply *associates*.

EXAMPLES 1. Let $\mathscr{C}$ be a Grothendieck category and $M = \mathrm{Sp}(\mathscr{C}) \cup \{x\}$ where $x \notin \mathrm{Sp}(\mathscr{C})$. For any prelocalizing subcategory $\mathscr{P}$ of $\mathscr{C}$ we denote by $\Gamma : \mathscr{P} \to M$ the predecomposition theory defined as follows. For $X \in \mathrm{Ob}\,\mathscr{P}$ and $m \in M$, then $m \in \Gamma(X)$ if:

(a) $m \in \mathrm{Sp}(\mathscr{C})$ and there exists a coirreducible subobject X' of X whose injective envelope is injective indecomposable of type m.

(b) $m = x$ and X has a non-zero quasi-continuous subobject X'.

We invite the reader to verify that Γ is in fact a predecomposition theory on $\mathscr{P}$ which is called the *spectral predecomposition theory*.

(2) Let M be a set consisting of exactly one element. Then for any prelocalizing subcategory $\mathscr{P}$ there exists a unique predecomposition theory $\Gamma : \mathscr{P} \to M$ such that $\Gamma(X) = M$ for any non-zero object X of $\mathscr{P}$. We shall say that such a predecomposition theory is *trivial*. Any predecomposition theory which associates to a non-zero object of $\mathscr{P}$ the same non-empty subset of a given set is also called trivial.

It is clear that the spectral predecomposition theory on $\mathscr{C}$ is trivial if and only if any object of $\mathscr{C}$ is quasi-continuous or equivalently, if $\mathrm{So}(\mathscr{C}_E) = 0$.

Let $\Gamma : \mathscr{P} \to M$ be a predecomposition theory. An object X of $\mathscr{P}$ is called Γ-*coirreducible* if $\Gamma(X)$ contains exactly one element. If X is Γ-coirreducible then any non-zero subobject of X and any essential extension of X is also Γ-coirreducible.

LEMMA 4.1. *Let $\{X_i\}_{i \in I}$ be a direct complete set of Γ-coirreducible subobjects of an object X; then X is Γ-coirreducible and $\Gamma(X_i) = \Gamma(X)$ for any i.*

Proof. Indeed, if $i, j \in I$ then $X_i + X_j \subseteq X_k$ for some $k \in I$, so that

$$\Gamma(X_i) \subseteq \Gamma(X_k), \qquad \Gamma(X_j) \subseteq \Gamma(X_k).$$

In conclusion $\Gamma(X_i) = \Gamma(X_j)$ for any $i, j \in I$. To finish, we can use condition (P3). ∎

A subobject X' of an object X is called Γ-*irreducible* if X/X' is Γ-coirreducible. If Γ is the spectral predecomposition theory, then an object X of $\mathscr{C}$ is L-coirreducible if and only if X is a l.c.-object and the injective envelopes of any two coirreducible subobjects of X have the same type.

LEMMA 4.2. *The set of Γ-coirreducible subobjects of an object X containing a Γ-coirreducible subobject X' of X has maximal elements.*

Proof. It is clear that if X'' is Γ-coirreducible and $X' \subseteq X''$, then

$$\Gamma(X') = \Gamma(X'').$$

Now we can use Lemma 4.1, to see that the set of Γ-coirreducible subobjects containing X' is inductive. ∎

If $m \in \Gamma(X)$ we shall say that m is Γ-*represented*, or simply *represented*, if there exists a subobject X' of X such that $\Gamma(X') = \{m\}$. Then X' is Γ-coirreducible; sometimes we shall call it m-coirreducible. Generally, there exist several m-coirreducible subobjects of an object X with no precise connection between them. Of course, by the above consideration, we can verify the following result.

PROPOSITION 4.3. *Let $X \in \mathrm{Ob}\,\mathscr{P}$ and let $m \in \Gamma(X)$ be a represented element. There exists an m-coirreducible subobject X_m of X such that for any m-coirreducible subobject X' of X we have $X' \cap X_m \neq 0$.*

Proof. Let X', X'' be two Γ-coirreducible subobjects of X such that

$$X' \cap X'' = 0;$$

then $\Gamma(X' + X'') = \Gamma(X') = \Gamma(X'')$. Moreover, if $\{X_i\}_i$ is a set of m-coirreducible subobjects of X whose sum is direct, the $\sum_i X_i$ is also a m-coirreducible object [see (P2) and (P3)]. Then we shall choose X_m to be a maximal direct sum of m-coirreducible subobjects of X ∎

Now we are able to formulate a variant of a K.R.S.G.-decomposition which we call Γ-K.R.S.G.-decomposition.

PROPOSITION 4.4. *Let $\Gamma : \mathscr{P} \to M$ be a predecomposition theory such that any non-zero object X of $\mathscr{P}$ contains a Γ-coirreducible subobject. Then any object of $\mathscr{P}$ is an essential extension of a direct sum of Γ-coirreducible subobjects.*

Proof. The proof is similar to the proof of Theorem 3.3. Indeed, the set of the direct sum of Γ-coirreducible subobjects of X is inductive relative to inclusion. A maximal element of this set is the desired object. ∎

We shall say that a predecomposition theory satisfies *condition* (P) if for all $X \in \mathrm{Ob}\,\mathscr{P}$, every element $m \in \Gamma(X)$ is represented.

THEOREM 4.5. *A predecomposition theory $\Gamma : \mathscr{P} \to M$ satisfies condition (P) if and only if any non-zero object X of $\mathscr{P}$ contains a Γ-coirreducible subobject.*

Proof. First assume that Γ satisfies condition (P); then for any object $X \neq 0$, $\Gamma(X) \neq \varnothing$ so that X has Γ-coirreducible subobjects.

Conversely, according to Proposition 4.4, any object X is the essential extension of a direct sum of Γ-coirreducible subobjects, denoted X'. Then $\Gamma(X) = \Gamma(X')$. But $X' = \coprod_{i\in I} X_i$, such that $\Gamma(X_i) = m_i$ and $m_i \neq m_j$ if $i \neq j$.

If I is a finite set with n elements, then we proceed by induction with respect to n to prove that $\Gamma(X) = \Gamma(X') = \bigcup_i \Gamma(X_i)$. Furthermore, we shall use condition ($P3$) to check that $\Gamma(X) = \Gamma(X') = \bigcup_i \Gamma(X_i)$ (for any set I). Finally, it is clear that any element of $\Gamma(X)$ is represented. ∎

Let X' be a subobject of an object X of $\mathscr{P}$. We shall say that the set $\{X_i\}_{i\in I}$ of subobjects of X gives a Γ-*decomposition* of X' if the following conditions are true:

(D1) $\bigcap_{i\in I} X_i = X'$ and the intersection is reduced;

(D2) X_i is a Γ-irreducible subobject for any i;

(D3) $\Gamma(X/X_i) \cap \Gamma(X/X_j) = \varnothing$ if $i \neq j$;

(D4) $\Gamma(X_i/X') = \Gamma(X/X') - \Gamma(X/X_i)$;

(D5) $\Gamma(X/X') = \bigcup_i \Gamma(X/X_i)$.

The subobject X' has a *finite* Γ-*decomposition* if the set $\{X_i\}_i$ of the above definition is finite. Finally, a predecomposition theory $\Gamma : \mathscr{P} \to M$ is called a *decomposition theory* if any subobject X' of an object X of $\mathscr{P}$ has a Γ-decomposition.

COROLLARY 4.6. *Assume that the subobjects* $\{X_i\}_i$ *of an object* X, *give a* Γ-*decomposition of the subobject* X'. *Then the subobjects* $\{X_i/X'\}_i$ *give a* Γ-*decomposition of* 0 *in the quotient object* X/X'.

The proof is easy and is left to the reader. ∎

The above corollary gives us the possibility of reducing the study of the Γ-decompositions of any subobject to the study of Γ-decompositions of the null subobject of a suitable object.

LEMMA 4.7. *Assume that the subobjects* $\{X_i\}_{i\in I}$ *give a* Γ-*decomposition of* 0 *in* X. *Then any* $m \in \Gamma(X)$ *is represented.*

Proof. It is clear that $m = \Gamma(X/X_i)$ for a suitable i. Moreover, let us denote $Y_i = \bigcap_{j\neq i} X_j$. Then $Y_i \neq 0$ and $X_i \cap Y_i = 0$. Then Y_i can be considered as a subobject of X/X_i and that $\Gamma(Y_i) = \Gamma(X/X_i) = m$. ∎

COROLLARY 4.8. *Any decomposition theory satisfies condition* (P). ∎

THEOREM 4.9. *Assume that the set* $\{X_i\}_i$ *of subobjects of* X *gives a* Γ*-decomposition of* 0. *Then a subobject* X' *of* X *is* Γ*-coirreducible if and only if there exists a* Γ*-irreducible subobject* X'' *of* X *such that* $X' \cap X'' = 0$.

Proof. Assume that X' is Γ-coirreducible and $\Gamma(X') = \{m_i\} = \Gamma(X/X_i)$. According to (D4) it follows that $X' \cap X_i = 0$. On the other hand, we see from the proof of Lemma 4.7, that X_i is Γ-irreducible.

Conversely, assume that $X' \cap X'' = 0$, X'' being Γ-irreducible; then $X' \subseteq X/X''$ so that $\Gamma(X') \subseteq \Gamma(X/X'')$. Hence X' is Γ-coirreducible, as a non-zero subobject of a Γ-coirreducible object. ∎

Now we shall give the very useful necessary and sufficient condition for a predecomposition theory to be a decomposition theory.

THEOREM 4.10. *Let* $\Gamma : \mathscr{P} \to M$ *be a predecomposition theory. The following assertions are equivalent*:

(1) Γ *is a decomposition theory*;

(2) Γ *has condition* (*P*).

Proof. The implication (1) ⇒ (2) is in fact Corollary 4.8.

(2) ⇒ (1). Let X be a non-zero object of $\mathscr{P}$ and $m \in \Gamma(X)$. Denote by X_m a subobject of X as in Proposition 4.3. Let us write $Y_m' = \sum_{m' \neq m} X_{m'}$. Then $X_m \cap Y_m' = 0$, X_m being in fact a direct sum of Γ-coirreducible subobjects. Furthermore, denote by Y_m a subobject which contains Y_m' and maximal with the property that $Y_m \cap X_m = 0$. We claim that the set $\{Y_m\}_{m \in \Gamma(X)}$ gives a Γ-decomposition of 0 in X. Indeed, if $Y \subseteq \bigcap_m Y_m$, then for any $m \in \Gamma(X)$ we have that $Y \cap X_m = 0$. If $Y \neq 0$, then it contains a Γ-coirreducible subobject Y', say $\Gamma(Y') = \{m_0\}$. But $Y \cap X_{m_0} = 0$, in contradiction with the choice of X_{m_0}. Hence $\bigcap_m Y_m = 0$, and by definition of the elements Y_m this last intersection is reduced.

We also see that X/Y_m is an essential extension of X_m, i.e. a Γ-coirreducible subobject; in addition $\Gamma(X/Y_m) = \Gamma(X_m) = \{m\}$. Finally, we must prove that $\Gamma(Y_m) = \Gamma(X) - \{m\}$. In fact, from the exact sequence:

$$0 \to Y_m \to X \to X/Y_m \to 0$$

we see that if $m' \in \Gamma(X)$, $m' \neq m$, then $m' \in \Gamma(Y_m)$ [see (P2)]. Finally, if Y is an m-coirreducible subobject of Y_m, then $Y \cap X_m \neq 0$ which is a contradiction. Hence necessarily $\Gamma(Y_m) = \Gamma(X) - \{m\}$. ∎

We shall use the decomposition theories for the characterization of the l.c.-categories. We need the following result:

LEMMA 4.11. A coirreducible object is Γ-coirreducible for any predecomposition theory Γ. ∎

For a prelocalizing subcategory $\mathscr{P}$ of a Grothendieck category $\mathscr{C}$ we denote by $\mathrm{Sp}(\mathscr{P})$ the subset of $\mathrm{Sp}(\mathscr{C})$ consisting of all injective indecomposables Q such that $Q_{\mathscr{P}} \neq 0$. If $\Gamma : \mathscr{P} \to M$ is a predecomposition theory, then from Lemma 4.11 any coirreducible object X is also Γ-coirreducible hence the assignment $Q_{\mathscr{P}} \rightsquigarrow \Gamma(Q_{\mathscr{P}})$ defines a map $f_\Gamma : \mathrm{Sp}(\mathscr{P}) \to M$.

THEOREM 4.12. *Let $\mathscr{C}$ be a Grothendieck category. With the above notations, the following assertions are equivalent:*

(1) *$\mathscr{C}$ is a l.c.-category;*

(2) *The assignment $\Gamma \rightsquigarrow f_\Gamma$ is a bijection between the set of all decomposition theories $\Gamma : \mathscr{C} \to M$ and the functions $f : \mathrm{Sp}(\mathscr{C}) \to M$ (M being a non-empty set).*

Proof. (1) ⇒ (2). Let $f : \mathrm{Sp}(\mathscr{C}) \to M$ be a map; for any $X \in \mathrm{Ob}\,\mathscr{C}$ we denote by $\Gamma_f(X)$ the subset of M consisting of all elements $m \in M$ such that there exists a coirreducible subobject X' of X with $f(E(X')) = m$. We claim that the assignment $X \rightsquigarrow \Gamma_f(X)$ is a predecomposition theory having the property (P). Indeed, if $X \neq 0$, then $\Gamma_f(X) \neq \varnothing$ and by convention $\Gamma_f(0) = \varnothing$. Consider the exact sequence:

$$0 \to X' \to X \to X'' \to 0 \tag{2}$$

It is clear that if $m \in \Gamma_f(X')$, then there exists a coirreducible subobject X_1 of X', such that $f(E(X_1)) = m$. But then X_1 can be regarded as a subobject of X, so that $m \in \Gamma(X)$. Hence, $\Gamma_f(X') \subseteq \Gamma_f(X)$. Now let $m \in \Gamma_f(X)$ be such that $m \notin \Gamma_f(X')$; then there exists a coirreducible subobject X_1 of X such that $f(E(X_1)) = m$. By hypothesis, $X_1 \cap X' = 0$, hence X_1 can be regarded as a subobject of X''. Hence by the exact sequence (2) we can check the relations:

$$\Gamma_f(X') \subseteq \Gamma_f(X) \subseteq \Gamma_f(X') \cup \Gamma_f(X'').$$

Now let $\{X_i\}_i$ be a direct complete set of subobjects of an object X, and $m \in \Gamma_f(X)$. Then $m = f(E(X'))$ where X' is a coirreducible subobject of X. By condition Ab 5 we deduce that $X' \cap X_i \neq 0$ for a suitable i, and hence $E(X_i \cap X') = E(X')$. In other words, $m \in \Gamma_f(X_i)$ or equivalently, $\Gamma_f(X) \subseteq \bigcup_i \Gamma_f(X_i)$. We have in fact an equality, the converse inclusion being obvious.

Furthermore, assume that X is an essential extension of X'; then for any coirreducible subobject X_1 of X we see that $X_1 \cap X' \neq 0$ hence $\Gamma_f(X) = \Gamma_f(X')$. Finally, by definition, any element $m \in \Gamma_f(X)$ is represented and it follows that Γ_f is in fact a predecomposition theory having the property (P), as claimed.

Now we shall prove that Γ_f determines the map f or equivalently, $f_{\Gamma_f} = f$. Indeed, if Q is an injective indecomposable of $\mathscr{C}$, then by definition

$$\Gamma_f(Q) = f(Q).$$

Conversely, for any decomposition theory $\Gamma : \mathscr{C} \to M$ we have $\Gamma_{f_\Gamma} = \Gamma$. In fact, if X is a coirreducible object, then $\Gamma(X)$ consists of one element, according to Lemma 4.11. Furthermore, by Theorem 3.3, an object X is an essential extension of a direct sum $\coprod_{i \in I} X_i$ of coirreducible objects, and $\Gamma(X) = \Gamma(\coprod_i X_i)$. Furthermore $\Gamma(\coprod_i X_i) = \bigcup_F \Gamma(\coprod_{i \in F} X_i)$, where F runs over all finite subsets of I. Hence, in order to prove that $\Gamma_{f_\Gamma} = \Gamma$ it will be sufficient to prove that $\Gamma_{f_\Gamma}(\coprod_{i \in F} X_i) = \Gamma(\coprod_{i \in F} X_i)$ for any F. But this follows from the equality $\Gamma_{f_\Gamma}(X_i) = \Gamma(X_i)$ for any i, and induction relative to the number of elements in F.

(2) $\Rightarrow$ (1). Two cases are possible:

(a) $\mathrm{So}(\mathscr{C}_E) = 0$, or equivalently, $\mathscr{C}$ does not contain any coirreducible subobjects. Then $\mathrm{Sp}(\mathscr{C}) = \varnothing$, hence for a suitable set M there exists exactly one map $f : \varnothing \to M$. But, obviously, we can define two distinct decomposition theories Γ_1, Γ_2 of $\mathscr{C}$ into M as follows. Let x, y be two distinct elements of M. Then, let us put $\Gamma_1(X) = \{x\}$, $\Gamma_2(X) = \{y\}$ for any non-zero object X. We see that Γ_1 and Γ_2 are trivial predecomposition theories having the property (P). The contradiction is obvious.

(b) Now let us assume that $\mathrm{So}(\mathscr{C}) \neq \varnothing$, and consider the set $M = \mathrm{Sp}(\mathscr{C}) \bigcup \{u, v\}$, where u and v are two distinct elements which do not belong to $\mathrm{Sp}(\mathscr{C})$. Denote by $f : \mathrm{Sp}(\mathscr{C}) \to M$ the map defined as follows: $f(Q) = Q$ for any $Q \in \mathrm{Sp}(\mathscr{C})$. If $\mathscr{C}$ is not a l.c.-category, then it has quasi-continuous objects. We shall define two distinct decomposition theories $\Gamma_1, \Gamma_2 : \mathscr{C} \to M$ such that $f_{\Gamma_1} = f_{\Gamma_2} = f$. The definition of Γ_1 is the following: if X is a quasi-continuous object, then $\Gamma_1(X) = \{u\}$; if X is a l.c. object then X is an essential extension of a direct sum $\coprod_i X_i$ of coirreducible objects and thus we shall define $\Gamma_1(X) = \bigcup_i \Gamma(X_i)$, where $\Gamma(X_i) = f(E(X_i))$ for all i. Finally any element X of $\mathscr{C}$ is an essential extension of a direct sum $X_1 \amalg X_2$ where X_1 is a l.c.-object and X_2 a quasi-continuous object. Then by definition $\Gamma_1(X) = \Gamma_1(X_1) \cup \Gamma_1(X_2)$. As above, by a slight computation we can prove that Γ_1 is a predecomposition theory having the property (P). The decomposition theory Γ_2 is defined in same way as Γ_1 with the only difference that $\Gamma_2(X) = \{v\}$ for any non-zero quasi-continuous object. The verification that $f_{\Gamma_1} = f_{\Gamma_2} = f$ is obvious.

In every case we have a contradiction. Hence, the assumption that $\mathscr{C}$ is not a l.c.-category is false. ■

COROLLARY 4.13. *Let $\mathscr{P}$ be a closed subcategory of a l.c.-category $\mathscr{C}$ and M a non-empty set. Then the mapping $\Gamma \rightsquigarrow f_\Gamma$ is a bijection between the decomposition theory $\Gamma : \mathscr{P} \to M$ and the set of all mapping $f : \mathrm{Sp}(\mathscr{P}) \to M$. Moreover, the family of all decomposition theories $\Gamma : \mathscr{P} \to M$ is in fact a set.*

The proof is obvious, $\mathscr{P}$ being also a l.c.-category. ■

Sometimes, the following result will be useful.

PROPOSITION 4.14. *Let $\mathscr{P}$ be a prelocalizing subcategory of a l.c.-category and $\Gamma : \mathscr{P} \to M$ a decomposition theory. For an object $X \in \mathrm{Ob}\,\mathscr{P}$ the following assertions are equivalent:*

(1) *X is Γ-coirreducible;*

(2) *X is the essential extension of a direct sum $\coprod_i X_i$ of coirreducible objects, such that $\Gamma(X_i) = \Gamma(X_j)$ for every $i, j \in I$.*

The proof is left to the reader. ■

Exercises

As always $\Gamma : \mathscr{P} \to M$ is a predecomposition theory, where $\mathscr{P}$ is a prelocalizing subcategory of a Grothendieck category.

1. Let Y, Z be subobjects of $X \in \mathrm{Ob}\,\mathscr{P}$ and $X' = Y \cap Z$. If Y is Γ-irreducible, then X' is Γ-irreducible in Z and $\Gamma(Z/X') = \Gamma(X/Y)$.

2. Assume that X' has a finite Γ-decomposition in X and that Y/X' is a non-zero subobject of X/X'. If X' is Γ-irreducible in Y, then there is a Γ-irreducible subobject X'' of X such that $X'' \cap Y = X'$ and

$$\Gamma(X/X'') = \Gamma(X/X').$$

3. Let Γ, $\Gamma' : \mathscr{P} \to M$ be two predecomposition theories. We shall say that Γ is stronger than Γ', written $\Gamma \geqslant \Gamma'$ if for all $X \in \mathrm{Ob}\,\mathscr{P}$, $\Gamma(X) \supseteq \Gamma'(X)$. Then any Γ-irreducible object is also Γ'-irreducible. Moreover if the subobject X' of X has a finite Γ-decomposition, then $\Gamma'(X/X') = \Gamma(X/X')$.

References

Bourbaki [26], [27]; Bucur [29]; Bucur–Deleanu [31]; Gabriel [59]; Năstăsescu [139]; Popescu [170]; Riley [186], [187].

5.5 Semi-noetherian categories

Let $\mathscr{C}$ be a Grothendieck-category. We shall define the *Krull–Gabriel filtration* of $\mathscr{C}$ as follows. For any ordinal number α we shall denote by $\mathscr{C}_\alpha$ the localizing subcategory of $\mathscr{C}$ defined in the following manner:

$\mathscr{C}_{-1}$ is the zero category.

Let us assume that $\alpha = \beta + 1$. Denote by $T_\beta : \mathscr{C} \to \mathscr{C}/\mathscr{C}_\beta$ the canonical functor and by S_β a right adjoint of T_β. Then an object X of $\mathscr{C}$ will be contained in $\mathscr{C}_\alpha$ if and only if $T_\beta(X) \in \mathrm{Ob}(\mathscr{C}/\mathscr{C}_\beta)_0$.

If α is a limit ordinal, then $\mathscr{C}_\alpha$ is the localizing subcategory generated by all localizing subcategories $\mathscr{C}_\beta$ with $\beta \leqslant \alpha$. It is clear that if $\alpha \leqslant \alpha'$ then $\mathscr{C}_\alpha \subseteq \mathscr{C}_{\alpha'}$. Because $\mathscr{C}$ has a set of localizing subcategories (Chapter 4, Theorem 14.2), there exists an ordinal α such that $\mathscr{C}_\alpha = \mathscr{C}_\tau$ for any $\tau \geqslant \alpha$. Let us put $\mathscr{C}_\tau = \bigcup_\alpha \mathscr{C}_\alpha$.

Thus $\mathscr{C}_0$ is the smallest localizing subcategory containing all simple objects.

Lemma 5.1. *With the above notations, $\mathscr{C}/\mathscr{C}_\tau$, does not contain simple objects.*

Proof. By definition, it is clear that τ is the smallest ordinal such that $\mathscr{C}_\tau = \mathscr{C}_\alpha$ for any $\alpha \geqslant \tau$. If $\mathscr{C}/\mathscr{C}_\tau$ has a simple object, then $(\mathscr{C}/\mathscr{C}_\tau)_0 \neq 0$ so that $\mathscr{C}_{\tau+1} \neq \mathscr{C}_\tau$, a contradiction. ∎

Frequently, we shall say that the localizing subcategories $\{\mathscr{C}_\alpha\}_\alpha$ define the Krull–Gabriel filtration of $\mathscr{C}$. We shall say that an object X of $\mathscr{C}$ has the *Krull–Gabriel dimension* defined or X is *semi-noetherian* if $X \in \mathrm{Ob}\,\mathscr{C}_\tau$. Then the smallest ordinal α so that $X \in \mathrm{Ob}\,\mathscr{C}_\alpha$ is denoted by $KG(X)$ and is called the *Krull–Gabriel dimension* of X. It is clear that $KG(X)$ has a predecessor.

We shall say that $\mathscr{C}$ is a *semi-noetherian category* or that the *Krull–Gabriel dimension of $\mathscr{C}$ is defined* if $\mathscr{C} = \mathscr{C}_\tau$, or equivalently any object of $\mathscr{C}$ is semi-noetherian. Usually, we shall write *s.n.-category* instead of semi-noetherian category.

Lemma 5.2. *The following assertions are equivalent for $\mathscr{C}$:*

(1) *$\mathscr{C}$ is a s.n.-category;*

(2) *For any localizing proper subcategory $\mathscr{F}$ of $\mathscr{C}$, the quotient category $\mathscr{C}/\mathscr{F}$ contains a simple object.*

Proof. (1) ⇒ (2). Let α be smallest ordinal such that $\mathscr{C}_\alpha \nsubseteq \mathscr{F}$ (there exists such an α because $\mathscr{F} \neq \mathscr{C}$). Obviously, $\alpha = \beta + 1$. Then, there exists an object X in $\mathscr{C}_\alpha$ such that $X \notin \mathrm{Ob}\,\mathscr{F}$. We can pick X so that $T_\beta(X)$ is simple in $\mathscr{C}/\mathscr{C}_\beta$ and X is $\mathscr{C}_\beta$-closed. Then, for any non-zero subobject X' of X, we see that $X/X' \in \mathrm{Ob}\,\mathscr{C}_\beta$, and so $X/X' \in \mathrm{Ob}\,\mathscr{F}$ by the definition of X. Hence $T(X)$ will be a simple object of $\mathscr{C}/\mathscr{F}$, where $T : \mathscr{C} \to \mathscr{C}/\mathscr{F}$ is the canonical functor.

The converse implication follows by Lemma 5.1 and the hypothesis. ∎

Corollary 5.3. *Let $\mathscr{F}$ be a localizing subcategory of a Grothendieck category $\mathscr{C}$. Then $\mathscr{C}$ is a s.n.-category if and only if $\mathscr{F}$ and $\mathscr{C}/\mathscr{F}$ are s.n.-categories.*

The proof is a direct consequence of the above lemma. ∎

For a s.n.-category $\mathscr{C}$ we shall denote by $KG(\mathscr{C})$ the smallest ordinal α such that $\mathscr{C}_\alpha = \mathscr{C}$. $KG(\mathscr{C})$ is called the *Krull–Gabriel dimension of* $\mathscr{C}$.

COROLLARY 5.4. *If $\mathscr{F}$ is a localizing subcategory of a s.n.-category $\mathscr{C}$, then:*

$$\sup(KG(\mathscr{F}), KG(\mathscr{C}/\mathscr{F})) \leqslant KG(\mathscr{C}) \leqslant KG(\mathscr{F}) \oplus KG(\mathscr{C}/\mathscr{F})$$

where "$\oplus$" is the sum of ordinal numbers.

The proof is simple and is left to the reader. ∎

THEOREM 5.5. *Any s.n.-category $\mathscr{C}$ is a l.c.-category.*

Proof. Let Q be a non-zero injective of $\mathscr{C}$ and α the smallest ordinal such that $Q_{\mathscr{C}_\alpha} \neq 0$. Then $\alpha = \beta + 1$ and Q is $\mathscr{C}_\beta$-closed. But it is clear that $T_\beta(Q_{\mathscr{C}_\alpha})$ contains a simple object U of $\mathscr{C}/\mathscr{C}_\beta$, and $S_\beta(U)$ is coirreducible and is contained in Q. Now, we shall utilize Theorem 3.3. ∎

Let $\mathscr{C}$ be a semi-noetherian category; for any ordinal $\alpha \leqslant KG(\mathscr{C})$ we denote by $\mathrm{Sp}_\alpha(\mathscr{C})$ the subset of $\mathrm{Sp}(\mathscr{C})$ consisting of all indecomposable injectives Q, which are $\mathscr{C}_\alpha$-closed and $Q_{\mathscr{C}_{\alpha+1}} \neq 0$. It is clear that for $\alpha \neq \beta$, $\mathrm{Sp}_\alpha(\mathscr{C}) \cap \mathrm{Sp}_\beta(\mathscr{C}) = \varnothing$ and $\bigcup_\alpha \mathrm{Sp}_\alpha(\mathscr{C}) = \mathrm{Sp}(\mathscr{C})$.

Now, let $Q \in \mathrm{Sp}_\alpha(\mathscr{C})$; then $T_\alpha(Q)$ contains a simple subobject U of $\mathscr{C}/\mathscr{C}_\alpha$ and thus $S_\alpha(U) = Q^\alpha$ is a coirreducible object. The objects U and Q^α are uniquely defined (up to isomorphism).

PROPOSITION 5.6. *Let $\mathscr{C}$ be a s.n.-category. Then, for any ordinal $\alpha \leqslant KG(\mathscr{C})$, the mapping*

$$Q \rightsquigarrow T_\alpha(Q^\alpha)$$

defines a bijection between $\mathrm{Sp}_\alpha(\mathscr{C})$ and the set of all types of simple objects of $\mathscr{C}/\mathscr{C}_\alpha$.

Proof. Let Q, $Q' \in \mathrm{Sp}_\alpha(\mathscr{C})$ such that $T_\alpha(Q^\alpha) \simeq T_\alpha(Q'^\alpha)$; then

$$S_\alpha T_\alpha(Q^\alpha) \simeq S_\alpha T_\alpha(Q'^\alpha)$$

so that $Q \simeq Q'$, being the injective envelopes of the isomorphic objects. Hence the map considered is an injection. Furthermore, if V is a simple object of $\mathscr{C}/\mathscr{C}_\alpha$ then $S_\alpha(V)$ is coirreducible and $E(S_\alpha(V))$ is an indecomposable injective, whose type is in $\mathrm{Sp}_\alpha(\mathscr{C})$. Clearly, $T_\alpha\{E[S_\alpha(V)]\} = V$, so that the map considered is also surjective. ∎

We have seen (Corollary 3.5) that an object of a l.c.-category has a unique K.R.S.G.-decomposition. For an s.n.-category we have the following more precise result:

THEOREM 5.7. *Let $\mathscr{C}$ be a s.n.-category and X a non-zero object of $\mathscr{C}$. Then for any $\alpha \leqslant KG(\mathscr{C})$ there exists a morphism $f: Y_\alpha \to X$ such that the following conditions are satisfied*:

(a) *Y_α is $\mathscr{C}_\alpha$-torsionfree*;

(b) *$T_\alpha(Y_\alpha)$ is a semisimple object of $\mathscr{C}/\mathscr{C}_\alpha$*;

(c) *The morphism $f = \coprod_\alpha f_\alpha = \coprod_\alpha Y_\alpha \to X$ is an essential monomorphism.*

Moreover, for any set $\{f_\alpha : Y_\alpha \to X\}_{\alpha \leqslant KG(\mathscr{C})}$ of morphisms of $\mathscr{C}$ which verifies the conditions (a)–(c) *we have that $T_\alpha(f_\alpha)$ defines an isomorphism between $T_\alpha(Y_\alpha)$ and* $\mathrm{So}(T_\alpha(X))$.

Proof. Let $\alpha < KG(\mathscr{C})$ be an ordinal number; we denote by X_α the greatest subobject of X in $\mathscr{C}_\alpha$. Also, let Z_α be a complement of X_α in $X_{\alpha+1}$ and $i_\alpha : Z_\alpha \to X$ the natural inclusion. Then the morphism $j = \coprod_\alpha i_\alpha : \coprod_\alpha Z_\alpha \to X$ is an essential monomorphism. In fact, this is clear if $KG(\mathscr{C}) = -1$ (i.e. $\mathscr{C}$ consists only of null objects). Assume the result for the s.n.-categories $\mathscr{C}'$ such that $KG(\mathscr{C}') < KG(\mathscr{C})$. Two cases are possible:

(1) $KG(\mathscr{C}) = \beta + 1$; then the inclusion morphism $h: X_\beta \amalg Z_\beta \to X$ is essential; on the other hand, j is equal to the composition

$$\coprod_\alpha Z_\alpha \xrightarrow{\ i\ } X_\beta \amalg Z_\alpha \xrightarrow{\ h\ } X,$$

where i is canonically defined. By assumption, i is essential, hence j is essential, as a composition of essential morphisms.

(2) $KG(\mathscr{C})$ is a limit ordinal; then $X = \Sigma X_\beta$ where β runs over the set of all the cardinal numbers $\beta < KG(\mathscr{C})$.

For any $\beta < KG(\mathscr{C})$, consider the morphism

$$j_\beta = \coprod_{\gamma<\beta} i_\gamma : \coprod_{\gamma<\beta} Z_\gamma \to X_\beta.$$

By hypothesis, any j_β is an essential monomorphism so that j is an essential monomorphism by Lemma 2.7 of Chapter 4.

Now the construction of the family $f_\alpha : Y_\alpha \to X$ is easy. Y_α will be a subobject of Z_α such that $T_\alpha(g_\alpha)$ is an isomorphism of $T_\alpha(Y_\alpha)$ onto $\mathrm{So}(T_\alpha(Z_\alpha))$

$$g_\alpha : Y_\alpha \to Z_\alpha$$

being the canonical inclusion. If P is a non-zero subobject of Z_α, then $T_\alpha(P)$ contains a simple subobject, hence $T_\alpha(P) \cap T_\alpha(Y_\alpha) \neq 0$. Therefore, necessarily $P \cap Y_\alpha \neq 0$, or equivalently, g_α is an essential monomorphism. Let us put $f_\alpha = i_\alpha g_\alpha$. The above conditions (a)–(c) are easily verified.

Conversely, assume that $\{f_\alpha\}$ is a family of morphisms which satisfies the above conditions (a)–(c). Then it is clear that $\mathrm{im}(T_\alpha(f_\alpha))$ is the greatest subobject of $\mathrm{im}(T_\alpha(f))$ in $\mathscr{C}_{\alpha+1}/\mathscr{C}_\alpha$. In fact, if H is a simple subobject of $T_\alpha(X)$, then $H \cap \mathrm{im}(T_\alpha(f)) \neq 0$, hence $H \cap \mathrm{im}(T_\alpha(f)) \subseteq \mathrm{im}(T_\alpha(f_\alpha))$. In conclusion, any simple subobject of $T_\alpha(X)$ is contained in $\mathrm{im}(T_\alpha(f_\alpha))$ as claimed. ∎

Now we shall give a new characterization of s.n.-categories using the prime localizing subcategories.

First we have the following useful result.

LEMMA 5.8. *Let $\mathscr{C}$ be a s.n.-category and $\{U_i\}_{i\in I}$ the set of all types of simple objects. For any $i \in I$, denote by Q_i the injective envelope of U_i. Then for an object X of $\mathscr{C}$, the following assertions are equivalent*:

(1) $X = 0$;

(2) $\mathrm{Hom}_{\mathscr{C}}(X, Q_i) = 0$ *for any* $i \in I$.

Proof. (2) ⇒ (1). Let $\mathscr{F}$ be the localizing subcategory of $\mathscr{C}$ consisting of all objects Y such that $\mathrm{Hom}_{\mathscr{C}}(Y, Q_i) = 0$ for any $i \in I$. If $\mathscr{F} \neq 0$, then it contains a simple object, by Lemma 5.3, a contradiction. ∎

THEOREM 5.9. *Let $\mathscr{C}$ be a Grothendieck category and* Spel($\mathscr{C}$) *the set of all prime localizing subcategories of $\mathscr{C}$. The following assertions are equivalent*:

(1) *$\mathscr{C}$ is a s.n.-category*;

(2) Spel($\mathscr{C}$) *is non-void and every localizing subcategory $\mathscr{F}$ of $\mathscr{C}$ can be represented as a reduced intersection*

$$\mathscr{F} = \bigcap_{i\in I} \mathscr{P}_i$$

where $\mathscr{P}_i \in$ Spel($\mathscr{C}$) *for any $i \in I$.*

Proof. (1) ⇒ (2). Let $T: \mathscr{C} \to \mathscr{C}/\mathscr{F}$ be the canonical functor, S a right adjoint of T, $\{U_i\}_{i\in I}$ a representative set of simple objects of $\mathscr{C}/\mathscr{F}$ and $Q_i = S(E(U_i))$. Write $\mathscr{P}_i$ the localizing subcategory of $\mathscr{C}$ cogenerated by Q_i. Thus by Theorem 20.6 of Chapter 4, we see that $\mathscr{P}_i$ is a prime localizing subcategory of $\mathscr{C}$ for any $i \in I$. Also it is clear that $\mathscr{F} \subseteq \mathscr{P}_i$ for every $i \in I$, so that $\mathscr{F} \subseteq \bigcap_{i\in I} \mathscr{P}_i$. Conversely, if X is an object of $\bigcap_{i\in I} \mathscr{P}_i$, then
$\mathrm{Hom}_{\mathscr{C}}(X, Q_i) = 0$, or equivalently, $\mathrm{Hom}_{\mathscr{C}/\mathscr{F}}(T(X), E(U_i)) = 0$ for any $i \in I$. Thus by Lemma 5.8, $T(X) = 0$, so that $X \in \mathrm{Ob}\,\mathscr{F}$. Hence

$$\mathscr{F} = \bigcap_{i\in I} \mathscr{P}_i.$$

Now we shall prove that this intersection is reduced. Indeed, let $i_0 \in I$, and $I' = I - i_0$. Then for any $i' \in I'$, one has: $\mathrm{Hom}_{\mathscr{C}/\mathscr{F}}(U_{i_0}, E(U_{i'})) = 0$. Thus if X is an object of $\mathscr{C}$ such that $T(X) \simeq U_{i_0}$, then $\mathrm{Hom}_{\mathscr{C}}(X, Q_{i'}) = 0$

for all $i' \in I'$, hence $X \in \mathrm{Ob}(\bigcap_{i' \in I'} \mathscr{P}_{i'})$ and obviously $X \notin \mathrm{Ob}\,\mathscr{F}$. Therefore $\bigcap_{i' \in I'} \mathscr{P}_{i'} \neq \bigcap_{i \in I} \mathscr{P}_i = \mathscr{F}$. In other words the above representation of $\mathscr{F}$ as an intersection of elements of Spel($\mathscr{C}$) is reduced.

(2) ⇒ (1). We intend to utilize Lemma 5.2. To this end, let $\mathscr{F}$ be a proper localizing subcategory of $\mathscr{C}$. Thus $\mathscr{F} = \bigcap_{i \in I} \mathscr{P}_i$ where $\mathscr{P}_i$ are elements of Spel($\mathscr{C}$) and the intersection is reduced. Let $i_0 \in I$, $T : \mathscr{C} \to \mathscr{C}/\mathscr{P}_{i_0}$ the canonical functor, S a right adjoint of T, U_{i_0} a simple object of $\mathscr{C}/\mathscr{P}_{i_0}$ and

$$K_{i_0} = S(U_{i_0}).$$

Thus $\mathscr{P}_{i_0}$ is cogenerated by $E(K_{i_0})$ and for any non-zero subobject X of K_{i_0}, $K_{i_0}/X \in \mathrm{Ob}\,\mathscr{P}_{i_0}$.

By hypothesis, $\mathscr{F}' = \bigcap_{i \in I - i_0} \mathscr{P}_i \neq \mathscr{F}$, so that there exists a non-zero object Y of $\mathscr{C}$ such that $Y \in \mathrm{Ob}\,\mathscr{F}'$ and $Y \notin \mathrm{Ob}\,\mathscr{F}$. Thus necessarily,

$$\mathrm{Hom}_{\mathscr{C}}(Y, E(K_{i_0})) \neq 0,$$

so that we can find a quotient object Y' of Y such that $Y' \subseteq E(K_{i_0})$. Now let $Z = Y' \cap K_{i_0}$. Thus Z is an object of $\mathscr{F}'$, by hypothesis and $Z \notin \mathrm{Ob}\,\mathscr{F}$. Finally, we see that for any non-zero subobject Z_1 of Z, $Z/Z_1 \in \mathrm{Ob}\,\mathscr{F}$. Thus necessarily the image of the object Z in the category $\mathscr{C}/\mathscr{F}$ will be a simple object. ■

A ring A is called *left semi-noetherian* (sometimes we write s.n. ring) if the category Mod A is a s.n.-category. The following characterization of a left semi-noetherian rings is in fact a paraphrase of Theorem 5.9. However we give a brief independent proof.

THEOREM 5.10. *For a ring A the following assertions are equivalent*:

(1) *A is a left semi-noetherian ring*;

(2) *Any element $F \in L(A)$ is a reduced intersection of the form*

$$F = \bigcap_{i \in I} F_i,$$

where F_i are left primes of A;

(3) *For any $F \in L(A)$, there exists a set $\{\mathfrak{a}\}_{i \in I}$ of left super-prime ideals, such that $F = \bigcap_{i \in I} F_{\mathfrak{a}_i}$ and this intersection is reduced*;

(4) *For any $F \in L(A)$ and any left ideal $\mathfrak{a}$ such that $\mathfrak{a} \notin F$ there exists a left ideal $\mathfrak{a}'$ and an element $x \in A$, $x \notin \mathfrak{a}$, such that $(\mathfrak{a} : x) \subseteq \mathfrak{a}'$, $\mathfrak{a}' \notin F$, and $\mathfrak{a}$ is maximal with these properties.*

Proof. The equivalences (1) ⇔ (2) ⇔ (3) follow by Theorem 5.9 and Theorem 20.9 of Chapter 4.

(1) ⇒ (4). Let $\{U_i\}_{i\in I}$ be the set of all types of simple objects of Mod $A/\mathscr{F}$; for any $i \in I$, let E_i be the injective envelope of U_i and $Q_i = S(E_i)$, where S is the inclusion functor Mod $A/\mathscr{F} \to$ Mod A. By Theorem 20.6 of Chapter 4, $F_i = F(Q_i)$ is a left prime of A. Obviously, $F \subseteq F_i$, for any i, so that

$$F \subseteq \bigcap_i F_i.$$

Now let $\mathfrak{a} \in \bigcap_i F_i$; then for any i, $\mathrm{Hom}_A(A/\mathfrak{a}, Q_i) = 0$ hence

$$\mathrm{Hom}_{\mathrm{Mod}\, A/\mathscr{F}}(T(A/\mathfrak{a}), E_i) = 0.$$

Then by Lemma 5.8 we see that $T(A/\mathfrak{a}) = 0$, or equivalently $A/\mathfrak{a} \in \mathrm{Ob}\,\mathscr{F}$, T being a left adjoint of S. Hence $\bigcap_i F_i = F$. Furthermore, let $\mathfrak{a}$ be a left ideal of A which does not belong to F. Then there exists $i \in I$ and a non-zero morphism $f: A/\mathfrak{a} \to Q_i$. Choose an element $x \in A$ such that $0 \neq f(\bar{a}) \in S(U_i)$, where $\bar{x}$ is the image of x in $A/\mathfrak{a}$. Then $Af(\bar{x}) \subseteq S(U_i)$, and $Af(\bar{x}) \simeq A/(\ker f : \bar{x})$. But it is clear that $(\mathfrak{a} : x) \subseteq (\ker f : \bar{x}) = \mathfrak{a}'$. Since $TS(U_i) \simeq U_i$, we see that $T(A/\mathfrak{a}') \simeq U_i$ so that every proper quoitient object of $A/\mathfrak{a}'$ belongs to $\mathscr{F}$. Therefore, it is clear that $\mathfrak{a}' \notin F$ and $\mathfrak{a}'$ is maximal with these properties.

(4) ⇒ (1). Let $G \in L(A)$; if G does not contain all the left ideals of A, then a left ideal $\mathfrak{a}$ can be chosen such that $\mathfrak{a} \notin G$ and is maximal with this property. Then for any left ideal $\mathfrak{b}$ such that $\mathfrak{a} \subseteq \mathfrak{b}$ and $\mathfrak{a} \neq \mathfrak{b}$ one has that $\mathfrak{b} \in G$ or equivalently $T(A/\mathfrak{b}) = 0$ where $T : \mathrm{Mod}\, A \to \mathrm{Mod}\, A/G$ is canonical. Thus it is clear that $T(A/\mathfrak{a})$ is simple. Now we use Lemma 5.2. ■

COROLLARY 5.11. *Let A be a ring, such that any system $F \in L(A)$ is finitely generated. Then A is a left s.n. ring.*

Proof. Let F be a system and $\mathfrak{a}$ a left ideal such that $\mathfrak{a} \notin F$. Denote by M the set of all left ideals $\mathfrak{a}'$ such that $\mathfrak{a} \subseteq \mathfrak{a}'$ and $\mathfrak{a}' \notin F$. Now the point is to prove that M is inductive. In fact, if $\{\mathfrak{a}_i'\}_i$ is a totally ordered subset of M, then $\bigcup_i \mathfrak{a}_i' = \mathfrak{a}'$ is again an element of M. If this is not the case, then $\mathfrak{a}' \in F$ so that $\mathfrak{a}'$ contains a finitely generated left ideal $\mathfrak{a}'' \in F$. It is clear that an index i can be found such that $\mathfrak{a}'' \subseteq \mathfrak{a}_i$, a contradiction. Now the proof is completed by Theorem 5.10, (4). ■

Now we shall say some words about semi-noetherian commutative rings.

Let A be a commutative ring and M a subset of Spec(A). Denote by F_M the set of all ideals $\mathfrak{a}$ such that $V(\mathfrak{a}) \subseteq M$ (recall that $V(\mathfrak{a})$ denotes the set of all prime ideals which contain $\mathfrak{a}$).

LEMMA 5.12. *With the above notation F_M is a localizing system.*

Proof. Indeed, if $\mathfrak{a}' \in F_M$ and $\mathfrak{a}$ is such that $(\mathfrak{a}:x) \in F_M$ for any $x \in \mathfrak{a}'$ and $\mathfrak{p} \in V(\mathfrak{a})$, then for any $x \in \mathfrak{a}'$, we have:

$$(\mathfrak{a}:x)\,x \subseteq \mathfrak{a} \subseteq \mathfrak{p}.$$

It follows that $\mathfrak{a}' \subseteq \mathfrak{p}$ or $(\mathfrak{a}:x) \subseteq \mathfrak{p}$ and in both cases we get $\mathfrak{p} \in M$. Hence $V(\mathfrak{a}) \subseteq M$ and $\mathfrak{a} \in F_M$. ■

For any $F \in L(A)$ we denote by $V(F)$ the set $F \cap \operatorname{Spec}(A)$. It is clear that for any F one has $F \subseteq F_{V(F)}$.

THEOREM 5.13. *Let A be a commutative ring. The following assertions are equivalent*:

(1) *A is semi-noetherian*;

(2) *Any localizing system F on A is an intersection of the form*

$$F = \bigcap_{i \in I} F_{\mathfrak{p}_i}$$

such that, for any i, $\mathfrak{p}_i$ is a prime ideal and if $\mathfrak{a} \supset \mathfrak{p}_i$ and $\mathfrak{a} \neq \mathfrak{p}_i$, then $\mathfrak{a} \in F$;

(3) *For any $F \in L(A)$ and any ideal $\mathfrak{a}$ such that $\mathfrak{a} \notin F$ there exists an ideal $\mathfrak{a}'$ such that $\mathfrak{a} \subseteq \mathfrak{a}'$, $\mathfrak{a}' \notin F$ and maximal with such properties*;

(4) *Any element $F \in L(A)$ is a reduced intersection of the form $F = \bigcap_i F_{\mathfrak{p}_i}$ where $\mathfrak{p}_i$ are prime ideals*;

(5) *For any $F \in L(A)$ one has $F = F_{V(F)}$ and any non-empty set of prime ideals has a maximal element.*

Proof. The equivalences (1) ⇔ (3) ⇔ (4) are in fact the same as in Theorem 5.10.

(2) ⇒ (4). Let $F = \bigcap_{i\in I} F_{\mathfrak{p}_i}$ be an intersection as in (2) and $\mathfrak{p}_i \neq \mathfrak{p}_j$ for any $i, j \in I$, $i \neq j$. We claim that this intersection is reduced. Indeed, assume that $F = \bigcap_{i \in I - \{i'\}} F_{\mathfrak{p} i}$ for some $i' \in I$. Then, $\mathfrak{p}_{ii} \notin F$ so that $\mathfrak{p}_{i'} \notin F_{\mathfrak{p}_i}$ for any some i. Then $\mathfrak{p}_{i'} \subset \mathfrak{p}_i$ so that $\mathfrak{p}_i \notin F$, by hypothesis, a contradiction. Hence the intersection is reduced, as claimed (see Exercise 15).

(4) ⇒ (2). Let $F = \bigcap_i F_{\mathfrak{p}_i}$ be a reduced intersection and $\mathfrak{p}_i \subset \mathfrak{a}$ for some i such that $\mathfrak{p}_i \neq \mathfrak{a}$. If $\mathfrak{a} \notin F$, then $\mathfrak{a} \notin F_{\mathfrak{p}_{i'}}$ for a suitable i', hence $\mathfrak{a} \subseteq \mathfrak{p}_{i'}$. In other words, $\mathfrak{p}_i \subseteq \mathfrak{a} \subseteq \mathfrak{p}_{i'}$. But then $F_{\mathfrak{p}_i} \supseteq F_{\mathfrak{p}_{i'}}$ and the intersection is obviously not reduced, a contradiction.

(3) ⇒ (5). Let $\mathfrak{a} \in F_{V(F)}$; then $V(\mathfrak{a}) \subseteq V(F)$. Assume that $\mathfrak{a} \notin F$ and let $\mathfrak{a}'$ be calculated as in (3). Then, by Chapter 4, Exercise 10 of Section 4.9 we see that $\mathfrak{a}'$ is prime. The contradiction is obvious; therefore, necessarily $F = F_{V(F)}$.

Now let M be a non-empty set of prime ideals and $M' = \mathrm{Spec}(A) - M$, i.e. the complementary set of M. It is clear that for any $\mathfrak{p} \in M$ one has $\mathfrak{p} \notin F_{M'}$. Again, using (3) and Exercise 10 of Section 4.9, Chapter 4, we see that M has a maximal element.

(5) $\Rightarrow$ (4). First we observe that for any $F \in L(A)$ there exist prime ideals $\mathfrak{p}$ such that $\mathfrak{p} \notin F$. In fact, if $V(F) = \mathrm{Spec}(A)$ then F is the improper system by (5). Let M be the set of prime ideals $\mathfrak{p}$ which are not in F and M' the set of maximal elements of M. It is clear by Exercise 15, that

$$F \subseteq \bigcap_{\mathfrak{p} \in M} F_{\mathfrak{p}} = \bigcap_{\mathfrak{p} \in M'} F_{\mathfrak{p}} = F'.$$

Now we claim that the last intersection is reduced and the inclusion is an equality. Indeed, if $\bigcap_{\mathfrak{p} \in M''} F_{\mathfrak{p}} = \bigcap_{\mathfrak{p} \in M'} F_{\mathfrak{p}}$ where $M'' = M' - \{\mathfrak{p}'\}$, then $\mathfrak{p}' \notin \bigcap_{\mathfrak{p} \in M''} F_{\mathfrak{p}}$ so that $\mathfrak{p}' \subseteq \mathfrak{p}$ for a suitable $\mathfrak{p} \in M''$, a contradiction. Finally, if $\mathfrak{a} \in F'$, then $V(\mathfrak{a}) \subseteq V(F')$, and $V(F') \subseteq V(F)$ by the construction of F'. Hence $\mathfrak{a} \in F$ therefore $F = F'$. The proof is now complete. ∎

COROLLARY 5.14. *Let A be a commutative semi-noetherian ring. The following statements are equivalent for $F \in L(A)$:*

(1) *F is a maximal element of $L(A)$;*

(2) *$F = F_{\mathfrak{p}}$ where $\mathfrak{p}$ is a minimal prime ideal of A.*

Proof. (1) $\Rightarrow$ (2). Let F be maximal in $L(A)$. Then $F = \bigcap_{\mathfrak{p}} F_{\mathfrak{p}}$. But it is clear that $F \subseteq F_{\mathfrak{p}}$ for any $\mathfrak{p}$ so that $F = F_{\mathfrak{p}}$ for every $\mathfrak{p}$. Hence, F is of the form $F_{\mathfrak{p}}$ for some $\mathfrak{p} \in \mathrm{Spec}(A)$. If this $\mathfrak{p}$ is not minimal, then it contains a proper prime ideal $\mathfrak{p}'$ so that $F_{\mathfrak{p}'} \supseteq F_{\mathfrak{p}}$, a contradiction (see Exercise 15).

(2) $\Rightarrow$ (1). Let $\mathfrak{p}$ be a minimal prime ideal and F an element of $L(A)$, finer than $F_{\mathfrak{p}}$. Let $\mathfrak{a} \notin F$; by Theorem 5.13, (3) and Chapter 4, Exercise 10 of Section 4.9, there is a prime ideal $\mathfrak{p}'$ with $\mathfrak{a} \subseteq \mathfrak{p}'$ and $\mathfrak{p}' \notin F$. Let us have $\mathfrak{p}' \neq \mathfrak{p}$. From the minimality of $\mathfrak{p}$ we derive that $\mathfrak{p}'$ contains an element of the complement of $\mathfrak{p}$ and thus $\mathfrak{p}'$ is contained in $F_{\mathfrak{p}}$. But this is false. Thus, we may conclude that $\mathfrak{p} = \mathfrak{p}'$ and consequently, $\mathfrak{p} \notin F$. Moreover, we derive that any ideal from F contains an element of the complement of $\mathfrak{p}$, and thus, it is in $F_{\mathfrak{p}}$. Finally, we have $F = F_{\mathfrak{p}}$. ∎

Let X be an A-module; an element $F \in \mathrm{Spel}(A)$ is said to be *associated to* X, if there exists a submodule X' of X such that $F(X') = F$. Frequently we shall denote by $\mathrm{Ass}(X)$ the set of all left primes of A which are associated to X.

LEMMA 5.15. *An element $F \in \mathrm{Spel}(A)$ is associated to X if and only if there exists an element $x \in X$ such that $\mathrm{Ann}(x)$ is a left super-prime ideal and $F_{\mathrm{Ann}(x)} = F$.*

Proof. Let us assume that $F \in \mathrm{Ass}(X)$ and let X' be a submodule of X such that $F(X') = F$. Then $T(E(X'))$ is an injective cogenerator of $\mathrm{Mod}\, A/\mathscr{F}$, so that if U is a simple object of this category, then $U \subseteq T(E(X'))$. But then $K_F \subseteq ST(E(X')) = E(X')$, so that $X' \cap K_F \neq 0$. It is clear by Theorem 20.9 of Chapter 4, that for any $x \in X' \cap K_F$ one has that $\mathrm{Ann}(x)$ is left super-prime and $F_{\mathrm{Ann}(x)} = F$.

The converse is obvious. ∎

As always, if $\mathscr{C}$ is a Grothendieck category, we denote by $\mathscr{C}_\tau$ the localizing subcategory of $\mathscr{C}$, consisting of all objects X such that $KG(X)$ is defined. This is equivalent to the definition of $\mathscr{C}_\tau$ given at the beginning of this paragraph.

THEOREM 5.16. *The following assertions are equivalent for a ring A:*

(1) *A is left s.n.;*

(2) *For any non-zero module X, one has* $\mathrm{Ass}(X) \neq \varnothing$ *and $KG(A/\mathfrak{a})$ is defined for any left super-prime ideal* $\mathfrak{a}$.

Proof. (1) ⇒ (2). If $X \neq 0$, then $\mathrm{Mod}\, A/\mathscr{F}(X)$ is a non-zero Grothendieck category. Let U be a simple object of this latter category. Then U can be considered as a subobject of $T(X)$ (because $T(E(X))$ is an injective cogenerator of $\mathrm{Mod}\, A/\mathscr{F}(X)$). Then $S(U) \subseteq ST(X)$, so that $X \cap S(U) \neq 0$. By Theorem 20.6 of Chapter 4, one can check that $F(S(U))$ is a left prime of A and for any $x \in X \cap S(U)$ one has $F_{\mathrm{Ann}(x)} = F(S(X))$. Now, we utilize Lemma 5.15.

(2) ⇒ (1). If $(\mathrm{Mod}\, A)_\tau \neq \mathrm{Mod}\, A$, then a $(\mathrm{Mod}\, A)_\tau$-torsionfree non-zero A-module X can be found. But this is a contradiction with the hypothesis. Hence, necessarily $(\mathrm{Mod}\, A)_\tau = \mathrm{Mod}\, A$, or equivalently, A is left s.n. ∎

Let A be a commutative ring and X an A-module. We shall say that a prime ideal $\mathfrak{p}$ *supports* X if $X_\mathfrak{p} \neq 0$. Denote by $\mathrm{Supp}(X)$ the set of all prime ideals p which support X. It is clear that $\mathrm{Supp}(X) \neq 0$ whenever $X \neq 0$, and $\mathrm{Ass}(X) \subseteq \mathrm{Supp}(X)$.

PROPOSITION 5.17. *Let A be a commutative s.n. ring. The smallest localizing subcategory of* $\mathrm{Mod}\, A$ *which contains an A-module X is exactly $\mathscr{F}_{\mathrm{Supp}(X)}$.*

Proof. First, by Exercise 11(c), one checks that $\mathscr{F}_{\mathrm{Supp}(X)}$ is a localizing subcategory which contains X. Now let $\mathscr{F}$ be an another localizing subcategory which contains X. Then, by Exercise 11(b), we deduce that for any

$$\mathfrak{p} \in \mathrm{Supp}(X)$$

one has $A/\mathfrak{p} \in \text{Ob}\,\mathscr{F}$. Hence if Y is an object of $\mathscr{F}_{\text{Supp}(X)}$, that is, if $\text{Supp}(Y) \subseteq \text{Supp}(X)$ then, $A/\mathfrak{p} \in \text{Ob}\,\mathscr{F}$ for any $\mathfrak{p} \in \text{Supp}(Y)$. If $y \in Y$ is an element such that $Ay = A/\text{Ann}(y) \notin \text{Ob}\,\mathscr{F}$, then $\text{Ann}(y) \notin F$ (the localizing system associated to $\mathscr{F}$) and by Theorem 5.13, (3), a prime ideal $\mathfrak{p}$ can be calculated such that $\text{Ann}(y) \subseteq \mathfrak{p}$ and $\mathfrak{p} \notin F$. But it is clear that $\mathfrak{p} \in \text{Supp}(Y)$ [Exercise 11, (b)], a contradiction. ∎

PROPOSITION 5.18. *Let A be a commutative s.n. ring and X an A-module. If $a \in A$ then the homothety $m_a : X \to X$ is a monomorphism if and only if $a \notin \mathfrak{p}$ for any $\mathfrak{p} \in \text{Ass}(X)$. In particular, the set of zero divisors of A is $\bigcup_{\mathfrak{p} \in \text{Ass}(A)} \mathfrak{p}$.*

Proof. If m_a is a monomorphism, then for any $x \in X$ such that $\text{Ann}(x) = \mathfrak{p}$ is a prime ideal one has $ax \neq 0$, or equivalently $a \notin \mathfrak{p}$. Conversely, if $a \notin \mathfrak{p}$ for every $\mathfrak{p} \in \text{Ass}(A)$, then necessarily $\text{Ass}(\ker m_a) = \varnothing$, so that m_a is a monomorphism (see Theorem 5.16). ∎

A localizing subcategory $\mathscr{F}$ of a Grothendieck $\mathscr{C}$ category is called *stable* if the injective envelope in $\mathscr{C}$ of any object of $\mathscr{F}$ is also an object of $\mathscr{F}$.

PROPOSITION 5.19. *Let A be a commutative semi-noetherian ring and $M = \{\mathfrak{p}_1, \ldots, \mathfrak{p}_n\}$ a finite set of prime ideals such that $\mathscr{F}_{V(\mathfrak{p}_i)}$ is stable for every i. Denote by $\mathscr{F}$ the smallest localizing subcategory of* Mod A *containing $A/\mathfrak{p}_i$ for each i. Then the following assertions are true:*

(1) *For any A-module X such that* $\text{Ass}(X) \subseteq M$ *it follows that* $X \in \text{Ob}\,\mathscr{F}$;

(2) *For any A-module X such that* $\text{Ass}(X) \subseteq M$ *it follows that* $\text{Ass}(X) = \text{Ass}_f(X)$;

(3) *If any element of M is maximal, then an A-module X belongs to $\mathscr{F}$ if and only if* $\text{Ass}(X) \subseteq M$.

Proof. (1) Let X be an A-module such that $\text{Ass}(X) \subseteq M$; then $E(X)$ is a finite direct sum of $\mathfrak{p}$-isotypic injectives. Because $\mathscr{F}_{V(\mathfrak{p}_i)} \subseteq \mathscr{F}$ for each i, one checks that $E(X) \in \text{Ob}\,\mathscr{F}$, hence $X \in \text{Ob}\,\mathscr{F}$.

(2) Let X be an A-module such that $\text{Ass}(X) \subseteq M$, $x \in X$, and $\mathfrak{p}$ a prime ideal, minimal in the set $V(\text{Ann}(x))$. If for any $\mathfrak{q} \in \text{Ass}(X)$ one has $\mathfrak{q} \nsubseteq \mathfrak{p}$ and $\mathfrak{q} \neq \mathfrak{p}$, then $\mathfrak{q} \in F_{\mathfrak{p}}$, so that $A/\mathfrak{q} \in \mathscr{F}_{\mathfrak{p}}$. Hence $F_{\mathfrak{q}} \subseteq F_{\mathfrak{p}}$ so that $E(X) \in \text{Ob}\,\mathscr{F}_{\mathfrak{p}}$ by hypothesis. But then $Ax \in \text{Ob}\,\mathscr{F}_{\mathfrak{p}}$, a contradiction. Therefore, necessarily $\mathfrak{q} \subseteq \mathfrak{p}$ for some $\mathfrak{q} \in \text{Ass}(X)$, and finally $\text{Ass}(X) = \text{Ass}_f(X)$.

(3) We shall prove that the full subcategory of Mod A consisting of all modules X such that $\text{Ass}(X) \subseteq M$ is localizing. To do this, it will be enough

to check that if $\mathrm{Ass}(X) \subseteq M$, and X' is a quotient object of X, then

$$\mathrm{Ass}(X') \subseteq M.$$

Indeed, $\mathrm{Supp}(X') \subseteq \mathrm{Supp}(X)$ and for every $\mathfrak{p} \in \mathrm{Supp}(X')$ one has $\mathfrak{p} \supseteq \mathfrak{q}$, with $\mathfrak{q} \in \mathrm{Ass}_f(X)$ (see Exercise 11). But then, according to (2) one has $\mathfrak{q} \in \mathrm{Ass}(X)$, and finally $\mathfrak{q} = \mathfrak{p} \in M$, by hypothesis. ■

COROLLARY 5.20. *Let A be a commutative s.n. ring and M a finite set of maximal ideals of A, such that $F_{V(\mathfrak{p})}$ is stable for every $\mathfrak{p} \in M$. Then the smallest localizing subcategory of* Mod *A containing all objects $A/\mathfrak{p}$, $\mathfrak{p} \in M$ is stable.* ■

THEOREM 5.21. *Let $\mathfrak{p}$ be a maximal ideal of a commutative semi-noetherian ring A. Assume that every non-zero prime ideal of A is maximal. Then the localizing subcategory $\mathscr{F}_{V(\mathfrak{p})}$ is stable.*

Proof. Let X be a non-zero object of $\mathscr{F}_{V(\mathfrak{p})}$. By Proposition 5.17 we check that $\mathrm{Ass}(X) = \{\mathfrak{p}\}$, so that $\mathrm{Ass}(E(X)) = \{\mathfrak{p}\}$. Then by Exercise (13) one checks that $\bigcup_{\mathfrak{q} \in \mathrm{Ass}_f(E(X))} \mathfrak{q} = \bigcup_{\mathfrak{q} \in \mathrm{Ass}(E(X))} \mathfrak{q} = \mathfrak{p}$, hence $\mathfrak{q} = \mathfrak{p}$ for every $\mathfrak{q} \in \mathrm{Ass}_f(X)$. Then by Exercise (11) we have that $\mathrm{Supp}(E(X)) \subseteq V(\mathfrak{p})$. Now we utilize Proposition 5.17. ■

PROPOSITION 5.22. *Let A be a commutative s.n. ring and $\mathfrak{p}$ a prime ideal which is not a minimal prime. There exists a prime ideal $\mathfrak{q}$ such that $\mathfrak{q} \subset \mathfrak{p}$, and for any prime ideal $\mathfrak{q}'$ such that $\mathfrak{q} \subseteq \mathfrak{q}' \subseteq \mathfrak{p}$ one has $\mathfrak{q} = \mathfrak{q}'$, or $\mathfrak{q}' = \mathfrak{p}$.*

Proof. Let $A_\mathfrak{p}$ be the local ring associated with $\mathfrak{p}$ (see Chapter 4, Section 4.20). Now, by Corollary 5.3 we check that $A_\mathfrak{p}$ is also s.n. ring. Let us put $M = \{\bar{\mathfrak{p}}\}$ where $\bar{\mathfrak{p}}$ is the unique maximal ideal of $A_\mathfrak{p}$. Then $\bar{\mathfrak{p}} \in F_M$ and F_M does not contain all ideals of $A_\mathfrak{p}$, by hypothesis. By Theorem 5.13 one checks that there exists a prime ideal $\bar{\mathfrak{q}}$ such that $\bar{\mathfrak{q}} \notin F_M$ and maximal with this property. It is clear that $\bar{\mathfrak{q}} \subset \bar{\mathfrak{p}}$ and for any prime ideal $\mathfrak{q}'$ such that $\bar{\mathfrak{q}} \subseteq \mathfrak{q}' \subseteq \bar{\mathfrak{p}}$ one has $\bar{\mathfrak{q}} = \mathfrak{q}'$ or $\mathfrak{q}' = \bar{\mathfrak{p}}$. Finally we use Exercise (7) of Section 4.20, Chapter 4. ■

Exercises

1. Let $\mathscr{C}$ be a s.n.-category. Then any localizing subcategory of $\mathscr{C}$ can be uniquely represented as a reduced intersection of prime localizing subcategories. Moreover $\mathrm{card}(L(\mathscr{C})) = 2^{\mathrm{card}(\mathrm{Spel}(\mathscr{C}))}$.

2. Any factor ring and any finite direct product of left s.n. rings is also a left s.n. ring.

3. Let A be a commutative s.n. ring. For any ordinal α we denote by $\mathrm{Spec}_\alpha(A)$ the subset of $\mathrm{Spec}(A)$ consisting of all elements $\mathfrak{p}$ such that

$$KG(A/\mathfrak{p}) \leqslant \alpha.$$

Then:

(a) $\bigcup_\alpha \mathrm{Spec}_\alpha(A) = \mathrm{Spec}(A)$;

(b) For any ordinal α, $\mathrm{Spec}_\alpha(A)$ is defined as follows:

(i) $\mathrm{Spec}_0(A)$ is the set of all maximal ideals of A;

(ii) If $\alpha = \beta + 1$, then $\mathrm{Spec}_\alpha(A)$ consists of all prime ideals $\mathfrak{p}$ such that for any prime ideal $\mathfrak{p}'$ such that $\mathfrak{p} \subset \mathfrak{p}'$ one has $\mathfrak{p}' \in \mathrm{Spec}_\beta(A)$.

(iii) If α is a limit ordinal, then $\mathrm{Spec}_\alpha(A) = \bigcup_{\beta<\alpha} \mathrm{Spec}_\beta(A)$.

Then $KG(\mathrm{Mod}\, A)$ is the smallest ordinal α such that $\mathrm{Spec}_\alpha(A) = \mathrm{Spec}(A)$.

4. For any left semi-noetherian ring one has $KG(\mathrm{Mod}\, A) = KG(A)$. Moreover $KG(A)$ has a predecessor.

5. Let A be a commutative ring and n a finite ordinal. The following assertions are equivalent:

(a) A is an s.n. ring and $KG(A) \leqslant n$;

(b) For any non-zero A-module X one has $\mathrm{Ass}(X) \neq \varnothing$ and any increasing sequence of prime ideals has no more than $n+1$ elements.

6. Let A be a commutative s.n. ring, $\mathfrak{p}$ a prime ideal of A and n a finite ordinal. The following statements are equivalent:

(a) $KG(A/\mathfrak{p}) \leqslant n$;

(b) Any increasing sequence of prime ideals

$$\mathfrak{p} \subset \mathfrak{p}_1 \subset \ldots \subset \mathfrak{p}_k \subset \ldots$$

has no more than $n+1$ elements.

7. Let A be a commutative ring and $F \in L(A)$. The following assertions are equivalent:

(a) There exists a set H of prime ideals such that

$$F = \bigcap_{\mathfrak{p}\in H} F_\mathfrak{p};$$

(b) For any ideal a not in F, there exists a prime ideal $\mathfrak{p}$ such that $\mathfrak{a} \subseteq \mathfrak{p}$ and $\mathfrak{p} \notin F$;

(c) $F = F_{V(F)}$.

A system F as above is called an *f-system*. If any system F on A is a f-system, then A is called an *f-ring*. It is clear that any s.n. ring is a f-ring.

8. A commutative ring A is called a *C.P.-ring* (compact packet) if for any ideal $\mathfrak{a}$ of A and any set $\{\mathfrak{p}_i\}_i$ of prime ideals such that $\mathfrak{a} \subseteq \bigcup_i \mathfrak{p}_i$ one has

that $\mathfrak{a} \subseteq \mathfrak{p}_i$ for a suitable i. A ring A is a *C.P.*-ring if and only if for any f-system $F \in L(A)$ there exists a multiplicative system S of elements of A such that $F = F_S$.

(For a finite system of prime ideals this property holds in every commutative ring, cf. [26] Chapter 2).

9. For a commutative ring A, the following assertions are equivalent:

(a) A is a f-ring and a C.P.-ring;

(b) For any $F \in L(A)$ there exists a multiplicative system S of elements on A, such that $F = F_S$.

(c) A is an f-ring and any prime ideal is the radical of a principal ideal. (Note that the *radical* $\sqrt{\mathfrak{a}}$ *of an ideal* $\mathfrak{a}$ is defined as an intersection of all prime ideals containing $\mathfrak{a}$).

Moreover, a ring A which satisfies the above equivalent conditions is a s.n. ring. Also a ring A such that any prime ideal is the radical of a principal ideal is a C.P.-ring.

10. Let A be a commutative s.n. ring. The following statements are equivalent:

(a) A is a C.P.-ring;

(b) For any $F \in L(A)$ there exists a multiplicative system S of elements of A such that $F = F_S$;

(c) Any prime ideal of A is the radical of a principal ideal.

Using these results, give examples of s.n. rings which are not *C.P.* rings.

11. (a) Let A be a commutative ring. Then for any exact sequence of A-modules:

$$0 \to X' \to X \to X'' \to 0$$

one has: $\mathrm{Supp}(X) = \mathrm{Supp}(X') \cup \mathrm{Supp}(X'')$.

(b) Further we shall say that a prime ideal $\mathfrak{p}$ is *weakly associated* with X if there exists an element $x \in X$ such that $\mathfrak{p}$ is minimal in the set of all prime ideals which contain $\mathrm{Ann}(x)$. Denote by $\mathrm{Ass}_f(X)$ the set of all primes weakly associated with X. Then $\mathrm{Ass}(X) \subseteq \mathrm{Ass}_f(X) \subseteq \mathrm{Supp}(X)$; moreover, a prime ideal $\mathfrak{p}$ is contained in $\mathrm{Supp}(X)$ if and only if $\mathfrak{p} \supseteq \mathfrak{p}'$, where

$$\mathfrak{p}' \in \mathrm{Ass}_f(X).$$

(c) If M is a set of prime ideals, then the full subcategory $\mathscr{F}_M$, consisting of all A-modules X such that $\mathrm{Supp}(X) \subseteq M$ is localizing. Moreover, the associated localizing system consists of all ideals $\mathfrak{a}$ such that $V(\mathfrak{a}) \subseteq M$ (and denoted by F_M).

12. Let A be a commutative s.n. ring, X an A-module and H the maximal elements of $\mathrm{Ass}(X)$. Then $F(X) = \bigcap_{\mathfrak{p}\in H} F_{\mathfrak{p}}$ and this intersection is reduced.

13. For any module X over a commutative s.n. ring, one has:

$$\bigcup_{\mathfrak{p}\in\mathrm{Ass}(X)} \mathfrak{p} = \bigcup_{\mathfrak{q}\in\mathrm{Ass}_f(X)} \mathfrak{q}.$$

14. Let A be a commutative non-discrete Archimedean valuation ring. Then Mod A is not a s.n.-category, whereas it is a l.c.-category. (Hint: Let $\mathfrak{m}$ be the maximal ideal of A; then $\mathfrak{m}^2 = \mathfrak{m}$, so that $F = \{\mathfrak{m}, A\}$ is a bilocalizing system on A. Then Mod $A/\mathscr{F}$ does not contain simple objects.)

15. Let A be a commutative ring, $F \in L(A)$ and $\mathfrak{p} \in \mathrm{Spec}(A)$. Then $\mathfrak{p} \notin F$ if and only if $F \subseteq F_{\mathfrak{p}}$. Moreover, if $\mathfrak{p}, \mathfrak{p}' \in \mathrm{Spec}(A)$, then $\mathfrak{p} \subset \mathfrak{p}'$ if and only if $F_{\mathfrak{p}'} \subset F_{\mathfrak{p}}$.

16. Let A be a commutative s.n. ring. Then the ring $A[X]$ is also an s.n. ring. Moreover, $KG(\mathrm{Mod}\, A[X]) = KG(\mathrm{Mod}\, A) + 1$.

References

Bourbaki [26]; Dickson [43]; Gabriel [59]; Goldman [76]; Năstăsescu [139]; Popescu [172]; Samuel–Zariski [202].

5.6 Semi-artinian categories

Let $\mathscr{C}$ be a Grothendieck category. Denote by $\mathrm{So}(\mathscr{C})$ the full subcategory of $\mathscr{C}$ generated by all semisimple objects. It is clear that any subobject and any quotient object of a semisimple object is also a semisimple object. (The zero object is considered to be a semisimple object). Also the direct sum of a family of semisimple objects is also a semisimple object. Thus $\mathrm{So}(\mathscr{C})$ is in fact a prelocalizing subcategory of $\mathscr{C}$, called *socle prelocalizing category*. If X is an object of $\mathscr{C}$, then by $\mathrm{So}(X)$ we shall denote the greatest subobject of X in $\mathrm{So}(\mathscr{C})$; $\mathrm{So}(X)$ is in fact the sum of all its simple subobjects (or zero otherwise), the *socle of* X. Thus the assignment $X \rightsquigarrow \mathrm{So}(X)$ defines a functor $\mathrm{So} : \mathscr{C} \to \mathrm{So}(\mathscr{C})$, called the *socle functor*.

According to Corollary 8.17 of Chapter 4, there exists a smallest localizing subcategory of $\mathscr{C}$ which contains $\mathrm{So}(\mathscr{C})$, or generated by $\mathrm{So}(\mathscr{C})$, that is the closure of $\mathrm{So}(\mathscr{C})$. This localizing subcategory will be denoted by $\overline{\mathrm{So}}(\mathscr{C})$ and is called sometimes the *socle localizing subcategory*. It is clear that with the notation at the beginning of Section 5.5 one has $\overline{\mathrm{So}}(\mathscr{C}) = \mathscr{C}_0$.

Recall that for any Grothendieck category $\mathscr{C}$ there exists a spectral category $\mathscr{C}_E$ and a canonical functor $P : \mathscr{C} \to \mathscr{C}_E$ (see Section 4.2 of Chapter 4).

THEOREM 6.1. *The following assertions are equivalent for a Grothendieck category* $\mathscr{C}$:

(1) $\mathscr{C}$ *is a s.n.-category and* $KG(\mathscr{C}) = 0$;

(2) *Any non-zero object of* $\mathscr{C}$ *contains a simple subobject*;

(3) *The canonical functor* $\mathrm{So} : \mathscr{C} \to \mathrm{So}(\mathscr{C})$ *defines an equivalence between* $\mathscr{C}_E$, *the spectral category associated with* $\mathscr{C}$ *and* $\mathrm{So}(\mathscr{C})$, *the prelocalizing socle subcategory*;

(4) *The canonical functor* $P : \mathscr{C} \to \mathscr{C}_E$ *has a fully and faithful left adjoint, and* $\mathscr{C}$ *is a l.c.-category*;

(5) $\mathscr{C}$ *is an s.n.-category and the canonical topology of* $\mathrm{Sp}(\mathscr{C})$ *is discrete.*

Proof. (1) ⇒ (2). It is clear that $\mathscr{C}$ is the smallest localizing subcategory generated by $\mathrm{So}(\mathscr{C})$, that is

$$\mathscr{C} = \bigcup_\alpha (\mathrm{So}(\mathscr{C}))^\alpha$$

(see Chapter 4, Lemma 8.13).

Let $X \in \mathrm{Ob}(\mathscr{C})$ be a non-zero object and α the smallest ordinal such that X does not contain non-zero subobjects of $(\mathrm{So}(\mathscr{C}))^\alpha$ (it is clear that $\alpha = 0$ if X contains simple objects). Also let Y be a non-zero subobject of X in $(\mathrm{So}(\mathscr{C}))^{\alpha+1} = (\mathrm{So}(\mathscr{C}))^\alpha \circ (\mathrm{So}(\mathscr{C}))$. By the definition of composition of any two prelocalizing subcategories (see Section 4.8 of Chapter 4) one checks that Y contains a simple subobject.

(2) ⇒ (3). The canonical functor So assigns to each object X its socle $\mathrm{So}(X)$ and has a left adjoint functor $I : \mathrm{So}(\mathscr{C}) \to \mathscr{C}$ which is full and faithful; moreover $I(\mathrm{So}(X)) = \mathrm{So}(X)$, viewed as an object of $\mathscr{C}$ (i.e. I is the inclusion functor).

Now if $f : X \to Y$ is an essential monomorphism then $\mathrm{So}(f)$ (which is in fact the restriction of f to $\mathrm{So}(X)$) is an isomorphism, by hypothesis. Then, by the definition of a spectral category (Chapter 4, Section 4.2) there exists a unique functor $H : \mathscr{C}_E \to \mathrm{So}(\mathscr{C})$ such that $HP = \mathrm{So}$. We claim that H is an equivalence of categories. In fact, if $X \in \mathrm{Ob}(\mathscr{C})$, then the inclusion $\mathrm{So}(X) \to X$ is an essential monomorphism, so that $H(P(X)) \simeq H(\mathrm{So}(X)) = \mathrm{So}(X)$, i.e. H is "onto" on the class of objects. To finish, it will be sufficient to prove that H is full and faithful. Indeed, let $t : P(X) \to P(Y)$ be a morphism of $\mathscr{C}_E$; then $t = P(f)P(s)^{-1}$ where $f : Z \to X$, $s : Z \to Y$ are morphisms of $\mathscr{C}$ and s is an essential monomorphism. Then $H(t) = HP(f)(HP(s))^{-1} = \mathrm{So}(f)\,\mathrm{So}(s)^{-1}$, and the relation $H(t) = 0$ is equivalent to $\mathrm{So}(f) = 0$, so that $\mathrm{So}(Z) \subseteq \ker f$. But then necessarily $P(f) = 0$, because $\mathrm{So}(Z)$ is essential in Z, therefore $t = 0$. Hence H is faithful. We leave it to the reader to prove that H is also full.

(3) $\Rightarrow$ (4). It is obvious that the functor IH is the claimed left adjoint of P.

(4) $\Rightarrow$ (2). Let I be a full and faithful left adjoint of P; then for any simple object U of $\mathscr{C}_E$ one has that the object $I(U)$ is also simple. Indeed, first we see that $I(U)$ is coirreducible because if for any two non-zero subobjects X, Y of $I(U)$ such that $X \cap Y = 0$ one has clearly that $P(X)$ and $P(Y)$ are non-zero subobjects of $PI(U) \simeq U$ and $P(X) \cap P(Y) = 0$, a contradiction. Now let X be a non-zero subobject of $I(U)$; then the natural inclusion $i: X \to I(U)$ is an essential monomorphism, so that $P(i)$ is an isomorphism, and by the commutative diagram:

$$\begin{array}{ccc} IP(X) & \xrightarrow{t_X} & X \\ \| & & \downarrow i \\ IPI(U) & \longrightarrow & I(U) \end{array}$$

in which t_X is produced by an arrow of adjunction, one checks that i is a right invertible monomorphism, i.e. an isomorphism. Hence, $I(U)$ is necessarily simple.

Now let X be a non-zero object of $\mathscr{C}$; then, by Theorem 3.3, $P(X)$ is semisimple and by hypothesis, the functorial morphism $t_X: IP(X) \to X$ is not zero. By the above considerations $IP(X)$ is also semisimple, so that $\operatorname{im} t_X$ is again semisimple. Hence X contains a simple subobject.

(1) $\Rightarrow$ (5). Let M be a subset of $\operatorname{Sp}(\mathscr{C})$; then any indecomposable injective $Q \in M$ contains a simple subobject. Let $\mathscr{F}$ be the localizing subcategory of $\mathscr{C}$ consisting of all objects X such that $\operatorname{Hom}_{\mathscr{C}}(X, Q) = 0$ for every $Q \in M$. If $Q' \in \overline{M}$, the closure of M in the canonical topology, and if $Q' \notin M$, then Q' contains a simple object U' and it is obvious that $U' \in \operatorname{Ob} \mathscr{F}$. This is false because Q' is $\mathscr{F}$-closed. Hence $M = \overline{M}$ so that the canonical topology of $\operatorname{Sp}(\mathscr{C})$ is the discrete topology.

(5) $\Rightarrow$ (1). Let $\mathscr{C}_0$ be the socle localizing subcategory (i.e. the smallest localizing subcategory which contains all simple objects). If $\mathscr{C}_0 \neq \mathscr{C}$, then $\mathscr{C}/\mathscr{C}_0 \neq 0$ so that $\operatorname{Sp}(\mathscr{C}/\mathscr{C}_0) \neq 0$ and by the definition of the canonical topology (see Section 5.3), this set is a closed subset of $\operatorname{Sp}(\mathscr{C})$ and also an open set by hypothesis, or equivalently, $\operatorname{Sp}(\mathscr{C}_0)$, its complementary set in $\operatorname{Sp}(\mathscr{C})$, is closed. By the definition of the closed sets in the canonical topology (see Corollary 3.7) and by Lemma 5.8, we find that $\overline{\operatorname{Sp}(\mathscr{C}_0)} = \operatorname{Sp}(\mathscr{C})$, a contradiction. Hence, necessarily $\mathscr{C}_0 = \mathscr{C}$. The proof is complete. ∎

A Grothendieck category $\mathscr{C}$ verifying the conditions of the above theorem is called a *semi-artinian* category. Frequently, we shall write s.a.-category instead of semi-artinian category.

COROLLARY 6.2. *Let $\mathscr{C}$ be a Grothendieck category and $\{\mathscr{C}_\alpha\}_\alpha$ its Krull–Gabriel filtration. Then for any ordinal $\alpha = \beta + 1$, one has that $\mathscr{C}_\alpha/\mathscr{C}_\beta$ is a s.a.-category.*

Proof. It will be sufficient to observe that $\mathscr{C}_0$ is a s.a.-category, according to the above theorem. ∎

COROLLARY 6.3. *Any localizing subcategory and any quotient category of a s.a.-category is also a s.a.-category.*

The proof will result from Corollary 5.4 and Theorem 6.1. ∎

A ring A is called *left semi-artinian* (we shall sometimes write left s.a.-ring) if the category Mod A is a semi-artinian category.

Recall that a left prime F of a ring A is maximal if $F = F_{\mathfrak{m}}$ where $\mathfrak{m}$ is a maximal left ideal.

THEOREM 6.4. *The following assertions are equivalent for a ring A:*

(1) *A is left s.a. ring;*

(2) *For any non-zero A-module X, one has* $\mathrm{Ass}(X) \neq \varnothing$ *and any left prime of A is maximal.*

Proof. (1) ⇒ (2). If U is a simple module, then $U \simeq A/\mathfrak{m}$ for a suitable maximal ideal $\mathfrak{m}$, so that by Lemma 5.15 and Theorems 6.1 and 20.6 of Chapter 4 one checks that for any non-zero A-module X, one has that $\mathrm{Ass}(X) \neq \varnothing$. Also, if F is a left prime, then $F = F(K_F)$ and K_F contains a simple A-module U; it is clear that $F = F(U)$, or equivalently, F is maximal.

(2) ⇒ (1). The proof is a direct consequence of the Lemma 5.15 and Theorem 6.1. ∎

Now, we shall give a criterion for a ring to be semi-artinian.

THEOREM 6.5. *The following assertions are equivalent for a ring A:*

(1) *A is left semi-artinian;*

(2) *$\mathfrak{R}(A)$ is right T-nilpotent and $A/\mathfrak{R}(A)$ is left semi-artinian.*

Proof. (1) ⇒ (2). To check the first assertion we shall use transfinite induction. Write $\mathfrak{R}_0 = 0$; if α is an ordinal such that it has a predecessor β then $\mathfrak{R}_\alpha$ is defined such that $\mathfrak{R}_\alpha/\mathfrak{R}_\beta$ is the socle of $\mathfrak{R}(A)/\mathfrak{R}_\beta$. If α is a limit ordinal, then $\mathfrak{R}_\alpha = \bigcup_{\beta<\alpha} \mathfrak{R}_\beta$. It is clear that $\mathfrak{R}_\alpha$ is an ideal of A for any ordinal α. A being semi-artinian, by Theorem 6.1, we check that $\mathfrak{R}(A) = \mathfrak{R}_{\alpha'}$ for an ordinal α'. Furthermore if $a \in \mathfrak{R}(A)$ we define $h(a)$ as the smallest ordinal α such that $a \in \mathfrak{R}_\alpha$. It is clear that $h(a)$ is not a limit ordinal, i.e. $h(a) = \beta + 1$ for a suitable ordinal β.

If U is a simple module, then $\mathfrak{R}(A)U = 0$, by Nakayama's Lemma (Chapter 3, Section 3.5, Exercise 11), so that from the exact sequence:

$$0 \to \mathfrak{R}_\beta \to \mathfrak{R}_{\beta+1} \to \mathfrak{R}_{\beta+1}/\mathfrak{R}_\beta \to 0$$

we deduce that $\mathfrak{R}(A)\,\mathfrak{R}_{\beta+1} \subseteq \mathfrak{R}_\beta$. Hence, for any $a, b \in \mathfrak{R}$, one has $h(ba) < h(a)$.

Now, let $a_1, a_2, \ldots, a_n, \ldots$ be a sequence of elements of $\mathfrak{R}(A)$ such that $a_n a_{n-1} \ldots a_1 \neq 0$ for every n. Then the set of ordinals $\{h(a_n a_{n-1} \ldots a_1)\}_n$ is an infinite decreasing sequence of ordinals, a contradiction. Hence $\mathfrak{R}(A)$ is right T-nilpotent. The last part follows by Exercise 2(b).

(2) $\Rightarrow$ (1). Let X be an A-module such that $\mathfrak{R}(A)X' \neq 0$ for any non-zero submodule of X. In particular, $\mathfrak{R}(A)X \neq 0$, so that an element $a_1 \in \mathfrak{R}(A)$ can be found for which $a_1 X \neq 0$. Let $X_1 = Aa_1 X$; by hypothesis, an element $a_2' \in \mathfrak{R}(A)$ can be chosen such that $a_2' X_1 \neq 0$ hence there exists a suitable $a_2 \in \mathfrak{R}(A)$ with $a_2 a_1 X \neq 0$. By the same procedure, for any natural number n, an element $a_n \in \mathfrak{R}(A)$ can be calculated so that

$$a_n a_{n-1} \ldots a_1 X \neq 0.$$

Hence, by hypothesis, for any non-zero A-module X, a non-zero submodule X' can be chosen such that $\mathfrak{R}(A)X' = 0$. Then X' is an $A/\mathfrak{R}(A)$-module and consequently it contains a simple $A/R(A)$-module. A slight computation proves that this simple $A/\mathfrak{R}(A)$-module, viewed as A-module, is also simple. ∎

COROLLARY 6.6. *Any left perfect ring is left semi-artinian.*

The proof follows by Theorem 6.5. ∎

The following result gives a characterization of commutative perfect rings in the class of commutative semi-artinian rings.

THEOREM 6.7. *The following assertions are equivalent for a commutative ring A:*

(1) *A is perfect;*

(2) *A is semi-artinian and* Spec(A) *is a finite set.*

Proof. (1) $\Rightarrow$ (2). By Theorem 6.4 and Corollary 6.6, any prime of A is maximal, hence Spec(A) is bijective with the set of types of simple objects (see also Lemma 10.2). Now it is clear that the canonical morphism

$$A \to A/\mathfrak{R}(A)$$

establishes a bijection between the set of maximal ideals of A and the set of all maximal ideals of $A/\mathfrak{R}(A)$. Now the result follows by the definition of perfect ring.

(2) ⇒ (1). Let $\mathfrak{p}_1, \ldots, \mathfrak{p}_n$ be all elements of $\mathrm{Ass}(A)$. By Exercise 13 of Section 5.5 one has

$$\bigcup_{\mathfrak{q}\in\mathrm{Spec}(A)} \mathfrak{q} = \bigcup_{i=1}^{n} \mathfrak{p}_i$$

because any prime ideal of A is obviously an element of $\mathrm{Ass}_f(A)$, so that

$$\mathfrak{q} \subseteq \bigcup_{i=1}^{n} \mathfrak{p}_i$$

for every prime ideal of A. Thus $\mathfrak{q} = \mathfrak{p}_i$ for a suitable i, hence any prime ideal of A belongs to $\mathrm{Ass}(A)$. We leave it to the reader to prove that $A/\mathfrak{R}(A)$ is a semisimple ring. The result now follows by Theorem 6.5. ■

NOTE. In [46] V. Dlab gave an example of a left semi-artinian ring A such that the set $\mathrm{Spel}(A)$ is finite and A is not a left perfect ring. Thus an analogue of the above theorem for the non-commutative case is not valid.

Examples of semi-artinian rings. Let B be a reduced ring having minimal left ideals. Then $\mathfrak{b}^2 = \mathfrak{b}$, for any such left ideal, so that an idempotent $e \in \mathfrak{b}$ can be calculated such that $\mathfrak{b} = Be$ (see Chapter 3, Section 3.11, Exercise 9). Let $\mathfrak{a}$ be the left socle of B and A the smallest subring of B containing $\mathfrak{a}$ and the identity element of B. If $\mathfrak{b}$ is a minimal left ideal of B, then it is also minimal viewed as a left ideal in A; because $\mathfrak{b} = Be$ and $BeB = AeA$ is a division ring (see Exercise 1). Any element of A is of the form $n.1 + x$ where $x \in \mathfrak{a}$ and n an integer. If B has as characteristic a prime number p, then $A/\mathfrak{a} \simeq \mathbf{Z}_p$ so that A is clearly a left semi-artinian ring.

There exists a particular case of the example considered. Let p be a prime number and $B = \prod \mathbf{Z}_p$ the direct product of an infinite set of copies of $\mathbf{Z}_p$. Let $\mathfrak{a}$ be the ideal $\coprod \mathbf{Z}_p$, consisting of all elements $x = (x_i) \in B$ such that $x_i = 0$ for all i, but a finite number of indices. Denote by A the subring of B generated by $\mathfrak{a}$ and the unit-element. As above, it is clear that A is a semi-artinian ring. Moreover A is not noetherian, but it is a regular ring.

For any Grothendieck category $\mathscr{C}$, we denote by $\mathrm{Sim}(\mathscr{C})$ the set of all types of simple objects. Now, if $\mathscr{C}$ is a s.a.-category, then the assignment

$$U \rightsquigarrow E(U)$$

gives a bijection between the sets $\mathrm{Sim}(\mathscr{C})$ and $\mathrm{Sp}(\mathscr{C})$.

For any $U \in \mathrm{Sim}(\mathscr{C})$, we denote by $\mathscr{L}(U)$ the smallest localizing subcategory containing U, and by $\mathscr{F}(U)$ the localizing subcategory generated by all objects X such that $\mathrm{Hom}_{\mathscr{C}}(X, E(U)) = 0$. Also for any object Y of $\mathscr{C}$ we denote by Y_U the greatest subobject of Y in $\mathscr{L}(U)$.

LEMMA 6.8. *If U, U' are non-isomorphic simple objects of $\mathscr{C}$, then*

$$\mathscr{L}(U) \subseteq \mathscr{F}(U').$$

Proof. First, we see that $U \in \mathrm{Ob}\,\mathscr{F}(U')$. Now if $X \in \mathrm{Ob}\,\mathscr{L}(U)$, then any non-zero quotient object of X contains U as subobject, so that if

$$f: X \to E(U')$$

is a non-zero morphism, then $U \subseteq \operatorname{im} f \subseteq E(U')$, a contradiction. Hence, necessarily $\mathscr{L}(U) \subseteq \mathscr{F}(U')$.

Now let $\mathscr{C}' = \prod_{U \in \mathrm{Sim}(\mathscr{C})}^{\mathscr{L}(U)}$ and $V: \mathscr{C} \to \mathscr{C}'$ the functor $V(Y) = \{Y_U\}_U$; it is obvious that the functor $H: \mathscr{C}' \to \mathscr{C}$ defined as: $H(\{Y_U\}_U) = \coprod_U Y_U$ is a left adjoint of V. ■

THEOREM 6.9. *Let $\mathscr{C}$ be a s.n.-category; with the above notations, the following assertions are equivalent:*

(1) *The functor $V: \mathscr{C} \to \mathscr{C}' = \Pi\, \mathscr{L}(U)$ is an equivalence of categories;*

(2) *For any simple object U, the localizing subcategory $\mathscr{F}(U)$ is stable.*

Proof. Let $u: HV \to \mathrm{Id}\,\mathscr{C}$ be an arrow of adjunction, chosen so that for each object X, $u_X: HV(X) = \coprod_U X_U \to X$ is in fact the sum morphism. By hypothesis, this morphism is an isomorphism. Now, let $X \in \mathrm{Ob}\,\mathscr{F}(U)$ and $Q = E(X)$; then $Q_U = 0$ (otherwise $Q_U \cap X \neq 0$, a contradiction), so that $Q = \coprod_{U' \neq U} Q_{U'}$. Thus $Q \in \mathrm{Ob}\,\mathscr{F}(U)$ by Lemma 6.8.

(2) ⇒ (1). We shall prove that H is an equivalence of categories. First, we shall show that $\mathscr{L}(U)$ is stable. Indeed, if $X \in \mathrm{Ob}\,\mathscr{L}(U)$ and Q is its injective envelope in $\mathscr{C}$, then for any $U' \neq U$, one has $Q \in \mathrm{Ob}\,\mathscr{F}(U')$, by Lemma 6.8 and the hypothesis. Now if $Q \notin \mathrm{Ob}\,\mathscr{L}(U)$, then $Q/Q_U \neq 0$, and so $(Q/Q_U)_{U'} \neq 0$ for a suitable simple object U' non isomorphic with U, a contradiction. Furthermore, let $\{Q_U\}_U$ be an injective object of $\mathscr{C}'$ (i.e. Q_U is injective in $\mathscr{L}(U)$ for each U), and $Q = \coprod_U Q_U = H(\{Q_U\}_U)$. If Q is not an injective object of $\mathscr{C}$, then $E(Q)/Q \neq 0$ and let R be a suitable subobject of $E(Q)$, which contains Q, so that $R/Q = U_0$ is a simple object. Since Q_{U_0} is injective in $\mathscr{C}$, then $R = Q_{U_0} \amalg Y$, where Y is a suitable supplement of Q_{U_0}, which contains $\coprod_{U \neq U_0} Q_U$ as an essential subobject. By Lemma 6.8 and the hypothesis, one verifies that $Y \in \mathrm{Ob}\,\mathscr{F}(U_0)$, so that

$$Y/\textstyle\coprod_{U \neq U_0} Q_U \simeq U_0 \in \mathrm{Ob}\,\mathscr{F}_{U_0},$$

a contradiction. Therefore, one has that $Q = E(Q)$ or equivalently, Q is an injective object. In other words, we have checked that $u_Q: HV(Q) \to Q$ is an isomorphism for any injective object of $\mathscr{C}$. Now let $X \in \mathrm{Ob}\,\mathscr{C}$; then one has the commutative diagram:

$$\begin{array}{ccccccc} 0 & \longrightarrow & HV(X) & \longrightarrow & HV(Q) & \longrightarrow & HV(Q') \\ & & \downarrow{\scriptstyle u_X} & & \downarrow{\scriptstyle u_Q} & & \downarrow{\scriptstyle u_{Q'}} \\ 0 & \longrightarrow & X & \longrightarrow & Q & \longrightarrow & Q' \end{array}$$

in which the bottom row is exact and Q and Q' are injectives. By left exactness of V and H, one checks that the top row is also exact, and by the above considerations u_Q and $u_{Q'}$ are isomorphisms. Finally, we see that u_X is also an isomorphism.

We leave it to the reader to check that H is a full and faithful functor. ∎

LEMMA 6.10. *Let M be a finite set of prime ideals of a commutative semi-artinian ring. Then the smallest localizing subcategory containing all objects $A/\mathfrak{p}$, $\mathfrak{p} \in M$ is stable.*

The proof follows by Theorem 5.21 and Corollary 5.20 (because any prime ideal of a semi-artinian ring is maximal and minimal simultaneously, by Theorem 6.4). ∎

COROLLARY 6.11. *Any commutative perfect ring is in a unique way a direct product of a finite set of commutative local perfect rings.*

The proof will result by Theorem 6.7, Lemma 6.10 and Theorem 6.9. ∎

Now let X be an object of a Grothendieck category $\mathscr{C}$. For any ordinal α we shall define a subobject X_α as follows: $X_0 = \mathrm{So}(X)$, the socle of X. If $\alpha = \beta + 1$, then X_α is defined such that $X_\alpha/X_\beta = \mathrm{So}(X/X_\beta)$. Finally, if α is a limit ordinal, then $X_\alpha = \bigcup_{\beta<\alpha} X_\beta$. Thus we have defined an increasing sequence of subobjects of X, called the *Loewy sequence* of X. The subobject $\bigcup_\alpha X_\alpha$ is in fact the greatest subobject of X in the localizing subcategory $\overline{\mathrm{So}}(\mathscr{C})$. It is clear that $\mathscr{C}$ is a s.a.-category if and only if for every X, one has $\bigcup_\alpha X_\alpha = X$. We shall say that the height of X is defined if $\bigcup_\alpha X_\alpha = X$.

Thus, the smallest ordinal α such that $X_\alpha = X$ is denoted by $h(X)$, and is called *height of* X. Also, if $\mathscr{C}$ is a s.a.-category, we shall define $h(\mathscr{C}) = \sup_{X \in \mathrm{Ob}\mathscr{C}} (h(X))$. The reader is invited to prove the following result.

PROPOSITION 6.12. *If $\{U_i\}_i$ is a set of generators of a s.a.-category $\mathscr{C}$, then $h(\mathscr{C}) = \sup_i (h(U_i))$.* ∎

Exercises

1. Let A be a ring. A left ideal Ae of A is minimal if and only if AeA is a division ring.

2. The following assertions are true:

(a) A left semi-artinian domain is a division ring;

(b) If $\mathfrak{a}$ is a two-sided ideal in a left semi-artinian ring A, then the ring $A/\mathfrak{a}$ is also a left semi-artinian ring;

(c) Any finite direct product of left semi-artinian rings is a left semi-artinian ring;

(d) If $u: A \rightarrow B$ is a left flat epimorphism of rings and A is left semi-artinian, then B is too.

3. We shall say that a ring A satisfies *condition* (H) if any non-zero A-module has a maximal submodule. The following assertions are equivalent for a commutative ring A:

(a) A is semi-artinian;

(b) $\mathfrak{R}(A)$ is T-nilpotent and $A/\mathfrak{R}(A)$ is semi-artinian and regular;

(c) A is a l.c.-ring and satisfies condition (H);

(d) Any prime ideal is maximal and for any non-zero A-module X, one has $\mathrm{Ass}(X) \neq \varnothing$.

(Hint: use the following result: If A is a reduced commutative ring and any prime ideal is maximal, then A is regular).

4. A commutative ring A is called *saturated* if any bimorphism $u: A \rightarrow B$ is in fact an isomorphism; A is called *strongly saturated* if any homomorphic image of A is saturated. Prove that the following assertions are equivalent for a commutative ring A:

(a) A is semi-artinian;

(b) $\mathfrak{R}(A)$ is T-nilpotent and A is a l.c.-ring and strongly saturated.

5. Let A be a commutative semi-artinian ring. The following assertions are equivalent:

(a) A is reduced;

(b) Any minimal ideal is projective;

(c) Any minimal ideal is injective;

(d) $J(A) = 0$, i.e. A is non-singular;

(e) A is a subring of a direct product of fields.

6. Let $\mathscr{C}$ be a semi-artinian category. Then there exists a bijection between the sets $\mathrm{Spel}(\mathscr{C})$ and $\mathrm{Sim}(\mathscr{C})$.

7. Let A be a left semi-artinian ring. There exists a natural bijection between $\mathrm{Spel}(A)$ and $\mathrm{Spel}(A/\mathfrak{R}(A))$.

8. Let A be a commutative ring. The following assertions are equivalent:

(a) A is a l.c.- and regular ring;

(b) A is a semi-artinian and regular ring;

(c) A is a semi-artinian and reduced ring;

(d) Any ideal of A is a reduced intersection of maximal ideals;

(e) Any submodule of an A-module is a reduced intersection of maximal ideals.

9. Let A be a commutative semi-artinian regular ring. Then A satisfies condition (H). Moreover any A-module X is an essential extension of a direct sum $P \amalg Q$ where P is a semi-simple projective module and Q a direct sum of simple injective modules.

10. Let $0 \to X' \to X \to X'' \to 0$ be an exact sequence in a Grothendieck category $\mathscr{C}$. Then the height of X is defined if and only if the heights of X' and X'' are defined. Moreover, one has:

$$\sup(h(X'), h(X'')) = h(X); \; h(X) \leqslant h(X') \oplus h(X'').$$

11. Let A be a left semi-artinian ring such that $h(A) \leqslant n$, where n is a natural number. Then A satisfies condition (H).

12. A Grothendieck category $\mathscr{C}$ is a s.a.-category if and only if $\mathscr{C}$ is a l.c.-category and the canonical functor $P: \mathscr{C} \to \mathscr{C}_E$ commutes with arbitrary direct products.

References

Bass [16]; Dlab [45], [46]; Gabriel [59]; Năstăsescu [139]; Năstăsescu–Popescu [143]; Popescu [172], [173].

5.7 Noetherian and artinian categories

Let $\mathscr{C}$ be an abelian category. We shall say that an object X of $\mathscr{C}$ is *noetherian* if any increasing sequence $X_1 \subseteq X_2 \subseteq \ldots \subseteq X_n \subseteq \ldots$ of subobjects of X is stationary. An object X of $\mathscr{C}$ will be called *artinian* if the object X^0 of $\mathscr{C}^0$ (the dual of $\mathscr{C}$) is noetherian, i.e. any descending sequence

$$X_1 \supseteq X_2 \supseteq X_3 \supseteq \ldots \supseteq X_n \supseteq \ldots$$

of subobjects of X becomes stationary.

LEMMA 7.1. *An object X of $\mathscr{C}$ is noetherian if and only if any set of subobjects of X (ordered by inclusion) has a maximal element.*

Moreover, an object X of $\mathscr{C}$ is artinian if and only if any set of subobjects of X (ordered by inclusion) has a minimal element.

Proof. Let X be noetherian and M a set of subobjects of X. Pick an element X_1 of M; if X_1 is not maximal, then we can find another element X_2 of M so that $X_1 \subset X_2$. By induction, for any integer $n > 0$ we can choose an element X_n of M such that $X_1 \subset X_2 \subset \ldots \subset X_{n-1} \subset X_n$. Since X is noetherian, this process will stop, that is X_n will be maximal for a certain n. The converse implication and the second part of the lemma are obvious. ∎

For an abelian category $\mathscr{C}$, we shall denote by $\mathscr{N}(\mathscr{C})$ (respectively, by $\mathscr{A}(\mathscr{C})$) or simply $\mathscr{N}$ (respectively, $\mathscr{A}$) if there is no danger of confusion, the full subcategory of $\mathscr{C}$ consisting of all noetherian (respectively, artinian) objects.

PROPOSITION 7.2. *The subcategories $\mathscr{N}$ (respectively, $\mathscr{A}$) are dense for any abelian category $\mathscr{C}$.*

Proof. Let us consider in $\mathscr{C}$ the exact sequence:

$$0 \longrightarrow X' \longrightarrow X \xrightarrow{p} X'' \longrightarrow 0$$

in which $X \in \mathrm{Ob}\,\mathscr{N}$. If $\{X_n'\}_n$ is an increasing sequence of subobjects of X' then it can be considered as a sequence of subobjects of X so that

$$X_n' = X'_{n+1} = \ldots$$

for a suitable n. Furthermore, if $\{X_n''\}_n$ is an increasing sequence of subobjects of X'', then the sequence $\{p^{-1}(X_n'')\}_n$ is increasing in X, so that it becomes stationary; but then, the sequence $\{X_n''\}_n$ also becomes stationary, hence X' and X'' are noetherian objects.

Conversely, assume that X' and X'' are noetherian and let $\{X_n\}_n$ be an increasing sequence of subobjects of X. Then the sequence $\{p(X_n)\}_n$ is increasing in X'' so that $p(X_{n_0}) = p(X_{n_0} + 1) = \ldots$ i.e.

$$X' + X_{n_0} = X' + X_{n_0+1} = \ldots$$

Now, the sequence $\{X_n \cap X'\}_n$ is necessarily stationary in X', so that $X' \cap X_{n_1} = X' \cap X_{n_1+1} = \ldots.$ Let us put $n' = \sup(n_0, n_1)$. Then for $n \geqslant n'$ one has:

$$p(X_n) \simeq (X_n + X')/X' \simeq X_n/X' \cap X_n \xrightarrow{f} X_{n+1}/X' \cap X_{n+1}$$

where the isomorphism f is chosen such that we have the commutative diagram, with exact rows:

$$\begin{array}{ccccccccc}
0 \rightarrow & X' \cap X_n & \longrightarrow & X_n & \longrightarrow & X_n/X' \cap X_n & \longrightarrow & 0 \\
& \| & & \downarrow i & & \downarrow f & & \\
0 \rightarrow & X' \cap X_{n+1} & \rightarrow & X_{n+1} & \rightarrow & X_{n+1}/X' \cap X_{n+1} & \rightarrow & 0
\end{array}$$

in which i is the inclusion. Now, it is clear that i is necessarily an isomorphism, so that X is also noetherian.

The second part of the proposition follows by duality. ∎

An object X of $\mathscr{C}$ is called *finite* if it is at the same time noetherian and artinian. We shall denote by $\mathscr{F}in(\mathscr{C})$ or simply $\mathscr{F}in$ the full subcategory consisting of all finite objects. From Proposition 7.2, we derive that $\mathscr{F}in$ is a dense subcategory. Moreover, $\mathscr{F}in = \mathscr{N} \cap \mathscr{A}$.

A *composition series* for an object X of $\mathscr{C}$ is a sequence

$$0 = X_0 \subset X_1 \subset \ldots \subset X_{n-1} \subset X_n = X$$

of subobjects of X such that X_{i+1}/X_i are simple objects for every $0 \leqslant i < n$. The number n is called the *length* of the considered composition series of X.

The following result gives a characterization of finite objects.

THEOREM 7.3. *An object X of $\mathscr{C}$ is finite if and only if it has a composition series.*

Proof. First, assume that X is finite. Then X is also artinian, so that the set of its non-zero subobjects (of course, we assume that $X \neq 0$) has a minimal one, say X_1. According to Proposition 7.2, X/X_1 is again an artinian object and if it is not zero, then it contains a minimal non-zero subobject, say Y_2. Let us put $X_2 = p^{-1}(Y_2)$, where $p : X \to X/X_1$ is canonical. In this way, for any natural number n we define a subobject X_n of X, such that $X_{n-1} \subset X_n$ and X_n/X_{n-1} is simple for every $n > 0$.

Since X is noetherian, then a natural number n' can be found such that $X_{n'} = X_{n'+1} = \ldots$, or equivalently $X_{n'} = X$. Now, define n as being the smallest integer n' so that $X_{n'} = X$. It is now obvious that the sequence $0 \subset X_1 \subset \ldots \subset X_n = X$ is a composition series of X.

Conversely, assume that X has a composition series

$$0 = X_0 \subset X_1 \subset \ldots \subset X_n = X.$$

We shall prove that X is noetherian, using the induction relative to n. If $n = 1$, then X is simple and all is clear. Assume that $n > 1$ and the result valid for an object X' having a composition series of length $<n$. Now, let $X_1' \subseteq X_2' \ldots \subseteq X_m' \subseteq \ldots$ an increasing sequence of subobjects of X, so that $X_1' \neq 0$ and let t be the smallest natural number for which $X_1' \subseteq X_t$ and $X_1' \nsubseteq X_{t-1}$. Then $X_1'/X_1' \cap X_{t-1} \subseteq X_t/X_{t-1}$ and because X_t/X_{t-1} is a simple object it follows that the last inclusion is in fact an equality, or equivalently $X_1' + X_{t-1} = X_t$. If for any natural number m we have

$$X_m' \subseteq X_t,$$

then necessarily $X_m' + X_{t-1} = X_t$ and $\{X_m' \cap X_{t-1}\}_m$ is an increasing sequence of subobjects of X_{t-1}. Because $t-1<t\leqslant n$, by the induction hypothesis, it follows that $X'_{m_0}\cap X_{t-1}=X'_{m_0+1}\cap X_{t-1}=\ldots$ for a suitable m_0. But then from the commutative diagram:

$$\begin{array}{ccccccccc} 0 & \to & X'_{m_0+1}\cap X_{t-1} & \to & X'_{m_0+1} & \to & X'_{m_0+1}/(X'_{m_0+1}\cap X_{t-1}) & \to & 0 \\ & & & & \downarrow{\scriptstyle i} & & & & \\ 0 & \to & X'_{m_0}\cap X_{t-1} & \to & X'_{m_0} & \to & X'_{m_0}/(X'_{m_0+1}\cap X_{t-1}) & \to & 0 \end{array}$$

with the exact row and i the canonical inclusion, we deduce that i is an isomorphism, so that $X'_{m_0}=X'_{m_0+1}$. By iteration, one has

$$X'_{m_0+1}=X'_{m_0+2}=\ldots.$$

Otherwise, let s be the smallest natural number such that $X_s'\not\subseteq X_t$, and t_1 the smallest natural number (obviously $\leqslant n$) such that $X_s'\subseteq X_{t_1}$. Then $t_1>t>0$ and the sequence $\{X_m'\cap X_{t_1-1}\}_m$ is an increasing sequence of X_{t_1-1} so that it becomes stationary by the induction hypothesis, since $t_1-1<n$. As above, we see that all X_m' are included in X_{t_1} and thus, the sequence X_m' is stationary, or an index m can be found such that $X_m'\not\subseteq X_{t_1}$. Again, as above we choose a natural number t_2 such that $t_1<t_2\leqslant n$ and $X_m'\subseteq X_{t_2}$. It is clear that after finitely many steps we find that $\{X_m'\}_m$ is a stationary sequence and finally X is noetherian.

In the same manner, we check that the object X^0 in $\mathscr{C}^0$ is noetherian, or equivalently, X is artinian. Hence X is a finite object. ∎

LEMMA 7.4. *Any noetherian object is a finite direct sum of indecomposable objects. The same result holds for any artinian or finite object.*

Proof. Let X be a noetherian object; for any natural number n we shall define a subobject X_n of X such that X_{n-1} is a direct summand of X_n and X_n a direct summand of X. In fact, $X_1=X$ if X is indecomposable, and X_1 is a proper non-zero direct summand, otherwise. Assume that X_{n-1} has been defined, so that $X=X_{n-1}\amalg Y_{n-1}$, then $X_n=X$ if Y_{n-1} is indecomposable and $X_n=X_{n-1}\amalg Z_n$ where Z_n is a non-zero proper direct summand of Y_{n-1} otherwise. By hypothesis, one has $X_n=X$ for a suitable n, so that $X_n=X_{n-1}\amalg Z_n$ and Z_n is indecomposable. In conclusion, any noetherian object contains indecomposable direct summands.

Again, for a natural number n, we shall define a direct summand X_n of X, such that X_n is a direct sum of n indecomposable objects. Assuming X_n defined, then $X=X_n\amalg Y_n$ hence $X_{n+1}=X_n\amalg Z_n$ where Z_n is an indecomposable direct summand of Y_n. It is clear that $\{X_n\}_n$ is an increasing sequence of subobjects of X so that $X_n=X$ for a suitable n, i.e. X is a direct sum of indecomposable objects. ∎

THEOREM 7.5. *Any finite object has a unique K.R.S.-decomposition.*

Proof. By the previous lemma, we find that a finite object has a K.R.S.-decomposition. Now, if X is an indecomposable finite object, then as in the proof of Lemma 1.5 we see that $\mathrm{End}_{\mathscr{C}}(X)$ is a local ring. Finally, the uniqueness of a K.R.S.-decomposition follows by Theorem 1.3. ∎

By a *noetherian* (respectively, *artinian*; respectively, *finite*) category, we mean an abelian category in which any object is noetherian (respectively, artinian; respectively, finite). It is clear that the dual of a noetherian category is artinian and conversely. Moreover, a category $\mathscr{C}$ is finite if and only if both it and its dual are noetherian.

If A is a ring, then the category $\mathscr{N}(\mathrm{Mod}\,A)$ is always non-zero, because any simple module is noetherian. Also, the category $\mathscr{A}(\mathrm{Mod}\,A)$ is non-zero since any simple module is also artinian. By Exercise 9 we see that the categories $\mathscr{N}(\mathrm{Mod}\,A)$ and $\mathscr{A}(\mathrm{Mod}\,A)$ are generally different. However, they coincide for a commutative semi-artinian ring.

PROPOSITION 7.6. *Let X be an A-module, where A is a commutative semi-artinian ring. The following assertions are equivalent:*

(1) *X is noetherian;*

(2) *X is artinian;*

(3) *X is a finite object.*

Proof. (1) ⇒ (3). Let $X_0 \subset X_2 \subset \ldots \subset X_n \subset \ldots$ be the Loewy sequence of X. Since X is noetherian, then $X_n = X$ for a suitable n and X_m/X_{m-1} is a finite direct sum of simple objects. Now, it is clear that X has a composition series, so that X is finite by Theorem 7.3.

(2) ⇒ (3). Let X^0 be a maximal submodule of X (which exists according to Section 5.6, Exercise 3). Furthermore, for any natural number n we shall denote by X^n a maximal submodule of X^{n-1}, $n \geqslant 1$. By hypothesis, $X^n = 0$ for a suitable n, so that X has a composition series.

The remaining assertions are obvious. ∎

Exercises

1. Any noetherian object is of finite type.

2. Any two composition series of an object X have the same length. Moreover, if $0 = X_0 \subset X_1 \subset \ldots \subset X_n = X$ and $0 = Y_0 \subset Y_1 \subset \ldots \subset Y_{n'} = X$, are

two composition series of an object X, then $n = n'$ and $X_{i+1}/X_i \simeq Y_{i+1}/Y_i$, $i = 0, \ldots, n-1$. Finally, if an object X has a composition series and

$$0 \subset X_1 \subset \ldots \subset X_m \subset X$$

is a sequence of subobjects, then it can be refined to a composition series.

3. Let $f: X \to X$ be an endomorphism. If X is noetherian and f an epimorphism, then f is an isomorphism. Dually, if X is artinian and f a monomorphism, then f is an isomorphism. Moreover, if X is finite and indecomposable, then any element of $\mathrm{End}_{\mathscr{C}}(X)$ is either nilpotent or an automorphism (i.e. $\mathrm{End}_{\mathscr{C}}(X)$ is a local ring). (Fitting's lemma).

4. Any noetherian A-module X is finitely generated. Moreover X is noetherian if and only if all its submodules are finitely generated.

5. Show that an abelian group X is finite (as an object in $\mathscr{A}\ell$) if and only if its underlying set is finite.

6. Let $0 = X_0 \subset X_1 \subset \ldots \subset X_n = X$ be a sequence of subobjects of an object X of an abelian category such that X_i/X_{i-1} is a noetherian (respectively, artinian, respectively, finite) object for every $i \geqslant 1$. Then X is also noetherian (respectively, artinian, respectively, finite) object.

7. Let A be a semisimple ring. Then $\mathscr{N}(\mathrm{Mod}\,A) = \mathscr{A}(\mathrm{Mod}\,A) = \mathscr{F}in$ $(\mathrm{Mod}\,A)$.

8. Any noetherian (respectively, artinian, respectively, finite) object of a Grothendieck category is an essential extension of a finite direct sum of coirreducible subobjects. Moreover $\mathscr{N}(\mathscr{C}), \mathscr{A}(\mathscr{C})$ and $\mathscr{F}in(\mathscr{C})$ are subcategories of $\mathscr{C}_\tau$ (using the notations of Section 5.5).

9. Let $\mathbf{Q}_p$ be the abelian group defined as in Chapter 3, Section 3.3, Example 2. Then $\mathbf{Q}_p$ is artinian object but not noetherian.

References

Amdal–Ringdal [8]; Bourbaki [21], [23]; Gabriel [59]; Matlis [125], [126]; Năstăsescu–Popescu [143]; Vidal [233].

5.8 Locally noetherian categories

By a *locally noetherian* (respectively, *locally artinian*; respectively, *locally finite*) category we mean a Grothendieck category having a set of noetherian (respectively, artinian, respectively, finite) generators. Usually, we shall write l.n.-category (respectively, l.a.-category; respectively, l.f.-category).

COROLLARY 8.1. *Any l.a.-category is semi-artinian. Moreover, any l.f.-category is a l.a.-category and hence is semi-artinian.*

The proof follows by Theorem 6.1. ■

THEOREM 8.2. *A category $\mathscr{C}$ is a l.f.-category if and only if it is a l.n. and s.a.-category.*

Proof. Assume that $\mathscr{C}$ is l.n. and s.a.-category. We shall prove that any noetherian object is finite. Indeed, let X be a noetherian object and

$$X_0 \subset X_1 \subset \dots \subset X_m \subset \dots$$

the Loewy sequence of X. The hypothesis claims that $X_{m'} = X$ for a suitable m' and X_r/X_{r-1} is a semisimple object, precisely, a direct sum of n_r simple objects, for any $1 \leqslant r \leqslant m'$.

Now, it is obvious that we can define a composition series of X having length $n_0 + n_1 + \dots + n_{m'}$.

The converse result is obvious. ■

LEMMA 8.3. *Let $\mathscr{C}$ be an abelian category, $\mathscr{F}$ a localizing subcategory and $T : \mathscr{C} \to \mathscr{C}/\mathscr{F}$ the canonical functor. If X is a noetherian (respectively, artinian; respectively, finite) object of $\mathscr{C}$, then so is $T(X)$ in $\mathscr{C}/\mathscr{F}$.*

Proof. By Proposition 7.2, we can assume that X is $\mathscr{F}$-torsion free. Let $\{Y_n\}_n$ be an increasing sequence of subobjects of $T(X)$ and S a right adjoint of T. Then the sequence $\{S(Y_n)\}_n$ of subobjects of $ST(X)$ is also increasing, so that $X \cap S(Y_n) = X \cap S(Y_{n+1}) = \dots$ for some n. (We can regard X as a subobject of $ST(X)$ since it is $\mathscr{F}$-torsion free). But then we have the commutative diagram canonically constructed:

$$\begin{array}{ccccccccc}
0 \to & X \cap S(Y_n) & \longrightarrow & S(Y_n) & \longrightarrow & S(Y_n)/X \cap S(Y_n) & \longrightarrow & 0 \\
& \| & & \downarrow{\scriptstyle i} & & \downarrow{\scriptstyle p} & & \\
0 \to & X \cap S(Y_{n+1}) & \to & S(Y_{n+1}) & \to & S(Y_{n+1})/X \cap S(Y_{n+1}) & \to & 0
\end{array}$$

By the "snake Lemma" we see that coker $i \in \operatorname{Ob}\mathscr{F}$, so that $T(i)$ is an isomorphism and finally $Y_n = TS(Y_n) = TS(Y_{n+1}) = Y_{n+1} = \dots.$

The other assertions are checked in the same way. ■

COROLLARY 8.4. *Let $\mathscr{F}$ be a localizing subcategory of a l.n.-category $\mathscr{C}$. Then $\mathscr{F}$ and $\mathscr{C}/\mathscr{F}$ are also l.n.-categories. The same result remains true for l.a. and l.f.-categories.*

The proof follows by the previous lemma. ■

THEOREM 8.5. *Any l.n.-category $\mathscr{C}$ is a s.n.-category. Moreover, if $\{\mathscr{C}_\alpha\}_\alpha$ is the Krull–Gabriel filtration of $\mathscr{C}$, then for any ordinal α, such that $\alpha = \beta + 1$, the category $\mathscr{C}_\alpha/\mathscr{C}_\beta$ is a l.f.-category.*

Proof. If $\mathscr{C}_\tau = \bigcup_\alpha \mathscr{C}_\alpha \neq \mathscr{C}$, then by Corollary 8.4 the category $\mathscr{C}/\mathscr{C}_\tau$ is also a l.n.-category, and by hypothesis has a non-zero noetherian object X. The set of proper subobjects of X has a maximal one hence the quotient object is simple, which contradicts the Lemma 5.1. Hence $\mathscr{C}_\tau = \mathscr{C}$, in other words, $\mathscr{C}$ is a s.n.-category.

Now, if $\alpha = \beta + 1$ is an ordinal, then by Corollary 6.2 $\mathscr{C}_\alpha/\mathscr{C}_\beta$ is a s.a.-category and thus, by Theorem 8.2 and Corollary 8.4, is in addition a l.f.-category. ∎

PROPOSITION 8.6. *The set of noetherian subobjects of an object X in a l.n.-category $\mathscr{C}$, is direct and complete. The same result remains true, in a suitable formulation for l.a.- and l.f.-categories.*

The proof is checked by Proposition 7.2. ∎

By virtue of Exercise 1 of Section 5.7 it is clear that any l.n.-category is locally of finite type, that is, it has a set of generators of finite type. Now we shall characterize the categories locally of finite type which are l.n.-categories.

THEOREM 8.7. *Let $\mathscr{C}$ be a Grothendieck category locally of finite type. The following assertions are equivalent:*

(1) *Any object of finite type is noetherian;*

(2) *$\mathscr{C}$ is a l.n.-category;*

(3) *Any direct sum of injective objects is again injective;*

(4) *If $\{Q_i, f_{ij}\}$ is a direct system of injective objects, then its inductive limit is again injective.*

Proof. The implications (1) ⇒ (2) and (4) ⇒ (3) are obvious.

(2) ⇒ (1). If X is an object of finite type, then the set $\{X_i\}_i$ of its noetherian subobjects is directed and complete by Proposition 8.6, so that $X = X_i$ for a suitable i, or equivalently X is noetherian.

(2) ⇒ (4). Let

$$f_i : Q_i \to \varinjlim_i Q_i = Q$$

be the structural morphisms and a morphism $g' : X' \to Q$ of a subobject X' of an object X into Q. Denote by M the set of all pairs (g'', X'') where X'' is an object of X which contains X' and $g'' : X'' \to Q$ a morphism which extends g'. It is clear that M, ordered in a natural manner, is an inductive set, so that it has a maximal element, say (g', X'). We claim that $X' = X$. If that is not the case, there exists a noetherian subobject Y of X such that $X' + Y \neq X'$. Denote $Y' = X' \cap Y$ and $h' : Y' \to Q$ the restriction of g' to Y'. Now, $h'(Y')$ is a noetherian object and $Q = \sum_i f_i(Q_i)$, so that $h'(Y') \subseteq f_i(Q_i)$ for a suitable i, because the set $\{f_i(Q_i)\}_i$ of subobjects of Q is complete. But Q_i is the inductive limit of its noetherian subobjects, so that a noetherian subobject H of Q_i can be chosen such that $f_i(H) = h'(Y')$ and denote $H' = H \cap \ker f_i$. By Chapter 2, Section 2.8, Exercise 6 we deduce that $H' = \sum_{j \geqslant i} H' \cap \ker f_{ij}$ and by noetherianness of H' one has

$$H' = H' \cap \ker f_{ij}$$

for a suitable j, or equivalently, if we write $P = f_{ij}(H)$, then f_j defines an isomorphism of P onto $h'(Y')$, denoted by t. Furthermore, let $h_1 : Y \to Q_j$ be a morphism which extends $t^{-1} h'$ and $h = f_j h_1$. It is obvious that h prolongs h'. Now by the cocartesian diagram:

$$\begin{array}{ccc} X' & \longrightarrow & X' + Y \\ \uparrow & & \uparrow \\ Y' & \longrightarrow & Y \end{array}$$

we can construct a morphism $g : X' + Y \to Q$ which extends g', a contradiction. Therefore, necessarily $X' = X$, so that Q is injective.

(3) $\Rightarrow$ (1). Let X be an object of finite type, and $X_1 \subseteq X_2 \subseteq \ldots \subseteq X_n \subseteq \ldots$ an increasing sequence of subobjects. Let us put $X' = \sum_n X_n$ and let

$$t_n : X'/X_n \to Q_n$$

be a monomorphism with Q_n injective and $Q = \coprod Q_n$. Denote by u_n the composition of morphisms:

$$X_n \longrightarrow X' \xrightarrow{\ s\ } \textstyle\coprod_{m<n} X'/X_m \xrightarrow{\ \bar{t}_n\ } \coprod_{m<n} Q_m \longrightarrow Q$$

where $\bar{t}_n$ is induced by t_m, $m < n$, and s by the natural epimorphisms

$$X' \to X'/X_m.$$

There exists a morphism $f : X' \to Q$ which extends u_n and a morphism $g : X \to Q$ whose restriction to X' is f. Because X is of finite type, hence small, g and consequently f has a factorization:

$$X' \to Q_1 \amalg \ldots \amalg Q_r \to Q.$$

Hence the canonical epimorphism $X' \to X'/X_n$ is null if $n > r$, so that $X_r = X_{r+1} = \ldots$, or equivalently X is noetherian. ■

Now we shall give a neat characterization of l.n.-categories, due to Gabriel. First we prove:

PROPOSITION 8.8. *Let $\mathscr{C}$ be a small artinian category. Then the category* Lex $\mathscr{C}$ *is a l.n.-category.*

Proof. By Theorem 11.3 of Chapter 4 it suffices to prove that for any object X of $\mathscr{C}$, the object $\bar{h}(X)$ is noetherian. In fact, let Y be a subobject of $\bar{h}(X)$; by Proposition 11.11 of Chapter 4, one has $Y = \Sigma \bar{h}(X'')$, where X'' runs through all quoitient objects of X such that $\bar{h}(X'') \subseteq V$. But a quotient set $\{\bar{h}(X'')\}$ considered above has a maximal element $\bar{h}(X_0'')$ and necessarily $Y = \bar{h}(X_0'')$. Hence any subobject of $\bar{h}(X)$ is also of the form $\bar{h}(X')$ with X' a suitable quotient object of X. The conclusion is now obvious. ■

Now let $\mathscr{C}$ be a l.n.-category and $\mathscr{C}' = \mathscr{N}(\mathscr{C})$ its full subcategory consisting of all noetherian objects. We invite the reader to check that $\mathscr{C}'$ is in fact equivalent with a small noetherian category. Furthermore, we consider the category $\mathscr{C}'' = (\mathscr{C}')^0$ and $\bar{h}: \mathscr{C}'' \to \operatorname{Lex} \mathscr{C}''$, the canonical functor (see Chapter 4, Section 11). Also let $t: \mathscr{C}'' \to \mathscr{C}$ be the functor ID where

$$D: \mathscr{C}'' \to \mathscr{C}'$$

is the duality functor and $I: \mathscr{C}' \to \mathscr{C}$ the inclusion functor. If X is an object of $\mathscr{C}$ then by Proposition 8.6 $X = \sum_i X_i$, where X_i runs through all noetherian subobjects of X. If Y is an object of $\mathscr{C}''$ then we put:

$$H(X)(Y) = \varinjlim_i \operatorname{Hom}_{\mathscr{C}''}(X_i^0, Y) = \varinjlim_i \operatorname{Hom}_{\mathscr{C}}(Y^0, X_i),$$

where X_i^0 is the dual of X_i and Y^0 the dual of Y (respectively in $\mathscr{C}''$ and $\mathscr{C}'$). It is not difficult to verify that the assignment $Y \rightsquigarrow H(X)(Y)$ is in fact a covariant left exact functor of $\mathscr{C}''$ into $\mathscr{A}b$, denoted by $H(X)$, and the assignment $X \rightsquigarrow H(X)$ defines as before a covariant functor $H: \mathscr{C} \to \operatorname{Lex} \mathscr{C}''$ such that $Ht = \bar{h}$. The Gabriel characterization of l.n.-categories is:

THEOREM 8.9. (Gabriel [59]). *Let $\mathscr{C}$ be a l.n.-category; with the above notations, the functor $H: \mathscr{C} \to \operatorname{Lex} \mathscr{C}''$ is an equivalence of categories.*

Proof. First we shall prove that any object of $\operatorname{Lex} \mathscr{C}''$ is of the form $H(X)$ with $X \in \operatorname{Ob} \mathscr{C}$. If F is a noetherian object of $\operatorname{Lex} \mathscr{C}''$, as in Proposition 8.8, one checks that $F = \bar{h}(X'')$ for a suitable $X'' \in \operatorname{Ob} \mathscr{C}''$, hence $F = H(X')$, where $X' = (X'')^0 \in \operatorname{Ob} \mathscr{C}'$. Now assume that $F \in \operatorname{Ob} \operatorname{Lex} \mathscr{C}''$ is not necessarily noetherian. Then, by Proposition 8.8, $F = \sum_i F_i$ where F_i runs through all its noetherian subobjects, and by the previous considerations one has $F_i = \bar{h}(X_i'')$. If $F_i \subseteq F_j$, then a unique epimorphism $u_{ij} : X_j'' \to X_i''$ can be chosen such that $\bar{h}(u_{ij})$ is the considered inclusion. Let us put $X_i = (X_i'')^0$ and $v_{ij} = (u_{ij})^0$.

Therefore one has in $\mathscr{C}'$ the inductive system $\{X_i, v_{ij}\}$ in which the v_{ij} are monomorphisms. If X is the inductive limit of this system, then

$$H(X) \simeq F,$$

because, for any $Y \in \operatorname{Ob}\mathscr{C}''$ one has:

$$H(X)(Y) = \varinjlim_i \operatorname{Hom}_{\mathscr{C}''}(X_i{}^0, Y) \simeq \varinjlim_i \operatorname{Hom}_{\mathscr{C}''}(X_i'', Y) \simeq$$

$$\varinjlim_i \operatorname{Hom}_{\operatorname{Lex}\mathscr{C}''}(\bar{h}(Y), \bar{h}(X_i'')) \simeq \operatorname{Hom}_{\operatorname{Lex}\mathscr{C}''}(\bar{h}(Y), \varinjlim_i \bar{h}(X_i'')) \simeq$$

$$\operatorname{Hom}_{\operatorname{Lex}\mathscr{C}''}(\bar{h}(Y), F) \simeq F(Y).$$

and all considered isomorphisms are functorial. Hence $H(X) \simeq F(Y)$ as claimed.

Now the point is to prove that H is a fully faithful functor.

In fact, if $f : X \to Y$ is a non-zero morphism X and is noetherian, then f has the factorization

$$X \xrightarrow{f'} Y' \xrightarrow{u} Y$$

where u is a monomorphism and Y' noetherian; if $H(f) = 0$, then $H(f') = 0$. But it is clear that $H(f') = \bar{h}((f')^0) \neq 0$. Hence $H(f)$ is necessarily non-zero. Now if $f : X \to Y$ is a non-zero morphism, and X not necessarily noetherian, then we can construct the commutative diagram:

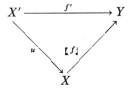

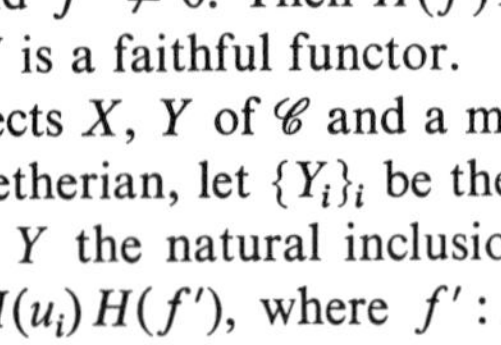

where X' is noetherian and $f' \neq 0$. Then $H(f)H(u) = H(f') \neq 0$ so that $H(f) \neq 0$; in conclusion H is a faithful functor.

Let us consider two objects X, Y of $\mathscr{C}$ and a morphism $f : H(X) \to H(Y)$. First, assume that X is noetherian, let $\{Y_i\}_i$ be the set of all noetherian subobjects of Y and $u_i : Y_i \to Y$ the natural inclusion. Then $\operatorname{im} f \subseteq H(Y_i)$ for an index i, so that $f = H(u_i)H(f')$, where $f' : X \to Y_i$ is a suitable morphism, because $Ht = \bar{h}$. If X is not noetherian, then $X = \sum_i X_i$, where again X_i runs through all its noetherian subobjects. If $u_i : X_i \to X$ are canonical inclusions, then for any i we can construct the following commutative diagram:

$$\begin{array}{ccc} \mathrm{Hom}_{\mathscr{C}}(X_i, Y) & \xrightarrow{k_i} & \mathrm{Hom}_{\mathrm{Lex}\,\mathscr{C}''}(H(X_i), H(Y)) \\ \uparrow{\scriptstyle \bar{u}_i} & & \uparrow{\scriptstyle \overline{H(u_i)}} \\ \mathrm{Hom}_{\mathscr{C}}(X, Y) & \xrightarrow{k} & \mathrm{Hom}_{\mathrm{Lex}\,\mathscr{C}''}(H(X), H(Y)) \end{array}$$

where the vertical morphisms are induced by u_i and the horizontal morphisms by H. But the previous considerations, k_i is an isomorphism, so that passing to the inductive limit by i, we find that k is also an isomorphism. ∎

Now we shall give a converse of Proposition 11.10 of Chapter 4 for the noetherian case.

PROPOSITION 8.10. *Let $\mathscr{C}$ be small artinian category. Then an object F of* Lex $\mathscr{C}$ *is injective if and only if it is exact.*

Proof. Let F be an exact covariant functor $F : \mathscr{C} \to \mathscr{Ab}$. We have to prove that F is an injective object of Lex $\mathscr{C}$. In fact, if $X \in \mathrm{Ob}\,\mathscr{C}$, then every subobject F' of $\bar{h}(X)$ is of the form $\bar{h}(X'')$, where X'' is a quotient object of X, so that one has an epimorphism $p : X \to X''$, such that $\bar{h}(p) : \bar{h}(X'') \to \bar{h}(X)$ is the natural inclusion. The morphism p induces the commutative diagram:

$$\begin{array}{ccccc} \mathrm{Hom}_{\mathrm{Lex}\,\mathscr{C}}(\bar{h}(X), F) & \longrightarrow & \mathrm{Hom}_{\mathrm{Lex}\,\mathscr{C}}(\bar{h}(X''), F) & & \\ \downarrow \wr & & \downarrow \wr & & \\ F(X) & \longrightarrow & F(X'') & \longrightarrow & 0 \end{array}$$

from which we conclude that the upper morphism is an epimorphism. Now we can use Theorem 8.9 and Proposition 8.6. ∎

NOTE 1. An immediate consequence of Proposition 8.10 is that any inductive direct limit of injective objects of a l.n.-category is also injective.

THEOREM 8.11. *Any injective object of a l.n.-category is in a unique way a direct sum of injective indecomposable objects, or equivalently, has a unique K.R.S.-decomposition.*

The proof is a direct consequence of Theorem 8.7 and Theorem 1.3. ∎

PROPOSITION 8.12. *Let $\mathscr{F}$ be a localizing subcategory of a l.n.-category $\mathscr{C}$. Then, the canonical functor $S : \mathscr{C}/\mathscr{F} \to \mathscr{C}$ commutes with inductive direct limits. Particularly, S commutes with direct sums.*

Proof. Let $\{X_i, u_{ij}\}$ be a direct system of objects of $\mathscr{C}/\mathscr{F}$, and X its limit. Then

$$X = T(\varinjlim_i S(X_i))$$

so that to complete the proof it will be sufficient to prove that the canonical morphism:

$$\varinjlim S(X_i) \to ST(\varinjlim S(X_i))$$

is an isomorphism, or equivalently, that

$$\varinjlim_i S(X_i)$$

is $\mathscr{F}$-closed. According to Lemma 4.1, Chapter 4, let $f: P \to Q$ be a morphism such that ker f and coker f are objects of $\mathscr{F}$, and P, Q are noetherian objects. Then one has the commutative diagram:

$$\begin{array}{ccc} \mathrm{Hom}_{\mathscr{C}}(P, \varinjlim S(X_i)) & \xrightarrow{\sim} & \varinjlim \mathrm{Hom}_{\mathscr{C}}(P, S(X_i)) \\ \uparrow{\scriptstyle \bar{f}} & & \uparrow\wr \\ \mathrm{Hom}_{\mathscr{C}}(Q \varinjlim S(X_i)) & \xrightarrow{\sim} & \varinjlim \mathrm{Hom}_{\mathscr{C}}(Q, S(X_i)) \end{array}$$

in which the horizontal morphisms are isomorphisms (because P and Q are finitely presented by Exercise 2). Because $S(X_i)$ is closed, the right morphism is an isomorphism, so that $\bar{f}$ is also an isomorphism. Now we can utilise Proposition 8.6. ∎

PROPOSITION 8.13. *If $\mathscr{C}$ is a l.n.-category, then any localizing subcategory $\mathscr{F}$ of $\mathscr{C}$ is determined by $\mathscr{N}(\mathscr{F})$. Moreover, there exists a natural bijection between the set of all localizing subcategories of $\mathscr{C}$ and all dense subcategories of $\mathscr{N}(\mathscr{C})$.*

This result is another consequence of Proposition 8.6. ∎

Exercises

1. Prove that the assertions of Theorem 8.7 are also equivalent with the following statements:

(a) An inductive direct limit of essential extensions is also an essential extension;

(b) An inductive direct limit of injective envelopes is also an injective envelope.

2. Any noetherian object is finitely presented.

3. A Grothendieck category $\mathscr{C}$ is l.n.-category if and only if $\mathscr{C}$ is a l.c.-category and the canonical functor $P: \mathscr{C} \to \mathscr{C}_E$ commutes with arbitrary

filtered inductive limits. Moreover, $\mathscr{C}$ is a l.f.-category if and only if $\mathscr{C}$ is a s.n.-category and the canonical functor $P : \mathscr{C} \to \mathscr{C}_E$ commutes with arbitrary direct products.

4. The following conditions on a Grothendieck category $\mathscr{C}$ are equivalent:

(a) There exists an injective object Q such that every injective of $\mathscr{C}$ is a direct sum of objects which are direct summands in Q;

(b) Every direct sum of injective objects is still injective;

(c) There is an object D of $\mathscr{C}$, such that every object of $\mathscr{C}$ is a subobject of a suitable direct sum of copies of D.

5. A Grothendieck category $\mathscr{C}$ is a l.n.-category if and only if any injective object is a direct sum of indecomposable injective objects.

6. The map:

$$\mathscr{C} \rightsquigarrow \operatorname{Lex} \mathscr{C}$$

defines a bijection between the (equivalence classes of) small abelian categories and the (equivalence classes of) locally coherent Grothendieck categories. Furthermore, $\mathscr{C}$ is naturally equivalent to the category of coherent objects of $\operatorname{Lex}\mathscr{C}$, and $\mathscr{C}$ is naturally embedded into $\operatorname{Lex}\mathscr{C}$, by means of the canonical functor $\bar{h} : \mathscr{C} \to \operatorname{Lex}\mathscr{C}$. (See Chapter 4, Section 4.11 and Exercise 9 of Section 4.11).

(Hint: The proof proceeds in the same way as that of Theorem 8.9).

7. Let $\mathscr{N}$ be a small noetherian category. Then the category $\operatorname{Lex}\mathscr{N}$ is a locally artinian category. Moreover, one has

$$KG(\operatorname{Lex}\mathscr{N}) \leqslant h(\operatorname{Lex}\mathscr{N}).$$

References

Bass [16]; Bucur–Deleanu [31]; Chase [37]; Faith–Walker [53]; Gabriel [59]; Matlis [125], [126]; Nouazé [153]; Papp [161]; Pareigis [162]; Roos [197], [198]; Stenström [214].

5.9 Noetherian and artinian rings

THEOREM 9.1. *The following assertions are equivalent for a ring A*:

(1) *A is a noetherian object of* Mod *A*;

(2) *Any left ideal of A is finitely generated*;

(3) *Any submodule of a finitely generated module is also finitely generated*;

(4) Mod *A is a l.n.-category.*

Proof. (1) ⇒ (2) results by Proposition 7.2 and Exercise 4 of Section 5.7.

(2) ⇒ (1). If $\mathfrak{a}_1 \subseteq \mathfrak{a}_2 \subseteq \ldots \subseteq \mathfrak{a}_n \subseteq \ldots$ is an increasing sequence of left ideals of A, then $\bigcup_n \mathfrak{a}_n$ is a finitely generated left ideal so that it is equal to $\mathfrak{a}_n$ for some n.

(4) ⇒ (1). From Proposition 8.6, $A = \sum \mathfrak{a}_i$, where $\mathfrak{a}_i$ runs through all noetherian left ideals of A, so that $A = \mathfrak{a}_i$ for a suitable i, since A is finitely generated.

The remaining equivalences are easily established, by Theorem 8.7. ∎

A ring A, as in Theorem 9.1, is called *left noetherian*. There exist classical examples of left noetherian rings which are not right noetherian, i.e. Mod A is a l.n.-category, but Mod A^0 is not (see [33] Chapter 1).

LEMMA 9.2. *Let X be a projective noetherian object of an abelian category $\mathscr{C}$. Then the ring* $\mathrm{End}_{\mathscr{C}}(X) = A$ *is right noetherian.*

Proof. Let $\mathfrak{a}$ be a right ideal of A and $f_1, \ldots, f_n \in \mathfrak{a}$; then $f_i : X \to X$ is an endomorphism of X and we write

$$[f_1, \ldots, f_n] = \sum_{i=1}^{n} \mathrm{im}\, f_i.$$

If $\{f_1, \ldots, f_n\}$ runs through all finite subsets of $\mathfrak{a}$ then the set of subobjects $[f_1, \ldots, f_n]$ of X has a maximal element, say $X' = [f_1, \ldots, f_n]$. Let

$$u_i : X \to X^n$$

be the structural morphisms and $f : X^n \to X$ the unique morphism such that $fu_i = f_i$, $i = 1, \ldots, n$. It is clear that $X' = \mathrm{im}\, f$ so that f has the factorization:

$$X^n \xrightarrow{p} X' \xrightarrow{j} X$$

where j is canonical inclusion and p an epimorphism. If $g : X \to X$ is a morphism of $\mathfrak{a}$, then $\mathrm{im}\, g \subseteq X'$ hence $g = jg'$. Since X is projective, there exists a morphism $h : X \to X^n$ such that $ph = g'$. If p_i is the projection associated with u_i, then $h = \sum_i u_i p_i h$ hence $g = jg' = jph = \sum_i jpu_i p_i h = \sum_i fu_i p_i h = \sum_i f_i (p_i h)$. Now, $p_i h \in A$ therefore g is an element of the right ideal generated by $f_1, \ldots, f_n$, i.e. this right ideal coincides with $\mathfrak{a}$. ∎

COROLLARY 9.3. *The ring of all $n \times n$-matrices over a left noetherian ring is a left noetherian ring.* ∎

COROLLARY 9.4. *Let $\mathscr{C}$ be an abelian noetherian category. The following assertions are equivalent*:

(1) *$\mathscr{C}$ is equivalent with $\mathscr{N}$*(Mod A^0), *where A is a right noetherian ring.*

(2) *$\mathscr{C}$ has a projective noetherian generator U.*

Proof. It will be sufficient to prove that (2) $\Rightarrow$ (1). In fact, let $A = \mathrm{End}_{\mathscr{C}}(U)$ and let $S: \mathscr{C} \to \mathrm{Mod}\, A^0$ be the functor: $S(X) = \mathrm{Hom}_{\mathscr{C}}(U, X)$. It is clear that S is a faithful functor and by Lemma 9.2, A is right noetherian. To complete the proof, it will be enough to show that S is full. First we shall prove that any object X of $\mathscr{C}$ is a quotient of U^n for some natural number n. To do this, let us consider the set of all subobjects X' of X such that $X' = \mathrm{im}\, f$ for a morphism $f: U^n \to X$ and let X' be a maximal element of this set. If $X' \neq X$, then a morphism $g: U \to X$ can be calculated so that $\mathrm{im}\, g \nsubseteq X'$ and so, the choice of X' is obviously violated. Finally, for any object X of $\mathscr{C}$, one has the exact sequence:

$$U^m \to U^n \to X \to 0$$

and the morphism $S(U, U^n): \mathrm{Hom}_{\mathscr{C}}(U, U^n) \to \mathrm{Hom}_A(S(U), S(U^n))$ is an isomorphism for any n.

We may leave it to the reader to fill in the details (see also Lemma 11.5 of Chapter 4). ∎

COROLLARY 9.5. *Let $\mathscr{C}$ be a l.n.-category having a projective noetherian generator U. Then $\mathscr{C}$ is equivalent with the category* $\mathrm{Mod}(\mathrm{End}_{\mathscr{C}}(U))^0$. ∎

A ring A is called *left artinian* if it is an artinian object of the category Mod A.

THEOREM 9.6. *Let A be a left artinian ring. Then $\mathfrak{R} = \mathfrak{R}(A)$ is a nilpotent ideal. Moreover $A/\mathfrak{R}(A)$ is a semisimple ring.*

Proof. The sequence $\{\mathfrak{R}^n\}_n$ becomes stationary so that $\mathfrak{R}^p = \mathfrak{R}^{p+1} = \dots = \mathfrak{a}$ for a suitable p. Let us assume that $\mathfrak{a} \neq 0$. There exists a left ideal $\mathfrak{b}$ so that $\mathfrak{a}\mathfrak{b} \neq 0$, for example $\mathfrak{b} = A$. Let $\mathfrak{b}$ be a minimal element in the set of such ideals. Then $\mathfrak{a}(\mathfrak{R}\mathfrak{b}) = (\mathfrak{a}\mathfrak{R})\mathfrak{b} = \mathfrak{a}\mathfrak{b} \neq 0$; since $\mathfrak{R}\mathfrak{b} \subseteq \mathfrak{b}$ by hypothesis one has that $\mathfrak{R}\mathfrak{b} = \mathfrak{b}$. We claim that $\mathfrak{b}$ is a finitely generated left ideal. Indeed, if $x \in \mathfrak{b}$ is such that $\mathfrak{a}x \neq 0$, then $\mathfrak{a}(Ax) \neq 0$ and $Ax \subseteq \mathfrak{b}$, i.e. $Ax = \mathfrak{b}$ by hypothesis. Finally, by the Lemma of Nakayama, $\mathfrak{b} = 0$, a contradiction. Hence $\mathfrak{R}^p = 0$.

The ring $A' = A/\mathfrak{R}(A)$ is obviously a left artinian ring, and $\mathfrak{R}(A') = 0$ .If $\mathfrak{a}$ is a minimal left ideal of A', then $\mathfrak{a} \nsubseteq \mathfrak{m}$ for a suitable maximal left ideal $\mathfrak{m}$, so that $\mathfrak{a} + \mathfrak{m} = A'$ and $\mathfrak{a} \cap \mathfrak{m} = 0$, or equivalently, any minimal left ideal is a direct summand. Furthermore, since A' is left artinian, there exists, in the set of left ideals which are not direct summands, a minimal element $\mathfrak{b}$ (in case this set is not empty). If $\mathfrak{b}$ contains a proper non-zero left ideal $\mathfrak{b}'$, then $\mathfrak{b}'$ is a direct summand of A', hence there exists a morphism $f: A' \to \mathfrak{b}'$ such that $fi = 1$, where $i: \mathfrak{b}' \to A'$ is the canonical inclusion. Thus $\mathfrak{b}'$ is also a direct summand in $\mathfrak{b}$ and we have $\mathfrak{b} = \mathfrak{b}' \amalg \mathfrak{b}''$. But $\mathfrak{b}''$ is also a direct summand of A' and as before $\mathfrak{b}$ is a direct summand of A'. Hence if $\mathfrak{b}$ is not a direct summand of A', then $\mathfrak{b}$ must be a simple left ideal, a contradiction. Therefore, the set of left ideals of A' which are not direct summands is empty, or equivalently A' is semisimple (see Chapter 3, Proposition 11.2). ∎

THEOREM 9.7. *Let A be a left artinian ring. The following assertions are equivalent for an A-module X:*

(1) *X is artinian;*

(2) *X is noetherian;*

(3) *X is finite.*

Moreover, any left artinian ring is also left noetherian.

Proof. (2) ⇒ (3). Because X is necessarily finitely generated, then X is a quotient module of A^n for a suitable n, so that X is artinian by Proposition 7.2. Now, from Theorem 7.3, X is a finite object.

(1) ⇒ (3). Let us assume that $R(A)^p = 0$ (see Theorem 9.6). Then one has the sequence

$$0 = \mathfrak{R}^p X \subseteq \mathfrak{R}^{p-1} X \subseteq \ldots \subseteq \mathfrak{R}X \subseteq X$$

of submodules of X. We shall prove that $X/\mathfrak{R}(X)$, $\mathfrak{R}(X)/\mathfrak{R}^2(X)$, ..., $\mathfrak{R}^{p-1} X$ are finite objects. It is clear that $\mathfrak{R}(\mathfrak{R}^i X/\mathfrak{R}^{i+1} X) = 0$ for every $0 \leqslant i \leqslant p$, so that any such module can be considered as an A/R-module, and by hypothesis, and Exercise 7 of Section 5.7 one has that it is finite (as A/R-module). Now it is obvious that this module is also finite as A-module, so that X is a finite object.

The last part of the theorem is now obvious. ∎

It is a known result (see [202]), that a commutative ring is artinian if and only if it is noetherian and any prime ideal is maximal. We shall give a generalization of this result using the notion of left prime of a ring.

THEOREM 9.8. *The following assertions are equivalent for a ring A.*

(1) *A is left artinian;*

(2) *A is left noetherian and any left prime of A is maximal.*

Proof. (1) ⇒ (2). From Theorem 9.7, A is left noetherian. Now, if F is a left prime of A, then K_F contains a simple module $U \simeq A/\mathfrak{m}$ so that $F = F_{\mathfrak{m}}$, where $\mathfrak{m}$ is a suitable maximal left ideal of A.

(2) ⇒ (1). By Theorem 6.4 one has that A is left semi-artinian. If

$$X_0 \subset X_1 \subset \ldots \subset X_n \subset \ldots$$

is the Loewy sequence of A, then X_n/X_{n-1} is a finite module and $X_n = A$ for a suitable n. Now it is clear that A is finite, hence artinian. ∎

THEOREM 9.9. *Let A be a commutative noetherian ring. Then any localizing subcategory of* Mod A *is stable.*

Proof. Let $\mathscr{F}$ be a localizing subcategory of Mod A and $X \in \mathrm{Ob}\,\mathscr{F}$. We want to check that E, the injective envelope of X, is also an object of $\mathscr{F}$. By Theorem 8.11, one checks that $E = \coprod_i Q_i$ where Q_i is an indecomposable injective object for any i, so that $Q_i = Q_{\mathfrak{p}_i}$, for a suitable prime ideal $\mathfrak{p}_i$ of A. Then $A/\mathfrak{p}_i \subseteq Q_i$, hence $(A/\mathfrak{p}_i) \cap X \neq 0$ and thus there exists an element $x \in X$ with $\mathrm{Ann}(x) = \mathfrak{p}_i$. Hence it will be sufficient to prove that for every prime ideal $\mathfrak{p}$ with $A/\mathfrak{p} \in \mathrm{Ob}\,\mathscr{F}$ one has $Q_\mathfrak{p} \in \mathrm{Ob}\,\mathscr{F}$. Before, we shall make some useful observations.

An element a of A will be called *almost-nilpotent* relatively to an A-module X if for any noetherian submodule X' of X there exists a natural number n such that $a^n X' = 0$. We shall use the following lemma:

LEMMA 9.10. *Let A be commutative noetherian ring and Q an indecomposable injective A-module. Then for any element $a \in A$, the homothety $m_a : Q \to Q$ is an isomorphism or a is almost-nilpotent relative to Q.*

Proof. If $\ker m_a = 0$, then m_a is an isomorphism because Q is indecomposable. Now, let us assume that $\ker m_a \neq 0$, and let M be a noetherian submodule of Q. Let us put $M_n = M \cap \ker m_{a^n}$ and $P_n = M \cap \mathrm{im}\, m_{a^n}$. Because $\{M_n\}$ is an increasing sequence of submodules of M, one has that

$$M_n = M_{n+1} = \ldots$$

for a suitable n. We claim that $M_n \cap P_n = 0$. In fact if $x \in M_n \cap P_n$, then $x = a^n y$ so that $a^n x = a^{2n} y = 0$, i.e. $y \in M_{2n} = M_n$, or equivalently $a^n y = x = 0$. Since Q is injective indecomposable, we have $P_n = 0$, because $M_n \supseteq M_1 \neq 0$. Therefore a is almost nilpotent relative to Q.

Now we continue the proof of the theorem. By Theorem 5.13 we see that $F = \cap F_\mathfrak{p}$ so that it will be enough to consider $F = F_\mathfrak{p}$ for some prime ideals $\mathfrak{p}$.

Therefore, let $\mathfrak{p}, \mathfrak{q} \in \mathrm{Spec}(A)$ such that $\mathfrak{q} \in F_\mathfrak{p}$. Then for every element $a \in A$ such that $a \notin \mathfrak{q}$ we see that $m_a : Q_\mathfrak{q} \to Q_\mathfrak{q}$ is an isomorphism, since $(\ker m_a) \cap (A/\mathfrak{q}) = 0$. If $a \in \mathfrak{q}$, then $A/\mathfrak{q} \subseteq \ker m_a$ therefore a is almost nilpotent relative to $Q_\mathfrak{q}$, so that for any noetherian submodule M of $Q_\mathfrak{q}$ one has $a^n M = 0$ for a suitable n. Now let $s \in \mathfrak{q} \cap (A - \mathfrak{p})$; then for any noetherian submodule M of $Q_\mathfrak{q}$ and any $x \in M$ one has $s^n x = 0$ for a suitable n, or equivalently $s^n \in \mathrm{Ann}(x)$. The proof is now complete, because of Proposition 8.6 and the definition of $F_\mathfrak{p}$. ∎

COROLLARY 9.11. *Any commutative artinian ring is a direct product (in a unique way) of finitely many artinian local rings.*

The proof is a direct consequence of Theorems 9.9 and 6.9. We note that Corollary 9.11 is also a direct consequence of Corollary 6.11. ∎

Another interesting consequence of Theorem 9.9 is the following:

COROLLARY 9.12. *Let A be a commutative noetherian ring and M the set of all maximal ideals of A. Then the subcategory $\overline{\mathrm{So}}(\mathrm{Mod}\, A)$, the smallest localizing subcategory containing $\mathrm{So}(\mathrm{Mod}\, A)$ is canonically equivalent with the direct product*

$$\prod_{\mathfrak{m} \in M} (\overline{\mathrm{So}})_{\mathfrak{m}} (\mathrm{Mod}\, A),$$

where $(\overline{\mathrm{So}})_{\mathfrak{m}} (\mathrm{Mod}\, A)$ is the smallest localizing subcategory of $\mathrm{Mod}\, A$ containing $A/\mathfrak{m}$.

The proof follows again by Theorems 9.9 and 6.9. ∎

In fact $\overline{\mathrm{So}}(\mathrm{Mod}\, A)$ is a l.f.-category so that the set of all finite subobjects of an object of this category is a direct complete set. We leave it to the reader to check the following result due to Matlis.

COROLLARY 9.13. (Matlis [126]). *If $\mathfrak{m}$ is a maximal ideal of a commutative noetherian ring, then $E(A/\mathfrak{m})$ is an artinian module. Particularly, any indecomposable A-module, where A is commutative and artinian, is finite.* ∎

Now we shall make some observations on left primes and on left noetherian rings.

LEMMA 9.14. *If $\mathfrak{p}$ is a prime two-sided ideal of a left noetherian ring then $A/\mathfrak{p}$ is a left order of a simple ring. Moreover, the module $E(A/\mathfrak{p})$ is in fact isotypic as A-module.*

Proof. The first part of the lemma will result from Chapter 4, Theorem 19.5. In fact, $E'(A/\mathfrak{p})$, the injective envelope of $A/\mathfrak{p}$ in $\mathrm{Mod}(A/\mathfrak{p})$, is isomorphic with $Q(A/\mathfrak{p})$ and thus it is a direct sum of simple $Q(A/\mathfrak{p})$-modules. Furthermore, we see that $E'(A/\mathfrak{p})$ is in fact a direct sum of indecomposable injective isomorphic $A/\mathfrak{p}$-modules. Finally, any coirreducible $A/\mathfrak{p}$-module is also coirreducible viewed as an A-module, so that $E(A/\mathfrak{p})$ is a direct sum of isomorphic indecomposable injective modules. ∎

Usually we shall denote by $Q_{\mathfrak{p}}$ an indecomposable injective contained in $E(A/\mathfrak{p})$.

COROLLARY 9.15. *For any prime ideal* $\mathfrak{p}$ *of a left noetherian ring* A, *the system* $F_{\mathfrak{p}}$ *is a left prime of* A.

The proof will follow from Lemma 10.2 and the previous lemma. ■

A two-sided ideal $\mathfrak{p}$ of a ring A is called *completely prime* if the ring $A/\mathfrak{p}$ is a domain.

THEOREM 9.16. *The following assertions are equivalent for a prime two-sided ideal* $\mathfrak{p}$ *of a left noetherian ring* A:

(1) $\mathfrak{p}$ *is completely prime*;

(2) $\mathfrak{p}$ *is irreducible*;

(3) $\mathfrak{p}$ *is left super-prime.*

Proof. (1) ⇒ (2). By Lemma 9.14 one checks that $Q(A/\mathfrak{p})$ is a simple ring and any element of $A/\mathfrak{p}$ becomes invertible in $Q(A/\mathfrak{p})$. Now it is clear that $Q(A/\mathfrak{p})$ is in fact a division ring, so that $A/\mathfrak{p}$ is coirreducible as $A/\mathfrak{p}$-module and finally as A-module.

(2) ⇒ (1). Then $E(A/\mathfrak{p})$, the injective envelope of $A/\mathfrak{p}$ in $\mathrm{Mod}(A/\mathfrak{p})$, is indecomposable injective so that by Lemma 9.14. $Q(A/\mathfrak{p})$ is a division ring, or equivalently $A/\mathfrak{p}$ is a left Ore domain.

(2) ⇒ (3). By Corollary 9.15, $F_{\mathfrak{p}}$ is a left prime of A so that $F_{\mathfrak{p}} = F_{\mathfrak{p}'}$ for a suitable left super prime ideal $\mathfrak{p}'$ of A. Now, we can choose $\mathfrak{p}'$ such that $\mathfrak{p}' = (\mathfrak{p} : x)$ where $x \notin \mathfrak{p}$, and obviously $\mathfrak{p} \subseteq \mathfrak{p}'$. If $y \in \mathfrak{p}'$, then $yx \in \mathfrak{p}$ so that $y \in \mathfrak{p}$ by (1). Finally, $\mathfrak{p} = \mathfrak{p}'$, or equivalently $\mathfrak{p}$ is left super prime.

The implication (3) ⇒ (1) will follow from Chapter 4, Section 4.20, Exercise 9. ■

Here there is a conjecture, namely, if all above conditions of Theorem 9.16 are also equivalent with:

(4) $F_{\mathfrak{p}}$ is a perfect left prime.

Obviously, one has the implication (4) ⇒ (3).

Exercises

1. Any left noetherian domain is a left Ore domain.

2. Let A be a domain such that $A/\mathfrak{a}$ is artinian for each left non-zero ideal. Then A is a left Ore domain. Can you say whether A is left noetherian? (We observe that A is necessarily a left semi-noetherian ring by Corollary 5.3).

3. Recall that a two-sided ideal $\mathfrak{a}$ of a ring A is called *prime* if for any two elements $x, y \in A$, the inclusion $xAy \subseteq \mathfrak{a}$ implies $x \in \mathfrak{a}$ or $y \in \mathfrak{a}$. An ideal $\mathfrak{a}$ is prime if and only if for any two left ideals $\mathfrak{b}$, $\mathfrak{c}$ of A the inclusion $\mathfrak{b}\,\mathfrak{c} \subseteq \mathfrak{a}$ implies that $\mathfrak{b} \subseteq \mathfrak{a}$ or $\mathfrak{c} \subseteq \mathfrak{a}$ (the same equivalence is valid for right ideals). Moreover, an ideal $\mathfrak{a}$ is prime if and only if there exists a non-zero submodule X such that $\mathrm{Ann}(X') = \mathfrak{a}$ for every non-zero submodule X' of X. So, if A is left noetherian, then for any non-zero module X, a maximal element of the set $\{\mathrm{Ann}(X')\}$ where X' runs through all non-zero submodules of X is a prime ideal.

4. If A is left noetherian, then any small object of $\mathrm{Mod}\,A$ is finitely generated.

5. (Faith and Walker, [53]). The following assertions are equivalent for a ring A:

(a) A is left noetherian;

(b) There exists a cardinal number c such that each injective A-module is a direct sum of modules, each generated by c elements;

(c) There exists an object H of $\mathrm{Mod}\,A$ such that every object in $\mathrm{Mod}\,A$ is isomorphic to a subobject of a direct sum of copies of H.

6. If A is left noetherian ring, and if the injective envelope of cyclic (respectively, finitely generated) modules in Mod A are finitely generated, then A is left artinian. Moreover if A is a ring such that each A-module is contained in a direct sum of finitely generated modules, then A is left artinian. (Hint: use the results of Kaplansky, given in Chapter 3, Section 3.9).

7. Let A be a commutative ring. Then A is artinian if and only if each injective A-module is a direct sum of finitely generated modules.

8. An artinian ring A is a cogenerator in $\mathrm{Mod}\,A$ if and only if A is a QF-ring.

9. The following assertions are equivalent for a ring A:

(a) A is a QF-ring;

(b) A is left artinian and left self-injective.

10. Let A be a left noetherian ring. The following assertions are equivalent:

(a) Any left prime of A is a special left prime;

(b) For each two-sided ideal $\mathfrak{a}$ of A, any left prime of $A/\mathfrak{a}$ is two-sided;

(c) For each prime ideal $\mathfrak{p}$ of A any left prime of $A/\mathfrak{p}$ is two-sided;

(d) For every injective indecomposable A-module Q there exists a prime ideal $\mathfrak{p}$ of A such that $Q \simeq E(A/\mathfrak{p})$.

11. Let A be a commutative noetherian ring. The following assertions are equivalent for an A-module X:

(a) X is artinian;

(b) X is a submodule of

$$\coprod_{i=1}^{n} E(A/\mathfrak{m}_i),$$

where $\mathfrak{m}_i$ are maximal ideals of A;

(c) $X \in \mathrm{Ob}\,\overline{\mathrm{So}}(\mathrm{Mod}\,A)$ and its socle is finitely generated.

Also the following assertions are equivalent:

(a′) $X \in \mathrm{Ob}\,\overline{\mathrm{So}}(\mathrm{Mod}\,A)$;

(b′) X is an essential extension of $\mathrm{So}(X)$;

(c′) X is a submodule of $\coprod_i E(A/\mathfrak{m}_i)$ where $\mathfrak{m}_i$ are maximal ideals of A;

(d′) Every finitely generated submodule of X is finite;

(e′) X is the sum of its artinian submodules.

12. Let A be a commutative noetherian ring and $Q = \amalg E(A/\mathfrak{m})$ where $\mathfrak{m}$ runs over all maximal ideals of A. Then $\mathrm{End}_A(Q) \simeq \Pi\, \mathrm{End}_A(E(A/\mathfrak{m}))$.

References

Bourbaki [23]; Faith–Walker [53]; Gabriel [59]; Krause [109]; Lam [111]; Matlis [125], [126]; Michler [130]; Papp [161]; Pareigis [162]; Samuel–Zariski [202].

5.10 Primary decomposition theory

Throughout, $\mathscr{C}$ is a Grothendieck category and $\mathscr{P}$ a prime localizing subcategory of $\mathscr{C}$. We shall denote by $Q_{\mathfrak{P}}$ an indecomposable injective object such that $\mathscr{F}(Q_{\mathfrak{P}}) = \mathscr{P}$. Moreover, if $\mathfrak{p}$ is a prime ideal of a commutative ring and $\mathscr{P} = \mathscr{F}_{\mathfrak{p}}$ then $Q_{\mathfrak{p}} = E(A/\mathfrak{p}) = Q_{\mathfrak{P}}$.

An object X of $\mathscr{C}$ will be called *$\mathscr{P}$-coprimary*, if $X \neq 0$ and $\mathscr{F}(X') = \mathscr{P}$ for any non-zero subobject X' of X. A subobject X' of X will be called *$\mathscr{P}$-primary* if X/X' is $\mathscr{P}$-coprimary. Finally, an object X will be called *$\mathscr{P}$-isotypic* if $E(X)$ is an essential extension of a direct sum $\coprod_i Q_i$ where $Q_i \simeq Q_{\mathfrak{P}}$ for every i.

Sometimes we shall write simply coprimary, primary and isotypic if there is no danger of confusion. Now recall that an element $\mathscr{P} \in \mathrm{Spel}(\mathscr{C})$ is associated with an object X if $\mathscr{F}(X') = \mathscr{P}$ for a suitable coirreducible subobject X' of X. The set of all elements of $\mathrm{Spel}(\mathscr{C})$ which are associated with X will be denoted by $\mathrm{Ass}(X)$.

THEOREM 10.1. *An object X of $\mathscr{C}$ is $\mathscr{P}$-coprimary if and only if it is $\mathscr{P}$-isotypic.*

Proof. If X is $\mathscr{P}$-isotypic then by Exercise 2 any non-zero subobject of X is also $\mathscr{P}$-isotypic. Now, the fact that X is $\mathscr{P}$-coprimary follows from Exercise 1.

Conversely, let us assume that X is $\mathscr{P}$-coprimary. We shall prove that any non-zero subobject X' of X contains a coirreducible subobject Y whose injective envelope is isomorphic with $Q_{\mathfrak{P}}$. In fact, if $Q' = E(X')$, then $\mathscr{F}(X') = \mathscr{F}(Q') = \mathscr{P}$ so that Q' is $\mathscr{P}$-closed. Now let U be a simple object of $\mathscr{C}/\mathscr{P}$ and $T: \mathscr{C} \to \mathscr{C}/\mathscr{P}$ the canonical functor. Then $T(Q')$ is an injective cogenerator of $\mathscr{C}/\mathscr{P}$ so that $U \subseteq T(Q')$. But then $S(U) \subseteq ST(Q') \simeq Q'$ so that $Y = S(U) \cap X' \neq 0$. It is clear that $E(Y) = E(S(Y)) = Q_{\mathfrak{P}}$. The proof now follows by an argument similar to the proof of Corollary 3.2. ∎

LEMMA 10.2. *Let $\mathscr{C}$ be s.n.-category. Then for every injective indecomposable Q, one has that $\mathscr{F}(Q)$ is a prime localizing subcategory. Moreover the assignment $Q \leadsto F(Q)$ defines a bijection between* $\mathrm{Sp}(\mathscr{C})$ *and* $\mathrm{Spel}(\mathscr{C})$.

Proof. It is clear that $\mathscr{C}/\mathscr{F}(Q)$ has simple objects and $T(Q)$ is an injective indecomposable cogenerator of $\mathscr{C}/\mathscr{F}(Q)$, or equivalently $\mathscr{F}(Q)$ is a prime localizing subcategory (see Chapter 4, Theorem 20.6). The remaining assertions are easy and are left to the reader. ∎

NOTE 1. One sees that an inverse of the given function is the assignment $\mathscr{P} \leadsto Q_{\mathfrak{P}}$, where $\mathscr{P} \in \mathrm{Spel}(\mathscr{C})$.

THEOREM 10.3. *Let $\mathscr{C}$ be s.n.-category. The following assertions are equivalent for an object X of $\mathscr{C}$:*

(1) *X is $\mathscr{P}$-isotypic;*

(2) *X is $\mathscr{P}$-coprimary;*

(3) $\mathrm{Ass}(X) = \{\mathscr{P}\}$.

Proof. By Theorem 10.1 it will be enough to prove that (1)⇔(3).

(1) ⇒ (3). It is clear that $Q_{\mathfrak{P}}$ can be considered as a subobject of $E(X)$ so that $X' = X \cap Q_{\mathfrak{P}} \neq 0$; also one see that $\mathscr{F}(X') = \mathscr{F}(Q_{\mathfrak{P}}) = \mathscr{P}$ so that $\mathscr{P} \in \mathrm{Ass}(X)$. If $\mathscr{P}' \in \mathrm{Ass}(X)$, then $\mathscr{P}' = \mathscr{F}(X')$ for a suitable coirreducible subobject of X so that $E(X') \simeq Q_{\mathfrak{P}}$, (see Chapter 4, Theorem 20.8). Therefore necessarily $\mathscr{P}' \equiv \mathscr{P}$ by hypothesis.

(3) ⇒ (1). The object $E(X)$ is an essential extension of a direct sum $\coprod_i Q_i$ of injective indecomposable objects. It is clear that for any i, $\mathscr{F}(Q_i)$ is associated with X so that $\mathscr{F}(Q_i) = \mathscr{P}$ for every i, or equivalently $Q_i \simeq Q_{\mathfrak{P}}$ (see Lemma 10.2). Hence X is $\mathscr{P}$-isotypic. ∎

Let $\mathscr{C}$ be a s.n.-category and M a non-empty set. By Theorem 4.12 a decomposition theory of $\mathscr{C}$ is completely defined by a mapping

$$f: \mathrm{Sp}(\mathscr{C}) \to M.$$

Now we set $M = \mathrm{Spel}(\mathscr{C})$ and let $f: \mathrm{Sp}(\mathscr{C}) \to \mathrm{Spel}(\mathscr{C})$ be the map defined in Lemma 10.2. In this way, we find a decomposition theory (constructed following the proof of Theorem 4.12) which is called *primary decomposition theory* and denoted by $\pi: \mathscr{C} \to \mathrm{Spel}(\mathscr{C})$. We invite the reader to prove the following result.

PROPOSITION 10.4. *Let $\mathscr{C}$ be a s.n.-category. An object X of $\mathscr{C}$ is π-coirreducible if and only if it is coprimary. A subobject X' of X is π-irreducible if and only if it is primary.* ■

COROLLARY 10.5. *Let $\mathscr{C}$ be a s.n.-category. Any irreducible subobject of an object X of $\mathscr{C}$ is π-irreducible. Also an element $\mathscr{P} \in \mathrm{Spel}(\mathscr{C})$ is associated with an object X if and only if the injective envelope of X contains an injective indecomposable object isomorphic with $Q_{\mathfrak{P}}$.* ■

Now we shall make a few remarks on the primary decomposition theory in commutative rings.

An ideal $\mathfrak{a}$ of a commutative ring A is called *classical primary* if for every two elements $x, y \in A$ such that $xy \in \mathfrak{a}$ and $x \notin \mathfrak{a}$ it follows that $y^n \in \mathfrak{a}$ for a suitable natural number n. It is easy to see that the radical $\sqrt{\mathfrak{a}}$ of a classical primary ideal is a prime ideal.

THEOREM 10.6. *Any classical primary ideal $\mathfrak{a}$ of a commutative semi-noetherian ring A is primary, more exactly $\mathfrak{p}$-primary, where $\mathfrak{p} = \sqrt{\mathfrak{a}}$.*

Proof. First we shall prove that $F_{\mathfrak{a}} = F_{\mathfrak{p}}$. Indeed, if $\mathfrak{b} \notin F_{\mathfrak{a}}$ then an element $x \in A$ can be found such that $\mathfrak{b}x \subseteq \mathfrak{a}$ and $x \notin \mathfrak{a}$. Then by hypothesis,

$$\mathfrak{b} \subseteq \mathfrak{p} = \sqrt{\mathfrak{a}}$$

so that $\mathfrak{b} \notin F_{\mathfrak{p}}$. Hence $F_{\mathfrak{p}} \subseteq F_{\mathfrak{a}}$. By Theorem 5.13 $F_{\mathfrak{a}} = \bigcap_i F_{\mathfrak{p}_i}$ so that

$$F_{\mathfrak{p}} \subseteq F_{\mathfrak{p}_i}$$

for every i, or equivalently $\mathfrak{p} \supseteq \mathfrak{p}_i$. But the relation $F_{\mathfrak{a}} \subseteq F_{\mathfrak{p}_i}$ shows that $\mathfrak{a} \subseteq \mathfrak{p}_i$, so that $\sqrt{\mathfrak{a}} \subseteq \mathfrak{p}_i$, in other words $\mathfrak{p} \subseteq \mathfrak{p}_i$. Thus we find that $\mathfrak{p} = \mathfrak{p}_i$ for every i.

Now let X be a non-zero submodule of $A/\mathfrak{a}$; then $F(X) \supseteq F_{\mathfrak{a}} = F_{\mathfrak{p}}$. By Theorem 5.13, $F(X) = \bigcap_i F_{\mathfrak{p}_i}$. From Exercise 12 of. Section 5.5, we can assume that $\mathfrak{p}_i \in \mathrm{Ass}(X)$ for every i, so that $\mathfrak{p}_i \in \mathrm{Ass}(A/\mathfrak{a})$ or equivalently

$\mathfrak{p}_i = \mathrm{Ann}(x)$ for a suitable element $x \in A/\mathfrak{a}$. But $x = \bar{y}$ with $y \in A$ and so, $\mathfrak{p}_i = (\mathfrak{a} : y) \supseteq \mathfrak{a}$. Hence necessarily $\mathfrak{p} \subseteq \mathfrak{p}_i$. If $\mathfrak{p} \neq \mathfrak{p}_i$, then $\mathfrak{p}_i \in F_{\mathfrak{p}} = F_{\mathfrak{a}}$, i.e. $\mathrm{Hom}_A(A/\mathfrak{p}_i, E(A/\mathfrak{a})) = 0$, a contradiction with the hypothesis. Hence, necessarily $\mathfrak{p}_i = \mathfrak{p}$ for any i, so that $F(X) = F_{\mathfrak{p}}$. In conclusion, $\mathfrak{a}$ is primary, namely $\sqrt{\mathfrak{a}} = \mathfrak{p}$-primary. ■

The following result is in some sense a converse of the previous theorem.

THEOREM 10.7. *Let $\mathfrak{p}$ be a prime ideal of a commutative semi-noetherian ring. The following assertions are equivalent*:

(1) *The localizing subcategory $\mathscr{F}_{V(\mathfrak{p})}$ (the smallest localizing subcategory containing $A/\mathfrak{p}$) contains any $\mathfrak{p}$-isotopic module,*

(2) *Any $\mathfrak{p}$-primary ideal $\mathfrak{a}$ is classical primary.*

Proof. (1) ⇒ (2). By hypothesis, $\mathscr{F}_{V(\mathfrak{p})} = \mathscr{F}$ contains any $\mathfrak{p}$-isotypic module. Now let $\mathfrak{a}$ be a $\mathfrak{p}$-primary ideal. First we shall prove that $\sqrt{\mathfrak{a}} = \mathfrak{p}$. In fact, let $\mathfrak{q}$ a prime ideal containing $\mathfrak{a}$. If $\mathfrak{p}$ is not contained in $\mathfrak{q}$, then $\mathfrak{p} \in F_{\mathfrak{q}}$, so that $A/\mathfrak{p} \in \mathrm{Ob}\,\mathscr{F}_{\mathfrak{q}}$. But then $\mathscr{F}_{V(\mathfrak{p})}$, being the smallest localizing subcategory which contains $A/\mathfrak{p}$ is necessarily included in $\mathscr{F}_{\mathfrak{q}}$. Therefore $A/\mathfrak{a} \in \mathrm{Ob}\,\mathscr{F}_{\mathfrak{q}}$, or equivalently $\mathrm{Hom}_A(A/\mathfrak{a}, E(A/\mathfrak{q})) = 0$, a contradiction. Hence, necessarily $\sqrt{\mathfrak{a}} = \mathfrak{p}$.

Now let $x, y \in A$ such that $xy \in \mathfrak{a}$ and $x \notin \mathfrak{a}$. If $y \notin \mathfrak{p}$, then $Ay \in F_{\mathfrak{p}}$ and the relation $Ayx \subseteq \mathfrak{a}$ proves that $x \in \mathfrak{a}$, a contradiction. Finally, we see that $\mathfrak{a}$ is classical primary.

(2) ⇒ (1). It will be enough to prove that any $\mathfrak{p}$ isotypic injective object Q is contained in $\mathscr{F}_{V(\mathfrak{p})}$, or equivalently $\mathrm{Supp}(Q) \subseteq V(\mathfrak{p})$ (see Proposition 5.17). Indeed, let $x \in Q$ and $\mathfrak{q}$ a prime ideal such that $\mathrm{Ann}(x) \subseteq \mathfrak{q}$. Now it is clear that Ax is also $\mathfrak{p}$-isotypic, or equivalently $\mathrm{Ann}(x)$ is $\mathfrak{p}$-primary, hence $\sqrt{\mathrm{Ann}(x)} = \mathfrak{p}$. Finally one has $\mathfrak{q} \supseteq \mathfrak{p}$, or equivalently $\mathrm{Supp}(Q) \subseteq V(\mathfrak{p})$, hence $Q \in \mathrm{Ob}\,\mathscr{F}_{V(\mathfrak{p})}$. ■

THEOREM 10.8. *Let A be a commutative noetherian ring and $\mathfrak{a}$ an ideal of A. The following assertions are equivalent*:

(1) $\mathfrak{a}$ *is $\mathfrak{p}$-primary*;

(2) $\mathfrak{a}$ *is classical primary and* $\sqrt{\mathfrak{a}} = \mathfrak{p}$.

The result is a consequence of the Theorem 9.9 and 10.7. ■

NOTE 2. In ([76], Theorem 6.1) there is a proof of Theorem 10.8 using the classical decomposition theory for noetherian rings. Hence we obtain another proof of Theorem 9.9 using the primary decomposition theory.

THEOREM 10.9. *Let* $\mathfrak{a}$ *be an ideal of a commutative semi-artinian ring* A. *The following assertions are equivalent*:

(1) $\mathfrak{a}$ *is* $\mathfrak{p}$-*primary*;

(2) $\mathfrak{a}$ *is classical primary*;

(3) $F_{\mathfrak{a}} = F$ *where* $\mathfrak{p} \in \operatorname{Spec}(A)$,

Proof. The implications (1) $\Leftrightarrow$ (2) and (2) $\Rightarrow$ (3) are direct consequences of Theorem 10.7, Corollary 5.20 and the definition of a primary ideal.

(3) $\Rightarrow$ (1). We claim that $A/\mathfrak{a}$ is isotypic. In fact, if $\mathfrak{p}' \in \operatorname{Ass}(A/\mathfrak{a})$, then $\mathfrak{p}' = \{\mathfrak{a} : x\}$, hence $\mathfrak{p}' \notin F_{\mathfrak{a}} = F_{\mathfrak{p}}$; in other words $\mathfrak{p}' \subseteq \mathfrak{p}$, or equivalently $\mathfrak{p}' = \mathfrak{p}$. ■

NOTE 3. The previous theorem is also valid for a semi-noetherian integral domain A, such that $KG(A) = 1$.

Exercises

1. Let X be an essential extension of a direct sum $\coprod_i X_i$. Then

$$\mathscr{F}(X) = \bigcap_i \mathscr{F}(X_i).$$

2. Prove that a non-zero subobject of a $\mathscr{P}$-isotypic object is also $\mathscr{P}$-isotypic. (Hint: use Corollary 3.2).

3. Let X', X'' be $\mathscr{P}$-primary subobjects of X. Then $X' \cap X''$ is also $\mathscr{P}$-primary.

4. Let $\mathscr{C}$ be a l.n.-category and Γ, $\Gamma' : \mathscr{C} \to M$ be two predecomposition theories and suppose that Γ is stronger than Γ'. Then if Γ is a decomposition theory, so is Γ', and moreover $\Gamma = \Gamma'$.

5. Prove that all results of this paragraph valid for left s.n.-rings are also valid for each left l.h.-ring. Moreover a commutative ring A is a l.h.-ring f and only if $\operatorname{Ass}(X) \neq \varnothing$ for each non-zero A-module.

References

Bourbaki [26]; Gabriel [59]; Goldman [76]; Hudry [95]; Michler [130]; Năstăsescu [139]; Nguen Trong Kham [145]; Popescu [168], [170], [172], [173], [174].

5.11 Decomposition theories on l.n.-categories

THEOREM 11.1. *Let $\mathscr{C}$ be a l.n.-category and $\Gamma : \mathscr{C} \to M$ a predecomposition theory. Then Γ is a decomposition theory if and only if the null subobject of any noetherian object has a Γ-decomposition.*

Proof. Assume that the null subobject of a noetherian object X has a Γ-decomposition. Then we claim that Γ satisfies condition (P) (see Section 5.4). In fact, let X be any object of $\mathscr{C}$; then the set $\{X_i\}_i$ of its noetherian subobjects is directed and complete, hence $\Gamma(X) = \bigcup_i \Gamma(X_i)$. Now if $m \in \Gamma(X)$, then $m \in \Gamma(X_i)$ for a suitable i, so that $\{m\} = \Gamma(X')$, where X' is a suitable subobject of X_i. Therefore the condition (P) holds, as claimed. ∎

Since every l.n.-category is a s.n.-category, all results of Section 5.10 about s.n.-categories can be applied to l.n.-categories. We shall examine some special decomposition theories on Mod A, where A is a left noetherian ring.

Throughout A will be a left noetherian ring. By Spep(A) we shall denote the set of all prime two-sided ideals of A. Let X be an A-module; then the set $\{\text{Ann}(X')\}$ where X' runs through all non-zero submodules of X has a maximal element Ann(X'); thus, for any non-zero submodule X'' of X' one has $\text{Ann}(X') = \text{Ann}(X'')$ and this last ideal is a prime ideal called *ter-associated* with X; generally the set of all elements of Spep(A) which are ter-associated with X is denoted by Ter(X). It is easy to see that for an indecomposable injective Q, the set Ter(Q) contains exactly one element. In this way we have defined a map:

$$\text{Ter} : \text{Sp}(\text{Mod } A) \to \text{Spep}(A).$$

LEMMA 11.2. *There exists a map $s : \text{Spep}(A) \to \text{Sp}(\text{Mod } A)$ such that* Ter *composed with s is the identity.*

Proof. Indeed, by Lemma 9.14, for every $\mathfrak{p} \in \text{Spep}(A)$, $E(A/\mathfrak{p})$ is a finite direct sum of isomorphic indecomposable injectives. An indecomposable injective contained in $E(A/\mathfrak{p})$ is denoted by $Q_{\mathfrak{p}}$. Now it is easy to see that $s(\mathfrak{p}) = Q_{\mathfrak{p}}$ and $\text{Ter}(Q_{\mathfrak{p}}) = \mathfrak{p}$. ∎

The sets $\mathfrak{p}$ and $Q_{\mathfrak{p}}$ are related by the following important result:

THEOREM 11.3. *Let $\mathscr{F}$ be a localizing subcategory of* Mod A *and $\mathfrak{p}$ a prime ideal of A. The following assertions are equivalent:*

(1) $A/\mathfrak{p} \in \text{Ob}\,\mathscr{F}$;

(2) $Q_{\mathfrak{p}} \in \text{Sp}(\mathscr{F})$.

Proof. The assertion (2) means that $Q_{\mathfrak{p}}$ contains a non-zero submodule of $\mathscr{F}$ (see Section 5.3). Thus the implication (1) ⇒ (2) is obvious. Now we shall prove that (2) ⇒ (1). Since $Q_{\mathfrak{p}}$ contains non-zero submodules of $\mathscr{F}$, one checks that $A/\mathfrak{p}$ contains a non-zero submodule X of $\mathscr{F}$ and

$$X \subseteq (A/\mathfrak{p}) \cap Q_{\mathfrak{p}}.$$

In the first place we may assume that $\mathrm{Ann}(X) = \{\mathfrak{p}\}$.

Let Y be a subset of X; we denote:

$l(Y) = \{a \in A/\mathfrak{p} \mid aY = 0\}$;

$\bar{l}(Y) = \{b \in Q(A/\mathfrak{p}) \mid bY = 0\}$.

Since $Q(A/\mathfrak{p})$ is a simple ring and $A/\mathfrak{p}$ is a left order of this ring (see Chapter 4, Theorem 19.5) and $l(Y) = \bar{l}(Y) \cap (A/\mathfrak{p})$, we check that the set of all left ideals $\{l(Y)\}$ when Y runs through all finite subsets of X has descending chain condition. Because X is the union of its finite subsets, then $l(X) = l(Y)$, where $Y = \{x_1, \ldots, x_n\} \subseteq X$. Hence:

$$l(X) = 0 = \bigcap_{1 \leqslant i \leqslant n} l(x_i)$$

so that

$$\mathrm{Ann}(X) = \mathfrak{p} = \bigcap_{1 \leqslant i \leqslant n}^{n} \mathrm{Ann}(x_i).$$

Finally we see that $A/\mathfrak{p}$ is a submodule of

$$\prod_{i=1}^{n} A/\mathrm{Ann}(x_i) \simeq \textstyle\prod_i Ax_i,$$

i.e. an object of $\mathscr{F}$. ■

The decomposition theory associated with the mapping $\mathrm{Ter} : \mathrm{Sp}(\mathrm{Mod}\, A) \to \mathrm{Spep}(A)$ is called the *tertiary decomposition theory* and is denoted by

$$\mathrm{Ter} : \mathrm{Mod}\, A \to \mathrm{Spep}(A).$$

Usually a Ter-coirreducible module is called *cotertiary* and any Ter-irreducible submodule of a module is called *tertiary*.

We shall say that a left noetherian ring A *has enough prime ideals* if the mapping $\mathrm{Ter} : \mathrm{Sp}(\mathrm{Mod}\, A) \to \mathrm{Spep}(A)$ is a bijection.

THEOREM 11.4. *The following assertions are equivalent for a left noetherian ring A:*

(1) *A has enough prime ideals;*

(2) *Any finitely generated cotertiary module is coprimary;*

(3) *Any cotertiary module is coprimary.*

Proof. (1) ⇒ (3). Let X be a cotertiary module and $\coprod_i Q_i = E(X)$, where Q_i are indecomposable injectives. Then for every i, j one has $\mathrm{Ter}(Q_i) = \mathrm{Ter}(Q_j)$, so that $Q_i \simeq Q_j$. Hence $E(X)$ is isotypic, or equivalently X is coprimary (see Theorem 10.1).

(2) ⇒ (1). Let Q, Q' be two indecomposable injectives so that $\mathrm{Ter}(Q) = \mathrm{Ter}(Q')$. Since any cotertiary coirreducible module is obviously coprimary, we find that Q (respectively, Q') contains a finitely generated submodule X (respectively, X'), which are cotertiary and $\mathrm{Ter}(X) = \mathrm{Ter}(X')$. Then $X \amalg X'$ is also cotertiary, hence coprimary. Finally $E(X \amalg X') = Q \amalg Q'$ is isotypic or equivalently $Q \simeq Q'$.■

Now let M be a non-empty set of prime ideals of A. Denote by F_M the set of all left ideals $\mathfrak{a}$ which contains a product $\mathfrak{p}_1 \dots \mathfrak{p}_n$ of elements of M. A slight computation proves that F_M is a localizing system on A. A noetherian object X belongs to F_M if and only if $\mathrm{Ann}(X)$ contains a product $\mathfrak{p}_1 \dots \mathfrak{p}_n$ of elements of M.

If $\mathfrak{p}$ is a prime ideal of A we shall denote by $C(\mathfrak{p})$ the set of all prime ideals of $F_{\mathfrak{p}}$.

LEMMA 11.5. *Let $\mathfrak{p}$ be a prime ideal. Then an ideal $\mathfrak{a}$ of A belongs to $F_{\mathfrak{p}}$ if and only if the relation $\mathfrak{a}x \subseteq \mathfrak{p}$ implies $x \in \mathfrak{p}$. Moreover a prime ideal $\mathfrak{q}$ belongs to $C(\mathfrak{p})$ if and only if $\mathfrak{q}$ contains an element of the complement of $\mathfrak{p}$.*

Proof. Let $\mathfrak{a} \in F_{\mathfrak{p}}$ and $x \in A$ such that $\mathfrak{a}x \subseteq \mathfrak{p}$; if $x \notin \mathfrak{p}$, then an element $y \in A$ can be found such that $yx \notin \mathfrak{p}$ and $y \in \mathfrak{a}$, a contradiction (see Chapter 4, Section 4.9 for the definition of $F_{\mathfrak{p}}$).

Conversely, assume that $\mathfrak{a}$ is an ideal such that the condition $\mathfrak{a}x \subseteq \mathfrak{p}$ implies $x \in \mathfrak{p}$. Let $x_1 \notin \mathfrak{p}$ and $x_2 \notin \mathfrak{a}$ such that $\mathfrak{a} \subseteq (\mathfrak{a} : x_2) \subseteq (\mathfrak{p} : x_1)$; then $\mathfrak{a}x_1 \subseteq \mathfrak{p}$ hence $x_1 \in \mathfrak{p}$, a contradiction. Therefore, necessarily $\mathfrak{a} \in F_{\mathfrak{p}}$.

Finally, let $\mathfrak{q} \in \mathrm{Spep}(A)$ and $s \in \mathfrak{q} \cap (A - \mathfrak{p})$; also let $x \in A$ such that $\mathfrak{q}x \subseteq \mathfrak{p}$. If $x \notin \mathfrak{p}$ then from relation $sAx \subseteq \mathfrak{p}$ it follows that $s \in \mathfrak{p}$, a contradiction. Hence by the first part of the Lemma we have that $\mathfrak{q} \in F_{\mathfrak{p}}$.■

NOTE 1. We observe that the fact that A is left noetherian ring is not needed here.

LEMMA 11.6. *Let Q be an indecomposable injective object and $\mathfrak{p} = \mathrm{Ter}(Q)$. Then $F_{\mathfrak{p}} \supseteq F(Q)$. Equality holds if and only if $A/\mathfrak{p}$ is Q-isotypic.*

Proof. Let $x \in Q$ be such that $F_{\mathrm{Ann}(x)} = F(Q)$. It is clear that $\mathrm{Ann}(x)$ is a left super-prime ideal and we can choose x such that $\mathrm{Ann}(Ax) = \mathfrak{p}$. Hence $\mathfrak{p} \notin F(Q)$. Thus, by Theorem 11.3 we see that $Q_{\mathfrak{p}}$ is $F(Q)$-closed so that $F_{\mathfrak{p}} = F(Q_{\mathfrak{p}}) \supseteq F(Q)$.

Assume that $F_{\mathfrak{p}} = F(Q)$. Then $Q_{\mathfrak{p}}$ is an indecomposable injective of definition of $F(Q)$, so that $Q_{\mathfrak{p}} \simeq Q$ by Theorem 20.8 of Chapter 4. The converse is obvious. ■

PROPOSITION 11.7. *The following assertions are equivalent for a ring A:*

(1) *A has enough prime ideals;*

(2) *The mapping $\mathfrak{p} \rightsquigarrow F_{\mathfrak{p}}$ establishes a bijection between* Spep(A) *and* Spel(A);

(3) *For any indecomposable injective Q one has $F(Q) = F_{\mathrm{Ter}(Q)}$.*

It suffices to apply Lemmas 9.14, 10.2 and 11.6. ■

LEMMA 11.8. *For any indecomposable injective Q one has*

$$V(F(Q)) = V(F_{\mathrm{Ter}(Q)}),$$

(where $V(F(Q))$ is the set of all prime ideals contained in $F(Q)$ and $V(F_{\mathrm{Ter}(Q)})$ the set of all prime ideals which contain $F_{\mathrm{Ter}(Q)}$).

Proof. The inclusion $F(V(Q) \subseteq V(F_{\mathrm{Ter}(Q)})$ holds by Lemma 11.6. Now let $\mathfrak{q} \in F_{\mathrm{Ter}(Q)}$; then by Lemma 11.5 it follows that $\mathfrak{q} \subseteq \mathrm{Ter}(Q)$. We claim that $\mathfrak{q} \in V(F(Q))$. Indeed, let $\mathfrak{a}$ be a super-prime left ideal of A such that $F_{\mathfrak{a}} = F(Q)$ (see Lemma 10.2). If $\mathfrak{q} \notin F_{\mathfrak{a}}$, then there are $x_1, x_2 \in A$ such that $x_1 \notin \mathfrak{q}$ $x_2 \notin \mathfrak{a}$ and $\mathfrak{q} \subseteq (\mathfrak{q} : x_1) \subseteq (\mathfrak{a} : x_2) = \mathfrak{a}'$. Now $\mathfrak{a}'$ is also a left super-prime ideal and $F_{\mathfrak{a}'} = F(Q)$. Furthermore, we see that $\mathfrak{q} \subseteq \mathrm{Ann}(X)$, for any non-zero submodule X of $A/\mathfrak{a}' \subseteq Q$. But for a suitable non-zero submodule X of $A/\mathfrak{a}'$, one has $\mathrm{Ter}(Q) = \mathrm{Ann}(X)$ and finally $\mathfrak{q} \subseteq \mathrm{Ter}(Q)$ a contradiction. Hence, necessarily $\mathfrak{q} \in F_{\mathfrak{a}}$, as claimed. ■

THEOREM 11.9. *The following assertions are equivalent for a (left noetherian) ring A:*

(1) *A has enough prime ideals;*

(2) *For any $F \in L(A)$ one has $F = F_{V(F)}$.*

Proof. (1) ⇒ (2). It is clear that $F_{V(F)} \subseteq F$, because the product of any two left ideals of a localizing system F is also an element of F. Conversely, let $\mathfrak{a} \in F$, and $X = A/\mathfrak{a}$. Also let $\mathfrak{p} \in \mathrm{Ter}(X)$; by the definition of $\mathrm{Ter}(X)$, there exists a non-zero submodule X_1 of X such that $\mathrm{Ann}(X_1) = \mathfrak{p}_1$ is a prime ideal. Then we see that $Q_{\mathfrak{p}_1} \in \mathrm{Sp}(\mathscr{F})$ by hypothesis, so that $\mathfrak{p}_1 \in F$ according to Theorem 11.3. Now for any natural number $n > 1$ we define a submodule X_n of X such that $X_{n-1} \subseteq X_n$ and $\mathrm{Ter}(X_n/X_{n-1}) = \mathfrak{p}_n$. Again we check that $\mathfrak{p}_n \in F$. Since A is left noetherian we see that $X_n = X$ for a suitable n. Then $(\mathfrak{p}_1 \mathfrak{p}_2 \ldots \mathfrak{p}_n) X = 0$ so that $\mathfrak{p}_1 \mathfrak{p}_2 \ldots \mathfrak{p}_n \subseteq \mathfrak{a}$, i.e., $\mathfrak{a} \in F_{V(F)}$.

(2) ⇒ (1). By Lemma 11.8 we find that $F(Q) = F_{\mathrm{Ter}(Q)}$ for any indecomposable injective Q. Thus the proof will result by Proposition 11.7. ■

NOTE 2. If A has enough prime ideals, then for any $\mathfrak{p}, \mathfrak{p}' \in \mathrm{Spep}(A)$ one has $F_{\mathfrak{p}} \subseteq F_{\mathfrak{p}'}$ if and only if $\mathfrak{p}' \subseteq \mathfrak{p}$. Particularly, if $\mathfrak{p}$ is a maximal prime ideal, then $F_{\mathfrak{p}}$ is minimal in $\mathrm{Spel}(A)$.

Now we shall give some manageable conditions for a left noetherian ring to have enough prime ideals. First we consider the following condition (G) given by Gabriel [59].

(G) For any left ideal $\mathfrak{a}$ such that $\mathrm{Ann}(A/\mathfrak{a}) = \mathfrak{p}$ is a prime ideal, there exist a finite number of elements $x_1, \ldots, x_r \in A/\mathfrak{a}$ such that:

$$\mathfrak{p} = \bigcap_{i=1}^{r} \mathrm{Ann}(x_i).$$

LEMMA 11.10. *If condition* (G) *is satisfied, then* A *has enough prime ideals.*
Proof. Let Q be an indecomposable injective, $\mathfrak{p} = \mathrm{Ter}(Q)$ and $x \in Q$ an element such that $\mathfrak{p} = \mathrm{Ann}(A/\mathrm{Ann}(x))$. Then

$$\mathfrak{p} = \bigcap_{i=1}^{r} \mathrm{Ann}(x_i),$$

with $x_i \in A/\mathrm{Ann}(x)$. But then $A/\mathfrak{p}$ is contained as submodule of

$$\prod_{i=1}^{r} (A/\mathrm{Ann}(x_i)).$$

Since $A/\mathrm{Ann}(x_i)$ is coirreducible for any i, we see that $Q_{\mathfrak{p}} \simeq Q$. Finally, if Q' is an another indecomposable injective such that $\mathrm{Ter}(Q') = \mathfrak{p}$, then

$$Q' \simeq Q_{\mathfrak{p}} \simeq Q. \blacksquare$$

If α is an ordinal number, then we shall denote by E_α the set of all prime ideals $\mathfrak{p}$ such that $KG(A/\mathfrak{p}) \leqslant \alpha$. Then we have the following result:

THEOREM 11.11. *Let* A *be a (left noetherian) ring having enough prime ideals. Using the above results, the sets* E_α *can be defined as follows:*

E_0 *is the set of maximal prime ideals of* A.

If $\alpha = \beta + 1$, *then* E_α *is consisting of all prime ideals* $\mathfrak{p}$ *such that for any prime ideal* $\mathfrak{q}$ *such that* $\mathfrak{p} \subset \mathfrak{q}$, *it follows that* $\mathfrak{q} \in E_\beta$.

If α *is a limit ordinal, then* $E_\alpha = \bigcup_{\beta < \alpha} E_\beta$.

Moreover $\mathrm{Spep}(A) = \bigcup_\alpha E_\alpha$ *and the smallest ordinal* α *such that*

$$E_\alpha = \mathrm{Spep}(A)$$

is exactly $KG(\mathrm{Mod}\, A)$.

Proof. If $KG(A/\mathfrak{p}) = 0$, then for every prime ideal $\mathfrak{q}$ such that $\mathfrak{p} \subseteq \mathfrak{q}$ one checks that $KG(A/\mathfrak{q}) = 0$. Then $F_{\mathfrak{p}}$ is a maximal left prime so that we can find a maximal ideal $\mathfrak{m}$ such that $F_{\mathfrak{p}} = F_{\mathfrak{m}}$ and $\mathfrak{m} = (\mathfrak{p} : x)$, $x \in A$. Now it is clear that $\mathfrak{m} \subseteq (\mathfrak{q} : x)$. If $\mathfrak{q} \neq \mathfrak{p}$ then necessarily $\mathfrak{q} \in F_{\mathfrak{m}}$, so that $\mathfrak{m} \neq (\mathfrak{q} : x)$; in other words $(\mathfrak{q} : x) = A$. But then $\mathfrak{q} = A$. Therefore $\mathfrak{p}$ is necessarily maximal.

Conversely, assume that $\mathfrak{p}$ is maximal. Then $F_{\mathfrak{p}}$ is necessarily minimal in Spel(A) so that $F_{\mathfrak{p}} = F_{\mathfrak{m}}$ where $\mathfrak{m}$ is a maximal left ideal. Now, it is clear that $A/\mathfrak{p}$ contains simple modules and for every left ideal $\mathfrak{a}$ which contains $\mathfrak{p}$, one checks also that $A/\mathfrak{a}$ contains simple modules (since $\mathfrak{m} = (\mathfrak{p} : x)$ and thus $(\mathfrak{a} : x) = \mathfrak{m}$). Hence we deduce that the height of $A/\mathfrak{p}$ is defined and thus it is a finite object, being noetherian. Finally, $KG(A/\mathfrak{p}) = 0$.

Now, let $\alpha = \beta + 1$. We can reduce the proof to the case $\beta = 0$ and leave to the reader to supply the details. The remaining assertions are also easy and are again left to the reader. ■

Before passing to the next part of this paragraph, we shall make some further observations.

Let $\mathscr{F}$ be a prelocalizing subcategory of Mod A, where A is any ring. For every A-module X, the submodules X' of X such that $X/X' \in \mathrm{Ob}\,\mathscr{F}$ define a fundamental system of neighbourhoods for a topology $T_F(X)$ on X. Relative to this topology X becomes a topological group. On the other hand the left ideals of a prelocalizing system F define on A a structure of a topological ring. In this way any module X becomes canonically a topological module. Moreover any morphism $f: X \to Y$ of A-modules, becomes naturally a morphism of topological modules. Particularly, if X' is a submodule of X, then the topology $T_F(X')$ is finer than the restriction of topology $T_F(X)$ to X'. We have the following important result on the relation between these topologies.

Proposition 11.12. *Let A be a ring (not necessarily left noetherian) and F a prelocalizing system on A. The following assertions are equivalent:*

(1) *For any A-module X and any submodule X' of X, the restriction to X' of the topology $T_F(X)$ coincides with the topology $T_F(X')$;*

(2) *For any A-module $X \in \mathrm{Ob}\,\mathscr{F}$, $E(X)$ is also an object of $\mathscr{F}$.*

Proof. (1) ⇒ (2). First we observe that $X \in \mathrm{Ob}\,\mathscr{F}$ if and only if $T_F(X)$ is the discrete topology. Now the topology $T_F(E(X))$ induces on X the discrete topology i.e. there exists an open submodule Y of $E(X)$ such that $Y \cap X = 0$. But then $Y = 0$, hence $T_F(E(X))$ is the discrete topology, or equivalently $E(X) \in \mathrm{Ob}\,\mathscr{F}$.

(2) ⇒ (1). Let X'' be a submodule of X' such that $X'/X'' \in \mathrm{Ob}\,\mathscr{F}$. We must find an open submodule Y of X such that $Y \cap X' \subseteq X''$. In fact, we

choose Y maximal with the property $Y \cap X' = X''$. Then X/Y is an essential extension of X'/X'', i.e. an object of $\mathscr{F}$ by hypothesis. ∎

COROLLARY 11.13. *Let F be a prelocalizing system on A such that the conditions of previous proposition are true. Also let X be an A-module and*

$$X'' = \bigcap X'$$

where X' runs over all open submodules of X. Then for every proper submodule X_1'' of X'', X''/X_1'' does not belong to $\mathscr{F}$ (or equivalently $T_F(X'')$ is the coarsest topology. ∎

Again A will be a left noetherian ring. Let $\mathfrak{a}$ be an ideal in A. We will say that A has the *Artin–Rees property* for $\mathfrak{a}$ if given any finitely generated A-module X, any submodule X' of X and a non-negative integer n, there exists $h(n)$ such that $\mathfrak{a}^{h(n)} X \cap X' \subseteq \mathfrak{a}^n X'$. Let F be localizing system consisting of all left ideals $\mathfrak{b}$ containing $\mathfrak{a}^n$, for a natural number n. Then by Proposition 11.12 we deduce that A has the Artin–Rees property for $\mathfrak{a}$ if and only if $\mathscr{F}$ is stable. (For the proof we utilize also Proposition 8.6).

By a *classical ring* we shall mean a left noetherian ring A having the Artin–Rees-property for every ideal $\mathfrak{a}$ of A. By Proposition 11.12 it follows that A is a classical ring if any localizing subcategory of Mod A is stable. Particularly, any commutative (noetherian) ring is a classical ring. We shall prove that the converse is true, viz that if A is a classical ring, then all localizing subcategories of Mod A are stable. First, we say that a module X is *negligible* if it is Ter-negligible, meaning that each element of X is annihilated by a product of primes belonging to Ter(X), i.e. $X \in \mathrm{Ob}\,\mathscr{F}_{\mathrm{Ter}\,X)}$.

The following result is an important completion of Theorem 9.9.

THEOREM 11.14 (Riley [187]). *Let A be a left noetherian ring. The following assertions are equivalent*:

(1) *A is a classical ring*;

(2) *Indecomposable injectives A-modules are negligible*;

(3) *All A-modules are negligible*;

(4) *For each ideal $\mathfrak{a}$ in A, the localizing subcategory $\mathscr{F}_{V(\mathfrak{a})}$ is stable.*

Proof. (1) ⇒ (2). Let Q be an indecomposable injective and $y \in Q$. Also let $\mathfrak{p} = \mathrm{Ter}(Q)$ and X a non-zero noetherian submodule of Q such that

$$\mathrm{Ann}(X) = \mathfrak{p}.$$

Then $Y = X + Ay$ is also noetherian and by hypothesis, there is a natural number $h(1)$ such that

$$\mathfrak{p}^{h(1)} Y \cap X \subseteq \mathfrak{p}X = 0.$$

Thus, necessarily $\mathfrak{p}^{h(1)} Y = 0$, because Q is coirreducible.

(2) $\Rightarrow$ (3). According to Proposition 8.6 will be sufficient to check that any noetherian module is negligible. In fact, let X be noetherian and

$$E(X) = \coprod_{i=1}^{r} Q_i.$$

Then

$$\mathrm{Ter}(X) = \bigcup_{i=1}^{r} \mathrm{Ter}(Q_i).$$

Also let $\mathfrak{p}_i = \mathrm{Ter}(Q_i)$ and $x = x_1 + \ldots + x_n$ an element of X. Then $\mathfrak{p}_i^{n_i} x_i = 0$ for some n_i, each $i = 1, \ldots, r$. Then $(\mathfrak{p}_1^{n_1} \ldots \mathfrak{p}_r^{n_r})\, x = 0$ and X is negligible.

(3) $\Rightarrow$ (4). Let $\mathscr{F}$ be a localizing subcategory of Mod A and Q an indecomposable injective such that $Q_{\mathscr{F}} \neq 0$, that is $Q \in \mathrm{Sp}(\mathscr{F})$. Then, by Exercise 3 and Theorem 11.3 one checks that $\mathfrak{p} = \mathrm{Ter}(Q)$ is an element of F, hence for any $x \in Q$, $\mathrm{Ann}(x) \supseteq \mathfrak{p}^n$, so that $Q \in \mathrm{Ob}\, \mathscr{F}$. Now, the proof follows since any injective is a direct sum of indecomposable injectives.

The implication (4) $\Rightarrow$ (1) follows by Proposition 11.12. ∎

COROLLARY 11.15. *A left artinian ring is a direct product of local left artinian rings if and only if A is a classical ring.*

The proof follows by Theorems 11.14 and 6.9. ∎

Exercises

1. Let $\Gamma : \mathscr{C} \to M$ be a decomposition theory, where $\mathscr{C}$ is a Grothendieck category. If X is an essential extension of a finite direct sum of coirreducible objects, the $\Gamma(X)$ is a finite set. This is the case for a noetherian (respectively, artinian or finite) object.

2. Any coprimary module is cotertiary; any primary submodule is tertiary. Moreover, a prime ideal $\mathfrak{a}$ is an element of $\mathrm{Ter}(X)$ is and only if X has a submodule X' such that $\mathfrak{a} = \mathrm{Ann}(X'')$ for every non-zero submodule X'' of X'.

3. If all indecomposable injectives A-modules are negligible, then A has enough prime ideals.

4. Any left artinian ring A satisfies the condition (G).

5. If any left ideal of A is two-sided, then A satisfies the condition (G).

6. Let A be a left noetherian ring such that $Z(A)$ is noetherian and A is a noetherian $Z(A)$-module. Then A satisfies the condition (G).

7. Let A be a left noetherian ring having enough prime ideals, $\mathfrak{p} \in \mathrm{Spep}(A)$ and n a finite ordinal. The following assertions are equivalent:

(a) $KG(A/\mathfrak{p}) \leqslant n$;

(b) Any chain of prime ideals

$$\mathfrak{p} \subset \mathfrak{p}_1 \subset \mathfrak{p}_2 \subset \ldots$$

contains no more than $n + 1$ elements.

8. Let k be a field of characteristic zero and $k(X)$ the fields of rational functions in X over k. Also let $d: k(X) \to k(X)$ be the ordinary derivative (i.e. $d(P(X)) = P'(X)$). Finally, let A be the subring of $\mathrm{End}_{\mathbf{Z}}(k(X))$ generated by d and all homotheties of $k(X)$. Then A^0 is a left noetherian ring having not enough prime ideals.

9. Let A be a commutative noetherian ring and X a finitely generated A-module. Then for each natural number n, there exists a natural number $h(n)$ such that:

$$\mathfrak{a}^{h(n)} X \cap X' = \mathfrak{a}^n X'.$$

10. Let $\mathscr{C}$ be l.n.-category and $\Gamma, \Gamma' : \mathscr{C} \to M$ two predecomposition theories, such that $\Gamma(X) = \Gamma'(X)$ for each noetherian object. Then $\Gamma = \Gamma'$. In particularly, any predecomposition theory $\Gamma: \mathscr{N}(\mathscr{C}) \to M$ can be uniquely extended to a predecomposition theory $\Gamma': \mathscr{C} \to M$. Moreover, Γ is a decomposition theory if and only if Γ' is a decomposition theory.

11. Let A be a left noetherian ring. For each noetherian A-module X, denote by $\mathrm{Supp}(X)$ the set of prime ideals of A which contains $\mathrm{Ann}(X)$. Then the assignment: $X \rightsquigarrow \mathrm{Supp}(X)$ defines a predecomposition theory $\mathrm{Supp}: \mathscr{N}(\mathrm{Mod}\, A) \to \mathrm{Spep}(A)$. If Supp is a decomposition theory, then it will coincide with Ter. Moreover, the following assertions are equivalent:

(a) Supp is a decomposition theory on $\mathscr{N}(\mathrm{Mod}\, A)$;

(b) The prime ideals in A are maximal ideals and the primes containing $\mathrm{Ann}(X)$ are annihilating ideals for X for all objects X in $\mathscr{N}(\mathrm{Mod}\, A)$;

(c) A is a classical ring in which the prime ideals are maximal;

(d) A is a finite direct product of classical rings each of which has a single prime ideal.

12. An artinian ring A is a direct product of local rings if and only if the primes containing $\mathrm{Ann}(X)$ are annihilating ideals for X for all finitely generated A-modules X.

13. Let A be a classical ring and $\mathfrak{a}$ an ideal in A. Let $\Delta = \bigcap_n \mathfrak{a}^n$. Then $\mathfrak{a}\Delta = \Delta$. Thus if Δ is contained in $\mathfrak{R}(A)$, we have $\bigcap_n \mathfrak{a}^n = 0$.

14. Let A be a left noetherian ring in which the non-zero prime ideals are maximal. Then A is a classical ring if and only if for any two prime ideals $\mathfrak{p}$, $\mathfrak{q}$ one has $\mathfrak{p}\mathfrak{q} = \mathfrak{q}\mathfrak{p}$, i.e. the multiplication of prime ideals is commutative.

15. A left noetherian ring A is a classical ring if and only if for any finitely generated module X, the minimal prime containing the annihilator of X belongs to $\mathrm{Ter}(X)$.

16. If A is a left noetherian and classical ring, the multiplication of the maximal ideals is commutative.

17. Let A be a ring and S the set of all left ideals of A. A predecomposition theory $\Gamma : \mathrm{Mod}\, A \to S$ is called *normal* if for each A-module X, the elements of $\Gamma(X)$ are *annihilating* ideals for X (i.e. if $\mathfrak{a} \in \Gamma(X)$, then $\mathfrak{a}X' = 0$ for some non-zero submodule X' of X) and for each $\mathfrak{a} \in \Gamma(X)$ one has: $\mathrm{Ann}(X) \subseteq \mathfrak{a}$. If A is left noetherian then Ter is the unique normal decomposition theory on Mod A.

References

Bourbaki [26]; Bucur [29]; Bucur–Deleanu [31]; Croisot–Lesieur [40]; Gabriel [59]; Michler [130]; Popescu [168], [170]; Riley [186], [187]; Samuel–Zariski [202].

6. Duality

The duality theory for abelian groups is mainly due to Pontryagin [164]; he has shown in fact that the dual category of $\mathcal{A}b$ is the category of compact abelian groups. Pontryagin's result together with the description of Grothendieck categories given in [60] allows the formal description of the dual category of any Grothendieck category.

In his thesis [59], Gabriel has constructed the dual of any locally finite category (see Theorem 4.8). Later on, Roos [198] extended Gabriel's techniques for locally noetherian categories. Last Oberst [155], adapting the Gabriel–Roos techniques succeeded in constructing the dual of any Grothendieck category.

In the first three paragraphs of this chapter (in a slightly different manner) we describe Oberst's results. In Section 6.4 we study the dual of the locally noetherian and locally finite categories. In Section 6.5 we dualize the localization procedure in Grothendieck categories called "colocalization". We note that colocalization is not only the formal dual of "localization"; the reader will see that it represents a technique "in its own right" and with its own power. Originally "colocalization" has been introduced by Roos in [198], but the present general form given here is due to C. Vraciu [234].

6.1 Linearly compact subcategories

Let A be a ring; by a *finitely closed* subcategory of Mod A, we shall mean a full subcategory $\mathscr{S}$, closed under finite direct sums, kernels and cokernels, and which contains any isomorphic copies of its objects. We shall denote by $\hat{\mathscr{S}}$ the following category: an object of $\hat{\mathscr{S}}$ is an A-module X having a topology τ such that (X, τ) is a complete Hausdorff topological group and X has a fundamental system of neighbourhoods of 0 consisting of submodules X' such that $X/X' \in \operatorname{Ob} \mathscr{S}$. Moreover, a morphism $f: X \to Y$ of $\hat{\mathscr{S}}$ is a continuous morphism of A-modules.

Let X be an object of $\hat{\mathscr{S}}$. A submodule X' of X is called *special open* if it is open and $X/X' \in \operatorname{Ob} \mathscr{S}$. A submodule X' of X is called *special closed* if $X' \in \operatorname{Ob} \hat{\mathscr{S}}$ with the induced topology (particularly, X' is complete in the induced topology so that is closed in X).

LEMMA 1.1. *Let M be an object of $\hat{\mathscr{S}}$ and N an open submodule of M. The following assertions are equivalent:*

(a) *N is special open in M;*

(b) *N is special closed in M;*

Moreover, if N is special open in M and P special open in N then P is special open in M.

Proof. (a) $\Rightarrow$ (b). First we see that N' is also closed (since it is open) so that it is complete in the induced topology. Now, for any open submodule U of M, with $U \subseteq N$ there exists a special open submodule $V \subseteq M$ such that $V \subseteq U$, and one has the exact sequence of Mod A:

$$0 \to N/V \to M/V \to M/N \to 0,$$

in which M/V and M/N are objects of $\mathscr{S}$. Since $\mathscr{S}$ is closed under kernels we check that $N/V \in \text{Ob}\,\mathscr{S}$, hence N with induced topology has a fundamental system of neighbourhoods consisting of submodules V with $N/V \in \text{Ob}\,\mathscr{S}$, in other words, $N \in \text{Ob}\,\hat{\mathscr{S}}$.

(b) $\Rightarrow$ (a). Because N is open in M, there exists a special open submodule U in M such that $U \subseteq N$. Now U is open in N, so there exists a special open submodule V of N with $V \subseteq U$. Then one has the exact sequence in Mod A:

$$N/V \to M/U \to M/N \to 0,$$

in which the N/V and M/U are objects of $\mathscr{S}$. Finally, we check that

$$M/N \in \text{Ob}\,\mathscr{S},$$

hence N is special open. ■

LEMMA 1.2. *Let M be an object of $\hat{\mathscr{S}}$, N special closed and P special open in M. Then $N \cap P$ is special open in N and $N + P$ special closed in M. Moreover if N is open in M, then $N \cap P$ is special open in M.*

Proof. Since $N \cap P$ is an open submodule of N, there exists a special open submodule U of N such that $U \subseteq N \cap P$. We have the diagram in *Mod A*, canonically constructed:

$$\begin{array}{l} N/U \\ \downarrow \\ N/(N \cap P) \to M/P \to M/(N+P) \to 0 \\ \downarrow \\ 0 \end{array}$$

with N/U and M/P in $\mathscr{S}$. Then by definition of $\mathscr{S}$, one checks that

$$N/(N \cap P) \quad \text{and} \quad M/(N + P)$$

are also objects of $\mathscr{S}$. The second part of the lemma will follow from Lemma 1.1. ∎

LEMMA 1.3. *Let $f: M_1 \to M_2$ be a morphism of $\hat{\mathscr{S}}$. If N_2 is special open in M_2, then $f^{-1}(N_2)$ is special open in M_1.*

Proof. First, it is clear that $f^{-1}(N_2)$ is an open submodule of M_1, hence there exists a special open submodule N_1 of M_1 with $N_1 \subseteq f^{-1}(N_2)$. Consider in Mod A the following commutative diagram:

$$\begin{array}{ccc} M_1/N_1 & \xrightarrow{\quad f' \quad} & M_2/N_2 \\ \downarrow & & \uparrow \\ M_1/f^{-1}(N_2) & = & M_1/f^{-1}(N_2)' \end{array}$$

in which f' is induced by f. Since M_1/N_1 and M_2/N_2 are objects of $\mathscr{S}$, we have that $M_1/f^{-1}(N_2) \in \operatorname{Ob} \mathscr{S}$. ∎

LEMMA 1.4. *Let $f: M_1 \to M_2$ be a morphism in $\hat{\mathscr{S}}$. Then* $\ker f = \bigcap N_1$ *where N_1 runs over all special open submodules such that* $\ker f \subseteq N_1$.

Proof. $\bigcap N_2 = 0$, because M_2 is Hausdorff, where N_2 runs over all special open submodules of M_2. Thus by Lemma 1.3, one has:

$$\bigcap N_1 \supseteq \ker f = f^{-1}(0) = \bigcap f^{-1}(N_2) \supseteq \bigcap N_1. \blacksquare$$

THEOREM 1.5. (Bourbaki [25]). *Let I be an inverse set and $(X_i, f_{ji})_{i\in I}$ a projective system of non-empty sets. Assume that for any $i \in I$, there exists a set σ_i of subsets of X_i such that:*

(a) *Any intersection of elements of σ_i is also an element of σ_i;*

(b) *The intersection of any inverse filtered system of non-empty elements of σ_i is also a non-empty set;*

(c) *For every $i \leqslant j$, $i, j \in I$ and any $x_j \in X_j$, $A_i \in \sigma_i$, one has $f_{ji}{}^{-1}(x_j) \in \sigma_i$ and $f_{ji}(A_i) \in \sigma_j$.*

Let us write

$$X = \varprojlim_i X_i$$

and let $f_i: X \to X_i$ be the structural projections. Then:

(i) $f_i(X) = \bigcap_{j \leqslant i} f_{ij}(X_j)$;

(ii) *X is a non-empty set.*

Proof. Denote by Σ the set of all families $U = \{A_i\}_{i \in I}$ such that:

($*$) $A_i \in \sigma_i$ and $A_i \neq \emptyset$ for any $i \in I$;

($**$) $f_{ij}(A_j) \subseteq A_i$ for any $i, j \in I, j \leqslant i$.

We define on Σ the following ordering: if $U = \{A_i\}_i$, $U' = \{A_i'\}_i$, let us put $U \geqslant U'$ if and only if $A_i \subseteq A_i'$ for every $i \in I$.

The set Σ is inductive. Indeed, let $U^\lambda = \{A_i^\lambda\}_i$ be a linearly ordered subset of Σ. For every $i \in I$, write $A_i = \bigcap_\lambda A_i^\lambda$. It is easy to see that $U = \{A_i\}_i$ satisfies the condition ($**$) and by (a) and (b) it satisfies also condition ($*$), so that $U \in \Sigma$ and $U^\lambda \leqslant U$ for every λ.

Furthermore, let $U = \{A_i\}_i$ be a maximal element of Σ, then $A_i = f_{ij}(A_j)$ for every $i, j \in I$, $j \leqslant i$. In fact, let us put $A_i' = \bigcap_{j \leqslant i} f_{ij}(A_j)$ for every $i \in I$ and $U' = \{A_i'\}_i$. First we see that if $k \leqslant j \leqslant i$, then by ($**$) one has:

$$f_{ik}(A_k) = f_{ij}(f_{jk}(A_k)) \subseteq f_{ij}(A_j);$$

in addition $f_{ij}(A_j) \in \sigma_i$ by (c) and $f_{ij}(A_j) \neq \emptyset$ by ($*$). Thus by (a) and (b) we conclude that U' satisfies ($*$). Now we claim that U' also satisfies ($**$). Indeed, if $j \leqslant i$, then $f_{ij}(A_j) \subseteq \bigcap_{k \leqslant j} f_{ij}(f_{jk}(A_k)) = \bigcap_{k \leqslant j} f_{ik}(A_k)$; but for every $l \leqslant i$, there exists $k \in I$, such that $k \leqslant l, k \leqslant j$ and thus $f_{ik}(A_k) = f_{il}(f_{lk}(A_k)) \subseteq f_{il}(A_l)$ hence $\bigcap_{k \leqslant j} f_{ik}(A_k) = \bigcap_{l \leqslant i} f_{il}(A_l) = A_i'$. Thus U' satisfies ($**$) as claimed, so that $U' \in \Sigma$. Since $A_i' \subseteq A_i$, for every i, one has that $U' = U$ by the maximality of U. Finally one has $A_i = f_{ij}(A_j)$ for every $j \leqslant i$.

Now we shall prove that $U = \{A_i\}_i$ is a maximal element of Σ, thus A_i consists of exactly one element, for every i. Let $x_i \in A_i$. For any $j \leqslant i$ let us put $B_j = A_j \cap f_{ij}^{-1}(x_i)$; if $j \nleqslant i$, then we consider $B_j = A_j$. Also let $B = \{B_i\}_i$; then $B \in \Sigma$. To see this, let $k \leqslant j \nleqslant i$; then $f_{jk}(B_k) \subseteq f_{jk}(A_k) \subseteq A_j = B_j$; if $k \leqslant j \leqslant i$, then the equality $f_{ik}^{-1}(x_i) = f_{jk}^{-1}(f_{ij}^{-1}(x_i))$ implies

$$f_{jk}(f_{ik}^{-1}(x_i)) = f_{ij}^{-1}(x_i).$$

Since $f_{jk}(A_k) \subseteq A_j$, then as before $f_{jk}(B_k) \subseteq B_j$ so that B has the property ($**$). Furthermore, for $j \leqslant i$, $A_i = f_{ij}(A_j)$ so that $B_j \neq \emptyset$ for every $j \in I$, and finally according to (a) and (b) one has $B_j \in \sigma_j$ for every $j \in I$. We obtain that $B \in \Sigma$. We have $U = B$ because $B_j \subseteq A_j$ for every j, (by the maximality of U); particularly $A_i = \{x_i\}$.

Now we are able to establish point (i) of the theorem. Firstly, it is obvious that $f_i(X) \subseteq \bigcap_{j \leqslant i} f_{ij}(X_j)$. Conversely, let $x_i \in \bigcap_{j \leqslant i} f_{ij}(X_j)$. Denote $B_j = f_{ij}^{-1}(x_i)$ if $j \leqslant i$, and $B_j = X_j$, otherwise. Then by hypothesis one has that B_j is non-empty and $B_j \in \sigma_j$ for every j. Also it is obvious that for every $k \leqslant j$ one has $f_{jk}(B_k) \subseteq B_j$. Therefore $B = \{B_j\}_j$ is an element of Σ. Now let U be a maximal element of Σ. Then $U = \{A_j\}_j$ and $A_j = \{y_j\}$ for every j by the above considerations, so that the element $(y_j)_j = y$ of the cartesian product $\prod_j X_j$ is in fact an element of X. Obviously $f_i(y) = y_i = x_i$, or equivalently $f_i(X) = \bigcap_{j \leqslant i} f_{ij}(X_j)$.

To complete the proof, we shall prove that (i) $\Rightarrow$ (ii). Indeed, because $f_{ij}(X_j) \neq \varnothing$ for every $j \leqslant i$, we see that the set $\{f_{ij}(X_j)\}_{j \leqslant i}$ is a decreasing family of subsets of X, belonging to σ_i. Thus by (b) one has $\bigcap_{j \leqslant i} f_{ij}(X_i) \neq \varnothing$. In this way, by (i) we have that $f_i(X) \neq \varnothing$, hence $X \neq \varnothing$ as required. ∎

Let A be a ring and $\mathscr{S}$ a full subcategory of Mod A.

(L.C.1) An object X of Mod A is called *$\mathscr{S}$-linearly compact* if for any decreasing family $\{X_i\}_i$ of its submodules with $X_i \in \operatorname{Ob} \mathscr{S}$, for every i, it follows that $\bigcap_i X_i \in \operatorname{Ob} \mathscr{S}$ and the canonical morphism

$$X \longrightarrow \varprojlim_i X/X_i$$

is an epimorphism.

We observe that a family $\{X_i\}_i$ of submodules of X is a *decreasing* or *inverse* family if for each pair of indices i, j there exists an index k such that $X_k \subseteq X_i \cap X_j$.

(L.C.2) We shall say that $\mathscr{S}$ is *linearly compact* subcategory if it is finitely closed and any object of $\mathscr{S}$ is $\mathscr{S}$-linearly compact.

In the sequel $\mathscr{S}$ will denote a linearly compact subcategory of Mod A.

NOTATIONS. For any exact sequence

$$0 \longrightarrow C \longrightarrow D \xrightarrow{\ p\ } E \longrightarrow 0 \tag{1}$$

of Mod A, we shall consider C as a submodule of D. If $e \in E$ is an element then $e = p(x)$, with $x \in D$; usually it will be more convenient to consider $e = x + C$ as a subset of D. Assume that $C \in \operatorname{Ob} \mathscr{S}$; then for $e \in E$, denote by σ_e the set of all subsets of e of the form $x' + C'$ where C' is a submodule of C, and $C' \in \operatorname{Ob} \mathscr{S}$; also the empty set will be considered an element of σ_e. Throughout we shall consider the exact sequence (1) with $C \in \operatorname{Ob} \mathscr{S}$.

LEMMA 1.6. *Let* $e = x + C \in E$. *Then*:

(i) *Any intersection of elements of* σ_e *is also an element of* σ_e;

(ii) *The intersection of a decreasing family of non-empty elements of* σ_e *is also a non-empty element of* σ_e.

Proof. Any decreasing family of non-empty sets of σ_e is of the form

$$\{x_i + C_i\}_i \quad \text{where} \quad \{C_i\}_i$$

is a decreasing family of submodules of C, with $C_i \in \text{Ob}\,\mathscr{S}$ for any i. Also x_i are elements of D such that $x_i - x_j \in C_j$ for every $C_i \subseteq C_j$. Since

$$(x - x_i) - (x - x_j) - (x_j - x_i) \in C_j$$

we check that the elements $\{x - x_i\}_i$ of C define an element

$$\{x - x_i + C_i\}_i \quad \text{of} \quad \varprojlim_i C/C_i.$$

Furthermore, by (L.C.1) the canonical mapping

$$C \to \varprojlim_i C/C_i$$

is surjective, so that an element $c \in C$ can be found such that $x - x_i - c \in C_i$ for any i. Then:

$$\bigcap_i (x_i + C_i) = x - c + \bigcap_i C_i \subseteq x + C$$

and $\bigcap_i C_i \in \text{Ob}\,\mathscr{S}$, by (L.C.1).

We have checked above the statement (ii) and a special case of (i). Now, the statement (i) will follow if we prove that for any two non-empty elements $x_1 + C_1$ and $x_2 + C_2$ of σ_e, the intersection: $(x_1 + C_1) \cap (x_2 + C_2)$ is also in σ_e. Indeed, by hypothesis, C_1, C_2 are submodules of C in $\mathscr{S}$ and by the exact sequence:

$$0 \to C_1 \cap C_2 \to C \to C/C_1 \amalg C/C_2$$

we conclude that $C_1 \cap C_2 \in \text{Ob}\,\mathscr{S}$. Now there are two possibilities: either $(x_1 + C_1) \cap (x_2 + C_2) = \varnothing$ (in this case there is nothing to prove) or $(x_1 + C_1) \cap (x_2 + C_2) \neq \varnothing$. Thus, let y be a common element, then

$$(y - x) - (x_1 - x) \in C_1 \quad \text{and} \quad (y - x) - (x_2 - x) \in C_2,$$

so that the elements $y - x$, $x_1 - x$, $x_2 - x$ of C define an element

$$(y - x + C_0, x_1 - x + C_1, x_2 - x + C_2) \in \varprojlim_{i=0,1,2} C/C_i$$

where $C_0 = C_1 \cap C_2$. Because of (L.C.1) an element $c \in C$ can be found such that $c - (y - x) \in C_0$, $c - (x_1 - x) \in C_1$, $c - (x_2 - x) \in C_2$. By the first relation one has: $y \in c - x + C_0$ and by the second one obtains: $c - x + C_0 \subseteq (x_1 + C_1) \cap (x_2 + C_2)$. But for any $z \in (x_1 + C_1) \cap (x_2 + C_2)$ one has $z - y \in C_1 \cap C_2 = C_0$, so that $z \in c - x + C_0$. Finally:

$$(x_1 + C_1) \cap (x_2 + C_2) = c - x + C_0 \in \sigma_e. \blacksquare$$

NOTATIONS. Let

$$\begin{array}{ccccccccc} 0 & \longrightarrow & C & \longrightarrow & D & \longrightarrow & E & \longrightarrow & 0 \\ & & \Big\downarrow{\scriptstyle f} & & \Big\downarrow{\scriptstyle g} & & \Big\downarrow{\scriptstyle h} & & \\ 0 & \longrightarrow & C' & \longrightarrow & D' & \longrightarrow & E' & \longrightarrow & 0 \end{array}$$

be a commutative diagram of Mod A with C, $C' \in \operatorname{Ob} \mathscr{S}$. Also let $x \in D$, $x' \in D'$ with $g(x) - x' \in C'$. *Then* $g(x + C) \subseteq x' + C'$, we denote by $\bar{g} : e = x + C \to e' = x' + C'$ the mapping induced by g.

LEMMA 1.7. *With the above notations, one has*:

(i) *For every* $y' \in e', \bar{g}^{-1}(y') \in \sigma_e$;

(ii) *For every* $U \in \sigma_e, \bar{g}(U) \in \sigma_{e'}$

Proof. (i) Let $y' \in e'$. One has two possibilities: first $\bar{g}^{-1}(y') = \varnothing$ (in this case there is nothing to prove) or there exists $y \in x + C$ such that $\bar{g}(y) = y'$. In the last case for every $z \in x + C$, such that $\bar{g}(z) = y'$, we have

$$z - y \in \ker g \cap C = \ker f,$$

hence $\bar{g}^{-1}(y') \subseteq y + \ker f$. Since the contrary inclusion is obvious, finally we have:

$$\bar{g}^{-1}(y') = y + \ker f \in \sigma_e.$$

(ii) Let $U \in \sigma_e$. Assume that $U \neq \varnothing$, so that $U = x_0 + C_0$ where C_0 is a submodule of C in $\mathscr{S}$, and $x_0 \in C$. Thus $x - x_0 \in C$ and so $\bar{g}(U) = g(x_0) + f(C_0)$, since $x' - g(x_0) = x' - g(x) + f(x - x_0) \in C'$ (because $x' - g(x) \in C'$ and $f(C_0) \in \operatorname{Ob} \mathscr{S}$). Finally $\bar{g}(U) \in \sigma_{e'}$. ∎

LEMMA 1.8. *Let I be a left filtered* (*or inverse*) *set and*

$$(s_i)_i : (D_i, g_{ji}) \to (E_i, h_{ji})$$

a morphism of I-filtered systems of A-modules, such that $s_i : D_i \to E_i$ is an epimorphism and $C_i = \ker s_i \in \operatorname{Ob} \mathscr{S}$ for every $i \in I$. Then the limit morphism:

$$s = \varprojlim_i s_i : \varprojlim_i D_i \to \varprojlim_i E_i$$

is an epimorphism.

Proof. For every $i \leqslant j$ we have in Mod A a commutative diagram with exact rows:

$$\begin{array}{ccccccccc} 0 & \longrightarrow & C_i & \longrightarrow & D_i & \xrightarrow{s_i} & E_i & \longrightarrow & 0 \\ & & \Big\downarrow & & \Big\downarrow{\scriptstyle g_{ji}} & & \Big\downarrow{\scriptstyle h_{ji}} & & \\ 0 & \longrightarrow & C_j & \longrightarrow & D_j & \xrightarrow{s_j} & E_j & \longrightarrow & 0 \end{array}$$

Let $z \in \varprojlim E_i$; then $z = (z_i)_i$, such that $z_i \in E_i$ and $h_{ji}(z_i) = z_j$ for every $i \leqslant j$. Since all maps s_i are surjective, for any i, we can choose an element $x_i \in D_i$ such that $s_i(x_i) = z_i$. Then one has:

$$s_j(x_j) = z_j = h_{ji}(z_i) = (h_{ji}s_i)(x_i) = s_j(g_{ji}(x_i)),$$

so that $x_j - g_{ji}(x_i) \in \ker s_i = C_i$, $i \leqslant j$. Let us denote $e_i = x_i + C_i$ and $\sigma_i = \sigma_{e_i}$ and let $\bar{g}_{ji} = e_i \to e_j$ be the map induced by g_{ji}.

Then $(e_i, \bar{g}_{ji})$ is an inverse system of sets and mappings and by Lemmas 1.6 and 1.7 the conditions (a)–(c) of Theorem 1.5 are fulfilled. Therefore, by Theorem 1.5 the set $\varprojlim e_i$ is non-empty, and for every

$$y = (y_i)_i \in \varprojlim_i e_i \subseteq \varprojlim_i D_i,$$

one has $s_i(y_i) = s_i(x_i) = z_i$, hence finally $s(y) = z$, or equivalently s is surjective. ■

LEMMA 1.9. *Let M be an object of $\mathcal{S}$ and $\{N_i\}_i$ a decreasing family of its special open submodules. Then the canonical map*:

$$M \to \varprojlim_i M/N_i$$

is surjective.

Proof. Let $(P_\alpha)_\alpha$ be the decreasing family of all special open submodules of M (see Lemma 1.2). For any i and any α, one has the following exact sequence in Mod A:

$$0 \to (N_i + P_\alpha)/P_\alpha \to M/P_\alpha \to M/(N_i + P_\alpha) \to 0$$

in which M/P_α is an object of $\mathcal{S}$; thus, by Lemma 1.2,

$$(N_i + P_\alpha)/P_\alpha \simeq N_i/(N_i \cap P_\alpha) \in \operatorname{Ob} \mathcal{S},$$

for every i and α. Since $\mathcal{S}$ is linearly compact, we check that for every α the canonical mapping:

$$M/P_\alpha \to \varprojlim_i M/(N_i + P_\alpha)$$

is onto, and its kernel belongs to $\mathcal{S}$. Also, according to Lemma 1.8 the canonical mapping

$$\varprojlim_\alpha M/P_\alpha \xrightarrow{s} \varprojlim_{i,\alpha} M/(N_i + P_\alpha)$$

is onto. On the other hand, one has the following commutative diagram in Mod A with exact rows:

$$\begin{array}{ccccccc}
0 & \longrightarrow & M & \xrightarrow{p} & \varprojlim_{\alpha} M/P_\alpha & & \\
\downarrow & & \downarrow{\scriptstyle q} & & \downarrow{\scriptstyle s} & & \\
\varprojlim_{i,\alpha} ((N_i + P_\alpha)/N_i) & \longrightarrow & \varprojlim_{i} M/N_i & \xrightarrow{r} & \varprojlim_{i,\alpha} (M/(N_i + P_\alpha)) & \longrightarrow & 0
\end{array}$$

Indeed, since M is complete (and Hausdorff) p is a bijection. Also we have proved above that s is surjective, hence so is r. For any i, there exists an index α such that $P_\alpha = N_i$, hence $\varprojlim ((N_i + P_\alpha)/N_i) = 0$. In conclusion

$$\varprojlim_{i,\alpha} ((N_i + P_\alpha)/N_i) = 0,$$

or equivalently, r is a bijective mapping, as claimed. ∎

LEMMA 1.10. *Let M be an object of $\hat{\mathscr{S}}$, $\{P_i\}_i$ a decreasing set of special open submodules of M and N a special closed submodule. Then*:

$$N + \textstyle\bigcap_i P_i = \bigcap_i (N + P_i).$$

Proof. For each index i, one has in Mod A the following commutative diagram with exact rows:

$$\begin{array}{ccccccc}
0 \to N & \longrightarrow & M & \longrightarrow & M/N & \longrightarrow & 0 \\
\downarrow & & \downarrow & & \downarrow & & \\
0 \to N/(N \cap P_i) & \longrightarrow & M/P_i & \longrightarrow & M/(N + P_i) & \longrightarrow & 0
\end{array}$$

By Lemma 1.2, $\{N \cap P_i\}_i$ is a decreasing family of special open subobjects of N, so that $N/(N \cap P_i) \in \operatorname{Ob} \mathscr{S}$ for every i. Also by Lemma 1.8, passing to projective limits we have the following commutative diagram with exact rows in Mod A:

$$\begin{array}{ccccccc}
0 \to N & \longrightarrow & M & \xrightarrow{p} & M/N & \longrightarrow & 0 \\
\downarrow{\scriptstyle f} & & \downarrow{\scriptstyle k} & & \downarrow{\scriptstyle l} & & \\
0 \to \varprojlim_{i} (N/N \cap P_i) & \longrightarrow & \varprojlim_{i} M/P_i & \longrightarrow & \varprojlim_{i} (M/(N + P_i)) & \longrightarrow & 0
\end{array}$$

Now, by Lemma 1.9 the mapping f is surjective, so that $\ker l = p(\ker k)$ or equivalently:

$$\textstyle\bigcap_i (N + P_i)/N = \ker l = p(\ker k) = p(\bigcap_i P_i) = (N + \bigcap_i P_i)/N.$$

Now the conclusion of Lemma 1.10 is obvious. ∎

LEMMA 1.11. *Let* $f : M_1 \to M_2$ *be a morphism of* $\hat{\mathscr{S}}$. *Then* $\ker f$ *is a special closed submodule of* M_1.

Proof. First, it is clear that $\ker f = f^{-1}(0)$ is a closed, hence complete submodule of M_1 in the induced topology. A fundamental system of neighbourhoods of 0 in $\ker f$ consists of all open submodules $N \cap \ker f$, where N runs over all special open submodules of M_1. Let $\{P_i\}_i$ be the decreasing family of all special open submodules of M_1 containing $\ker f$. Then by Lemma 1.4, one has:

$$\ker f = \bigcap_i P_i.$$

Now, let N be special open in M_1. By Lemma 1.1 and 1.10 we have:

$$\ker f/(N \cap \ker f) \simeq (N + \ker f)/N \simeq (N + \bigcap_i P_i)/N \simeq (\bigcap_i (N + P_i))/N.$$

Finally, we see that $\{(N + P_i)/N\}_i$ is a decreasing family of submodules of $M/N \in \operatorname{Ob} \mathscr{S}$ and by Lemma 1.2, $(N + P_i)/N \simeq P_i/(N \cap P_i) \in \operatorname{Ob} \mathscr{S}$ for any i so that $\ker f/(N \cap \ker f) = \bigcap_i ((N + P_i)/N)$ since $\mathscr{S}$ is linearly compact. ■

LEMMA 1.12. *Let* M *be an object of* $\hat{\mathscr{S}}$ *and* N *a submodule of* M. *The following assertions are equivalent*:

(a) N *is special closed in* M;

(b) M/N *with the quotient topology is an object of* $\hat{\mathscr{S}}$;

(c) $N = \bigcap U$, *where* U *runs over all special open submodules of* M, *which contain* N.

Proof. (a) $\Rightarrow$ (b). Let $\{U_i\}_i$ be the set of all special open submodules of M. Then $N/N \cap U_i \in \operatorname{Ob} \mathscr{S}$ for every i (Lemma 1.2) and by Lemma 1.8, we have the commutative diagram with exact rows, in Mod A:

$$\begin{array}{ccccccc}
0 \to N & \longrightarrow & M & \longrightarrow & M/N & \longrightarrow & 0 \\
\downarrow f & & \downarrow k & & \downarrow l & & \\
0 \to \varprojlim_i (N/(N \cap {}_iU)) & \to & \varprojlim_i M/U_i & \to & \varprojlim_i (M/(N + U_i)) & \to & 0
\end{array}$$

The submodules $N \cap U$, where U runs over all special open submodules of M, define a fundamental system of neighbourhoods of 0 in N. Since N in the induced topology is an object of $\hat{\mathscr{S}}$, hence complete, we check that the mapping f is in fact an isomorphism; because k is also an isomorphism, we deduce that l is an isomorphism.

A fundamental system of neighbourhoods of M/N (topologized with the quotient topology) is formed by all submodules $(N + U)/U$, where U is a special open submodule of M, and moreover one has:

$$(M/N)/((N + U)/N) \simeq M/(N + U) \in \operatorname{Ob} \mathscr{S}$$

by Lemma 1.2. Finally, by the previous considerations we have that M/N with the quotient topology is an object of $\mathscr{S}$.

(b) $\Rightarrow$ (c). By the statement (b) the canonical mapping:

$$p : M \to M/N$$

is a morphism of $\mathscr{S}$ and by Lemma 1.4, one has:

$$N = \ker p = \bigcap U,$$

where U runs over all special open submodules of M containing N.

(c) $\Rightarrow$ (a). If (c) is satisfied, then one has in Mod A the exact sequence:

$$0 \longrightarrow N \longrightarrow M \xrightarrow{f} \textstyle\prod_U (M/U), \tag{2}$$

where f is canonically defined. Now assume that M/U is equipped with the discrete topology; then the direct product of (2) is an object of $\mathscr{S}$ (see Theorem 1.15 below) and the canonical mapping f is a morphism in $\mathscr{S}$. Thus by Lemma 1.11, $N = \ker f$ is a special closed submodule of M. ∎

NOTE. We may check directly (without using Theorem 1.15) that the direct product of (2) is an object of $\mathscr{S}$. We suggest that the reader fills in the details.

LEMMA 1.13. *Let M be an object of $\mathscr{S}$ and $\{N_i\}_i$ a decreasing family of special closed submodule of M. Then the canonical map*:

$$M \to \varprojlim_i M/N_i$$

is surjective.

Proof. Let $\{P_\alpha\}_\alpha$ be the decreasing set of all special open submodules of M. Consider the following diagram of Mod A, canonically constructed:

$$\begin{array}{ccc} M & \xrightarrow{\quad q \quad} & \varprojlim_i M/N_i \\ {\scriptstyle p}\downarrow & & \downarrow{\scriptstyle r} \\ \varprojlim_\alpha M/P_\alpha & \xrightarrow{\ s\ } & \varprojlim_{i,\alpha} (M/(N_i + P_\alpha)) \end{array}$$

Since M is complete, we have that p is surjective; also by Lemma 1.12, (b), the map:

$$M/N_i \to \varprojlim_\alpha (M/(N_i + P_\alpha))$$

is an isomorphism for each i, so that the mapping r is surjective. The map s, which appeared in the proof of Lemma 1.9 is surjective because N_i is special closed. Finally, one has that q is surjective. ∎

LEMMA 1.14. *Let* $f: M_1 \to M_2$ *be a morphism of* $\hat{\mathscr{S}}$. *Then* $\operatorname{im} f$ *is special closed in* M_2.

Proof. Let:

$$\begin{array}{ccc} M_1 & \xrightarrow{\ f\ } & M_2 \\ \downarrow & & \uparrow \\ M_1/\ker f & \xrightarrow{\ \bar{f}\ } & \operatorname{im} f \end{array}$$

be the canonical decomposition of f in Mod A. Assume that $M_1/\ker f$ is identified with $\operatorname{im} f$ through $\bar{f}$. According to Lemma 1.11 $\ker f$ is special closed in M_1, and by Lemma 1.12, (b) $M_1/\ker f = \operatorname{im} f$ equipped with the quotient topology is an object of $\hat{\mathscr{S}}$. Since $\bar{f}$ is obviously a continuous mapping, the induced topology of $\operatorname{im} f$ is finer than the quotient topology. Particularly $\operatorname{im} f$ is a complete (and Hausdorff) topological group in the induced topology.

A fundamental system of neighbourhoods of $\operatorname{im} f$ in the induced topology consists of all the submodules $\operatorname{im} f \cap V$, where V is special open in M_2. But then, for each special open V in M_2, $\operatorname{im} f \cap V$ is open also in the quotient topology of $\operatorname{im} f$, so that there exists a special open submodule U of $\operatorname{im} f$ in the quotient topology, such that $U \subseteq \operatorname{im} f \cap V$. Let $g: \operatorname{im} f/U \to M_2/V$ be the canonical mapping and the commutative diagram of Mod A, which gives the canonical decomposition of g in Mod A

$$\begin{array}{ccc} \operatorname{im} f/U & \longrightarrow & M_2/V \\ \downarrow & & \uparrow \\ \operatorname{im} f/(\operatorname{im} f \cap V) & = & \operatorname{im} f/(\operatorname{im} f \cap V) \end{array}$$

Since $\operatorname{im} f/U$ and M_2/U are objects of $\mathscr{S}$, one checks that

$$\operatorname{im} f/(\operatorname{im} f \cap V) \in \operatorname{Ob} \mathscr{S},$$

hence $\operatorname{im} f$ is a special closed submodule of M_2 as claimed. ∎

THEOREM 1.15. *The category* $\hat{\mathscr{S}}$ *is a preabelian category with projective limits. The kernels, cokernels and projective limits of* $\hat{\mathscr{S}}$ *are calculated as in* Mod A *and equipped respectively with the induced topology, the quotient topology and with the topology of projective limits* (*see* [24], *Chapter* 1).

Proof. Let $f: M_1 \to M_2$ be a morphism of $\mathscr{S}$. Then $\ker f$ in $\operatorname{Mod} A$, is special closed in M_1 (Lemma 1.11) so that equipped with the induced topology it is an object of $\mathscr{S}$, and obviously it is the kernel of f in $\mathscr{S}$. Furthermore, by Lemma 1.14 $\operatorname{im} f$ is special closed in M_2, so that according to Lemma 1.12 (b), $\operatorname{coker} f$ in $\operatorname{Mod} A$ equipped with quotient topology is an object of $\mathscr{S}$ and obviously $\operatorname{coker} f$ is in $\mathscr{S}$.

To finish the proof it will be sufficient (see Chapter 1, Theorem 4.1) to check that $\mathscr{S}$ has direct products. Indeed, if $\{M_i\}_{i \in I}$ is a set of objects of $\mathscr{S}$ then $\prod_i M_i$, the direct product equipped with product topology is a complete topological group. A fundamental system of neighbourhoods of $\prod_i M_i$ consists of the submodules of the form: $\prod_i N_i$, where N_i is special open in M_i for each i, and $N_i = M_i$ for all but a finite set of indices. For such a submodule of $\prod_i M_i$ we denote by I' the subset of I consisting of all indices i' such that $N_{i'} \neq M_{i'}$. Then I' is a finite subset of I and we have:

$$\textstyle\prod_i M_i / \prod_i N_i \xrightarrow{\sim} \prod_i (M_i/N_i) \xrightarrow{\sim} \prod_{i' \in I'} (M_{i'}/N_{i'}) \in \operatorname{Ob} \mathscr{S}.$$

Finally we see that ΠM_i is an object of $\mathscr{S}$. ■

LEMMA 1.16. *Let M be an object of $\mathscr{S}$ and N_1, N_2 two special closed submodules of M. Then $N_1 \cap N_2$ is special closed in M. Moreover, if M/N_1 and M/N_2 are objects of $\mathscr{S}$, then $M/(N_1 \cap N_2) \in \operatorname{Ob} \mathscr{S}$.*

Proof. Let $f: M \to M/M_1 \amalg M/M_2$ be the canonical mapping and its decomposition in $\operatorname{Mod} A$:

$$\begin{array}{ccc} M & \xrightarrow{\quad f \quad} & M/N_1 \amalg M/N_2 \\ \downarrow & & \uparrow \\ M/(N_1 \cap N_2) & = & M/(N_1 \cap N_2) \end{array}$$

We shall consider M/N_1 and M/N_2 as objects of $\mathscr{S}$, equipped with the quotient topology and f as a morphism of $\mathscr{S}$ [see Lemma 1.12 (b)]. Now we shall examine the canonical decomposition of f in $\mathscr{S}$. It is clear that $\operatorname{coim} f$ in $\mathscr{S}$ is in fact $\operatorname{coim} f$ in $\operatorname{Mod} A$, equipped with the quotient topology (Theorem 1.15); in other words $M/(N_1 \cap N_2)$ equipped with the quotient topology is an object of $\mathscr{S}$, and thus, by Lemma 1.12 (a), $N_1 \cap N_2$ is special closed in M. Also, by Theorem 1.15 $M/N_1 \cap N_2$ in the induced topology is a special closed submodule of $M/N_1 \amalg M/N_2$. Now, by Lemma 1.12, (c), any special closed submodule of an object of $\mathscr{S}$ is also an object of $\mathscr{S}$, so that particularly $M/N_1 \cap N_2 \in \operatorname{Ob} \mathscr{S}$. ■

LEMMA 1.17. *Let $f: M_1 \to M_2$ be a morphism of $\mathscr{S}$ and N_2 a special closed of M_2. Then $f^{-1}(N_2)$ is special closed in M_1. Moreover if $M_2/N_2 \in \operatorname{Ob} \mathscr{S}$, then $M_1/f^{-1}(N_2) \in \operatorname{Ob} \mathscr{S}$.*

Proof. Denote by $\bar{f}: M_1 \to M_2/N_2$ the canonical mapping induced by f and its canonical decomposition in Mod A:

$$\begin{array}{ccc} M_1 & \longrightarrow & M_2/N_2 \\ \downarrow & & \uparrow \\ M_1/f^{-1}(N_2) & = & M_1/f^{-1}(N_2) \end{array}$$

Now the proof is similar to the proof of Lemma 1.16. ∎

An object M of $\hat{\mathscr{S}}$ is called *strict* if each special closed submodule N of M such that $M/N \in \operatorname{Ob} \mathscr{S}$ is also open. Denote by $\tilde{\mathscr{S}}$ the full subcategory of $\hat{\mathscr{S}}$ consisting of all strict objects.

LEMMA 1.18. *The inclusion functor $I : \tilde{\mathscr{S}} \to \hat{\mathscr{S}}$ has a right adjoint S. Moreover $SI = \operatorname{Id} \tilde{\mathscr{S}}$.*

Proof. Let M be an object of $\hat{\mathscr{S}}$. The family $\{N_i\}_i$ of its special closed submodules with $M/N_i \in \operatorname{Ob} \mathscr{S}$ is a decreasing family, by Lemma 1.16. Thus, according to Lemma 1.14, the canonical mapping:

$$M \xrightarrow{p} \varprojlim_i N/N_i$$

is onto. Finally, according to Lemma 1.1, the set $\{N_i\}_i$ contains all special open submodules of M, so that p is in fact a bijection. Therefore, there exists an unique topology on the A-module M such that M becomes a topological group, having all submodules $\{N_i\}_i$ as fundamental system of neighbourhoods of zero. It is obvious that M, equipped with this topology is an object of $\hat{\mathscr{S}}$, denoted by $S(M)$. Now, since the original topology of M is finer than this topology, we see that the identical map:

$$\psi_M : S(M) \to M$$

is continuous, hence a morphism of $\hat{\mathscr{S}}$.

We claim that $S(M)$ is in fact an object of $\tilde{\mathscr{S}}$. Indeed, if N is special closed in $S(M)$, then it is special closed in M [see Lemma 1.12 (c)]; now, if

$$M/N \in \operatorname{Ob} \mathscr{S}$$

then N is open in $S(M)$ by the definition of the topology of $S(M)$, hence $S(M)$ is an object of $\tilde{\mathscr{S}}$ as claimed. Also a glance at the definition of $S(M)$ proves that $S(M) = M$ for each object M of $\tilde{\mathscr{S}}$.

Finally, let M' be an object of $\tilde{\mathscr{S}}$ and $f : M' \to M$ a morphism in $\hat{\mathscr{S}}$. Also let N be special closed in M' and $M/f^{-1}(N) \in \operatorname{Ob} \mathscr{S}$, hence $f^{-1}(N)$ is special open by hypothesis. Thus f can be considered as a morphism

$\bar{f}: M' \to S(M)$ and $\psi_M \bar{f} = f$. In conclusion S is a right adjoint of I and the morphism $\{\psi_M\}_M$ defines an arrow of adjunction of S with I.

Now, any object M of $\mathcal{S}$ will be considered as an object of $\tilde{\mathcal{S}}$ with the discrete topology. ■

THEOREM 1.19. *The category $\tilde{\mathcal{S}}$ is an abelian category with the condition* Ab 5*. *S is a full subcategory of $\tilde{\mathcal{S}}$, closed under subobjects, quotient objects and finite direct sums. Moreover, the objects of $\mathcal{S}$ form a class of cogenerators of $\tilde{\mathcal{S}}$.*

Proof. By Theorem 1.15 and Lemma 1.18, one checks that $\tilde{\mathcal{S}}$ is a preabelian category with projective limits.

Now let $f: M_1 \to M_2$ be a bimorphism of $\tilde{\mathcal{S}}$; then by Lemma 1.18, f is an isomorphism of A-modules. Also, if N_1 is special open in M_1, then according to Lemma 1.14, $f(N_1)$ is special closed in M_2, and

$$M_2/f(N_1) \simeq M_1/N_1 \in \operatorname{Ob} \mathcal{S},$$

so that $f(N_1)$ is also special open in M_2; hence f is necessarily a bicontinuous isomorphism of A-modules, or equivalently, an isomorphism of $\tilde{\mathcal{S}}$. Then, by Chapter 2, Proposition 3.1, $\tilde{\mathcal{S}}$ is abelian.

If M is an object in $\tilde{\mathcal{S}}$ and N a subobject of M in $\tilde{\mathcal{S}}$, then N is a submodule of M, and M/N is an object of $\tilde{\mathcal{S}}$, equipped with the quotient topology, hence by Lemma 1.14(a) N is special closed in M. Since $\tilde{\mathcal{S}}$ is abelian, then the topology of N is in fact the induced topology. Particularly a sequence

$$0 \to N \to M \to P \to 0$$

is exact in $\tilde{\mathcal{S}}$ if and only if the underlying sequence of A-modules is exact in Mod A.

The condition Ab 5* now follows from Lemma 1.13.

Finally, let $f: M_1 \to M_2$ be a non-zero morphism of $\tilde{\mathcal{S}}$; then an element $x_1 \in M_1$ can be found such that $f(x_1) \neq 0$; since M_2 is Hausdorff, there exists a special open submodule N_2 of M_2 such that $f(x_1) \notin N_2$. Thus $M_2/N_2 \in \operatorname{Ob} \mathcal{S}$ so that the composed morphism:

$$M_1 \xrightarrow{f} M_2 \xrightarrow{p} M_2/N_2$$

where p is canonical, is non-zero. Hence the objects of $\mathcal{S}$ form a class of cogenerators of $\tilde{\mathcal{S}}$. ■

6.2 Topologically linearly compact rings

Let A be a ring. An A-module M is called *algebraically linearly compact* (we shall write: a.l.c.) if for each decreasing family $\{M_i\}_i$ of finitely generated submodules M_i of M the module $\bigcap_i M_i$ is also finitely generated and the canonical map

$$M \to \varprojlim_i M/M_i$$

is surjective. The ring A is called *algebraically left linearly compact* (a.l.c.) if A considered as A-module is algebraically linearly compact.

LEMMA 2.1. *Let*

$$0 \longrightarrow N \xrightarrow{f} M \xrightarrow{g} P \longrightarrow 0$$

be an exact sequence of Mod A. *If M is a.l.c. and N finitely generated, then P is a.l.c.*

Proof. For any submodule P' of P, one has the canonically constructed exact sequence

$$0 \to N \to g^{-1}(P') \to P' \to 0.$$

Since N is finitely generated, $g^{-1}(P')$ is finitely generated whenever P' is finitely generated.

Now let $\{P_i\}_i$ be a decreasing family of finitely generated submodules of P; then $\{g^{-1}(P_i)\}_i$ is also a decreasing family of finitely generated submodules of M.

Since g is a surjective mapping, one has

$$P' = \bigcap_i P_i = g(g^{-1}(\bigcap_i P_i)) = g(\bigcap_i g^{-1}(P_i))$$

so that P' is finitely generated by hypothesis.

Now for any i one has the canonically constructed commutative diagram:

$$\begin{array}{ccc} M & \xrightarrow{\quad g \quad} & P \\ \downarrow & & \downarrow \\ M/g^{-1}(P_i) & \xrightarrow[\sim]{g_i} & P/P_i \end{array}$$

where g_i is an isomorphism. Passing to projective limits we obtain the commutative diagram:

$$\begin{array}{ccc} M & \xrightarrow{\quad g \quad} & P \\ {\scriptstyle t}\downarrow & & \downarrow{\scriptstyle t'} \\ \varprojlim_i (M/g^{-1}(P_i)) & \xrightarrow[\sim]{\bar{g}} & \varprojlim_i (P/P_i) \end{array}$$

in which t is surjective by hypothesis. Thus t' is also surjective. ■

Now we shall denote by Coh(A) the full subcategory of Mod A generated by all coherent objects (using the notations of Section 3.5, Chapter 3, one has $\mathrm{Coh}(A) = \mathrm{Coh}(\mathrm{Mod}\,A)$). Also we shall utilize frequently the results of Section 3.5, Chapter 3 concerning finitely generated and coherent objects.

We shall consider a prelocalizing subcategory $\mathscr{P}$ of Mod A; then $\mathscr{P}$ is a Grothendieck category and notions such as: object of finite type and coherent object in $\mathscr{P}$ are the same as in Mod A. In particular

$$\mathrm{Coh}(\mathscr{P}) = \mathscr{P} \cap \mathrm{Coh}(A).$$

LEMMA 2.2. *Let M be a coherent object of $\mathscr{P}$ and N a finitely generated submodule of M. If N and M/N are a.l.c. modules then M is a.l.c. too.*

Proof. Let $\{M_i\}_i$ be a decreasing family of finitely generated submodules of M. Then one has the commutative diagram canonically constructed:

$$\begin{array}{ccccccccc}
 & & 0 & & 0 & & 0 & & \\
 & & \downarrow & & \downarrow & & \downarrow & & \\
0 & \longrightarrow & \bigcap_i (N \cap M_i) & \longrightarrow & \bigcap_i M_i & \xrightarrow{n} & \bigcap_i ((N + M_i)/N) & \dashrightarrow & 0 \\
 & & \downarrow & & \downarrow & & \downarrow & & \\
0 & \longrightarrow & N & \longrightarrow & M & \xrightarrow{p} & M/N & \longrightarrow & 0 \\
 & & \downarrow f & & \downarrow k & & \downarrow l & & \\
0 & \longrightarrow & \varprojlim_i (N/(N \cap M_i)) & \longrightarrow & \varprojlim_i (M/M_i) & \xrightarrow{m} & \varprojlim_i (M/(N + M_i)) & \dashrightarrow & 0 \\
 & & \downarrow & & \downarrow & & \downarrow & & \\
 & & 0 & & 0 & & 0 & &
\end{array}$$

Since $N + M_i$ is finitely generated for each i, then $\{(N + M_i)/N\}_i$ is a decreasing family of finitely generated submodules of M/N. Furthermore, because M/N is a.l.c. then $\bigcap_i ((M_i + N)/N_i)$ is finitely generated and l is a surjective map. Also, since M is coherent, then $\{N \cap M_i\}_i$ is a decreasing family of finitely generated submodules of N, so that $\bigcap_i (N \cap M_i)$ is finitely generated and f is a surjection, by hypothesis. Thus one has: k is a surjective map, also n is a surjection and $\bigcap_i M_i$ is finitely generated; in conclusion M is a.l.c. ■

LEMMA 2.3. *Let $M \in \mathrm{Ob}\,\mathrm{Coh}(\mathscr{P})$ and N, P be two finitely generated a.l.c. submodules of M. Then $N + P$ is coherent and a.l.c.*

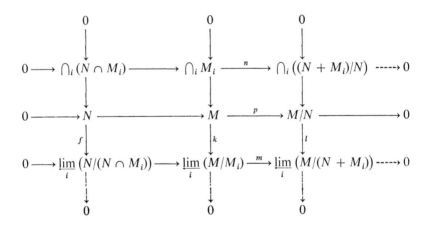

Proof. $N + P$ is finitely generated, thus coherent. Also, by the previous lemma $N \sqcap P$ is an a.l.c.-module and one has the following exact sequence:

$$0 \longrightarrow \ker s \longrightarrow N \sqcap P \xrightarrow{s} N + P \longrightarrow 0$$

where s is the mapping "sum". Since $N \sqcap P$ is finitely generated and $N + P$ coherent, then $\ker s$ is finitely generated, hence by Lemma 2.1 $N + P$ is a.l.c.-module. ∎

By a *left linear topological ring*, we mean a topological ring A having a fundamental system of neighbourhoods of zero consisting of left ideals. Thus the set of all open left ideals of A is in fact a left prelocalizing system P on A, and conversely, a left prelocalizing system on A defines a topology on A, such that A becomes a left linear topological ring. We shall refer to this topology as the P-topology. A left ideal $\mathfrak{a}$ of A is open in the P-topology if and only if $\mathfrak{a} \in P$. By a P-discrete module we mean an A-module X such that for each $x \in X$, $\operatorname{Ann}(x)$ is open, or equivalently $\operatorname{Ann}(x) \in P$. Obviously, a P-discrete module is in fact an object of $\mathscr{P}$ (for the notion of the prelocalizing subcategory associated with P; see Chapter 4, Theorem 9.1) and conversely.

By a *left linearly topological compact ring* A, we shall mean a left linearly topological ring A having a fundamental system of neighbourhoods consisting of ideals $\mathfrak{a}$ such that $A/\mathfrak{a}$ is a.l.c.

THEOREM 2.4. *Let P be a left prelocalizing system on A such that A becomes in the P-topology a left linearly topological compact ring. Then* Coh($\mathscr{P}$) *is a linearly compact subcategory of* Mod A.

Proof. By the results of Chapter 3, Section 3.5, we check that Coh($\mathscr{P}$) is closed under finite direct sums, kernels and cokernels, calculated in $\mathscr{P}$, hence in Mod A. Also, any A-module isomorphic with an object of Coh($\mathscr{P}$) is always in Coh($\mathscr{P}$). To prove that an object M of Coh($\mathscr{P}$) is Coh($\mathscr{P}$)-linearly compact, it will be enough to check that it is a.l.c. In fact, if

$$M \in \operatorname{Ob} \operatorname{Coh}(\mathscr{P}),$$

then M is finitely generated; moreover if M is cyclic, then $M \simeq A/\mathfrak{a}$, where $\mathfrak{a}$ is an element of P, hence there exists a left ideal $\mathfrak{a}' \subseteq \mathfrak{a}$ such that $A/\mathfrak{a}'$ is an a.l.c. module. Then, by the exact sequence:

$$0 \rightarrow \mathfrak{a}/\mathfrak{a}' \rightarrow A/\mathfrak{a}' \rightarrow A/\mathfrak{a} \rightarrow 0$$

and by Lemma 2.1, we have that $M \simeq A/\mathfrak{a}$ is a.l.c.

Generally, let $M = Ax_1 + \ldots + Ax_n$, where, as above, Ax_i is a.l.c. for each i. Then, by Lemma 2.3 one has that M is a.l.c. ∎

As always let P be a prelocalizing system on A as in the previous theorem; we shall denote by $\mathrm{TF}(\mathscr{P})$ the following category. An object of $\mathrm{TF}(\mathscr{P})$ is an A-module M equipped with a topology having a fundamental system of neighbourhoods of zero consisting of submodules M' such that M/M' is finitely generated; the morphisms of $\mathrm{TF}(\mathscr{P})$ are continuous morphisms of A-modules. By $\mathrm{TC}(\mathscr{P})$ we denote the full subcategory of $\mathrm{TF}(\mathscr{P})$ generated by all objects M having a fundamental system of neighbourhoods of zero consisting of submodules M' such that $M/M' \in \mathrm{Ob}\,\mathrm{Coh}(\mathscr{P})$.

LEMMA 2.5. *The inclusion functor*:

$$I_1 : \mathrm{TC}(\mathscr{P}) \to \mathrm{TF}(\mathscr{P})$$

has a left adjoint denoted by H_1.

Proof. Let $M \in \mathrm{Ob}\,\mathrm{TF}(\mathscr{P})$ and M_1, M_2 be two open submodules of M such that $M/M_i \in \mathrm{Ob}\,\mathrm{Coh}(\mathscr{P})$, $i = 1, 2$. Then $M_1 \cap M_2$ is an open submodule of M, so that there exists an open submodule M' of M, $M' \subseteq M_1 \cap M_2$ and M/M' is finitely generated. But then $M/M_1 \cap M_2$ is also finitely generated, as a quotient of a finitely generated module. Now by the inclusion

$$M/M_1 \cap M_2 \to M/M_1 \amalg M/M_2$$

we have that $M/M_1 \cap M_2$ is an object of $\mathrm{Coh}(\mathscr{P})$, since the direct sum is an object of $\mathrm{Coh}(\mathscr{P})$. Then there exists a unique topology on M such that M becomes a topological group, having a fundamental system of neighbourhoods of zero, consisting of all open submodules M' such that

$$M/M' \in \mathrm{Ob}\,\mathrm{Coh}(\mathscr{P}).$$

We shall denote this topological group by $H_1(M)$. Now it is clear that $H_1(M)$ is an object of $\mathrm{TC}(\mathscr{P})$ and since the topology of $H_1(M)$ is finer than the old topology of M, we find that the identical map (which is denoted by

$$\phi_M^1 : M \to H_1(M))$$

is continuous. Furthermore, let $f : M \to N$ be a morphism in $\mathrm{TF}(\mathscr{P})$ then for every open submodule N' of N such that $N/N' \in \mathrm{Ob}\,\mathrm{Coh}(\mathscr{P})$, as above, we can prove that $f^{-1}(N')$ is an open submodule of M such that $M/f^{-1}(N')$ is an object of $\mathrm{Coh}(\mathscr{P})$. Moreover, if $N \in \mathrm{Ob}\,\mathrm{TC}(\mathscr{P})$, then the morphism f can be obviously factored by a unique morphism $\bar{f} : H_1(M) \to N$. ∎

By Theorem 2.4, the full subcategory $\mathrm{Coh}(\mathscr{P})$ is a linearly compact subcategory of $\mathrm{Mod}\,A$, so that we can utilize all results of Section 6.1. Particularly $\widehat{\mathrm{Coh}}(\mathscr{P})$ is denoted by $\mathrm{CTC}(\mathscr{P})$. Moreover $\mathrm{CTC}(\mathscr{P})$ is the full subcategory of $\mathrm{TC}(\mathscr{P})$ generated by all complete and Hausdorff objects. Also, the full subcategory $\widetilde{\text{Coh}(\mathscr{P})}$ of $\text{Coh}(\hat{\mathscr{P}}) = \text{CTC}(\mathscr{P})$ is denoted by $\text{STC}(\mathscr{P})$; by Section 6.1, an object of $\text{CTC}(\mathscr{P})$ belongs to $\text{STC}(\mathscr{P})$ if and only if it is strict.

LEMMA 2.6. *The inclusion functor:*

$$I_2 : \text{CTC}(\mathscr{P}) \to \text{TC}(\mathscr{P})$$

has a left adjoint denoted by H_2.

Proof. Let M be an object of $\text{TC}(\mathscr{P})$ and $\{M_i\}_i$ the family of its open submodules M_i such that $M/M_i \in \text{Ob Coh}(\mathscr{P})$. Then M/M_i in the discrete topology is an object of $\text{CTC}(\mathscr{P})$, so that

$$\varprojlim_i (M/M_i)$$

equipped with the topology of projective limit is also an object of $\text{CTC}(\mathscr{P})$ (see Theorem 1.15), which is denoted by $H_2(M)$; also denote by

$$\phi_M^2 : M \to \varprojlim_i (M/M_i) = H_2(M)$$

the canonical morphism (which is obviously a morphism of $\text{TC}(\mathscr{P})$).

Now if $f : M \to N$ is a morphism of $\text{TC}(\mathscr{P})$, then for each open submodule N' of N, such that $N/N' \in \text{Ob Coh}(\mathscr{P})$, we see that $f^{-1}(N')$ is an open submodule of M and $M/f^{-1}(N') \in \text{Ob Coh}(\mathscr{P})$. Thus a unique morphism $\bar{f} : H_2(M) \to H_2(N)$ can be found such that the diagram:

$$\begin{array}{ccc} M & \xrightarrow{f} & N \\ {\scriptstyle \phi_M^2}\downarrow & & \downarrow{\scriptstyle \phi_N^2} \\ H_2(M) & \xrightarrow{\bar{f}} & H_2(N) \end{array}$$

is commutative. We put $\bar{f} = H_2(f)$. Finally, if $N \in \text{Ob CTC}(\mathscr{P})$, then $\phi^2{}_N$ is an isomorphism, so that the proof can be completed by Chapter 1, Theorem 5.1. ■

According to Lemmas 2.5 and 2.6 we check that the inclusion functor:

$$I = I_1 I_2 : \text{CTC}(\mathscr{P}) \to \text{TF}(\mathscr{P})$$

has a left adjoint $H = H_2 H_1$. We shall say that H is the functor of *coherent completion.* Also, by Lemma 1.18, the inclusion functor $J : \text{STC}(\mathscr{P}) \to \text{CTC}(\mathscr{P})$ has a right adjoint $S : \text{CTC}(\mathscr{P}) \to \text{STC}(\mathscr{P})$.

Now it is clear that the left linearly topological compact ring A (in the P-topology) is an object of $\text{TF}(\mathscr{P})$; thus $SH(A) \in \text{Ob STC}(\mathscr{P})$. We shall say that A is *strictly complete and coherent* if $A \in \mathrm{Ob}\,\mathrm{STC}(\mathscr{P})$, in the P-topology.

By a *co-Grothendieck category*, we shall mean an Ab 5* category having a set of cogenerators, i.e. a category whose dual is Grothendieck.

THEOREM 2.7. *Let P be a left prelocalizing system on the ring A such that A is left linearly topologically compact in the P-topology. Then* $\mathrm{STC}(\mathscr{P})$ *is a co-Grothendieck category having $SH(A)$ as a projective generator. The category* $\mathrm{Coh}(\mathscr{P})$ *is a full subcategory of* $\mathrm{STC}(\mathscr{P})$, *closed under subobjects, quotient objects and finite direct sums.*

Also, for any object X of $\mathrm{Coh}(\mathscr{P})$ *there exists a natural number $k \geqslant 1$ and an epimorphism $SH(A)^k \to X$ in* $\mathrm{STC}(\mathscr{P})$. *Moreover the objects of* $\mathrm{Coh}(\mathscr{P})$ *forms a set of congenerators of* STC $(\mathscr{P})$'

Proof. Firstly, we see that any morphism $h: A \to M$, where $M \in \mathrm{Ob}\,\mathrm{TC}(\mathscr{P})$, is a morphism of $\mathrm{TF}(\mathscr{P})$. Indeed, for each open submodule M' of M with $M/M' \in \mathrm{Ob}\,\mathrm{Coh}(\mathscr{P})$, there exists a monomorphism $A/h^{-1}(M') \to M/M'$ of A-modules, hence $A/h^{-1}(M')$ is an object of $\mathscr{P}$, or equivalently, $h^{-1}(M')$ is an open left ideal of A.

Now denote by $\phi: A \to H(A)$ and $\psi: SH(A) \to H(A)$ the corresponding morphism of adjunction. We claim that $SH(A)$ is a projective object of $\mathrm{STC}(\mathscr{P})$. For let $f: N \to M$ be an epimorphism of $\mathrm{STC}(\mathscr{P})$ and

$$g: SH(A) \to M$$

a morphism in $\mathrm{STC}(\mathscr{P})$. Thus we have the following diagram:

$$\begin{array}{ccccc} A & \xrightarrow{\phi} & H(A) & \underset{\psi^{-1}}{\overset{\psi}{\leftrightarrows}} & SH(A) \\ {\scriptstyle h}\downarrow & \swarrow{\scriptstyle k} & & & \downarrow{\scriptstyle g} \\ N & & \xrightarrow{f} & & M \longrightarrow 0 \end{array}$$

Since ψ is a bijective map and f is surjective, there exists a morphism h of A-modules such that $g\psi^{-1}\phi = fh$. By the above consideration, h is a morphism of $TF(\mathscr{P})$ so that there exists a morphism k such that $k\phi = h$. Thus one has:

$$fk\psi\psi^{-1}\phi = fk\phi = fh = g\psi^{-1}\phi$$

hence $fk\psi = g$, since $k\psi$ is a morphism of $\mathrm{STC}(\mathscr{P})$. Finally $SH(A)$ is a projective in $\mathrm{STC}(\mathscr{P})$, as claimed.

Furthermore, we shall prove that $SH(A)$ is a generator of $\mathrm{STC}(\mathscr{P})$. For that let $f: M \to N$ be a non-zero morphism of $\mathrm{STC}(\mathscr{P})$ and $x \in M$ such that $f(x) \neq 0$. Then in Mod A one has the commutative diagram:

$$\begin{array}{ccccccc} SH(A) & \xrightarrow{\psi} & H(A) & \xrightarrow{H(u_x)} & M & \xrightarrow{f} & N \\ & & \uparrow & \nearrow_{u_x} & & & \\ & & A & & & & \end{array}$$

Since $fu_x \neq 0$, one has that $fH(u_x) \neq 0$ hence $fH(u_x)\psi \neq 0$, and $H(u_x)\psi$ is a morphism of STC($\mathscr{P}$). Therefore $SH(A)$ is a generator of STC($\mathscr{P}$).

Finally let X be an object of Coh($\mathscr{P}$); then X is a finitely generated A-module and let $x_1, \ldots, x_k$ be its generators. Then the morphisms

$$H(u_{x_i})\psi : SH(A) \to X \quad \text{of} \quad \text{STC}(\mathscr{P}), \qquad i = 1, \ldots k,$$

define an epimorphism:

$$SH(A)^k \to X$$

in STC($\mathscr{P}$).

All other statements of the theorem will follow from Theorem 1.19. ■

6.3 The duality theorem for Grothendieck categories

A category $\mathscr{C}$ is called *skeletally small* if it is equivalent to a small category, i.e. if there exists a set of types of objects of $\mathscr{C}$. A full subcategory $\mathscr{C}'$ of a category $\mathscr{C}$ is called *generating subcategory* if its objects provide a class of generators of $\mathscr{C}$.

LEMMA 3.1. *If $\mathscr{C}$ is a Grothendieck category, then there is a full, skeletally small, finitely closed and generating subcategory $\mathscr{N}$ of $\mathscr{C}$ and an injective cogenerator E of $\mathscr{C}$ such that for each $X \in \mathrm{Ob}\,\mathscr{N}$ there is an exact sequence*:

$$0 \to X \to E^k \qquad k \in \mathbf{N}$$

Proof. Let D be a set of generators of $\mathscr{C}$ and $\mathscr{N}$ a full abelian subcategory of $\mathscr{C}$ containing D (see Chapter 4, Proposition 11.7). We can take as E an injective envelope of $\amalg X$, where X runs over all non-isomorphic objects of $\mathscr{N}$. ■

A Grothendieck category $\mathscr{C}$ being given, there are generally many pairs $(\mathscr{N}, E)$ of a full, skeletally small, finitely closed generating subcategory $\mathscr{N}$ of $\mathscr{C}$ and an injective cogenerator E such that each $X \in \mathrm{Ob}\,\mathscr{N}$ can be embedded into some E^k. For special $\mathscr{C}$'s however special pairs $(\mathscr{N}, E)$ are naturally associated with $\mathscr{C}$.

EXAMPLES 1. If $\mathscr{C}$ is a l.n. (respectively l.a. or l.f.)-category, one can take as $\mathscr{N}$ the full subcategory of all noetherian (respectively, artinian, respectively, finite) objects of $\mathscr{C}$. An appropriate choice for E is the direct sum of a representative system of indecomposable injective objects of $\mathscr{C}$ (see Chapter 5, Theorem 8.10).

2. Let $\mathscr{C}$ be a spectral Grothendieck category and U a generator of $\mathscr{C}$. Then U is also an injective cogenerator of $\mathscr{C}$. The full subcategory $\mathscr{N}$ of $\mathscr{C}$ of all subobjects of some U^k, $k \in \mathbf{N}$ is skeletally small, finitely closed and generates $\mathscr{C}$. The pair $(\mathscr{N}, U)$ satisfies the conditions of Lemma 3.1.

Now let $\mathscr{C}$ be a Grothendieck category and $(\mathscr{N}, E)$ a pair of a full, skeletally small, finitely closed and generating subcategory of $\mathscr{C}$ together with an injective cogenerator E such that for each $X \in \operatorname{Ob} \mathscr{N}$, there is an exact sequence:

$$0 \to X \to E^k \qquad k \in \mathbf{N}.$$

If $X \in \operatorname{Ob} \mathscr{C}$, then the abelian group $\operatorname{Hom}_{\mathscr{C}}(X, E)$ is canonically an $\operatorname{End}_{\mathscr{C}}(E) = R$-module. In fact, the assignment $X \rightsquigarrow \operatorname{Hom}_{\mathscr{C}}(X, E)$ defines a contravariant, faithful and exact functor:

$$G : \mathscr{C} \to \operatorname{Mod} R.$$

Sometimes instead of G, we shall consider the functor $G^0 : \mathscr{C}^0 \to \operatorname{Mod} R$, such that $G^0 D = G$, where $D : \mathscr{C} \to \mathscr{C}^0$ is the duality functor. It is clear that G commutes with projective limits.

If X' is a subobject of an object X of $\mathscr{C}$, we shall denote by $l_X(X')$ the image of the canonical monomorphism:

$$G(X/X') \to G(X)$$

(induced by the canonical epimorphism $X \to X/X'$). In fact

$$l_X(X') = \{f \in \operatorname{Hom}_{\mathscr{C}}(X, E), \quad \text{such that} \quad X' \subseteq \ker f\}.$$

It is obvious that $G(X)/l_X(X') \simeq G(X')$.

LEMMA 3.2. *For any $N \in \operatorname{Ob} \mathscr{N}$, $G(N)$ is a finitely generated R-module. Conversely, if $N \in \operatorname{Ob} \mathscr{N}$, then any finitely generated submodule of $G(N)$ is of the form $l_N(N')$ where $N' \in \operatorname{Ob} \mathscr{N}$.*

Proof. More generally, if $X \in \operatorname{Ob} \mathscr{C}$, $f_1, \ldots, f_n \in G(X)$ and $f : X \to E^n$, the unique morphism defined such that $p_i f = f_i$ (where $p_i : E^n \to E$ are structural projections) for each i, then we claim that $l_X(\ker f) = Rf_1 + \ldots + Rf_n$.

Indeed, let $j : \ker f \to X$ be the inclusion; thus $f_i j = p_i fj = 0$, so that $f_i \in l_X(\ker f)$, or equivalently $Rf_1 + \ldots + Rf_n \subseteq l_X(\ker f)$. Conversely, if

$g \in l_X(\ker f)$, one has the commutative diagram:

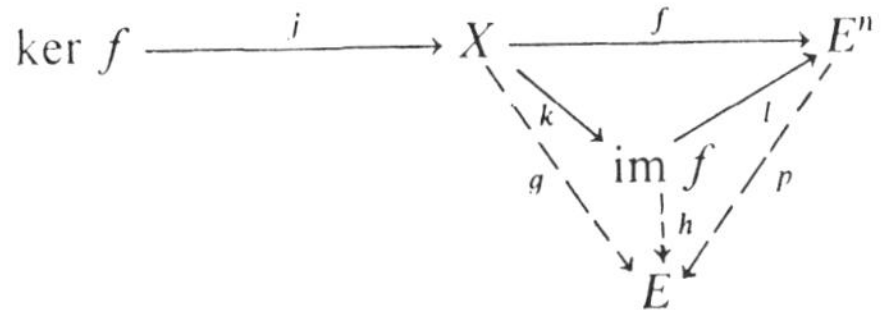

where $f = lk$ is the canonical decomposition of f, $hk = g$ (because $gj = 0$) and p is calculated by the injectivity of E, i.e. $pl = h$. Then $pf = plk = hk = g$. Also, if u_i are structural injections associated with p_i, then we have:

$$f = \textstyle\sum_i u_i p_i f = \sum_i u_i f_i.$$

Now, if $r_i = pu_i$, then one has:

$$g = pf = \textstyle\sum_i (pu_i) f_i = \sum_i r_i f_i \in Rf_1 + \ldots + Rf_n$$

as claimed.

Particularly, if $N \in \operatorname{Ob} \mathcal{N}$, then there is a monomorphism $f : N \to E^n$, $n \in \mathbf{N}$ so that:

$$l_N(\ker f) = Rf_1 + \ldots + Rf_n = l_N(0) = G(N),$$

where $f_i = p_i f$, $p_i : E^n \to E$ being structural projections. ∎

Note 1. Let $X \in \operatorname{Ob} \mathscr{C}$ and N, N' two of its subobjects in $\mathcal{N}$. Then,

$$l_X(N) \cap l_X(N') = l_X(N + N')$$

and obviously $N + N' \in \operatorname{Ob} \mathcal{N}$. Thus $G(X)$ has a unique topology such that $G(X)$ is a topological group and its submodules $l_X(N)$, $N \in \operatorname{Ob} \mathcal{N}$ is a fundamental system of neighbourhoods of zero. $G(X)$ will always be considered with this topology. Particularly R is a left linearly topological ring. We shall denote by $P(\mathcal{N})$ the prelocalizing system associated with this linear topology.

Lemma 3.3. *For any $N \in \operatorname{Ob} \mathcal{N}$, $G(N)$ is an a.l.c. R-module. Particularly, R is a left linearly topological compact ring.*

Proof. Let $\{F_i\}_i$ be a decreasing set of finitely generated submodules of $G(N)$. By Lemma 3.2 we have $F_i = l_N(N_i)$, $N_i \in \operatorname{Ob} \mathcal{N}$, for each i. Thus $\{N_i\}_i$ is direct set of subobjects of N and $\Sigma N_i \in \operatorname{Ob} \mathcal{N}$. Moreover, we have:

$$\textstyle\bigcap_i F_i = \bigcap_i l_N(N_i) = l_N(\sum_i N_i)$$

hence $\bigcap_i F_i$ is finitely generated by Lemma 3.2. Furthermore, we have:

$$\varprojlim_i (G(N)/F_i) = \varprojlim_i (G(N)/l_N(N_i)) \simeq \varprojlim_i G(N_i) =$$

$G(\varinjlim N_i) = G(N)/l_N(\varinjlim N_i) \simeq G(N)/ \bigcap_i F_i$.

In other words, the canonical map:

$$G(N) \to \varprojlim_i (G(N)/F_i)$$

is injective. ∎

As a direct consequence of Lemma 11.5 of Chapter 4 we obtain the following result:

LEMMA 3.4. *Let $N_1, N_2 \in \mathrm{Ob}\,\mathcal{N}$ and $g : G(N_1) \to G(N_2)$ a morphism of R-modules. Then there exists a unique morphism $f : N_2 \to N_1$ such that*

$$G(f) = g. \ ∎$$

LEMMA 3.5. *For any $N \in \mathrm{Ob}\,\mathcal{N}$, one has $G(N) \in \mathrm{Ob}\,\mathrm{Coh}(P(\mathcal{N}))$ and conversely for each object $T \in \mathrm{Ob}\,\mathrm{Coh}(P(\mathcal{N}))$, there exists $N \in \mathrm{Ob}\,\mathcal{N}$ such that $G(N) \simeq T$.*

Proof. For each $f \in G(N)$ we have: $\mathrm{Ann}(f) = l_E(\mathrm{im}\, f)$ and $\mathrm{im}\, f \in \mathrm{Ob}\,\mathcal{N}$, so that $\mathrm{Ann}(f)$ is an open left ideal of R, therefore $G(N) \in \mathrm{Ob}\,P(\mathcal{N})$. Moreover, it is known by Lemma 3.1 that $G(N)$ is finitely generated.

Now let T be a finitely generated object of $P(\mathcal{N})$. Thus there exist

$$N' \in \mathrm{Ob}\,\mathcal{N}$$

and an epimorphism of R-modules:

$$h : G(N') \to T.$$

Indeed, if $t_1, \dots t_n \in T$ are generators of T, then we have the canonical morphism of R-modules

$$\coprod_{i=1}^{n} (R/\mathrm{Ann}(t_i)) \xrightarrow{h_1} T.$$

Since $T \in \mathrm{Ob}\,P(\mathcal{N})$, there exists for each $i = 1, \dots, n$ a subobject N_i of E, $N_i \in \mathrm{Ob}\,\mathcal{N}$, such that $l_E(N_i) \subseteq \mathrm{Ann}(t_i)$. Then we have the canonical epimorphism of R-modules

$$G(N_i) = R/l_E(N_i) \xrightarrow{h_i} R/\mathrm{Ann}(t_i) \qquad i = 1, \dots n$$

and finally we get

$$h = \left(\coprod_{i=1}^{n} h_i\right).$$

Furthermore, if $f: T \to G(N)$ is a morphism of R-modules, then by Lemma 3.4 there exists a morphism $f': N \to N'$ in $\mathscr{N}$, such that $G(f') = fh$. Since E is an injective object, we have:

$$\ker(fh) = \ker G(f') = G(\operatorname{coker} f')$$

and $\operatorname{coker} f' \in \operatorname{Ob} \mathscr{N}$; thus, by Lemma 3.1, $\ker(fh)$ is finitely generated so that $\ker f = h(\ker fh)$ is also finitely generated. Therefore

$$G(N) \in \operatorname{Coh}(P(\mathscr{N})).$$

Hence, for $T \in \operatorname{Coh}(P(\mathscr{N}))$ we see that $\ker h$ is a finitely generated submodule of $G(N')$ so, by Lemma 3.2 $\ker h = l_{N'}(N'')$, $N'' \in \operatorname{Ob} \mathscr{N}$. Finally, since h is an epimorphism, one obtains

$$T \simeq G(N')/\ker h \simeq G(N')/1_{N'}(N'') \simeq G(N''). \blacksquare$$

NOTE 2. By Lemmas 3.4 and 3.5 we prove that G defines an equivalence between $\mathscr{N}^0$ and $\operatorname{Coh}(P(\mathscr{N}))$.

LEMMA 3.6. *For each $X \in \operatorname{Ob} \mathscr{C}$, $G(X)$ is an object of* $\operatorname{CTC}(P(\mathscr{N}))$, *and for each morphism $f: X \to Y$ of $\mathscr{C}$, the morphism $G(f): G(Y) \to G(X)$ is a morphism of* $\operatorname{CTC}(P(\mathscr{N}))$. *(In other words, the "image" of the functor G is contained in the subcategory* $\operatorname{CTC}(P(\mathscr{N}))$ *of* Mod R.)

Proof. By Lemma 3.5 and by the definition of the topology of $G(X)$ one checks that $G(X) \in \operatorname{Ob} \operatorname{TC}(P(\mathscr{N}))$.

Now the set $\{N_i\}_i$ of all subobjects of X in $\mathscr{N}$ is direct and complete by hypothesis, therefore one obtains:

$$G(X) = G(\varinjlim_i N_i) = \varprojlim_i G(N_i) \simeq \varprojlim (G(X)/l_X(N_i)),$$

in other words, $G(X)$ is a complete topological group (or equivalently $G(X) \in \operatorname{Ob} \operatorname{CTC}(P(\mathscr{N}))$.

If $f: X \to Y$ is a morphism of $\mathscr{C}$ and $N \subseteq X$ is a subobject of X in $\mathscr{N}$, then $G(f)^{-1}(l_X(N)) = l_Y(f(N))$, and $f(N) \in \operatorname{Ob} \mathscr{N}$, so that $G(f)$ is a continuous mapping or equivalently, $G(f)$ is a morphism of $\operatorname{CTC}(P(\mathscr{N}))$. $\blacksquare$

Henceforth we shall use the terminology of Section 6.1.

LEMMA 3.7. *Let $X \in \operatorname{Ob} \mathscr{C}$. Any special open submodule of $G(X)$ is of the form $l_X(N)$, where N is a subobject of X in $\mathscr{N}$.*

Proof. Let V be a special open submodule of $G(X)$; thus there exists a subobject N' of X, $N' \in \operatorname{Ob} \mathscr{N}$, $l_X(N') \subseteq V$. Since $G(X)/V \in \operatorname{Ob} \operatorname{Coh}(P(\mathscr{N}))$, by

Lemma 3.5 there exists $N \in \mathrm{Ob}\,\mathcal{N}$ such that $G(X)/V \simeq G(N)$. Then one obtains the following diagram:

$$G(N') \simeq G(X)/l_X(N') \xrightarrow{g} G(X)/V \xrightarrow{\sim} G(N),$$

in which g is a canonical epimorphism of R-modules. Now according to Lemma 3.4, $g = G(f)$, where $f : N \to N'$ is a morphism in $\mathscr{C}$. Finally f is necessarily a monomorphism of $\mathscr{C}$ and obviously $V = l_X(N)$. ∎

LEMMA 3.8. *Let* $X, Y \in \mathrm{Ob}\,\mathscr{C}$ *and* $f : G(Y) \to G(X)$ *a morphism of* $\mathrm{CTC}(P(\mathcal{N}))$. *Then there exists a unique morphism* $f' : X \to Y$ *of* $\mathscr{C}$ *such that* $G(f') = f$.

Proof. Let $\{N_i\}_i$ be the (directed and complete) set of all subobjects of X in $\mathcal{N}$. Then $f^{-1}(l_X(N_i))$ is a special open submodule of $G(Y)$ for each i (see Lemma 1.3) so that by Lemma 3.7 we have:

$$f^{-1}(l_X(N_i)) = l_Y(\overline{N}_i)$$

where $\overline{N}_i$ is a suitable subobject of Y in $\mathcal{N}$. Let $t_i : N_i \to X$ and $\bar{t}_i : \overline{N}_i \to Y$ be the canonical inclusions. Then the morphism f defines a unique morphism $f_i : G(\overline{N}_i) \to G(N_i)$ such that the following diagram, canonically constructed, is commutative:

$$\begin{array}{ccc} G(Y) & \xrightarrow{\quad f \quad} & G(X) \\ \downarrow{\scriptstyle G(\bar{t}_i)} & & \downarrow{\scriptstyle G(t_i)} \\ G(Y)/f^{-1}(l_X(N_i)) \simeq G(\overline{N}_i) & \xrightarrow{\ f_i\ } & G(N_i) \end{array}$$

Now by Lemma 3.4 there exists a unique morphism $f_i' : N_i \to \overline{N}_i$ such that $G(f_i') = f_i$. By a slight computation we prove that the morphism $\bar{t}_i f_i'$ defines in fact a morphism of the direct system $\{N_i\}_i$ to Y, so that there exists a unique morphism $f' : X \to Y$ such that $f' t_i = \bar{t}_i f_i'$. Finally we have $f = G(f')$. The uniqueness of f' is obvious. ∎

LEMMA 3.9. *For each* $X \in \mathrm{Ob}\,\mathscr{C}$, *one has* $G(X) \in \mathrm{Ob}\,\mathrm{STC}(P(\mathcal{N}))$. *Conversely, for each object* $M \in \mathrm{Ob}\,\mathrm{STC}(P(\mathcal{N}))$ *there exists an object* X *of* $\mathscr{C}$ *such that* $M \simeq G(X)$.

Proof. For each special closed submodule V of $G(X)$, there exists a directed family $\{N_i\}_i$ of subobjects of X such that

$$V = \textstyle\bigcap_i l_X(N_i) = l_X(\varinjlim N_i) \qquad \text{(see Lemma 1.12).}$$

Moreover, $G(X)/V \in \operatorname{Ob}\operatorname{Coh}(P(\mathcal{N}))$, so that according to Lemma 3.5, there exist $N \in \operatorname{Ob}\mathcal{N}$ and the diagram:

$$G(N) \xrightarrow{f} G(\varinjlim_i N_i) \overset{\sim}{\longleftarrow} G(X)/V$$

where f is an isomorphism of R-modules. Now, since $G(N)$ is an object of $\operatorname{CTC}(P(\mathcal{N}))$, (with the discrete topology), then by Lemma 3.8 there exists a morphism $f': \varinjlim N_i \to N$ of $\mathscr{C}$ such that $G(f') = f$. Finally we see that f' is necessarily an isomorphism, hence V is a special open submodule of $G(X)$.

Furthermore, let $M \in \operatorname{Ob}\operatorname{STC}(P(\mathcal{N}))$ and $\{M_i\}_i$ the set of its special open submodules. Then for each i, we have $M/M_i \in \operatorname{Ob}\operatorname{Coh}(P(\mathcal{N}))$ so that by Lemma 3.5 there exist an object N_i of $\mathcal{N}$ and an isomorphism

$$g_i : M/M_i \to G(N_i).$$

Now, if $M_i \subseteq M_j$, then we have the commutative diagram of R-modules

$$\begin{array}{ccc} M/M_i & \xrightarrow{p_{ij}} & M/M_j \\ {\scriptstyle g_i}\downarrow\wr & & {\scriptstyle g_j}\downarrow\wr \\ G(N_i) & \xrightarrow{G(u_{ij})} & G(N_j) \end{array}$$

where p_{ij} is the canonical epimorphism and u_{ij} is calculated by Lemma 3.4 in a canonical way. Thus we have in $\mathscr{C}$ the direct system $\{N_i, u_{ij}\}$, let

$$\varinjlim_i N_i$$

be its inductive limit. Finally we leave to the reader to check that

$$M \simeq \varprojlim_i (M/M_i) \simeq \varprojlim G(N_i) \simeq G(\varinjlim N_i). \ \blacksquare$$

Summarizing all previous lemmas, we have the following result.

THEOREM 3.10. *Let $\mathscr{C}$ be a Grothendieck category and $(\mathcal{N}, E)$ a pair consisting of a full, skeletally small, finitely closed and generating subcategory of $\mathscr{C}$ together with an injective cogenerator E such that $X \in \operatorname{Ob}\mathcal{N}$, X is isomorphic with a subobject of E^k, for some natural number k. The ring $R = \operatorname{End}_{\mathscr{C}}(E)$ is a strict complete topologically left coherent and linearly compact ring. Moreover the functor $G^0 : \mathscr{C}^0 \to \operatorname{Mod} R$ defines an equivalence of categories between $\mathscr{C}^0$ and $\operatorname{STC}(P(\mathcal{N}))$.* $\blacksquare$

THEOREM 3.11. *Let R be a strict complete topologically left coherent and linearly compact ring. Then there exist a Grothendieck category $\mathscr{C}$ and an injective cogenerator E of $\mathscr{C}$ such that $R \simeq \operatorname{End}_{\mathscr{C}}(E)$.* $\blacksquare$

Exercises

1. Let A be a ring, $\mathcal{N}$ the full subcategory of Mod A of all submodules of finitely generated modules and E an injective A-module. The following assertions are equivalent:

(a) The module E is a cogenerator of Mod A, and for each $X \in \operatorname{Ob} \mathcal{N}$, there exists an exact sequence:

$$0 \to X \to E^k \qquad (k \in \mathbf{N}),$$

(b) For each left ideal $\mathfrak{a}$ of A there are finitely many elements $x_i \in E$, such that $\mathfrak{a} = \bigcap_i \operatorname{Ann}(x_i)$.

2. Let A be a ring and $\mathcal{N}$ the full subcategory of Mod A, as in Exercise 1. The following assertions are equivalent:

(a) The module A is an injective cogenerator of Mod A, and for each $X \in \operatorname{Ob} \mathcal{N}$, there is an exact sequence

$$0 \to X \to A^k \qquad (k \in \mathbf{N});$$

(b) The ring A is left self-injective and for each left ideal $\mathfrak{a}$ of A there are finitely many elements x_i in A such that $\mathfrak{a} = \bigcap_i \operatorname{Ann}(x_i)$.

3. Let $\mathscr{C}$ be a Grothendieck category and R a strict complete topologically left coherent and linearly compact ring. Assume that E is an injective cogenerator of $\mathscr{C}$ and that $h: R \to \operatorname{End}_{\mathscr{C}}(E)$ is a ring isomorphism (no topology here). Identify $R = \operatorname{End}_{\mathscr{C}}(E)$, via h. Let $\mathcal{N}$ be the full subcategory of $\mathscr{C}$ of all those N such that $\operatorname{Hom}_{\mathscr{C}}(N, E) \in \operatorname{Coh}(\mathscr{P})$, (where $\mathscr{P}$ is the left prelocalizing system on R, which defines the given topology of R) and there is a monomorphism $N \to E^k$, for some natural number k.

(a) Assume that for each special open left ideal $\mathfrak{a}$ of R, there is a subobject $N \subseteq E$ with $\mathfrak{a} = \operatorname{Hom}_{\mathscr{C}}(E/N, E)$. Then the functor $X \rightsquigarrow \operatorname{Hom}_{\mathscr{C}}(X, E)$ induces an equivalence of categories

$$\mathcal{N}^0 \xrightarrow{\sim} \operatorname{Coh}(\mathscr{P});$$

(b) If $N \in \operatorname{Ob} \mathcal{N}$ and if $N' \subseteq N$ is a subobject of N, such that N/N' can be embedded into some E^k, where k is a natural number, then $N' \in \operatorname{Ob} \mathcal{N}$.

4. Assumptions as in Exercise 1. Assume, in addition, that for each subobject $X \subseteq E$, the left ideal $\operatorname{Hom}_{\mathscr{C}}(E/X, E)$ of $R = \operatorname{End}_{\mathscr{C}}(E)$ is special closed. Then $\mathcal{N}$ is finitely closed.

5. Let $\mathscr{C}$ be a Grothendieck category and R a strict complete topologically left coherent and linearly compact ring. The following assertions are equivalent:

(a) $\mathscr{C}$ and $\operatorname{STC}(\mathscr{P})$ are dual to each other (here $\mathscr{P}$ is a left prelocalizing system which defines the topology of R).

(b) There are an injective cogenerator E of $\mathscr{C}$ and a ring isomorphism $h: R \to \mathrm{End}_{\mathscr{C}}(E)$ (identify $R = \mathrm{End}_{\mathscr{C}}(E)$, via h) with the following properties:

(i) The subobjects X of E^k, k a natural number, form a set of generators of $\mathscr{C}$.

(ii) For each subobject $X \subseteq E$, the left ideal $\mathrm{Hom}_{\mathscr{C}}(E/X, E)$ of R is special closed and for each special open left ideal $\mathfrak{a}$ of R there is a subobject X of E with $\mathfrak{a} = \mathrm{Hom}_{\mathscr{C}}(E/X, E)$.

6. (Topological Morita theorem). Let R and R' be strict complete topologically right coherent and linearly compact rings. Then the following assertions are equivalent:

(a) The categories $\mathrm{STC}(\mathscr{P})$ and $\mathrm{STC}(\mathscr{P}')$ are equivalent. (Here $\mathscr{P}$ (respectively, $\mathscr{P}'$) is the left prelocalizing system of R (respectively, R') consisting of all open left ideals).

(b) There exist a projective generator U in $\mathrm{STC}(\mathscr{P}')$ and a ring isomorphism $h: R \to \mathrm{End}_{\mathrm{STC}(\mathscr{P}')}(U)$ (identify these two rings via h) with the following properties:

(i) The quotient objects of all U^k, k a natural number, form a family of cogenerators of $\mathrm{STC}(\mathscr{P}')$.

(ii) For each special closed submodule Y of U the right ideal $\mathrm{Hom}_{\mathrm{CTC}(\mathscr{P}')}(U, Y)$ of R is special closed, and for each special open right ideal $\mathfrak{a}$ of R, there is a special closed submodule Y of U, with

$$\mathfrak{a} = \mathrm{Hom}_{\mathrm{CTC}(\mathscr{P}')}(U, Y).$$

If (a) and (b) are satisfied, the equivalence is given by:

$$Y \rightsquigarrow \mathrm{Hom}_{\mathrm{STC}(\mathscr{P}')}(U, Y).$$

7. Let R be a ring and E an injective R^0-module such that for each right ideal $\mathfrak{a}$ of R, there is a finite family $x_1, \ldots, x_n$ of elements in E with

$$\mathfrak{a} = \bigcap_{i=1}^{n} \mathrm{Ann}(x_i).$$

Denote $S = \mathrm{End}_{R^0}(E)$ and by $\mathscr{N}$ the full subcategory of $\mathrm{Mod}\, R^0$, consisting of all submodules of finitely generated R^0-modules (see Exercise 1).

(a) With the linear topology defined by $P(\mathscr{N})$ the ring S is a strict complete topologically left coherent and linearly compact ring, and with the canonical S-module structure and the discrete topology E is in $\mathrm{STC}(P(\mathscr{N}))$.

(b) There are the inverse dualities:

$$\mathrm{Mod}\, R^0 \underset{\mathrm{Hom}_{STC(P(\mathscr{N}))}(?, E)}{\overset{\mathrm{Hom}_{R^0}(?, E)}{\rightleftarrows}} \mathrm{STC}(P(\mathscr{N})).$$

The topology on $\mathrm{Hom}_R(X, E)$, $X \in \mathrm{Ob}\,\mathrm{Mod}\,R^0$ is canonically defined. (See Note 1). The R^0-module structure of

$$\mathrm{Hom}_{\mathrm{STC}(P(\mathcal{N}))}(X, E),\ X \in \mathrm{Ob}\,\mathrm{STC}(P(\mathcal{N}))$$

is induced by that of E.

8. Let S be a strict complete topologically left coherent and linearly compact ring. Then $\mathrm{STC}(\mathscr{P})$ is dual to a module category $\mathrm{Mod}\,R^0$ if and only if $\mathrm{STC}(\mathscr{P})$ admits an injective cogenerator of cofinite type. If E is such a cogenerator and if $R = \mathrm{End}_{\mathrm{STC}(\mathcal{N})}(E)$, then a duality is given by

$$X \rightsquigarrow \mathrm{Hom}_{\mathrm{STC}(\mathcal{N})}(X, E), \qquad X \in \mathrm{Ob}\,\mathrm{STC}(\mathscr{P}).$$

(Here P is the left prelocalizing system on S which defines the topology of S).

9. Let R be a ring. The following assertions are equivalent:

(a) R is right injective ring and each right ideal of R is the annihilator of a finite set of elements of R (such a ring is called a *generalized right Quasi-Frobenius ring*).

(b) (i) R is algebraically left coherent and linearly compact.

(ii) If $\{\mathfrak{a}_i\}_i$ is an inverse family of finitely generated left ideals of R with $\bigcap_i \mathfrak{a}_i = 0$, then $\mathfrak{a}_i = 0$ for some i.

(iii) If $\mathfrak{a}$ is a finitely generated left ideal of R, then the natural inclusion $\mathfrak{a} \to R$ induces the epimorphism $\mathrm{Hom}_R(\mathfrak{a}, R) \to \mathrm{Hom}_R(R, R)$.

(iv) The right annihilator of any proper finitely generated left ideal $\mathfrak{a}$ of R is not zero. (Hint: use Exercise 2).

10. A regular ring R is right self-injective if and only if R is algebraically left linearly compact.

6.4 Duality theory for l.n. and l.f.-categories

Let A be a ring and $\mathscr{S}$ a finitely closed full subcategory of $\mathrm{Mod}\,A$.

PROPOSITION 4.1. *If $\mathscr{S}$ is an artinian category, then $\mathscr{S}$ is a linearly compact subcategory of* Mod A. *Moreover $\tilde{\mathscr{P}} = \mathscr{P}$ and $(\mathscr{P})^0$ is a l.n.-category.*

Proof. If $\mathscr{S}$ is artinian, then for each object $M \in \mathrm{Ob}\,\mathscr{S}$ and every inverse set $\{M_i\}_i$ of its subobjects with $M_i \in \mathrm{Ob}\,\mathscr{S}$ for any i, there exists an index i', such that $\bigcap_i M_i = M_{i'}$. Particularly $\mathscr{S}$ is a linearly compact subcategory. Now we shall prove that $\tilde{\mathscr{P}} = \mathscr{P}$. To do this, let $M \in \mathrm{Ob}\,\mathscr{P}$ and N a special closed submodule of M. By Lemma 1.12(b), M/N equipped with the quotient topology is an object of $\mathscr{P}$, therefore if $\{U_i\}_i$ is the set of all special open submodules of M, then $\{(N + U_i)/N\}_i$ is a family of subobjects of M/N belonging to $\mathscr{S}$ and the canonical mapping:

$$M/N \xrightarrow{f} \varprojlim_i (M/(N + U_i))$$

is an isomorphism of R-modules. Now if $M/N \in \mathrm{Ob}\,\mathscr{S}$, then there exists an index i' such that

$$(N + U_{i'})/N = \bigcap_i((N + U_i)/N) = \ker f = 0,$$

hence $N = N + U_{i'}$ is special open in M (see Lemma 1.1). Finally we have $M \in \mathrm{Ob}\,\tilde{\mathscr{S}}$ as claimed. ∎

An A-module M is called *coperfect* if for each inverse family $\{M_i\}_i$ of its finitely generated subobjects, there exists an index i' with $\bigcap_i M_i = M_{i'}$. Obviously, any coperfect A-module is algebraically linearly compact. Thus, as a direct consequence of Lemma 2.1 one has the following result:

LEMMA 4.2. *Let* $0 \to N \to M \to P \to 0$ *be an exact sequence of* Mod A *with* M *coperfect and* N *finitely generated. Then* P *is also coperfect.* ∎

LEMMA 4.3. *Let* $\mathscr{P}$ *be a prelocalizing subcategory of* Mod A. *If* M *is a coherent object of* $\mathscr{P}$ *and* N *a finitely generated submodule of* M, *such that* N *and* M/N *are coperfect modules then* M *is also coperfect.*

Proof. If $\{M_i\}_i$ is an inverse set of finitely generated submodules of M, one has the following exact sequence, canonically constructed:

$$0 \to \bigcap_i (N \cap M_i) \to \bigcap_i M_i \to \bigcap_i (N + M_i)/N \to 0$$

(see Lemma 1.12). Thus by hypothesis, there exists an index i' such that

$$\bigcap_i (N \cap M_i) = N \cap M_{i'} \quad \text{and} \quad \bigcap_i (N + M_i)/N = (N + M_{i'})/N.$$

By a slight computation, one checks that $\bigcap_i M_i = M_{i'}$. ∎

COROLLARY 4.4. *Let* $\mathscr{P}$ *and* M *be as in the previous lemma. If* N *and* P *are two finitely generated and coperfect submodules of* M, *then* $N + P$ *is coherent and coperfect.*

The proof follows by Lemma 4.3 and Lemma 2.3. ∎

A topological ring A is called *left coperfect* if it has a fundamental system of neighbourhoods of zero, consisting of left ideals $\mathfrak{a}$ with $A/\mathfrak{a}$ a coperfect module. Obviously, any topological left coperfect ring is linearly compact. A topological left coperfect ring A is complete and coherent if it is an object of CTC($\mathscr{P}$). (As always by $\mathscr{P}$ we shall denote the left prelocalizing system of all open left ideals of A).

THEOREM 4.5. *Let A be a topological left coperfect ring. Then the category* CTC($\mathscr{P}$) *is a l.a.-category. The full subcategory* Coh($\mathscr{P}$) *of* CTC($\mathscr{P}$) *is just the category* $\mathscr{A}$(CTC($\mathscr{P}$)), *the full subcategory of all artinian objects. Moreover, the object $H(A)$ is a projective generator of* CTC($\mathscr{P}$), *namely a direct product $\prod_i K_i$ of indecomposable projective objects and any indecomposable projective object K of* CTC($\mathscr{P}$) *is isomorphic with K_i for a suitable i.*

Proof. As in Theorem 2.7, using the previous lemmas we may verify that Coh($\mathscr{P}$) is an artinian category. Also by Proposition 4.1 and Theorem 2.7 we deduce that $(\mathrm{CTC}(\mathscr{P}))^0$ is a l.n.-category.

Now if M is an artinian object of CTC($\mathscr{P}$), and $\{M_i\}_i$ is the inverse set of its special open submodules, then $0 = \bigcap_i M_i = M_{i'}$ for a suitable index i', so that $M = M/M_{i'} \in \mathrm{Ob}\,\mathrm{Coh}(\mathscr{P})$.

Furthermore, let K be an indecomposable projective object of CTC($\mathscr{P}$) and U a special open submodule of P. Then U is a subobject of K in CTC($\mathscr{P}$) and one has a canonical epimorphism (in CTC($\mathscr{P}$))

$$K \xrightarrow{f} K/U \longrightarrow 0. \tag{3}$$

It is clear that if $K/U \neq 0$ then (3) is a projective envelope of K/U. But $K/U \in \mathrm{Ob}\,\mathrm{Coh}(\mathscr{P})$, so that by Theorem 2.7, there exists a natural integer $k \geqslant 1$ and the epimorphism

$$H(A)^k \to K/U \to 0.$$

Now it is clear that $H(A)^k$ is projective and K can be viewed as a direct summand of $H(A)^k$. If

$$H(A) = \textstyle\prod_i K_i$$

is a representation of $H(A)$ as a direct product of indecomposable projectives (see Chapter 5, Theorem 8.10), then one has:

$$H(A)^k \simeq \textstyle\prod_i K_i^k,$$

so that by Theorem 1.3, of Chapter 5, one has that $K = K_i$ for a suitable i. ∎

A topological ring A is called *left pseudo-compact*, if it is complete, Hausdorff and has a fundamental system of neighbourhoods of zero consisting of left ideals $\mathfrak{a}$ for which $A/\mathfrak{a}$ is a finite A-module. Obviously, a left pseudo-compact ring is left coperfect, complete and coherent, and according to Theorem 4.5, the category $(\mathrm{CTC}(\mathscr{P}))^0$ is a l.f.-category.

Now let $\mathscr{C}$ be a l.n.-category and $\mathscr{N} = \mathscr{N}(\mathscr{C})$ the full subcategory of all noetherian objects of $\mathscr{C}$. Obviously $\mathscr{N}$ is a skeletally small and finitely closed subcategory. An injective object E of $\mathscr{C}$ is called *big injective*, if for any indecomposable injective object Q of $\mathscr{C}$, there exists a monomorphism $Q \to E$. Obviously, a big injective object is a cogenerator of $\mathscr{C}$ and for any noetherian object N of $\mathscr{C}$, there exist a natural number k and an embedding

$$0 \to N \to E^k.$$

Further, we shall fix a big injective E of $\mathscr{C}$. Let $R = \operatorname{End}_{\mathscr{C}}(E)$ and

$$G : \mathscr{C} \to \operatorname{Mod} R$$

be the functor $G(X) = \operatorname{Hom}_{\mathscr{C}}(X, E)$. In the sequel we shall use the notation of Section 6.3. The following result is a direct consequence of Lemma 3.1.

LEMMA 4.6. *For each noetherian object N of $\mathscr{C}$, the R-module $G(N)$ is coperfect. Particularly R is canonically a topological left coperfect ring.* ■

By the previous results and general results of Section 6.3, one has the following important result:

THEOREM 4.7. *Let $\mathscr{C}$ be a l.n.-category $\mathscr{N} = \mathscr{N}(\mathscr{C})$ and E a big injective of $\mathscr{C}$. Then $R = \operatorname{End}_{\mathscr{C}}(E)$ is a topological left coperfect, complete and coherent ring and the functor*

$$G : \mathscr{C} \to \operatorname{CTC}(P(\mathscr{N})),$$

$G(X) = \operatorname{Hom}_{\mathscr{C}}(X, E)$ *defines an equivalence between* $\mathscr{C}^0$ *and* $\operatorname{CTC}(P(\mathscr{N}))$. *Moreover the restriction of functor G to $\mathscr{N}$ defines an equivalence between $\mathscr{N}^0$ and* $\operatorname{Coh}(P(\mathscr{N}))$.

Conversely, if R is a topological left coperfect, complete and coherent ring, then there exist a l.n.-category $\mathscr{C}$ and a big injective E of $\mathscr{C}$ such that

$$R \simeq \operatorname{End}_{\mathscr{C}}(E). \blacksquare$$

Finally let $\mathscr{C}$ be a l.f.-category and E a big injective object of $\mathscr{C}$. Then for each object $N \in \operatorname{Ob} \operatorname{Fin}(\mathscr{C})$ one has the exact sequence:

$$0 \to N \to E^k$$

for some natural number k. We shall consider the ring $R = \operatorname{End}_{\mathscr{C}}(E)$ canonically topologized and having as a fundamental system of neighbourhoods the left ideals $l_E(N)$, where $N \in \operatorname{Ob} \operatorname{Fin}(\mathscr{C})$. By Lemma 4.6 we see that R is a topological coperfect complete and coherent ring, and the functor

$$X \rightsquigarrow \operatorname{Hom}_{\mathscr{C}}(X, E)$$

defines an equivalence between $\mathscr{C}^0$ and $\operatorname{CTC}(P(\operatorname{Fin}(\mathscr{C})))$. Particularly, for each $N \in \operatorname{Ob} \operatorname{Fin}(\mathscr{C})$, the object $\operatorname{Hom}_{\mathscr{C}}(N, E)$ is a finite object of

$$\operatorname{CTC}(P(\operatorname{Fin}(\mathscr{C}))).$$

Moreover, one has the following result:

THEOREM 4.8. *Let $\mathscr{C}$ be a l.f.-category and E a big injective of $\mathscr{C}$. Then the ring $R = \mathrm{End}_{\mathscr{C}}(E)$ (topologized as above) is a topological left pseudo-compact ring. Conversely if R is a topological left pseudo-compact ring, there exists a l.f.-category $\mathscr{C}$ and a big injective E of $\mathscr{C}$ such that $R \simeq \mathrm{End}_{\mathscr{C}}(E)$.*

Proof. Assume that R is a topological left coperfect, complete and coherent. Also let $X \in \mathrm{Coh}(\mathscr{P})$ be a simple object of $\mathrm{CTC}(\mathscr{P})$, and X' a R-submodule of X. Then for each $x \in X'$ one has $Rx \simeq R/\mathrm{Ann}(x) \in \mathrm{Ob}\,\mathrm{Coh}(\mathscr{P})$. Since X is coherent in $\mathscr{P}$ and Rx a finitely generated submodule of $\mathscr{P}$, then

$$Rx \in \mathrm{Ob}\,\mathrm{Coh}(\mathscr{P}) \subseteq CTC(\mathscr{P}),$$

so that $Rx = 0$ or $Rx = X$, thus if $X' \neq 0$, then $X' = X$. Then we check that X is a finite object of $\mathrm{Coh}(\mathscr{P})$, so that X is also a finite object of Mod R.

Now, if $\mathscr{C}$ is a l.f.-category and $R = \mathrm{End}_{\mathscr{C}}(E)$ (E a big injective of $\mathscr{C}$), then for each $N \subseteq E, N \in \mathrm{Ob}\,\mathrm{Fin}(\mathscr{C})$, $R/l_E(N) \simeq \mathrm{Hom}_{\mathscr{C}}(N, E) \in \mathrm{Ob}\,\mathrm{Coh}(P(\mathrm{Fin}(\mathscr{C})))$ and this last object is a finite object of $\mathrm{CTC}(P(\mathrm{Fin}(\mathscr{C})))$. Therefore R is a topological left pseudo-compact ring. The other statements of the theorem are obvious. ∎

Exercises

1. Let A be a linearly topological left coherent, coperfect and complete ring. Then $A/\mathfrak{R}(A)$ is a regular left self-injective ring of the form $\prod_i \mathrm{End}_{K_i}(V_i)$ where the skew fields K_i and the K_i-vector spaces V_i are uniquely determined. (Hint: use Theorem 4.7 and Exercise 2).

2. Let Q be an injective object of a Grothendieck category $\mathscr{C}$ such that $h(Q)$ (the height of Q) is defined and $h(Q) \leqslant \aleph_0$. Let also $A = \mathrm{End}_{\mathscr{C}}(Q)$. Then:

(a) $\mathfrak{R}(A)$ is the subset of A consisting of all elements f, such that

$$\ker f \supseteq \mathrm{So}(Q);$$

(b) $\bigcap_n \mathfrak{R}(A)^n = 0$;

(c) $A/\mathfrak{R}(A)$ is a direct product (in a unique way) of rings of endomorphisms of vector spaces over certain skew fields.

Moreover, the above results are valid for each injective object of a finite category. Particularly $h(X) \leqslant \aleph_0$ for each object X of a l.f.-category.

3. Let $\mathscr{C}$ be a l.n.-category, Q an injective object of $\mathscr{C}$ and $\mathscr{F} = \mathscr{F}_Q$ (the localizing subcategory generated by all objects X such that $\mathrm{Hom}_{\mathscr{C}}(X, Q) = 0$). We shall equip $R = \mathrm{End}_{\mathscr{C}}(Q)$ with the topology having a fundamental system of neighbourhoods consisting of all left ideals of the form $l_Q(N)$

where N is a noetherian subobject of Q (denote by $\mathscr{P}$ the set of all left open ideals). Then R is a topological left coperfect, complete and coherent ring and the functor:

$$M : \mathscr{C} \to \mathrm{CTC}(\mathscr{P}),$$

defined as $M(X) = \mathrm{Hom}_{\mathscr{C}}(X, Q)$ define an equivalence between $(\mathscr{C}/\mathscr{F})^0$ and $\mathrm{CTC}(\mathscr{P})$.

4. By a *sober* ring we shall mean a topological left pseudo-compact ring A so that $A/\mathfrak{R}(A)$ is a product of skew fields. We shall say that two topological left pseudo-compact rings A and B are *equivalent* if the categories $\mathrm{CTC}(P(A))$ and $\mathrm{CTC}(P(B))$ are equivalent (here $P(A)$ and $P(B)$ are respectively, left prelocalizing system of all open left ideals on A respectively, on B).

Let $\mathscr{C}$ be a l.f.-category and $E = \coprod Q$ where Q runs over all types of indecomposable injectives of $\mathscr{C}$; then E is a big injective of $\mathscr{C}$ and the ring $\mathrm{End}_{\mathscr{C}}(E)$ (canonically topologized as in Theorem 4.8) is sober. Also prove that any topological left pseudo-compact ring is equivalent of a sober ring, and any two sober rings are equivalent if and only if they are isomorphic.

5. By a *quasi-Frobenius category* we shall mean a Grothendieck category $\mathscr{C}$ such that every injective object of $\mathscr{C}$ is also projective. Let $\mathscr{P}$ be a left prelocalizing system of a ring A so that A becomes a left linearly topological, complete, coherent and coperfect ring. The following assertions are equivalent:

(a) $(\mathrm{TC}(\mathscr{P}))^0$ is a quasi-Frobenius category;

(b) A is an injective object in $\mathrm{TC}(\mathscr{P})$;

(c) For every coherent, closed left ideal $\mathfrak{a}$ of A, every continuous A-linear map $\mathfrak{a} \to A$ is a right multiplication by an element of A.

6. Prove that a right perfect, left coherent and left self-injective ring is a QF-ring.

6.5 Colocalization

Let A be a strict complete topologically left coherent and linearly compact ring. We shall denote by $\mathrm{STC}(A)$ the category $\mathrm{STC}(\mathscr{P})$, where P is the prelocalizing system of all open left ideals of A.

A full subcategory $\mathscr{C}$ of $\mathrm{STC}(A)$ is called *precolocalizing* if its dual in $(\mathrm{STC}(A))^0$, is prelocalizing or equivalently if it is closed with respect to subobjects, quotient objects and direct products in $\mathrm{STC}(A)$.

If $\mathscr{C}$ is a precolocalizing subcategory of $\mathrm{STC}(A)$ we shall denote by C the set of all left ideals $\mathfrak{a}$ of A in $\mathrm{STC}(A)$, such that $A/\mathfrak{a} \in \mathrm{Ob}\,\mathscr{C}$.

LEMMA 5.1. *The set C has the following properties*:

(i) *If* $\mathfrak{a} \in C$ *and* $\mathfrak{b}$ *is a subobject of* A (*in* STC(A)) *and* $\mathfrak{a} \subseteq \mathfrak{b}$, *then* $\mathfrak{b} \in C$;

(ii) *If* $\{\mathfrak{a}_i\}_i$ *is a set of elements of* C, *then* $\bigcap_i \mathfrak{a}_i \in C$;

(iii) *If* $\mathfrak{a} \in C$ *then* $(\mathfrak{a} : x) \in C$ *for each* $x \in A$.

LEMMA 5.2. *Let* $\mathfrak{a}_C = \bigcap_{\mathfrak{a} \in C} \mathfrak{a}$. *Then*:

(i) $\mathfrak{a}_C \in C$, *and for each subobject* $\mathfrak{a}$ *of* A *in* STC(A), *we have* $\mathfrak{a} \in C$ *if and only if* $\mathfrak{a}_C \subseteq \mathfrak{a}$;

(ii) $\mathfrak{a}_C$ *is an ideal of* A. ■

THEOREM 5.3. *Let* $\mathfrak{a}$ *be a subobject of* A *in* STC(A), *such that* $\mathfrak{a}$ *is a two-sided ideal. Then the ring* $A/\mathfrak{a}$ *equipped with the quotient topology is a strict complete topologically left coherent and linearly compact ring. The category* STC($A/\mathfrak{a}$) *is canonically a precolocalizing subcategory of* STC(A).

Proof. It is easy to see that the assignment:

$$\mathfrak{b} \rightsquigarrow \mathfrak{b}/\mathfrak{a}$$

defines a bijection between the set of all left ideals of A which contain $\mathfrak{a}$ and the left ideals of $A/\mathfrak{a}$.

Now with the quotient topology $A/\mathfrak{a}$ is a linearly topologized ring, and as A-module $A/\mathfrak{a}$ belongs to STC(A).

It is clear that Mod($A/\mathfrak{a}$) is identified with the full subcategory of Mod A generated by all A-modules M such that $\mathfrak{a} \subseteq \operatorname{Ann}(x)$ for each $x \in M$. Also it is clear that an object M of Mod($A/\mathfrak{a}$) is an $A/\mathfrak{a}$-a.l.c. module if and only if it is a.l.c. considered as A-module. Thus, if $\mathfrak{b}$ is a left open ideal of A, with $\mathfrak{a} \subseteq \mathfrak{b}$, and $A/\mathfrak{b}$ is an a.l.c. A-module, then $\mathfrak{b}/\mathfrak{a}$ is a left open ideal of $A/\mathfrak{a}$ and $(A/\mathfrak{a})/(\mathfrak{b}/\mathfrak{a}) \simeq A/\mathfrak{b}$ is an a.l.c. $A/\mathfrak{a}$-module. In conclusion we check that $A/\mathfrak{a}$ is linearly topologized and left compact.

Furthermore, let M be an $A/\mathfrak{a}$-module; then for each $x \in M$ we have $\operatorname{Ann}_{A/\mathfrak{a}}(x) = \operatorname{Ann}_A(x)/\mathfrak{a}$, so that $\operatorname{Ann}_{A/\mathfrak{a}}(x)$ is open in $A/\mathfrak{a}$ if and only if $\operatorname{Ann}_A(x)$ is open in A. Thus we have verified that

$$\operatorname{Dis}(A/\mathfrak{a}) = \operatorname{Dis}(A) \cap \operatorname{Mod}(A/\mathfrak{a}),$$

where by Dis(A) we shall denote the prelocalizing subcategory associated with the prelocalizing system of all open left ideals. Also we have:

$$\operatorname{Coh}(\operatorname{Dis}(A/\mathfrak{a})) = \operatorname{Coh}(\operatorname{Dis}(A)) \cap \operatorname{Mod}(A/\mathfrak{a}).$$

From this last equality, it will follow that $M \in \operatorname{Ob}\operatorname{STC}(A/\mathfrak{a})$ if and only if $M \in \operatorname{Ob}\operatorname{STC}(A)$ and for each $x \in M$ we have $\mathfrak{a} \subseteq \operatorname{Ann}(x)$. Since

$$A/\mathfrak{a} \in \operatorname{Ob}\operatorname{STC}(A),$$

then $A/\mathfrak{a} \in \mathrm{Ob}\,\mathrm{STC}(A/\mathfrak{a})$ so that $A/\mathfrak{a}$ is strict complete and coherent. Finally it is obvious that $\mathrm{STC}(A/\mathfrak{a})$ is a precolocalizing subcategory of $\mathrm{STC}(A)$. ∎

THEOREM 5.4. *The mapping*

$$\mathscr{C} \rightsquigarrow \mathfrak{a}_C$$

is a bijection between the set of all precolocalizing subcategories of $\mathrm{STC}(A)$ *and the set of all two-sided ideals of* A *which are subobjects of* A *in* $\mathrm{STC}(A)$. *An inverse mapping is given by*:

$$\mathfrak{a} \rightsquigarrow \mathrm{STC}(A/\mathfrak{a}).$$

Proof. Let $\mathscr{C}$ be precolocalizing subcategory of $\mathrm{STC}(A)$. We shall prove that an object M of $\mathrm{STC}(A)$ belongs to $\mathscr{C}$ if and only if for each $x \in M$, we have $\mathfrak{a}_C \subseteq \mathrm{Ann}(x)$, or equivalently

$$\mathrm{STC}(A/\mathfrak{a}_C) \equiv \mathscr{C}.$$

Firstly we observe that for $M \in \mathrm{Ob}\,\mathrm{STC}(A)$ and $x \in M$, the mapping $u_x : A \to M$, is a morphism of $\mathrm{STC}(A)$. In other words, $\mathrm{Ann}(x)$ is a subobject of A in $\mathrm{STC}(A)$ and one obtains the monomorphism in $\mathrm{STC}(A)$:

$$0 \to A/\mathrm{Ann}(x) \to M.$$

Therefore, if $M \in \mathrm{Ob}\,\mathscr{C}$ then $A/\mathrm{Ann}(x) \in \mathrm{Ob}\,\mathscr{C}$ and thus $\mathfrak{a}_C \subseteq \mathrm{Ann}(x)$.

Conversely, let $M \in \mathrm{Ob}\,\mathrm{STC}(A)$ be such that for each $x \in M$, $\mathfrak{a}_C \subseteq \mathrm{Ann}(x)$. Then we must prove that $M \in \mathrm{Ob}\,\mathscr{C}$. Since $\mathscr{C}$ is closed under projective limits, it will be sufficient to assume that $M \in \mathrm{Ob}\,\mathrm{Coh}(\mathrm{Dis}(A))$. Thus, M is a finitely generated A-module and let $x_1, \ldots, x_n$ be its generators. Then we have the canonical epimorphism of $\mathrm{STC}(A)$:

$$\coprod_{i=1}^{n} A/\mathrm{Ann}(x_i) \to M \to 0$$

so that $M \in \mathrm{Ob}\,\mathscr{C}$, because $A/\mathrm{Ann}(x_i) \in \mathrm{Ob}\,\mathscr{C}$, for each i.

Now let $\mathfrak{a}$ be a two-sided ideal of A in $\mathrm{STC}(A)$ and $\mathscr{C} = \mathrm{STC}(A/\mathfrak{a})$. The proof will be complete if we prove that $\mathfrak{a} = \mathfrak{a}_C$. Indeed, since

$$A/\mathfrak{a} \in \mathrm{Ob}\,\mathrm{STC}(A/\mathfrak{a}) = \mathscr{C},$$

it will result that $\mathfrak{a} \in C$ so that $\mathfrak{a}_C \subseteq \mathfrak{a}$. Conversely, let $\mathfrak{b}$ be a subobject of A in $\mathrm{STC}(A)$, with $\mathfrak{b} \in C$, i.e. $A/\mathfrak{b} \in \mathrm{Ob}\,\mathscr{C}$. Thus $\mathfrak{a} \subseteq \mathrm{Ann}(1 + \mathfrak{b}) = \mathfrak{b}$, so that $\mathfrak{a} \subseteq \bigcap_{\mathfrak{b} \in C} = \mathfrak{a}_C$, as claimed. ∎

As always, $\mathfrak{a}$ will be a two-sided ideal of A in $\mathrm{STC}(A)$. For each

$$M \in \mathrm{Ob}\,\mathrm{STC}(A),$$

we shall consider the abelian group

$$\mathfrak{a} \otimes_A M$$

which is canonically an A-module (because $\mathfrak{a}$ is an (A, A^0)-bimodule). Also we shall consider the mapping

$$f_M : \mathfrak{a} \otimes_A M \to M$$

defined by $f_M(a \otimes x) = ax$, $a \in \mathfrak{a}$, $x \in M$.

Further let $\mathscr{C} = \mathrm{STC}(A/\mathfrak{a})$, and

$$E : 0 \longrightarrow C \xrightarrow{\alpha} Y \xrightarrow{\beta} M \longrightarrow 0$$

an exact sequence of STC(A), with $C \in \mathrm{Ob}\,\mathscr{C}$. If $\beta(y) = \beta(y')$, y, $y' \in Y$, then $y - y' \in \mathrm{im}\,\alpha \in \mathrm{Ob}\,\mathscr{C}$ so that for each $a \in \mathfrak{a}$, we have $ay = ay'$. Thus there exists a morphism of A-modules:

$$g_E : \mathfrak{a} \otimes_A M \to Y,$$

with $\beta g_E = f_M$, defined by $g_E(a \otimes x) = ay$, $a \in \mathfrak{a}, x \in M, y \in Y$ and $\beta(y) = x$.

LEMMA 5.5. *Consider the A-module* $\mathfrak{a} \otimes_A M$, *equipped with the topology having as a fundamental system of neighbourhoods of zero the set of all submodules* $\{g_E^{-1}(Y')\}_{Y'}$, *where* Y' *runs over all special open submodules of* Y. *With this topology* $\mathfrak{a} \otimes_A M$ *is an object of* TC(A).

Proof. We recall that an object of TC(A) is in fact an A-module X, equipped with a topology, such that X becomes a topological group having a fundamental system of neighbourhoods of zero consisting of all submodules X' with $X/X' \in \mathrm{Ob}\,\mathrm{Coh}(\mathrm{Dis}(A))$. Therefore we must prove that

$$(\mathfrak{a} \otimes_A M)/g_E^{-1}(Y') \in \mathrm{Ob}\,\mathrm{Coh}(\mathrm{Dis}(A)),$$

for each special open submodule Y' of Y. Since we have the canonical inclusion:

$$0 \to \mathfrak{a} \otimes_A M/g_E^{-1}(Y') \to Y/Y'$$

induced by g_E and $Y/Y' \in \mathrm{Ob}\,\mathrm{Coh}(\mathrm{Dis}(A))$, it will be sufficient to check that $\mathfrak{a} \otimes_A M/g_E^{-1}(Y')$ is finitely generated. Indeed, because Y/Y' is finitely generated, we may take $y_1 + Y', \ldots, y_m + Y'$ to be a generating set, where $y_1, \ldots, y_m \in Y$. Thus, for each $i = 1, \ldots, m$, the mapping

$$v_i : \mathfrak{a} \to Y$$

defined by $v_i(a) = ay_i$, $a \in \mathfrak{a}$, is a morphism of STC(A), so that $v_i^{-1}(Y')$ is a special open in $\mathfrak{a}$. Then $\mathfrak{a}/v_i^{-1}(Y') \in \mathrm{Coh}(\mathrm{Dis}(A))$; let

$$a_1^i + v_i^{-1}(Y'), \ldots, a_{n_i}^i + v_i^{-1}(Y')$$

be its generators (as A-module), $a_1^i, \ldots, a_{n_i}^i \in \mathfrak{a}$, $i = 1, \ldots, m$. We write $x_i = \beta(y_i)$, and for each $x \in M$, let $y \in Y$, with $\beta(y) = x$. There exist elements $b_1, \ldots b_m \in A$, such that $y - b_1 y_1 - \ldots - b_m y_m \in Y'$. Thus, for each $a \in \mathfrak{a}$, we have

$$a \otimes x = a \otimes (x - b_1 x_1 - \ldots - b_m x_m) + a \otimes b_1 x_1 + \ldots + a \otimes b_m x_m$$

and

$$\begin{aligned} a \otimes (x - b_1 x_1 - \ldots - b_m x_m) \\ = a \otimes \beta(y - b_1 y_1 - \ldots - b_m y_m) \in g_E^{-1}(Y'). \end{aligned}$$

For each $1 \leqslant i \leqslant m$, one obtains $ab_i \in \mathfrak{a}$, so that there exist elements

$$b_1^i, \ldots, b_{n_i}^i \in A$$

with

$$ab_i - b_1^i a_1^i - \ldots - b_{n_i}^i a_{n_i}^i \in v_i^{-1}(Y')$$

and

$$(ab_i - b_1^i a_1^i - \ldots - b_{n_i}^i a_{n_i}^i) \otimes x_i \in g_E^{-1}(Y').$$

Thus we have

$$a \otimes x - \sum_{i=1}^{m} \sum_{j=1}^{n_i} b_j^i (a_j^i \otimes x_i) \in g_E^{-1}(Y')$$

so that $\mathfrak{a} \otimes_A M / g_E^{-1}(Y')$ is generated by elements $a_j^i \otimes x_i + g_E^{-1}(Y')$, $1 \leqslant i \leqslant m$, $1 \leqslant j \leqslant n_i$. ∎

The object of $TC(A)$ defined by Lemma 5.5 will be denoted by $\mathscr{C}(M, g_E)$. The mapping $g_E : \mathscr{C}(M, g_E) \to M$ is obviously a morphism of TC(A).

Particularly, from the exact sequence:

$$E_0 : 0 \longrightarrow 0 \longrightarrow M \xrightarrow{1_M} M \longrightarrow 0$$

we shall obtain (according to Lemma 5.5) the object $\mathscr{C}(M, f_M)$, because $g_{E_0} = f_M$. This last object of TC(A) will be denoted by $\mathscr{C}(M)$ or $\mathfrak{a} \otimes_A M$, if there is no danger of confusion concerning the topologies on $\mathfrak{a} \otimes_A M$.

For any exact sequence:

$$E : 0 \to C \to Y \to M \to 0$$

with $C \in \mathrm{Ob}\mathscr{C}$, we see that $\mathscr{C}(M)$ and $\mathscr{C}(M, g_E)$ have the same underlying A-module, namely, $\mathfrak{a} \otimes_A M$ and the identity mapping:

$$\mathscr{C}(M, g_E) \to \mathscr{C}(M)$$

is a continuous mapping, that is a morphism of TC(A).

Recall (see Section 6.2) that the inclusion functor $I_1 : \mathrm{CTC}(A) \to \mathrm{TC}(A)$ has a left adjoint T, defined as follows: if $X \in \mathrm{Ob}\,\mathrm{TC}(A)$, and $\{X_i\}_i$ is a fundamental system of neighbourhoods of zero consisting of submodules of X, such that $X/X_i \in \mathrm{Ob}\,\mathrm{Coh}\,(\mathrm{Dis}(A))$, for each i, then $T(X) = \varprojlim X/X_i$ and an arrow of adjunction $\phi_X : X \to T(X)$ is defined canonically. Also the inclusion functor $J : \mathrm{STC}(A) \to \mathrm{CTC}(A)$ has a right adjoint S, so that for each object M of $\mathrm{CTC}(A)$, an arrow of adjunction $\psi_M : S(M) \to M$ is given by the identity of M (in fact $S(M)$ and M have the same underlying modules and the topology of $S(M)$ is finer than the topology of M).

Now for $M \in \mathrm{Ob}\,\mathrm{STC}(A)$, let $\mathfrak{a} \otimes_A M = \mathscr{C}(M)$ and let us put:

$$\mathfrak{a} \mathbin{\widehat{\otimes}}_A M = ST(\mathfrak{a} \otimes_A M) \in \mathrm{Ob}\,\mathrm{STC}(A).$$

Furthermore, we denote by $\bar{f}_M : T(\mathfrak{a} \otimes_A M) \to M$ the unique morphism of $\mathrm{CTC}(A)$ such that $\bar{f}_M \phi = f_M$. Also, we denote by $\hat{f}_M$ the composition

$$\mathfrak{a} \mathbin{\widehat{\otimes}}_A M \xrightarrow{\psi} T(\mathfrak{a} \otimes_A M) \xrightarrow{\bar{f}_M} M,$$

where ϕ and ψ are the corresponding arrows of adjunction.

In other words, we have the following commutative diagram:

$$\begin{array}{ccc} \mathfrak{a} \otimes_A M & \xrightarrow{f_M} & M \\ {\scriptstyle\phi}\downarrow & \overset{\bar{f}_M}{\nearrow} & \uparrow{\scriptstyle\hat{f}_M} \\ T(\mathfrak{a} \otimes_A M) & \longleftarrow & \mathrm{ST}(\mathfrak{a} \otimes_A M) = \mathfrak{a} \mathbin{\widehat{\otimes}}_A M \end{array}$$

Finally, we shall write

$$\overline{\mathfrak{a}M} = \mathrm{im}\ \hat{f}_M.$$

Lemma 5.6. *For each subobject N of M in* $\mathrm{STC}(A)$, *we have $M/N \in \mathrm{Ob}\,\mathscr{C}$ if and only if N contains all products of the form ax with $a \in \mathfrak{a}$ and $x \in M$. Moreover $\overline{\mathfrak{a}M}$ is the smallest subobject of M in* $\mathrm{STC}(A)$ *with this property. Particularly*, coker $\hat{f}_M = M/\overline{\mathfrak{a}M} \in \mathrm{Ob}\,\mathscr{C}$.

Proof. The first assertion results from the equality $\mathscr{C} = \mathrm{STC}(A/\mathfrak{a})$.

Furthermore, we have:

$$\mathrm{im}\ f_M = \mathrm{im}(\bar{f}_M \phi) \subseteq \mathrm{im}(\bar{f}_M) = \mathrm{im}\ \hat{f}_M,$$

so that $\overline{\mathfrak{a}M}$ contains all products $ax = f_M(a \otimes x)$ with $a \in \mathfrak{a}$ and $x \in M$. If M' is a subobject of M in $\mathrm{STC}(A)$ which contains all products ax, $a \in \mathfrak{a}$, $x \in M$, then f_M may be factored as $f_M = ig$ where $i : M' \to M$ is the canonical inclusion and $g : \mathfrak{a} \otimes_A M \to M'$ is defined in the obvious way and it is a morphism of $\mathrm{TC}(A)$. Thus there exists a unique morphism

$$h : T(\mathfrak{a} \otimes_A M) \to M'$$

of CTC(A) such that $h\phi = g$ and $ih\phi = ig = f_M = \bar{f}_M \phi$ implies $ih = \bar{f}_M$. In other words, we have the following commutative diagram:

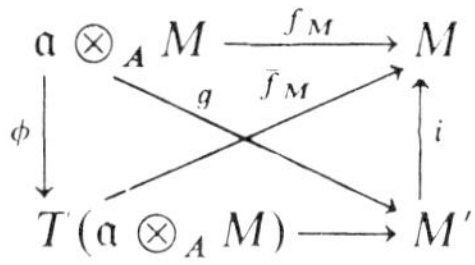

But then we have: im $\bar{f}_M \subseteq M'$, as required. ■

Proposition 5.7. *The inclusion functor* $J : \mathscr{C} \to$ STC(A) *has a left adjoint* $K :$ STC(A) $\to \mathscr{C}$ *defined by* $K(M) = M/\overline{\mathfrak{a}M}$.

Proof. We have the following exact sequence of STC(A):

$$0 \to \overline{\mathfrak{a}M} \to M \to M/\overline{\mathfrak{a}M} \to 0,$$

and $M/\overline{\mathfrak{a}M} \in \mathrm{Ob}\,\mathscr{C}$. For each morphism $f : M \to X$ with $X \in \mathrm{Ob}\,\mathscr{C}$, we have $f(ax) = af(x) = 0$, for every $x \in \mathfrak{a}$, $x \in M$, so that $\ker f$ is a subobject of M in STC(A) which contains all products ax, $a \in \mathfrak{a}$, $x \in M$. According to the previous lemma, we have $\overline{\mathfrak{a}M} \subseteq \ker f$, hence there exists a unique morphism $\bar{f} : M/\overline{\mathfrak{a}M} \to X$ of $\mathscr{C}$ such that the following diagram is commutative:

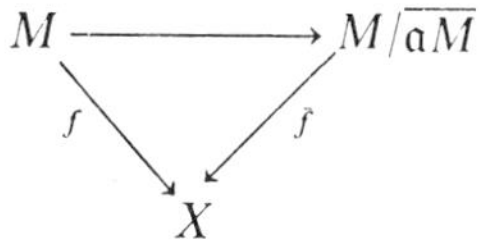

Lemma 5.8. *For any* $M \in \mathrm{Ob}\,$STC(A) *we have* $\ker \hat{f}_M \in \mathrm{Ob}\,\mathscr{C}$.

Proof. Let $\{M_i\}_{i \in I}$ be the decreasing family of all special open submodules of M. For every $i \in I$, we have the following commutative diagram:

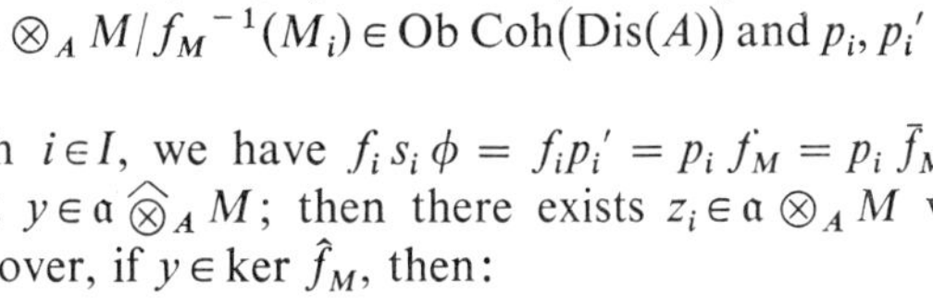

in which f_i is induced by M, s_i is the unique morphism of CTC(A) with $s_i \phi = p_i'$ (since $\mathfrak{a} \otimes_A M/f_M^{-1}(M_i) \in \mathrm{Ob}\,\mathrm{Coh}(\mathrm{Dis}(A))$ and p_i, p_i' are canonical epimorphisms.

Now, for each $i \in I$, we have $f_i s_i \phi = f_i p_i' = p_i f_M = p_i \bar{f}_M \phi$, therefore $f_i s_i = p_i \bar{f}_M$. Let $y \in \mathfrak{a} \hat{\otimes}_A M$; then there exists $z_i \in \mathfrak{a} \otimes_A M$ with $p_i'(z_i) = s_i(y)$, $i \in I$. Moreover, if $y \in \ker \hat{f}_M$, then:

$$0 = p_i(\bar{f}_M(y)) = f_i(s_i(y)) = f_i(p_i'(z_i)) = p_i(f_M(z_i)),$$

so that $f_M(z_i) \in M_i$ for each $i \in I$. Also, let $z_i = \sum_{j=1}^{n_i} a_j^{\,i} \otimes x_j^{\,i}$, $a_j^{\,i} \in \mathfrak{a}$, $x_j^{\,i} \in M$; then $f_M(z_i) = \sum_j a_j^{\,i} x_j^{\,i}$ and for each $a \in A$, one obtains:

$$a z_i = \textstyle\sum_j a a_j^{\,i} \otimes x_j^{\,i} = \sum_j a \otimes a_j^{\,i} x_j^{\,i} = a \otimes f_M(z_i) \in f_M^{-1}(M_i).$$

But then $s_i(ay) = p_i(az_i) = 0$, for each $i \in I$, hence by the definition of $T(\mathfrak{a} \otimes_A M)$ it follows that $ay = 0$. In other words, $\mathfrak{a} \subseteq \mathrm{Ann}(y)$, for each $i \in I$, hence $\ker \hat{f}_M \in \mathrm{Ob}\,\mathscr{C}$.

Now if $f: N \to M$ is a morphism of $\mathrm{STC}(A)$, then we can define the mapping $1 \otimes f: \mathfrak{a} \otimes_A N \to \mathfrak{a} \otimes_A M$. The equality

$$f_M(1 \otimes f) = f f_N$$

proves that $1 \otimes f$ is a morphism of $\mathrm{TC}(A)$, (because $\mathfrak{a} \otimes_A M$ is in fact $\mathscr{C}(M)$, i.e. an object of $\mathrm{TC}(A)$). ∎

Lemma 5.9. *If* $\hat{f}_M : \mathfrak{a} \mathbin{\hat{\otimes}}_A M \to M$ *is an epimorphism of* $\mathrm{STC}(A)$, *we have* $\hat{f}_{\mathfrak{a} \hat{\otimes}_A M} = ST(1 \otimes \hat{f}_M)$ *and this morphism is an epimorphism of* $\mathrm{STC}(A)$.

Proof. For brevity, we shall write $f = f_M$, $N = \mathfrak{a} \mathbin{\hat{\otimes}}_A M$, $g = f_N$ and $h = 1 \otimes \hat{f}$. Since an epimorphism of $\mathrm{STC}(A)$ or $\mathrm{CTC}(A)$ is in fact a surjective mapping (see Section 6.2), it will follow that h, $T(h)$ and $\mathrm{ST}(h)$ are surjections, therefore $ST(h)$ will be an epimorphism of $\mathrm{STC}(A)$. Now, we consider the following diagram:

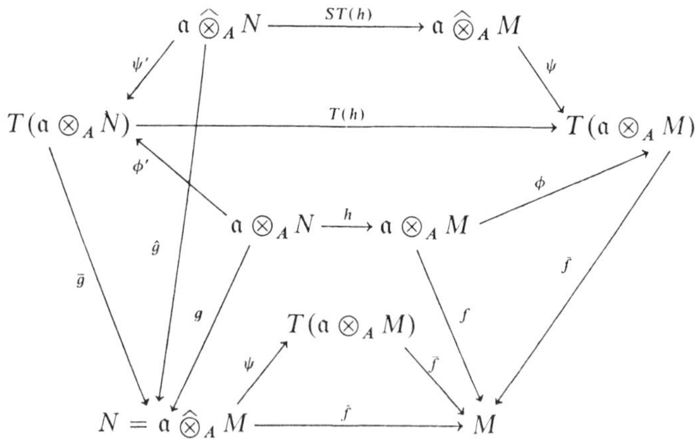

in which we have: $\bar{f}\psi g = \hat{f} g = f h = \bar{f}\phi h$, hence $\psi g = \phi h$, because $\bar{f}$ is surjective. But thus $\psi \bar{g} \phi' = \psi g = \phi h = T(h)\phi'$, hence $\psi \bar{g} = T(h)$ and $\psi \mathrm{ST}(h) = T(h)\psi' = \psi \bar{g} \psi' = \psi \hat{g}$, hence finally $\hat{g} = \mathrm{ST}(h)$, as claimed. ∎

Lemma 5.10. *For each* $X \in \mathrm{Ob}\,\mathscr{C}$ *we have* $\mathfrak{a} \mathbin{\hat{\otimes}}_A X = 0$.

Proof. For each $x \in X$, we have $\mathfrak{a} \subseteq \mathrm{Ann}(x)$, hence for every $a \in \mathfrak{a}$, $f_X(a \otimes x) = ax = 0$, so that $f_X = 0$. But then necessarily, $\mathfrak{a} \hat{\otimes}_A X = 0$. ∎

THEOREM 5.11. *Let $\mathfrak{a}$ be a two-sided ideal of A, subobject of A in* STC(A) *and $M \in \mathrm{Ob}\,\mathrm{STC}(A)$.*

(1) *The following assertions are equivalent:*

(a) *M has no non-zero quotient object in $\mathscr{C}$;*

(b) $\overline{\mathfrak{a}M} = M$;

(c) *$\hat{f}_M$ is an epimorphism of* STC(A);

(2) *The following assertions are also equivalent:*

(a) *A has no non-zero quotient object in $\mathscr{C}$, and any exact sequence $0 \to C \to Y \to M \to 0$ with $C \in \mathrm{Ob}\,\mathscr{C}$ is splitting;*

(b) *We have in STC(A) the following splitting exact sequence:*

$$0 \longrightarrow \ker f_M \longrightarrow \mathfrak{a} \hat{\otimes}_A M \xrightarrow{\hat{f}_M} M \longrightarrow 0;$$

(c) *The morphism $\hat{f}_M : \mathfrak{a} \hat{\otimes}_A M \to M$ is an isomorphism of* STC(A).

Proof. (1) The statement (1) will follow easily by Lemma 5.6 and the equality $\overline{\mathfrak{a}M} = \mathrm{im}\,\hat{f}_M$.

(2) (a) ⇒ (b) According to (1) $\hat{f}_M$ is an epimorphism of STC(A) so that by Lemma 5.8, $\ker \hat{f}_M \in \mathrm{Ob}\,\mathscr{C}$.

(b) ⇒ (c) By Lemma 5.9, $\hat{f}_{\mathfrak{a}\hat{\otimes}_A M}$ is an epimorphism of STC(A), hence by (1) any quotient object of $\mathfrak{a} \hat{\otimes}_A M$ in $\mathscr{C}$ is null. Particularly, $\ker \hat{f}_M = 0$.

(b) ⇒ (a) Let

$$E : 0 \longrightarrow C \xrightarrow{\alpha} Y \xrightarrow{\beta} M \longrightarrow 0$$

be an exact sequence of STC(A) with $C \in \mathrm{Ob}\,\mathscr{C}$. Then $f_M i = \beta g_E$, in TC(A), where $i : \mathscr{C}(M, g_E) \to \mathfrak{a} \otimes_A M$ is in fact the identity mapping. It is clear that there exists a morphism $j : T(\mathfrak{a} \otimes_A M) \to T(\mathscr{C}(M, g_E))$ of CTC(A) such that $T(i)\,j = 1$. Thus we have the following commutative diagram:

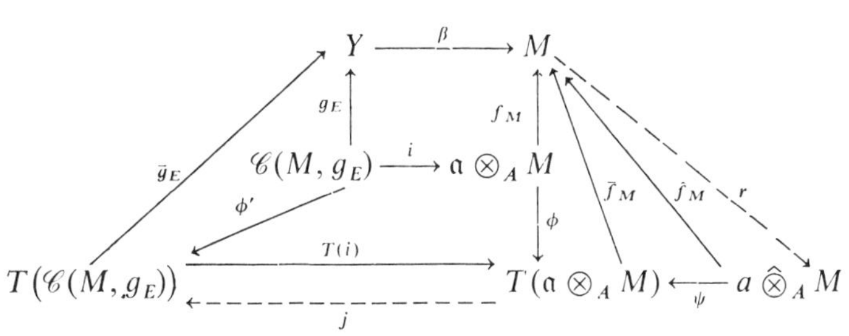

in which one obtains $\beta\bar{g}_E\phi' = \beta g_E = f_M i = \bar{f}_M \phi i = \bar{f}_M T(i)\phi'$, hence $\beta g_E = \bar{f}_M T(i)$. But then we have: $\beta g_{\bar{E}} j\psi = \bar{f}_M \psi = \hat{f}_M$. By hypothesis, there exists a morphism $r: M \to \mathfrak{a} \hat{\otimes}_A M$ of STC(A) with $\hat{f}_M r = 1$. Thus

$$\beta(\bar{g}_E j\psi r) = 1,$$

i.e. E is a splitting sequence of STC(A).

The implication (c) ⇒ (b) is obvious. ∎

Let $\mathscr{C}_1$, $\mathscr{C}_2$ be two precolocalizing subcategories of STC(A), say $\mathscr{C}_1 = \mathrm{STC}(A/\mathfrak{a}_1)$, $\mathscr{C}_2 = \mathrm{STC}(A/\mathfrak{a}_2)$ where $\mathfrak{a}_1$ and $\mathfrak{a}_2$ are two-sided ideals of A, subobjects of A in STC(A). We denote by $\mathscr{C}_1.\mathscr{C}_2$ the following full subcategory of STC(A). An object M of STC(A) belongs to $\mathscr{C}_1.\mathscr{C}_2$ if $\overline{\mathfrak{a}_2 M} = \bigcap M'$ belongs of $\mathscr{C}_1$, where M' runs through all subobjects of M (in STC(A)) such that $M/M' \in \mathrm{Ob}\,\mathscr{C}_2$. $\mathscr{C}_1.\mathscr{C}_2$ is called the *coproduct* of $\mathscr{C}_1$ with $\mathscr{C}_2$ and is a precolocalizing subcategory. We observe that $\mathscr{C}_2 \subseteq \mathscr{C}_1.\mathscr{C}_2$ and for each subobject $\mathfrak{a}$ of $\mathfrak{a}_2$ in STC(A), the condition $A/\mathfrak{a} \in \mathrm{Ob}\,\mathscr{C}_1.\mathscr{C}_2$ is equivalent with $\mathfrak{a}_2/\mathfrak{a} \in \mathrm{Ob}\,\mathscr{C}_1$.

A dense precolocalizing subcategory of STC(A) is called *colocalizing*. It is clear that a precolocalizing subcategory $\mathscr{C}$ of STC(A) is localizing if and only if $\mathscr{C}.\mathscr{C} = \mathscr{C}$.

If $\mathfrak{a}$ is a two-sided ideal of A, subobject of A in STC(A), we shall write $\mathfrak{a}^2 = \overline{\mathfrak{a}.\mathfrak{a}}$; also we shall say that $\mathfrak{a}$ is topologically *idempotent* if $\mathfrak{a}^2 = \mathfrak{a}$. We observe that we have already a notion of idempotent ideal which is generally different from the notion used here. We hope that the reader will not confuse these notions.

THEOREM 5.12. *Let $\mathscr{C}_1$, $\mathscr{C}_2$ be two precolocalizing subcategories of* STC(A) *and $\mathscr{C} = \mathscr{C}_1.\mathscr{C}_2$. Then*:

$$\mathfrak{a}_C = \overline{\mathfrak{a}_{C_1} \cdot \mathfrak{a}_{C_2}}.$$

In particular, a precolocalizing subcategory $\mathscr{C}$ is colocalizing if and only if $\mathfrak{a}_C$ is topologically idempotent. Thus the mapping $\mathscr{C} \rightsquigarrow \mathfrak{a}_C$ is a bijection between the set of colocalizing subcategories of STC(A) *and the set of all topologically idempotent two-sided ideals of A (which are subobjects of A in* STC(A)). *The inverse mapping is*: $\mathfrak{a} \rightsquigarrow \mathrm{STC}(A/\mathfrak{a})$.

Proof. Let $\mathscr{C}_1 = \mathrm{STC}(A/\mathfrak{a}_1)$, $\mathscr{C}_2 = \mathrm{STC}(A/\mathfrak{a}_2)$ and $\mathrm{STC}(A/\mathfrak{a}) = \mathscr{C} = \mathscr{C}_1.\mathscr{C}_2$. Since $\mathscr{C}_2 \subseteq \mathscr{C}$, one obtains $\mathfrak{a} \subseteq \mathfrak{a}_2$. Also $A/\mathfrak{a} \in \mathrm{Ob}\,\mathscr{C}$, hence $\mathfrak{a}_2/\mathfrak{a} \in \mathrm{Ob}\,\mathscr{C}_1$. Now let $x \in \mathfrak{a}_1$, $y \in \mathfrak{a}_2$. Then:

$$\mathfrak{a}_1 \subseteq \mathrm{Ann}(y + \mathfrak{a}) = \{a \in A \mid ay \in \mathfrak{a}\}.$$

Thus $xy \in \mathfrak{a}$, so that $\overline{\mathfrak{a}_1\,\mathfrak{a}_2} \subseteq \mathfrak{a}$.

Conversely, let $\mathfrak{b}$ be a subobject of $\mathfrak{a}_2$ in STC(A), which contains all products xy with $x \in \mathfrak{a}_1$, $y \in \mathfrak{a}_2$. In particular $\mathfrak{b}$ is a subobject of A in STC(A). Now we must prove that $\mathfrak{a} \subseteq \mathfrak{b}$, i.e. $A/\mathfrak{b} \in \operatorname{Ob} \mathscr{C}$, or equivalently

$$\mathfrak{a}_2/\mathfrak{b} \in \operatorname{Ob} \mathscr{C}_1.$$

Indeed, for each $x \in \mathfrak{a}_1$, $y \in \mathfrak{a}_2$ we have, by hypothesis $xy \in \mathfrak{b}$, hence

$$\mathfrak{a}_1 \subseteq \operatorname{Ann}(y + \mathfrak{b}),$$

therefore $\mathfrak{a}_2/\mathfrak{b} \in \operatorname{Ob} \mathscr{C}_1$. ∎

Lemma 5.13. *If* $\mathfrak{a}$ *is a topologically idempotent two-sided ideal of* A *in* STC(A), *then* $\bar{f}_{\overline{\mathfrak{a}M}}$ *is an epimorphism in* STC(A), *for each* $M \in \operatorname{Ob} \mathrm{STC}(A)$.

Proof. Let N be a subobject of $\overline{\mathfrak{a}M}$ of STC(A) which contains all products ax, with $a \in \mathfrak{a}$ and $x \in \overline{\mathfrak{a}M}$. In particular N is a subobject of M in STC(A). Let $x \in M$; we consider the mapping $u : \mathfrak{a} \to M$, with $u(a) = ax$, which is a morphism of STC(A). For each a, $a' \in \mathfrak{a}$, we have $a'x \in \overline{\mathfrak{a}M}$, hence $aa'x \in N$, so that $aa' \in u^{-1}(N)$, hence $\mathfrak{a}^2 \subseteq u^{-1}(N)$. Since $\mathfrak{a}$ is topologically idempotent two-sided ideal of A in STC(A), thus $\mathfrak{a} \subseteq u^{-1}(N)$, hence N contains all products ax, with $a \in \mathfrak{a}$, $x \in M$, i.e. $N = \overline{\mathfrak{a}M}$. The assertion now follows from Lemma 5.11. ∎

Let $\mathscr{C}$ be a colocalizing subcategory of STC(A); an object N of STC(A) as in Theorem 5.11, (2) is called *$\mathscr{C}$-coclosed*, or simple *coclosed* if there is no danger of confusion.

Lemma 5.14. *If* $M \in \operatorname{Ob} \mathrm{STC}(A)$ *is such that* $\bar{f}_M$ *is an epimorphism, then* $\mathfrak{a} \hat{\otimes}_A M$ *is* $\mathscr{C}$-*coclosed.*

Proof. For brevity, we write $N = \mathfrak{a} \hat{\otimes}_A M$, $f = \hat{f}_M$. Thus we have the exact sequence of STC(A):

$$0 \longrightarrow \ker f \longrightarrow N \xrightarrow{f} M \longrightarrow 0,$$

with $\ker f \in \operatorname{Ob} \mathscr{C}$ (see Lemma 5.8). Then the sequence:

$$\mathfrak{a} \otimes_A \ker f \longrightarrow \mathfrak{a} \otimes_A N \xrightarrow{1 \otimes f} \mathfrak{a} \otimes_A M \longrightarrow 0$$

is exact in Mod A. This sequence will be considered in TC(A) if $\mathfrak{a} \otimes_A M$ is not equipped with its natural topology, but with the topology defined such that the mapping $1 \otimes f$ is continuous. Thus $\mathfrak{a} \otimes_A M$ equipped with this new topology is obviously an object of TC(A), which will be denoted by $(\mathfrak{a} \otimes_A M)'$.

Now, it is clear that the identity mapping $i: (\mathfrak{a} \otimes_A M)' \to \mathfrak{a} \otimes_A M$ is a morphism of STC(A). According to Lemma 5.10, we see that:

$$\mathrm{ST}\big((1\otimes f)'\big) : \mathfrak{a} \,\hat{\otimes}_A N \to \mathrm{ST}\big((\mathfrak{a} \otimes_A M)'\big)$$

is an isomorphism of STC(A), where by $(1\otimes f)'$ we have denoted the mapping $1\otimes f$ considered as morphism $(1\otimes f)' : \mathfrak{a} \otimes_A N \to (\mathfrak{a} \otimes_A M)'$ of TC(A). Thus one obtains the following commutative diagram:

$$\begin{array}{ccc} \mathfrak{a}\,\hat{\otimes}_A N & \xrightarrow{\mathrm{ST}((1\otimes f)')} & \mathrm{ST}\big((\mathfrak{a}\otimes_A M)'\big) \\ & {\scriptstyle \mathrm{ST}(1\otimes f)}\searrow \qquad \nearrow{\scriptstyle S(j)} \qquad \swarrow{\scriptstyle \mathrm{ST}(i)} & \\ & \mathfrak{a}\,\hat{\otimes}_A M = N & \end{array}$$

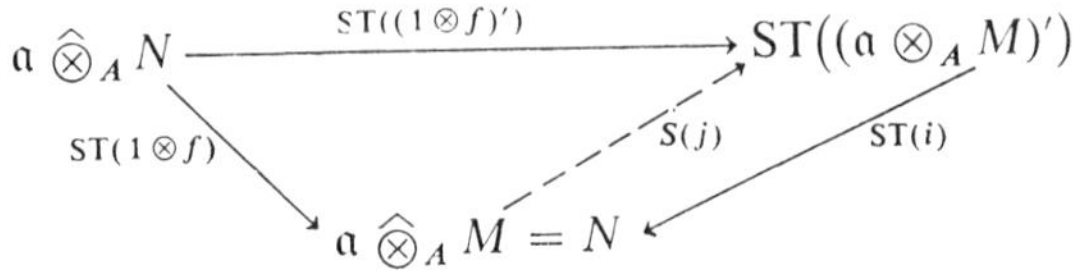

But there exists a morphism $j: T(\mathfrak{a} \otimes_A M) \to T\big((\mathfrak{a} \otimes_A M)'\big)$, of CTC($A$) such that $T(i)j = 1$, and by Lemma 5.9, $f_N = \mathrm{ST}(1\otimes f)$. Therefore the exact sequence:

$$0 \longrightarrow \ker \hat{f}_N \longrightarrow \mathfrak{a}\,\hat{\otimes}_A N \xrightarrow{\hat{f}_N} N \longrightarrow 0$$

is exact and splitting, so that N is $\mathscr{C}$-coclosed. ■

Now it is obviously the case that all the notions introduced above such as: precolocalizing and colocalizing subcategory and coclosed objects are dual to the corresponding notions of the localization theory. Also, if $\mathscr{C}$ is a colocalizing subcategory of STC(A), then we can define the notion of *coquotient* category and the notion of *$\mathscr{C}$-coenvelope* of an object M of STC(A): a morphism $f: N \to M$ of STC(A) is an $\mathscr{C}$-coenvelope of M if $\ker f \in \mathrm{Ob}\,\mathscr{C}$, $\mathrm{coker}\, f \in \mathrm{Ob}\,\mathscr{C}$ and N is $\mathscr{C}$-coclosed. Finally, the following result is true:

THEOREM 5.15. *Let $\mathscr{C}$ be a colocalizing subcategory of* STC(A). *Then* $\mathscr{C}$ = STC($A/\mathfrak{a}$), *where $\mathfrak{a}$ is a topologically idempotent two-sided ideal of A in* STC(A). *Also, for each object M of* STC(A), *the composition*:

$$\mathfrak{a}\,\hat{\otimes}_A \overline{\mathfrak{a}M} \xrightarrow{\hat{f}_{\overline{\mathfrak{a}M}}} \overline{\mathfrak{a}M} \longrightarrow M$$

is a $\mathscr{C}$-coenvelope of M. The coquotient category STC(A)/$\mathscr{C}$ *is canonically equivalent with the full subcategory of* STC(A), *generated by all objects M such that $\hat{f}_M : \mathfrak{a}\,\hat{\otimes}_A M \to M$ is an isomorphism. Finally, a right adjoint of the inclusion functor* STC(A)/$\mathscr{C} \to$ STC(A) *is the functor*

$$M \rightsquigarrow \mathfrak{a}\,\hat{\otimes}_A \overline{\mathfrak{a}M}. \ ■$$

Now for each ordinal α, we define a precolocalizing subcategory $\mathscr{C}_\alpha$ of STC(A) as follows:

$\mathscr{C}_{-1}$ is the zero category.

$\mathscr{C}_0$ is the full subcategory of STC(A) generated by all objects X which are direct products of simple objects of STC(A) (an object of $\mathscr{C}_0$ is called *co-semisimple*). $\mathscr{C}_0$ is in fact the smallest precolocalizing subcategory of STC(A) which contains all simple objects of STC(A).

If $\alpha = \beta + 1$, then $\mathscr{C}_\alpha = \mathscr{C}_0 \mathscr{C}_\beta$.

If α is a limit ordinal, then $\mathscr{C}_\alpha = \bigcup_{\beta<\alpha} \mathscr{C}_\beta$.

It is clear that there exists an ordinal α such that $\mathscr{C}_\alpha = \mathscr{C}_\gamma$ for each $\gamma \geqslant \alpha$. Thus $\bigcup_\alpha \mathscr{C}_\alpha = \mathscr{C}_\sigma$ is called the *cosocle* colocalizing subcategory of STC(A). $\mathscr{C}_\sigma$ is obviously the smallest colocalizing subcategory of STC(A) which contains all simple objects. It is clear that $(\mathrm{STC}(A))^0$ is semi-artinian if and only if $\mathscr{C}_\sigma = \mathrm{STC}(A)$ or equivalently, any non-zero object of STC(A) contains a maximal proper subobject.

Now for each $M \in \mathrm{Ob}\,\mathrm{STC}(A)$, we define

$$\mathrm{rad}(M) = \bigcap \ker h,$$

where f runs through all morphisms $h : M \to N$ such that $\ker h$ is maximal. Obviously, if $f : M \to N$ is a morphism of STC(A) then $f(\mathrm{rad}(M)) \subseteq \mathrm{rad}(N)$; in particular, if $\mathfrak{a}$ is a two-sided ideal of A, subobject of A in STC(A), then for each $x \in A$, the morphism $u_x : \mathfrak{a} \to \mathfrak{a}$ (i.e. $u_x(y) = yx$) is a morphism of STC(A) and $\mathrm{rad}(\mathfrak{a})\, x = u_x(\mathrm{rad}(\mathfrak{a}))$, so that $\mathrm{rad}(\mathfrak{a})$ is also a two-sided ideal of A.

THEOREM 5.16. *Let A be a complete topologically left coherent and linearly compact ring. For any ordinal α we define a two-sided ideal $\mathfrak{a}_\alpha$ of A as follows:*

$\mathfrak{a}_0 = \mathrm{rad}(A)$;

$\mathfrak{a}_{\alpha+1} = \mathrm{rad}(\mathfrak{a}_\alpha)$;

If *α is a limit ordinal, then* $\mathfrak{a}_\alpha = \bigcap_{\beta<\alpha} \mathfrak{a}_\beta$.

Finally we denote $\mathfrak{a}_\sigma = \bigcap_\alpha \mathfrak{a}_\alpha$.

Thus with the above notation we have:

$$\mathscr{C}_\alpha = \mathrm{STC}(A/\mathfrak{a}_\alpha).$$

Particularly $\mathscr{C}_\sigma = \mathrm{STC}(A/\mathfrak{a}_\sigma)$, *so that $\mathfrak{a}_\sigma$ is a topological idempotent two-sided ideal of A and a subobject of A in* STC(A). *Finally,* $(\mathrm{STC}(A))^0$ *is a semi-artinian category if and only if* $\mathfrak{a}_\sigma = 0$.

Proof. For each simple object S of STC(A), there exists a non-zero morphism $h: A \to S$, so that $\mathfrak{a}_0 \subseteq \ker h$ implies $S \simeq A/\ker h \in \operatorname{Ob} \operatorname{STC}(A/\mathfrak{a}_0)$, hence $\mathscr{C}_0 \subseteq \operatorname{STC}(A/\mathfrak{a}_0)$. On the other hand, $A/\mathfrak{a}_0 \in \operatorname{Ob} \mathscr{C}_0$, therefore

$$\operatorname{STC}(A/\mathfrak{a}_0) \subseteq \mathscr{C}_0.$$

Now let α be an ordinal such that $\mathscr{C}_\beta = \operatorname{STC}(A/\mathfrak{a}_\alpha)$ for each $\beta < \alpha$. If $\alpha = \beta + 1$, then $\mathfrak{a}_\beta/\mathfrak{a}_\alpha \in \operatorname{Ob} \mathscr{C}_0$, hence $A/\mathfrak{a}_\alpha \in \operatorname{Ob}(\mathscr{C}_0 . \operatorname{STC}(A/\mathfrak{a}_\beta)) = \mathscr{C}_0 . \mathscr{C}_\alpha = \mathscr{C}_\alpha$, i.e. $\operatorname{STC}(A/\mathfrak{a}_\alpha) \subseteq \mathscr{C}_\alpha$. Conversely, for each subobject $\mathfrak{a}$ of $\mathfrak{a}_\beta$ in STC(A), with $\mathfrak{a}_\beta/\mathfrak{a} \in \operatorname{Ob} \mathscr{C}_0$, there exists a family of morphisms $h: \mathfrak{a} \to S_h$, with S_h a simple object of STC(A), such that $\mathfrak{a} = \bigcap \ker h$, hence $\mathfrak{a}_\alpha \subseteq \mathfrak{a}$, i.e. $\mathfrak{a}_\alpha \subseteq \mathfrak{a}_0\, \mathfrak{a}_\beta = \mathfrak{a}_{C\alpha}$. Therefore $\mathscr{C}_\alpha \subseteq \operatorname{STC}(A/\mathfrak{a}_\alpha)$.

Finally, if α is a limit ordinal, then for each $\beta < \alpha$ we have $\mathfrak{a}_\alpha \subseteq \mathfrak{a}_\beta$, i.e. $\mathscr{C}_\beta = \operatorname{STC}(A/\mathfrak{a}_\beta) \subseteq \operatorname{STC}(A/\mathfrak{a}_\alpha)$, hence $\mathscr{C}_\alpha \subseteq \operatorname{STC}(A/\mathfrak{a}_\alpha)$. Also, for each $\beta < \alpha$, we have $A/\mathfrak{a}_\beta \in \operatorname{Ob} \mathscr{C}_\beta \subseteq \mathscr{C}_\alpha$, hence $A/\mathfrak{a}_\alpha \simeq \varprojlim A/\mathfrak{a}_\beta \in \operatorname{Ob} \mathscr{C}$, i.e. $\operatorname{STC}(A/\mathfrak{a}_\alpha) \subseteq \mathscr{C}_\alpha$. ■

Bibliography

1. Akiba, T., Remark on generalized rings of quotients, I. *Proc. Jap. Acad.* **40** (1964), 801–806
2. Akiba, T., Remark on generalized rings of quotients, II, *J. Math. Kyoto Univ.* **5** (1965), 39–44.
3. Akiba, T., Remark on generalized rings of quotients, III, *J. Math. Kyoto Univ.* **9** (1969), 205–212.
4. Alin, J. S., "Structure of Torsion Modules", Ph.D. Thesis, University of Nebraska, 1967.
5. Alin, J. S., A primary decomposition theory for torsion modules, *Proc. Amer. Math. Soc.* **27** (1971), 43–48.
6. Alin, J. S. and Dickson, S. E., Goldie's torsion theory and its derived functors, *Pacific J. Math.* **24** (1968), 195–203.
7. Almkvist, G., Fractional categories, *Arkiv. Math.* **7** (1968), 449–476
8. Amdal, I. K. and Ringdal, F., Catégories unisérielles, *C. R. Acad. Sci. Paris, Série A* **267** (1968), 85–87 and 247–249.
9. Atiyah, M. F., On the Krull–Schmidt theorem with applications to sheaves, *Bull. Soc. Math. France,* **84** (1956), 307–317.
10. Auslander, M., Coherent functors, *Proc. Conf. Cat. Algebra,* La Jolla (1965), 189–231.
11. Azumaya, G., Corrections and supplementaries to my paper concerning Krull–Remak–Schmidt's theorem, *Nagoya Math. J.* **1** (1950), 117–124.
12. Baer, R., Abelian groups that are direct summands of every containing Abelian group, *Bull. Amer. Math. Soc.* **46** (1940), 800–806.
13. Bănică, C. P. and Popescu, N., Quelques considérations sur l'exactitude des foncteurs, *Bull. Math. Soc. Sci. Math. Phys.* R.P.R. **7** (1963), 143–147.
14. Bănică, C. and Popescu, N., Categorii cit, *St. cerc. mat.* **17** (1965), 951–985.
15. Bănică, C. and Popescu, N., Sur les catégories préabeliennes, *Rev. Roum. Math. Pures Appl.* **10** (1965), 621–633.
16. Bass, H., Finistic dimension and a homological generalization of semi-primary rings, *Trans. Amer. Math. Soc.* **95** (1960), 466–488.
17. Bkouche, R., "Localizations et anneaux des sections", University of Brest, 1969.
18. Bkouche, R., "Couples spectraux et faisceaux associés. Applications aux anneaux de fonctions", University of Brest, 1970.
19. Bkouche, R., Couples spectraux et faisceaux associés. Applications aux anneaux de fonctions, *Bull. Soc. Math. France* **98** (1970), 253–295.
20. Bourbaki, N. "Théorie des ensembles", 2e éd., Hermann, Paris, 1963.
21. Bourbaki, N., "Algèbre", Chap. 2, 3e éd., Hermann, Paris, 1962.
22. Bourbaki, N., "Algèbre", Chap. 4–5, 2e éd., Hermann, Paris, 1959.
23. Bourbaki, N., "Algèbre", Chap. 8, Hermann, Paris, 1958.
24. Bourbaki, N., "Topologie générale", Chap. 1–2, 3e éd., Hermann, Paris, 1961.

25. Bourbaki, N., "Topologie générale", Chap. 3–4, 3e éd., Hermann, Paris, 1960,
26. Bourbaki, N., "Algèbre commutative", Chap. 1–2 and Chap. 3–4, Hermann, Paris, 1961.
27. Bourbaki, N., "Algèbre commutative", Chap. 5–6, Hermann, Paris, 1964.
28. Buchsbaum, D. A., Exact categories and duality, *Trans. Amer. Math. Soc.* **80** (1955), 1–34.
29. Bucur, I., Fonctions définies sur le spectre d'une catégorie et théorie de la décomposition, *Rev. Roum. Math. Pures Appl.* **9** (1964), 583–588.
30. Bucur, I., "Algebră omologică, Editura Didactică si Pedagogică, Bucuresti, 1965.
31. Bucur, I. and Deleanu, A., "Introduction to the Theory of Categories and Functors", Wiley, London, 1968.
32. Burmistrovich, I. E., Imbedding of an additive category in a category with direct products, *Dokl. Akad. Nauk. SSSR* (Russian) **132** (1960), 1235–1237.
33. Cartan, H. and Eilenberg, S., "Homological Algebra", Princeton University Press, Princeton, 1956.
34. Cateforis, V. C., Flat regular quotient rings, *Trans. Amer. Math. Soc.* **138** (1969), 241–250.
35. Cateforis, V. C., On regular self-injective rings, *Pacific. J. Math.*, **30** (1969), 39–45.
36. Cateforis, V. C. and Sandomierski, F., On modules of singular submodule zero, *Can. J. Math.* **23** (1971), 345–354.
37. Chase, S. U., Direct products of modules, *Trans. Amer. Math. Soc.* **97** (1960), 457–473.
38. Cohn, P. M., "Universal Algebra", Harper and Row, New York, London, Tokyo, 1965.
39. Cohn, P. M., "Free Rings and Their Relations", LMS Monographs No. 2, Academic Press (London & New York), 1971.
40. Croisot, R. and Lesieur, L., "Algèbre noethérienne non-commutative", Gauthier-Villars, Paris, 1963.
41. Desq, R., Sur les anneaux de quotients, *C. R. Acad. Sci. Paris, Série A* **270** (1971), 199–202.
42. Dickson, S. E., A torsion theory for abelian categories, *Trans. Amer. Math. Soc.* **121** (1966), 233–235.
43. Dickson, S. E., Decomposition of modules, *Math. Z.* **90** (1965), 9–13.
44. Dieudonné, J., Remarks on quasi-Frobenius rings, *Ill. J. Math.* **2** (1958), 346–354.
45. Dlab, V., On a class of perfect rings, *Can. J. Math.* **22** (1970), 822–826.
46. Dlab, V., A characterization of perfect rings, *Pacific J. Math.* **33** (1970), 79–88.
47. Duma, A., Sur les catégories des diagrammes, *Rev. Roum. Math. Pures Appl.* **10** (1965), 653–657.
48. Eckmann, B. and Schopf, A., Über injektive Moduln, *Archiv. Math.* **4** (1953), 75–78.
49. Elizarov, V. P., On quotient rings of associative rings, *Isv. Akad. Nauk SSSR* (Russian), **24** (1960), 153–170.
50. Elizarov, V. P., Flat extension of rings, *Soviet. Math. Dokl.* (Russian) **8** (1967), 905–907.
51. Endo, S., On flat modules over commutative rings, *J. Math. Soc. Japan,* **14** (1962), 284–291.
52. Faith. C., Rings with ascending chain condition on annihilators, *Nagoya Math. J.* **27** (1966), 179–191.

53. Faith, C. and Walker, E. A., Direct sum representations of injective modules, *J. Algebra,* **5** (1967), 203–221.
54. Findlay, G. D., Flat epimorphic extension of rings. *Math. Z.* **118** (1970), 281–288.
55. Fort, J., Sommes directes de sous-modules coirréductibles d'un module, *C. R. Acad. Sci. Paris* **262** (1966), 1239–1242.
56. Fort, J., Sommes directes de sous-modules coirréductibles d'un module, *Math. Z.* **103** (1968), 363–388.
57. Freyd, P., "Abelian Categories", Harper & Row, New York, 1964.
58. Freyd, P., Algebra-values functors in general categories and tensor product in particular, *Colloq. Math.* **14** (1966), 89–105.
59. Gabriel, P., Des catégories abéliennes, *Bull. Soc. Math. France* **90** (1962), 323–448.
60. Gabriel, P. and Popescu, N., Caractérisation des catégories abéliennes avec générateurs et limites inductives exactes, *C. R. Acad. Sci. Paris* **258** (1964), 4188–4191.
61. Gabriel, P. and Oberst, U., Spektralkategorien und reguläre Ringe im von Neumannschen Sinn, *Math. Z.* **92** (1966), 389–395.
62. Gabriel, P. and Zisman, M., "Category of fractions and homotopy theory", Ergebnisse der Math. Springer, Berlin, 1967.
63. Gabriel, P. and Rentschler, R., Sur la dimension des anneaux et ensembles ordonnés, *C. R. Acad. Sci. Paris, Série A* **265** (1967), 712–715.
64. Gabriel, P. and Ulmer, F., Lokal präsentierbare Kategorien, "Lecture Notes in Math". **221** Springer, (1971), 89–103.
65. Giraud, J., Analysis situs, *Séminaire Bourbaki,* **15** (1962/63), Exposé 256; 256–266.
66. Godement, R., "Théorie des faisceaux", Hermann, Paris, 1958.
67. Goblot, R., Catégories modulaires, *C. R. Acad. Sci. Paris, Série A* **267** (1968), 381–383.
68. Goblot, R., Catégories modulaires ayant assez d'objects projectifs, *C. R. Acad. Sci. Paris, série A* **267** (1968), 461–464.
69. Goblot, R., Catégories modulaires commutatives qui sont des catégories de faisceaux quasi-cohérents sur un schéma, *C. R. Acad. Sci. Paris* **268** (1969), 92–95.
70. Goblot, R., Sur les anneaux linéairement compacts, *C. R. Acad. Sci. Paris* **270** (1970), 1212–1215.
71. Goblot, R., "Sur deux classes de catégories de Grothendieck", Thèse. Univ. de Lille, 1971.
72. Goldie, A. W., The structure of prime rings under ascending chain conditions, *Proc. Lond. Math. Soc.* (3), **8** (1958), 589–608.
73. Goldie, A. W., Semi-prime rings with maximum condition, *Proc. Lond. Math. Soc.* (3) **10** (1960), 201–220.
74. Goldie, A. W., Torsion free modules and rings, *J. Algebra* **1** (1964), 286–297.
75. Goldie, A. W., Localization in non-commutative noetherian rings, *J. Algebra* **5** (1967), 89–105.
76. Goldman, O., Rings and modules of quotients, *J. Algebra,* **13** (1969), 10–48.
77. Gray, J. W., Sheaves with values in a category, *Topology* **3** (1965), 1–18.
78. Govorov, V. E., The rings over which the flat modules are free (Russian). *Sibir. Math. J.* **6** (1965), 300–304

79. Grothendieck, A., Sur quelques points d'algèbre homologique, *Tohôku Math. J.* **9** (1957), 119–221.
80. Grothendieck, A. and Dieudonné, J., Eléments de géométrie algébrique, *Publ. Inst. des Hautes Et. Sci.* **4** (1960).
81. Gruson, L., Complétion abélienne, *Bull. Soc. Math.* **90** (1966), 17–40.
82. Hacque, M., Remarques sur les épimorphismes d'anneaux, *C. R. Acad. Sci. Paris, Série A* **268** (1969), 1447–1450.
83. Hacque, M., Mono-sous-catégories d'une catégorie de modules, *Publ. Dept. Math. Univ. Lyon* **6** (1969), 13–48.
84. Hacque, M., Localisations exactes et localisations plates, *Publ. Dept. Math. Univ. Lyon* **6** (1969), 97–117.
85. Hacque, M. and Maury, G., "Séminaire d'Algèbre", Dept. Math. Univ. Lyon, Année 1969–70, Tome I, II and Annee 1970–71, Tome I, II.
86. Harada, M., Semi-primary Abelian categories, *Osaka J. Math.* **5** (1968), 189–198.
87. Harada, M., Note on quasi-injective modules, *Osaka J. Math.* **2** (1965), 351–356.
88. Heller, A., Homological algebra in Abelian categories, *Ann. of Math.* **68** (1958), 484–525.
89. Heller, A. and Rowe, K. A., On the category of sheaves, *Amer. J. Math.* **84** (1962), 205–216.
90. Huber, P. I., Standard constructions in Abelian categories, *Math. Ann.* **146** (1962), 321–325.
91. Hudry, A., Sous-modules Σ-clos, *C. R. Acad. Sci. Paris, Série A* **267** (1968), 789–791.
92. Hudry, A., Quelques remarques sur la notion d'extension rationnelle maximale d'un module, *Publ. Dept. Math. Lyon* **6** (1969).
93. Hudry, A., Quelques applications de la localisation de Gabriel. *C. R. Acad. Sci. Paris, Série A* **270** (1970), 8–11.
94. Hudry, A., "Applications de la thèorie de la localisation aux anneaux et aux modules" (These de 3ème cycle). Univ. Lyon, 1970.
95. Hudry, A., Sur les anneaux localement homogènes, *C. R. Acad. Sci. Paris, Série A* **271** (1970), 1214–1217.
96. Jacobson, N., Structure of rings, *Amer. Math. Soc. Coll. Publ.* Vol. 37, 1956.
97. Jans, J. P., On orders in quasi-Frobenius Rings, *J. Algebra* **7** (1967), 35–43.
98. Jensen, C. U., A remark on flat and projective modules, *Can. J. Math.* **18** (1966), 943–950.
99. Johnson, R. E., The extended centralizer of a ring over a module, *Proc. Amer. Math. Soc.* **2** (1951), 891–895.
100. Johnson, R. E. and Wong, E. T., Self-injective rings, *Can. Math. Bull.* **2** (1951), 167–173.
101. Kan, D. M., Adjoint functors, *Trans. Amer. Math. Soc.* **294** (1958), 294–329.
102. Kaplansky, I., Projective modules, *Ann. Math.* **68** (1958), 372–377.
103. Kasu, A. I., Closed classes of left Σ-modules and closed sets of left ideals of the ring Λ, *Mat. Zametki* (Russian) **5** (1969), 381–390.
104. Kato, T., Self-injective rings, *Tohôku Math. J.* **19** (1967), 485–495.
105. Kato, T., Torsionless modules, *Tohôku Math. J.* **20** (1968), 234–243.
106. Kleisli, H., Homotopy theory in Abelian categories, *Can. J. Math.* **14** (1962), 139–169.

107. Knight, J. T., On epimorphisms of non-commutative rings, *Proc. Camb. Phil. Soc.* **68** (1970), 589–601.
108. Koch. K., On almost maximal right ideals, *Proc. Amer. Math. Soc.* **25** (1970), 266–271.
109. Krause, G., On the Krull-dimension of left noetherian left Matlis-rings, *Math. Z.* **118** (1970), 207–214.
110. Kuroš, A. G., "Teoriya grupp", Moscow, 1969.
111. Lam, T. Y., The category of noetherian modules, *Proc. Nat. Acad. Sci. U.S.A.* **55** (1966), 1038–1040.
112. Lambek, J., On Utumi's ring of quotients, *Can. J. Math.* **15** (1963), 363–370.
113. Lambek, J., "Lectures on Rings and Modules", Blaisdell, Waltham, 1966.
114. Lambek, J., "Torsion Theories, Additive Semantics and Rings of Quotients", Springer Lecture Notes, **177**, 1971.
115. Lazard, D., Sur les modules plats, *C. R. Acad. Sci. Paris.* **258** (1964), 6313–6316.
116. Lazard, D., Epimorphismes plats d'anneaux. *C. R. Acad. Sci. Paris* **266** (1968), 314–316.
117. Lazard, D., Autour de la platitude, *Bull. Soc. Math. France* **97** (1968), 81–128.
118. Leclerc, C., "Catégories des *A*-modules non-singuliers", Thèse (3ème cycle) Univ. Lyon, 1969.
119. Lubkin, S., Imbedding of abelian categories, *Trans. Amer. Math. Soc.* **97** (1960), 410–417.
120. Maclane, S., "Homology", Springer, Berlin, 1963.
121. Maranda, J. M., Injective structures, *Trans. Amer. Math. Soc.* **110** (1964), 98–135.
122. Marot, J., Espace des localisations sur un anneau non-nécessairement commutatif, *C. R. Acad. Sci. Paris* **271** (1970), 1148–1151.
123. Marot, J., "Sur la localisation", Univ. Brest, 1969.
124. Marot, J., "Espaces des localisations sur un anneau", Univ. Brest, 1970.
125. Matlis, E., Injective modules over Noetherian rings, *Pacific J. Math.* **18** (1958), 511–528.
126. Matlis, E., Modules with descending chain conditions, *Trans. Amer. Math. Soc.* **97** (1960), 495–508.
127. Maury, G., Modules projectifs, modules injectives. Généralisations, *L'enseigmenent mathématique* **7** (1968), 257–261.
128. Mazet, P., Générateurs, relations et épimorphismes d'anneaux, *C. R. Acad. Sci. Paris* **266** (1968), 309–311.
129. Mewborn, A. C. and Winton, C. N., Orders in self injective semi-perfect rings, *J. Algebra* **13** (1969), 5–9.
130. Michler, G., Goldman's primary decomposition and the tertiary decomposition, *J. Algebra* **16** (1970), 129–137.
131. Mishina, A. P. and Skornjakov, L. A., "Abelian groups and Modules" (in Russian), Moscow, 1969.
132. Mitchell, B., The full imbedding theorem, *Amer. J. Math.* **86** (1964), 619–637.
133. Mitchell, B., "Theory of Categories", Academic Press, New York and London, 1965.
134. Morita, K., Duality for modules and its applications to the theory of rings with minimum condition, *Sci. Rep. Tokyo, Kyoiku Daigaku* **6** (1958–59), 83–142.
135. Morita, K., Category isomorphisms and endomorphism rings of modules, *Trans. Amer. Math. Soc.* **103** (1962), 451–469.
136. Morita, K., Localization in categories of modules, I, *Math. Z.* **114** (1970), 121–144.

137. Morita, K., Localization in categories of modules, II, *J. Reine Angew. Math.* **242** (1970), 163–169.
138. Morita, K., Localization in categories of modules, III, *Math. Z.* **119** (1971), 313–320.
139. Năstăsescu, C., Décomposition primaire dans les anneaux semi-artiniens, *J. Algebra* **14** (1970), 170–181.
140. Năstăsescu, C. and Nită, C., Sur le centre d'une catégorie de Grothendieck, *C. R. Acad. Sci. Paris* **265** (1967), 373–375.
141. Năstăsescu, C. and Popescu, N., Sur la structure des objets de certaines catégories abéliennes, *C. R. Acad. Sci. Paris* **262** (1966), 1295–1297.
142. Năstăsescu, C. and Popescu, N., Quelques observations sur les topos abéliens, *Rev. Roum. Math. Pures. Appl.* **12** (1967), 553–563.
143. Năstăsescu, C. and Popescu, N., Les anneaux semi-artinians, *Bull. Soc. Math. France* **96** (1968), 357–368.
144. Năstăsescu, C. and Popescu, N., On the localization ring of a ring, *J. Algebra* **15** (1970), 41–56.
145. Nguen Trong Kham, "*D*-anneaux", Thèse ,Univ. de Bucarest. 1972.
146. Nită, C., Sur les anneaux *A* tels que tout *A*-module simple est isomorphe à un idéal, *C. R. Acad. Sci. Paris,* **268** (1969), 88–91.
147. Nită, C., Remarques sur les anneaux *N*-reguliers, *C. R. Acad. Sci. Paris* **271** (1970), 345–348.
148. Nită, C., L'anneau des endomorphismes d'un module et *S*-anneau, *C. R. Acad. Paris Sci.* **272** (1971), 940–941.
149. Nită, C., Anneaux self injectifs de dimension finie, *Rev. Roum. Math. Pures Appl.* **15** (1970), 583–588.
150. Nită, C., Anneaux *N*-réguliers, *C. R. Acad. Sci. Paris* **268** (1969), 1241–1243.
151. Nită, C., Anneaux *N*-réguliers de dimension finie, In Press.
152. Northcott, D. G., "An introduction to Homological Algebra," Cambridge University Press, 1960.
153. Nouazé, Y., Catégories localement de type fini et catégories localement noetheriennes, *C. R. Acad. Sci. Paris* **257** (1963), 823–824.
154. Oberst, U., Duality theory for Grothendieck categories, *Bull. Amer. Math. Soc.* **75** (1969), 1401–1408.
155. Oberst, U., Duality theory for Grothendieck categories and linearly compact rings, *J. Algebra* **15** (1970), 473–542.
156. Olivier, J. P., Anneaux absolument plats universels et épimorphismes d'anneaux, *C. R. Acad. Aci. Paris* **266** (1968), 317–318.
157. Oort, F., On the definition of an abelian category, *Proc. Roy. Neth. Acad. Sci. A.* **70** (1970), 83–92.
158. Ore, O., Linear Equations in non-commutative fields, *Ann. Math.* **32** (1931), 463–477.
159. Osofsky, B. L., A generalization of quasi-Frobenius rings, *J. Algebra* **4** (1966), 373–387.
160. Ouzilou, R., Faisceaux additifs et applications catégoriques, *Publ. Dept. Math. Univ. Lyon* **3** (1966), 2–41.
161. Papp. Z., On algebraically closed modules, *Publ. Math. Debrecen* **6** (1959), 311–327.

162. Pareigis, B., "Categories and Functors", Academic Press, London and New York, 1970.
163. Pareigis, B., Radikale und kleine Moduln. *Bayerische Akad. Wissenschaft, München* (1966).
164. Pontryagin, L. S., "Topological Groups", (Russian), Moscow, 1954.
165. Pop Horia, The structure of rings of a single ideal. (In Press).
166. Popescu, A. and Sevici, C., The embedding of some additive categories. (In Press).
167. Popescu, N., Elemente de teoria fasciculelor, V, *St. Cerc. Mat.* **18** (1966), 945–991.
168. Popescu, N., Elemente de teoria fasciculelor, VI, *St. Cerc. Mat.* **19** (1967), 205–240.
169. Popescu, N., La localisation pour des sites, *Rev. Roum. Math. Pures. Appl.* **10** (1965), 1031–1044.
170. Popescu, N., La théorie générale de la décomposition, *Rev. Roum. Math. Pures. Appl.* **7** (1967), 1365–1371.
171. Popescu, N., Le spectre (à gauche) d'un anneau, *J. Algebra* **18** (1971), 213–228.
172. Popescu, N., Les anneaux semi-noethériens, *C. R. Acad. Sci. Paris* **272** (1971), 1439–1441.
173. Popescu, N., Sur les C.P. anneaux, *C. R. Acad. Sci. Paris.* **272** (1971), 1493–1496.
174. Popescu, N., Théorie de la décomposition primaire dans les anneaux semi-noétheriens, *J. Algebra,* **23** (1972), 482–492.
175. Popescu, N. and Spircu, T., Sur les épimorphismes plats d'anneaux, *C. R. Acad. Sci. Paris* **268** (1969), 376–379.
176. Popescu, N., and Spircu, T., Quelques observations sur les épimorphismes plats (à gauche) des anneaux, *J. Algebra* **16** (1970), 40–59.
177. Popescu, N. and Spulber, D., Les quasi-ordres (à gauche) des anneaux, *J. Algebra* **17** (1971), 474–481.
178. Radu, A., Théorème général du produit tensoriel, *Rev. Roum. Math. Pures Appl.* **8** (1968), 1153–1158.
179. Raicov, D. A., Semiabelian categories, *Dokl. Acad. Nauk. SSSR* (Russian) **188** (1969), 1006–1009.
180. Ravel, J., Injectivité. Généralisations et applications, extensions essentielles dans un treillis de Johnson, *Publ. Dept. Math. Univ. Lyon* **4** (1967), 1–74.
181. Reis, C. M. and Visvanathan, T. M., A compactness property for prime ideals in noetherian rings, *Proc. Amer. Math. Soc.* **25** (1970), 353–356.
182. Renault, G., Anneau associé à un module injectif, *Bull. Sci. Math.* **92** (1968), 53–58.
183. Rentschler, R., Sur les modules M tels que Hom(M,—) commute avec les sommes directes, *C. R. Acad. Sci. Paris* **268** (1969), 930–933.
184. Ribenboim, P., "Extensions epi-plates de fractions", Queen's Math. Preprint No. 1970–47.
185. Rieffel, M. A., A general Wedderburn theorem *Proc. Nat. Acad. Sci. U.S.A.* **54** (1965), 1513.
186. Riley, J. A., Axiomatic primary and tertiary decomposition theory, *Trans. Amer. Math. Soc.* **105** (1962), 177–201.
187. Riley, J. A., "Ideal theory in Noetherian rings", Dissertation Brandeis University, 1964.
188. Roby, N., Sur les épimorphismes de la catégorie des anneaux, *C. R. Acad. Sci. Paris,* **266** (1968), 213–313.
189. Roby, N., Diverses caractérisations des épimorphismes, *Exposé* 3, *Séminaire P. Samuel,* 1967/68.

190. Roos, J. E., Caractérisation des catégories qui sont quotient de catégories de modules par de sous-catégories bilocalisantes, *C. R. Acad. Sci. Paris* **261** (1965), 4954–4957.
191. Roos, J. E., Sur les foncteurs dérivés des produits infinis dans les catégories de Grothendieck. Exemples et contre-exemples, *C. R. Acad. Sci. Paris, Série A* **263** (1966), 895–898.
192. Roos, J. E., Sur la condition Ab 6 et ses variantes dans les catégories abéliennes, *C. R. Acad. Sci. Paris, Série A,* **264** (1967), 991–994.
193. Roos, J. E., Sur les catégories spectrales localement distributives, *C. R. Acad. Sci. Paris* **265** (1967), 14–17.
194. Roos, J. E., "Locally distributive Spectral Categories and Strongly Regular Rings", Reports of the Midwest Category Seminar, Springer, Berlin (1967), 156–181.
195. Roos, J. E., Sur l'anneau maximal de fractions des AW^*-algèbres et des anneaux de Baer, *C. R. Acad. Paris, Série A* **266** (1968), 120–123.
196. Roos, J. E., Sur la décomposition bornée des objets injectifs dans les catégories de Grothendieck, *C. R. Acad. Sci. Paris, Série A,* **266** (1968), 449–452.
197. Roos, J. E., Sur la structure des catégories abéliennes localement noethériennes, *C. R. Acad. Sci. Paris, Série A* **266** (1968), 701–704.
198. Roos, J. E., Locally noetherian categories and generalized strictly linearly compact rings. Applications. Category theory, homology theory and their applications, II ,"Battelle Institute Conference, 1968", Volume 2, Springer, Berlin, 1969, pp. 197–227.
199. Roos, J. E., On the structure of abelian categories with generators and exact limits. Applications (in press).
200. Roux, A., Sur une équivalence de catégories abéliennes, *C. R. Acad. Sci. Paris* **258** (1964), 5566–5569.
201. Samuel, P. Les épimorphismes d'anneaux, Séminaire, Paris 1967/68.
202. Samuel, P. and Zariski, O., "Commutative Algebra", Vols. I and II, van Nostrand, 1958, 1960.
203. Sandomierski, F. L., Semisimple maximal quotient rings, *Trans. Amer. Math. Soc.* **128** (1967), 112–120.
204. Sevici, C., A new proof of a theorem of Kaplansky (In press).
205. Silver, L., Non-commutative localization and applications, *J. Algebra* **7** (1967), 44–76.
206. Simson, D., Note on projective modules, *Bull. Acad. Pol. Sci. Serie des Sci. Math. Astr. et Phys.* **17** (1969), 355–359.
207. Skornjakov, L. A., Elizarov's quotient ring and the localization principle, *Math. Zametki* (Russian) **1** (1967), 263–268.
208. Spulber, D., Sur une classe des anneaux de Prüfer, *C. R. Acad. Sci. Paris, Série A,* **271** (1970), 365–367.
209. Stenström, B., Pure submodules, *Arkiv Math.* **7** (1967), 159–171.
210. Stenström, B., High submodules and purity, *Arkiv Math.* **7** (1967), 173–176.
211. Stenström, B., Purity in functor categories, *J. Algebra,* **8** (1968), 352–361.
212. Stenström, B. On the completion of modules in an additive topology, *J. Algebra* **16** (1970), 523–540.
213. Stenström, B., "Rings and Modules of Quotients", Springer Lecture Notes 237 (1971).

214. Stenström, B., Direct sum decompositions in Grothendieck categories, *Arkiv Math.* **7** (1968), 427–432.
215. Storrer, H. H., Epimorphismen von kommutativen Ringen, *Comment. Math. Helv.* **43** (1968), 378–401.
216. Storrer, H. H., A note on quasi-Frobenius rings and ring epimorphisms, *Can. Math. Bull.* **12** (1969), 287–292.
217. Swan, R. G., "Algebraic *K*-theory". Springer Lecture Notes No. 76 Berlin, 1968.
218. Tachikawa, H., On splitting of module categories, *Math. Z.* **111** (1969), 145–150.
219. Tachikawa, H., Double centralizers and dominant dimensions, *Math. Z.* **116** (1970), 79–88.
220. Tachikawa, H., Localization and artinian quotient rings, *Math. Z.* **119** (1971), 239–253.
221. Takashi, H., Adjoint pairs of functors on abelian categories, *Journal Fac. Sc. Univ. Tokyo* **13** (1962), 175–181.
222. Teply, M. L., Torsion-free injective modules, *Pacific J. Math.* **28** (1969), 441–453.
223. Teply, M. L., Some aspects of Goldie's torsion theory. *Pacific J. Math.* **29** (1969), 447–459.
224. Teply, M. L., Torsion-free projective modules, *Proc. Amer. Math. Soc.* **27** (1971), 29–34.
225. Tisseron, C., Objects injectifs dans une catégorie abélienne avec générateurs et limites inductives exactes, *Publ. Dept. Math. Lyon,* **4** (1967), 145–148.
226. Tisseron, C., "Quelques applications de la notion de localisation", Thèse (3ème cycle), Univ. Lyon, 1969.
227. Tisseron, C., Quelques applications de la localisation *Publ. Dept. Math. Lyon,* **4–6** (1969), 1–53.
228. Turnidge, D. R., Torsion theories and semihereditary rings, *Proc. Amer. Math. Soc.* **24** (1970), 137–143.
229. Ulmer, F., Properties of dense and relative adjoint functors, *J. Algebra* **8** (1968), 77–95.
230. Ulmer, F., Representable functors with values in arbitrary categories, *J. Algebra* **8** (1968), 96–129.
231. Ulmer, F., On the existence and exactness of the associated sheaf functor (to appear.)
232. Utumi, Y., On quotient rings, *Osaka Math. J.* **8** (1956), 1–18.
233. Vidal, R., Sur les catégories abéliennes localement artiniennes, *C. R. Acad. Sci. Paris, série A,* **272** (1971), 1293–1296.
234. Vraciu, C., Colocalization (to appear).
235. Walker, C. L., and Walker, E. A. Quotient categories of modules, "Proc. Conf. on Cat. Alg." La Jolla, 1965, Springer, Berlin, 1966, pp. 404–420.
236. Walker, C. L. and Walker, E. A., Quotient categories and rings of quotients (to appear in *Trans. Amer. Math. Soc.*).
237. Water, van der, A., A property of torsion-free modules over left Ore domains, *Proc. Amer. Math. Soc.* **25** (1970), 199–201.
238. Watts, C. E., Intrinsic characterization of some additive functors, *Proc. Amer. Math. Soc.* **11** (1960), 5–8.

Subject Index

D

E

F

G

H

I

J

K

L

S

T